Medical Biotechnology, Biopharmaceutics, Forensic Science and Bioinformatics

Multidisciplinary Applications and Advances in Biotechnology: Contributions from the Biotechnology Society of Nigeria Working Groups

Series Editors:

Benjamin Ewa Ubi

Department of Biotechnology / Biotechnology Research & Development Centre, Ebonyi State University Abakaliki, Nigeria

Abdulrazak B. Ibrahim

Department of Biochemistry, Ahmadu Bello University, Zaria, Nigeria / Forum for Agriculturl Research in Africa (FARA) , Accra, Ghana

Hajiya Mairo Inuwa

Department of Biochemistry, Ahmadu Bello University, Zaria, Nigeria

Ifeoma Maureen Ezeonu

Department of Microbiology, University of Nigeria, Nsukka, Nigeria

Charles Oluwaseun Adetunji

Edo State University Uzairue, Nigeria

Abubakar Gidado

Department of Biochemistry / North East Zonal Biotechnology Centre of Excellence, University of Maiduguri, Maiduguri, Bornu State, Nigeria

Emmanuel Olufemi Ekundayo

Department of Microbiology, Michael Okpara University of Agriculture, Umudike, Nigeria

Fermentation and Algal Biotechnologies for the Food, Beverage and Other Bioproduct Industries

James Chukwuma Ogbonna, Sylvia Uzochukwu, Emeka Godfrey Nwoba, Charles Oluwaseun Adetunji, Nwadiuto (Diuto) Esiobu, Abdulrazak B. Ibrahim, Benjamin Ewa Ubi

Agricultural Biotechnology, Biodiversity and Bioresources Conservation and Utilization

Olawole Obembe, Emmanuel Olufemi Ekundayo, Okoli Stanley Arinze, Abubakar Gidado, Charles Oluwaseun Adetunji, Abdulrazak B. Ibrahim, Benjamin Ewa Ubi

Medical Biotechnology, Biopharmaceutics, Forensic Science and Bioinformatics
Hajiya Mairo Inuwa, Ifeoma Maureen Ezeonu, Charles Oluwaseun Adetunji, Abubakar Gidado, Emmanuel Olufemi Ekundayo, Abdulrazak B. Ibrahim, Benjamin Ewa Ubi

Biosafety and Bioethics in Biotechnology: Policy, Advocacy, and Capacity Building
Sylvia Uzochukwu, Okoli Stanley Arinze, Nwadiuto (Diuto) Esiobu, Emeka Godfrey Nwoba, Christpeace Nwagbo Ezebuiro, Charles Oluwaseun Adetunji, Abdulrazak B. Ibrahim, and Benjamin Ewa Ubi

Microbiomes and Emerging Applications
Nwadiuto (Diuto) Esiobu, James Ogbonna, Charles Oluwaseun Adetunji, Olawole Obembe, Ifeoma Maureen Ezeonu, Abdulrazak B. Ibrahim, and Benjamin Ewa Ubi

Bioenergy and Environmental Biotechnology for Sustainable Development
Akinola Rasheed Popoola, Emeka Godfrey Nwoba, James Chukwuma Ogbonna, Charles Oluwaseun Adetunji, Nwadiuto (Diuto) Esiobu, Abdulrazak B. Ibrahim, and Benjamin Ewa Ubi

For more information about this series, please visit: https://www.routledge.com/Multidisciplinary-Applications-and-Advances-in-Biotechnology/book-series/MAABNIG

Medical Biotechnology, Biopharmaceutics, Forensic Science and Bioinformatics

Edited by
Hajiya Mairo Inuwa, Ifeoma Maureen Ezeonu,
Charles Oluwaseun Adetunji,
Emmanuel Olufemi Ekundayo, Abubakar Gidado,
Abdulrazak B. Ibrahim, and Benjamin Ewa Ubi

CRC Press
Taylor & Francis Group
Boca Raton London New York

CRC Press is an imprint of the
Taylor & Francis Group, an informa business

First edition published 2022
by CRC Press
6000 Broken Sound Parkway NW, Suite 300, Boca Raton, FL 33487-2742

and by CRC Press
4 Park Square, Milton Park, Abingdon, Oxon, OX14 4RN

CRC Press is an imprint of Taylor & Francis Group, LLC

Library of Congress Cataloging-in-Publication Data
Names: Ubi, Benjamin Ewa, 1968- editor. | Biotechnology Society of Nigeria, issuing body.
Title: Multidisciplinary applications and advances in biotechnology : contributions from the Biotechnology Society of Nigeria working groups, six-volume set / edited by Benjamin Ewa Ubi [and fourteen others]. Description: First edition. | Boca Raton : CRC Press, 2022. | Includes bibliographical references and index. |
Contents: Volume 1. Fermentation and Algal Biotechnologies for the Food, Beverage and Other Bioproduct Industries—Volume 2. Agricultural biotechnology, biodiversity and bio-resources conservation and utilization—Volume 3. Medical Biotechnology, Biopharmaceutics, Forensic Science and Bioinformatics— Volume 4. Biosafety and Bioethics in Biotechnology—Volume 5. Microbiomes and Emerging Applications—Volume 6. Bioenergy and environmental biotechnology for sustainable development, |
Summary: "This six volume book set examines a range of topics and applications related to biotechnology. Volumes include fermentation and algal biotechnologies; agricultural biotechnology; medical biotechnology, biopharmaceutics; biosafety, bioethics, biotechnology policy; microbiomes; bioenergy and environmental biotechnology for sustainable development"—Provided by publisher. Identifiers: LCCN 2021047039 (print) | LCCN 2021047040 (ebook) | ISBN 9780367766948 (v. 1 ; hardback) | ISBN 9781032013763 (v. 1 ; paperback) | ISBN 9780367766955 (v. 2 ; hardback) | ISBN 9781032014982 (v. 2 ; paperback) | ISBN 9780367766962 (v. 3 ; hardback) | ISBN 9781032015026 (v. 3 ; paperback) | ISBN 9780367767020 (v. 4 ; hardback) | ISBN 9781032015781 (v. 4 ; paperback) | ISBN 9780367767044 (v. 5 ; hardback) | ISBN 9781032018331 (v. 5 ; paperback) | ISBN 9780367767051 (v. 6 ; hardback) | ISBN 9781032018416 (v. 6 ; paperback) | ISBN 9781003178378 (v. 1 ; ebook) | ISBN 9781003178880 (v. 2 ; ebook) | ISBN 9781003178903 (v. 3 ; ebook) | ISBN 9781003179177 (v. 4 ; ebook) | ISBN 9781003180241 (v. 5 ; ebook) | ISBN 9781003180289 (v. 6 ; ebook)
Subjects: LCSH: Biotechnology—Nigeria.
Classification: LCC TP248.195.N6 M95 2022 (print) | LCC TP248.195.N6 (ebook) | DDC 660.609669—dc23
LC record available at https://lccn.loc.gov/2021047039
LC ebook record available at https://lccn.loc.gov/2021047040

ISBN: 978-0-367-76696-2 (hbk)
ISBN: 978-1-032-01502-6 (pbk)
ISBN: 978-1-003-17890-3 (ebk)

DOI: 10.1201/9781003178903

Typeset in Times
by codeMantra

Dedication

This book is dedicated to Professor Vincent Ado Tenebe, FBSN, Chairman of the Board of Trustees (BoT) of the Biotechnology Society of Nigeria (BSN) for his passion for Biotechnology and commitment to the cause of the Society.

Contents

PART 2 Working Group 07: Medical and Pharmaceutical Biotechnology

Handling Editor: Prof. Hajiya Mairo Inuwa

PART 5 Working Group 13: Medical Insect Biotechnology, Genomics and Molecular Genetics of Pests

Handling Editor: Dr. Abdulrazak Ibrahim

Preface

In recognition of biotechnology's pivotal role as a burgeoning technology in fostering a knowledge-driven bio-based economy, the Biotechnology Society of Nigeria (BSN) has – in the about 40 years of its founding – made enormous contributions towards expanding the frontiers of biotechnology in Nigeria for sustainable livelihoods. This book is one of the multi-volume series on the Multidisciplinary Applications and Advances in Biotechnology: Contributions from the BSN, with each volume covering specific contemporary topics targeted at providing scientists in the academia, industry, research institutes/agencies, policy makers, etc. with recent advances in the field to bridge the knowledge gap in the crucial areas of biotechnology.

This book is designed to document advances in biotechnology in Nigeria over the past decades and the emerging biotech sector as a key driver of the bioeconomy. It is a timely knowledge product detailing key issues on recent developments in the field of biotechnology particularly in Nigeria. It is written for use by various readers including postgraduate students, professionals in the area of research, industry players, policy makers, educationists, science advocacy groups, communication experts, etc. Additionally, the book engenders a deeper understanding of underlying developments in biotechnology, its historical perspectives and context in Nigeria, processes and science-policy nexus that have placed Nigeria as a leader in biotechnology in Africa.

This volume covers 5 of the 16 Working Groups of the Society, namely Molecular Diagnostics and DNA Forensics (WG04); Medical and Pharmaceutical Biotechnology (WG07); Omics and Bioinformatics (WG08); Medicinal and Underutilized Plants (WG12); and Medical Insect Biotechnology, Genomics and Molecular Genetics of Pests (WG13). It contains important topics on exploiting Nigeria's mega biodiversity for food security, improved nutrition and health; DNA forensics and molecular diagnostics; medical biotechnology and biopharmaceuticals; medicinal and underutilized plants; global trends, impact and mitigation of antibiotic resistance; bioinformatics applications; medical insect biotechnology, etc. in 23 chapters that bring to light the applications of available knowledge and deliverables in the thematic areas covered by the array of seasoned contributing authors.

Each chapter addresses the relevant points in the subject areas covered and discusses each of the thematic areas concisely. The chapters are co-authored by a team of different scientific experts in the field which guarantees in-depth scholarly output in a high level of interdisciplinary collaborative manner. We have leveraged the pool of expertise available in the five BSN Working Groups to ensure a higher level of sustainability on the deliverables expected from these thematic research areas.

This book is a useful reference material, a comprehensive description of the recent advances in the field, and a valuable guide to the subject matter's original literature. It is specifically targeted at scientists and researchers working in the different areas of biotechnology including DNA forensics and its applications, medical and biopharmaceuticals including the novel SARs-Cov 2, underutilized and medicinal plants used traditionally for the treatment of various ailments and boosting nutrition and the rural economy in Africa, molecular diagnostics of diseases and biosurveillance, bioinformatics, antibiotic resistance and mitigation measures, etc. This book will be valuable to professionals in universities, research institutions, industry or government agencies, and policy makers/science advocacy groups.

Overall, this volume is an excellent piece of scientific contributions that targets a wide range of audiences including policy makers, private sector, researchers, civil society, farmers and fishing communities, government and environmentalists, which could help in fostering an innovative pathway for future sustainable and healthy populations, molecular diagnostics, arthropod midgut components and intervention against infectious diseases, genomics-based vector surveillance for effective outbreak prevention and control, antimicrobial agents from indigenous plants, antibiotic resistance, bioinformatics, nutraceuticals, nanodrugs, etc., which are important to meet the health and economic needs of an increasing global population. It should also be of relevance to industrialists with interest in valorizing Africa's indigenous bioresources and biotech start-ups, medical biotechnology sector, professionals involved in biopharmaceuticals and nutraceutics, educators and researchers, and educational and research institutions. Readers will find this book a valuable resource containing useful information in the thematic areas contributed by a seasoned and diverse array of scholars within this professional Society.

The carefully selected editorial team and the authors of individual chapters drawn from several key institutions were chosen for their recognized international expertise and contributions to the various subject areas. Their willingness to make these invaluable contributions in their respective fields of expertise reflecting the different working groups and in a collaborative way to foster research networking is gratefully acknowledged. The editors wish to thank the different manuscripts' reviewers for their unceasing dedication and significant input. We thank Prof. Malachy O. Akoroda of the Department of Agronomy, University of Ibadan, who shared the initial idea that inspired this book initiative with me just after the conclusion of the 32nd Annual Conference of the Society held at the International Institute of Tropical Agriculture (IITA) Ibadan, Nigeria. Furthermore, we highly appreciate the foresight, diligent support, and understanding of the publisher, Taylor and Francis Group (USA), for publishing *Multidisciplinary Applications and Advances in Biotechnology – VI: Medical Biotechnology, Biopharmaceutics, Forensic Science and Bioinformatics* with their international hallmark of excellence. We return special thanks to Dr. Marc Gutierrez and Mr. Nick Mould, whose tremendous and painstaking efforts, constant suggestions, and guidance at all stages led to the successful publication.

Prof. Benjamin Ewa Ubi
President, Biotechnology Society of Nigeria (BSN)
& Chief Editor
25 March, 2021

Foreword

The Biotechnology Society of Nigeria (BSN) has been a progressive Society with a focus on shaping and advancing biotechnology research, policy and deployment in Nigeria. As expected, the Society provides veritable and unbiased arguments for the advancement of policy and research in all spheres of biotechnology. Besides, government agencies and departments such as the National Biotechnology Development Agency (NABDA) have always worked with the BSN due to the wealth of experience that could be tapped from the abundant expertise of its members as individuals, groups and the society as an entity.

This is particularly so because having been at a point, the Director-General and Chief Executive Officer of the National Biotechnology Development Agency in Nigeria, one is always challenged with the provision of the appropriate path for the development of the technology with emphasis on what areas and types of technology are relevant to the country now and in the future. The intellectual discourses of the Society have always provided some level of guidance at such crossroads, while the provision of a permanent site for the headquarters of the BSN which I did is sine qua non in its efforts at infrastructural development.

A typical example was developing curriculum for biotechnology education in Nigeria. This debate was an important one considering that education curriculum for biotechnology in Nigeria will form the bedrock of knowledge and shape what aspects of the Body of Knowledge the country would specialize in. Of course, this would also shape national proficiency and determine the depth to which the technology could be applied to national challenges such as maintenance of environmental integrity, promotion of health and medicine, agricultural revolution and food security, and the enhancement of industrial processes.

It is this quality of discussion with sector application of biotechnology in mind that has shaped the creation of this book. This book for the most part focuses on helping the reader to appreciate biotechnology as a concept, its application in meeting national challenges and the necessary policy requirements that are needed to create the connection between knowledge, need and solution.

For instance, our clearer understanding of the structures of DNA and RNA has presented various techniques for DNA research and manipulation, from the principles of nucleic acid extraction to sequencing techniques and genomic fingerprinting. Similarly, in molecular diagnostics and forensic science, several markers have been used to identity pathogens such as viruses, bacteria, parasites and fungi, monitor disease progression and outcome, and identify crime suspects, victims of air mishaps and the resolution of parental disputes. This book therefore comes at no better time than now when the transition to a green economy is an indispensable part of efforts aimed at mitigating the effect of climate change and the myriads of health, increased agricultural and environmental challenges that accompany it. This informed the emphasis on the deployment of relevant biotechnologies for improved healthcare delivery, massive increase in food production and environmental cleanliness through the use of bio-digesters to convert bio-degradable wastes to energy and cooking gas.

The content of this book addresses the regional needs of the continent, and especially emphasizes the dimensions and spread required of scientists and government to provide a functional integration of biotechnology into national development. In this regard, this book lays a great foundation and brings in completely new and unassailable ways, including frontline researchers in the field of biotechnology both within and outside the country. Of course, here you find the premium authors, researchers, policy analysts, etc., whose experiences could form the footing for further diversification of thoughts and efforts in mainstreaming biotechnology into national development agenda.

This is not necessarily the first time that the BSN is making a scientific publication, but this certainly is the first time emphasis is laid on a holistic approach to biotechnology development in Nigeria. This book differs from any scientific publication from the Society because it captures broadly the three dimensions of biotechnology that include research, application/deployment and policy. In this, it is not a substitute for the *Nigerian Journal of Biotechnology* but provides additional layers of application and policy dimensions as to how biotechnology should be applied and administered.

Finally, this is just the beginning of a great effort to bring to the knowledge of the public, government and researchers the need to think biotechnology and live biotechnology. This book will become the Body of Knowledge that will guide all biological scientists (irrespective of specialization), all policy players in the biotechniques sector and all development agencies, on what is relevant in mainstreaming biotechnology into everyday lives of Nigerians. I will end with the following remark: the future of biotechnology development in Nigeria and sub-Saharan Africa for that matter will be shaped to a good extent by the current and future dimensions of the discourses that this and subsequent editions of this book will bring on the table.

Prof. Alex U. Akpa FWHO, FRSTMH, FCAI, SRFIARSA
Formerly Ag. Director-General/Chief Executive Officer
National Biotechnology Development Agency, Abuja
Nigeria

Editors

Prof. Hajiya Mairo Inuwa received his Ph.D. in Biochemistry (Plant Gene Expression) from the University of Glasgow, Commonwealth Scholar; M.Sc. and B.Sc. in Biochemistry and Postgraduate Diploma in Computer Science from Ahmadu Bello University Zaria; and Postgraduate Diploma in Management from the University of Lagos. She is currently Director, Centre for Biotechnology Research and Training, Ahmadu Bello University. Her research interest is in eukaryotic genetic-based research such as in vitro erythropoietin action of some plant materials for potential treatment of sickle cell anemia and diabetes, and purification and characterization of enzymes such as DNA methylase, sialidase and cysteine protease. Her high index of academic proficiency is demonstrated in over several publications in peer-reviewed journals. Prof. Inuwa is an excellent academician, demonstrated in about 30 years of teaching and research, and mentorship and supervision of hundreds of students at undergraduate and postgraduate levels (Ph.D., M.Sc. and PGD) in Biochemistry, Biotechnology and Nutrition. She is a member of the team of experts that won the World Bank-assisted funding by the African Centre of Excellence in Neglected Tropical Diseases and Forensic Biotechnology (ACENTDFB), Ahmadu Bello University; member of the team of research experts on development of drought-tolerant maize and pest-resistant cotton with the Institute for Agricultural Research on an IITA-sponsored program. She has sound research-industry experience retiring from Glaxo Nigeria Plc after an award of a 10-year meritorious service as a Quality Assurance, Research and Development Manager (1991–2002).

Professor Ifeoma Maureen Ezeonu, B.Sc. (Nig.), M.Sc., Ph.D. (Atlanta), is a Professor of Medical Microbiology and Molecular Genetics in the Department of Microbiology, University of Nigeria, Nsukka, and currently the Director of the University of Nigeria Central Science Laboratory. She obtained her B.Sc. in Microbiology from the University of Nigeria, Nsukka, in 1987 and her M.Sc. and Ph.D. in Microbiology from Georgia State University, Atlanta, Georgia, USA, in 1991 and 1994, respectively. Her areas of research include: bacteriology, medical biotechnology, molecular biology, bacterial drug resistance, probiotics and prebiotics. She is the Coordinator of the Drug-Resistant Infections Research Group of the University of Nigeria. She has received several awards and grants including Georgia State University SGA Academic Affairs and Honor Society Council Certificate of Achievement, Georgia State University Regent Opportunity Scholarship, US National Institute of General Medical Sciences Pre-Doctoral Grant (F31GM016377) for study of Growth of Fungi on HVAC materials, University of Nigeria Senate Research Grant for study of use of plant extracts for inhibition of urinary catheter biofilms, and Tertiary Education Trust Fund Institutional Based Research Intervention for study of epidemiology of drug-resistant tuberculosis in Enugu, Nigeria. She has published several high-impact articles in peer-reviewed journals as can be seen in her Research Gate and Google scholar profiles.

Charles Oluwaseun Adetunji, Ph.D., is an Associate Professor of Microbiology and Biotechnology at the Microbiology Department, Faculty of Science, Edo State University Uzairue EDSU, Edo State, Nigeria, where he utilized the application of biological techniques and microbial bioprocesses for the actualization of sustainable development goals and Agrarian revolution, through quality teaching, research and community service. He is currently the Acting Director of Intellectual Property and Technology Transfer, Sub-Dean for Faculty of Science, and the Head of department of Microbiology at EDSU. He is a Visiting Professor and the Executive Director for the Center of Biotechnology, Precious Cornerstone University, Ibadan.. He has won several scientific awards and grants from renowned academic bodies like Council of Scientific and Industrial Research (CSIR), India; Department of Biotechnology (DBT), India; the World Academy of Science (TWAS), Italy; Netherlands Fellowship Programme (NPF), the Netherlands; the Agency for International Development Cooperation; Israel; and Royal Academy of Engineering, UK, among many others. He has filed several scientific patents on Bioherbicides, Biopesticides, Nanobiosurfactants, Nanobiopesticdes and many more. He has published over 300 scientific articles in refereed journals and conference proceedings both international and local journals. He was ranked among the top 500 prolific authors in Nigeria between 2019 till date by SciVal/SCOPUS. He is currently editing several biotechnology text books with Elsevier. He was recently appointed as the President and Chairman Governing Council of the Nigerian Bioinformatics and Genomics Network Society. He is presently a series editor with Taylor and Francis, USA editing several textbooks on Agricultural Biotechnology, Nanotechnology, Pharmafoods, and Environmental Sciences. The breadth of his scholarly contributions to research is evident from his contributions which cover several topics relating to Food Security, Agriculture and Environmental Sustainability. His research interest includes microbiology, biotechnology, bioresources and nanotechnology. He is an editorial board member of many international journals and serves as a reviewer to many double-blind peer-review journals like Elsevier, Springer, Taylor and Francis, Wiley, PLOS One, Frontiers, Nature, American Chemistry Society, Bentham Science Publishers, etc. He is a member of many scientific and professional bodies including the American Society for Microbiology, Nigerian Young Academy, Biotechnology Society of Nigeria and Nigerian Society for Microbiology. He has won a lot of international recognition and also acted as a keynote speaker delivering invited talks/position papers at various universities, research institutes and several centres of excellence which span across several continents of the globe. Over the last 15 years, he has built strong working collaborations with reputable research groups in numerous and leading universities across the globe. He is the convener for Recent Advances in Biotechnology, which is an annual international conference where renowned Microbiologists and Biotechnologists come together to share their latest discoveries (https://raibconference.wixsite.com/website). He is the Founder & CEO of BECTIK Biotechnology and Nanotechnology Company. His Biotechnology Company consists of a team of leading academics in their fields and thrives to deliver solutions all around Bio- and Nanotechnology. The company also delivers solutions that are tailored to your business needs and for any scale.

Abubakar Gidado, Ph.D., is a Professor of Biochemistry and the Director of North-East Zonal Biotechnology Centre of Excellence (NEZBCE) at the University of Maiduguri. He obtained his B.Sc. and M.Sc. (General Biochemistry) from the University of Maiduguri, Ph.D. (Biochemistry) from A.B.U. Zaria, Nigeria, and a post-doctoral research at Chiang Mai University, Thailand. His research interests include identifying and isolating compounds from natural products with medicinal properties and developing antimicrobials targeting microbes causing post-harvest losses. He has published more than 40 peer-reviewed journal articles and organized several training workshops on molecular biology techniques. Professor Gidado is a member of the Biotechnology Society of Nigeria (BSN) and Nigerian Society of Biochemistry and Molecular Biology (NSBM).

Emmanuel Olufemi Ekundayo, Ph.D., is an Associate Professor of Medical Microbiology and Microbial Genetics and the Vice President of the Biotechnology Society of Nigeria. He was the Director of Centre for Molecular Biosciences and Biotechnology, Michael Okpara University of Agriculture, Umudike from 2014 to 2020. Dr. Ekundayo graduated from the University of Jos, Nigeria, with Bachelor of Science degree (Zoology) in 1991 and obtained the Master of Science degree in Medical Microbiology from the same university in 1994. The research interests of Dr. Emmanuel Ekundayo have focused on development and evaluation of diagnostics for infectious diseases, antimicrobial drug resistance and antimicrobial properties of medicinal plants.

Abdulrazak B. Ibrahim, Ph.D., is an agricultural biotechnologist with experience and skills on plant biotechnology stewardship, stakeholder engagement, foresight analysis, development of transgenic crops, laboratory and field management of GM crops, biosafety analysis, adoption and discontinuation of events. He was trained in the Brazilian Agricultural Research Cooperation (EMBRAPA Center for Genetic Resources), from where he obtained a Doctorate degree in Molecular Biology from Universidade de Brasilia (UnB) and M.Sc. from Universidade Federal do Ceara (UFC), Fortaleza, and acquired best practices in agricultural biotechnology leading to the development and patenting of a transgenic technology for the control of white-fly using RNA interference technology within EMBRAPA, and the development of virus-resistant cowpea, under the Africa-Brazil Innovation Platform, implemented by EMBRAPA and Forum for Agricultural Research in Africa (FARA), with support from the Department for International Development (DFID), Bill & Melinda Gates Foundation (BMGF), International Fund for Agricultural Development (IFAD) and Foundation Arthur Bernardes (FUNARBE).

Benjamin Ewa Ubi, Ph.D., is a Professor of Plant Breeding and Biotechnology. He obtained his B.Sc. Honours (Agric.) Agronomy from the University of Calabar (1990) and M.Sc. Agronomy – Crop Science (Plant Breeding and Tissue Culture) (1994) and Ph.D. Agronomy (Plant Breeding and Biotechnology) (1998) from the University of Ibadan, Nigeria. His Ph.D. research was completed at the then Biotechnology Laboratory of the International Institute of Tropical Agriculture (IITA), Ibadan, where he was a graduate research fellow and became the President of the International Association of Research Scholars and Fellows (IARSAF) in 1996. Among several other research engagements, Prof. Ubi served as Science and Technology Agency (STA) Postdoctoral Fellow under the Japan International Science and Technology Exchange Centre (JISTEC) 1999–2001 at the Plant Breeding Department (Biotechnology Laboratory) of the National Institute for Livestock and Grassland Science (NARO, National Agricultural Research Organization), Nishinasuno, Japan; and as a three-term Visiting Scientist at the NARO National Institute of Fruit Tree Science (NIFTS), Tsukuba, Japan, under the auspices of the Inoue Research Foundation of Japan Fellowship (April 2003–April 2004) and the Japanese Society for the Promotion of Science (JSPS) visiting fellowships (October 2008–October 2009; May 2012–March 2013). Prof. Ubi served as a Visiting Professor at the Arid Land Research Centre (The Global Centre of Excellence for Dryland Science) (Molecular Breeding Laboratory), Tottori University, Japan; and Visiting Professor, University of Tokyo, JapanAmong several other academic engagements, Prof. Ubi has served as the Director of the Biotechnology Research and Development Centre, Ebonyi State University, Abakaliki (August 2006–March 2008) and the Head of the Department of Biotechnology, Ebonyi State University for several years; Dean, Faculty of Science, Ebonyi State University Abakaliki and Pioneer Editor-in-Chief of *Science Journal* <https://sciencejournal.ng>, a publication of the Faculty of Science, Ebonyi State University, Abakaliki; President, Biotechnology Society of Nigeria (BSN) and Leader of the Crops Genetics, Breeding and Genetics Working Group of the Biotechnology Society of Nigeria (BSN).

Contributors

Oluchi Cynthia Abiamere
Department of Environmental Biotechnology
National Biotechnology Development Agency (NABDA)
Abuja, Nigeria

I. Abraham
Department of Agricultural Biotechnology
and
Open Forum on Agricultural Biotechnology (OFAB) in Africa, Nigeria Chapter
National Biotechnology Development Agency (NABDA)
Abuja, Nigeria

I. J. Adeosun
Laboratory of Molecular Biology, Immunology and Bioinformatics
Department of Microbiology
Adeleke University
Ede, Nigeria

Oluwafemi Abiodun Adepoju
Department of Biochemistry
Ahmadu Bello University
Zaria, Nigeria

Charles Oluwaseun Adetunji
Applied Microbiology, Biotechnology and Nanotechnology Laboratory
Department of Microbiology
Edo State University Uzairue
Okpella, Nigeria

Juliana Bunmi Adetunji
Nutritional and Toxicological Research Laboratory
Department of Biochemistry Sciences
Osun State University
Osogbo, Nigeria

Folasade Muibat Adeyemi
Department of Microbiology
Osun State University
Osogbo, Nigeria

D. E. Agbonlahor
Department of Medical Laboratory Science, Faculty of Basic Medical Sciences
Edo State University
Uzairue, Nigeria

Muhammad Akram
Department of Eastern Medicine
Government College University Faisalabad
Faisalabad, Pakistan

Geoffrey Obinna Anoliefo
Applied Environmental Biosciences and Public Health Research Group
and
Environmental Biotechnology and Sustainability Research Group
Department of Plant Biology and Biotechnology, Faculty of Life Sciences
University of Benin
Benin City, Nigeria

Rumaisa Ansari
Department of Eastern Medicine
Government College University Faisalabad
Faisalabad, Pakistan

Emmanuel Oluwadare Balogun
Department of Biochemistry
Ahmadu Bello University
Zaria, Nigeria

O. O. Bamigboye
Biotechnology/Environmental Laboratory
Department of Microbiology
Adeleke University
Ede, Nigeria

Ruth Ebunoluwa Bodunrinde
Department of Microbiology
Federal University of Technology
Akure, Nigeria

E. A. Brisibe
Department of Genetics and Biotechnology
University of Calabar
Calabar, Nigeria

Genevieve Dable-Tupas
Clinical Epidemiologist, Director for Research and Publication, Head, Research Center
College of Medicine
Davao Medical School Foundation, Inc. (DMSFI)
Davao City, Philippines

Chukwuebuka Egbuna
Department of Biochemistry, Faculty of Natural Sciences
Chukwuemeka Odumegwu Ojukwu University
Uli Campus, Nigeria
and
Nutritional Biochemistry and Toxicology Unit, World Bank Africa Centre of Excellence
Centre for Public Health and Toxicological Research (PUTOR)
University of Port Harcourt
Port Harcourt, Nigeria

E. O. Ekundayo
Department of Microbiology
Michael Okpara University of Agriculture
Umudike, Nigeria

Temidayo Oluyomi Elufisan
Instituto Politécnico Nacional
Centro de Biotecnología Genómica
Reynosa, México

Godwill Azeh Engwa
Godfrey Okoye University
Thinkers Corner, Nigeria

D. Enoma
Covenant Applied Informatics and Communication African Center of Excellence
Covenant University
Ota, Nigeria

Onyekachukwu O. Enweani
Medical Centre
University of Nigeria
Enugu, Nigeria

Ifeoma B. Enweani-Nwokelo
Department of Medical Laboratory Science
Nnamdi Azikiwe University
Awka, Nigeria

A. O. Eremwanarue
Lahor Centre for Molecular Diagnostics
Benin City, Nigeria

D. Esiobu
Applied Biotech International Nigeria Ltd.
Abuja, Nigeria

Nwadiuto (Diuto) Esiobu
Applied Biotech International Nigeria Ltd
Wuye, Abuja
and
Biological Sciences Department
Florida Atlantic University
Boca Raton, USA

Ifeoma M. Ezeonu
Department of Microbiology
University of Nigeria
Nsukka, Nigeria

Anton Ficai
Applied Microbiology, Biotechnology and Nanotechnology Laboratory
Department of Microbiology
Edo University Iyamho
Auchi, Nigeria

Denisa Ficai
Faculty of Applied Chemistry and Materials Science
University Politehnica of Bucharest
Bucharest, Romania

Njidda Gadzama
Biological Sciences Department
University of Maiduguri
Maiduguri, Nigeria

G. S. George
Department of Medical Laboratory Science
Niger Delta University
Amassoma, Nigeria

Daniel Ingo Hefft
University Centre Reaseheath
Food and Agricultural Sciences, Reaseheath College
Nantwich, UK

Bashiru Ibrahim
Department of Biochemistry
Ahmadu Bello University
Zaria, Nigeria

Jonathan C. Ifemeje
Department of Medical Biotechnology
National Biotechnology Development Agency
Abuja, Nigeria

J. C. Igwe
Department of Pharmaceutical Microbiology and Biotechnology, Faculty of Pharmaceutical Sciences
Kaduna State University
Kaduna, Nigeria

Beckley Ikhajiagbe
Applied Environmental Biosciences and Public Health Research Group
and
Environmental Biotechnology and Sustainability Research Group
Department of Plant Biology and Biotechnology, Faculty of Life Sciences
University of Benin
Benin City, Nigeria

Obianuju P. Ilo
School of Agricultural, Earth and Environmental Sciences
University of KwaZulu-Natal
Durban, South Africa
and
Environmental Biotechnology and Bioconservation Department
National Biotechnology Development Agency
Abuja, Nigeria

I. R. Iroha
Department of Microbiology
Ebonyi State University
Abakaliki, Nigeria

M. O. Kaka
Laboratory of Molecular Biology, Immunology and Bioinformatics
Department of Microbiology
Adeleke University
Ede, Nigeria

Umme Laila
Department of Eastern Medicine
Government College University Faisalabad
Faisalabad, Pakistan

Scott E. Miller
National Museum of Natural History
Smithsonian Institution
Washington, D.C., United States

S. S. D. Mohammed
Department of Microbiology and Biotechnology, Faculty of Natural and Applied Sciences
Nile University of Nigeria
Abuja, Nigeria

Saheed Ibrahim Musa
Environmental Biotechnology and Sustainability Research Group
Department of Plant Biology and Biotechnology, Faculty of Life Sciences
University of Benin
Benin City, Nigeria
and
Department of Biological Sciences
Admiralty University of Nigeria
Ibusa, Nigeria

Wilson Nwankwo
Informatics and CyberPhysical Systems Laboratory
Department of Computer Science
Edo State University Uzairue
Auchi, Nigeria

Francisca Nwaokorie
Department of Medical Laboratory Science
University of Lagos
Lagos, Nigeria

Friday Nwalo Nweke
Alex Ekwueme Federal University
Ndufu-Alike, Nigeria

O. O. Obembe
UNESCO Chair on Plant Biotechnology, Plant Science Research Cluster
Department of Biological Science
Covenant University
Ota, Nigeria

Bamidele Oderinde
Department of Medical Laboratory Science
College of Medical Sciences
University of Maiduguri
Maiduguri, Nigeria

Ugoy Sonia A. Ogbonna
Department of Applied Microbiology and Brewing
Nnamdi Azikiwe University
Awka, Nigeria

Edokpolor Osazee Ohanmu
Environmental Biotechnology and Sustainability Research Group
Department of Plant Biology and Biotechnology, Faculty of Life Sciences
University of Benin
Benin City, Nigeria
and
Department of Environmental Management and Toxicology
Edo University
Iyamho, Nigeria

Ihekerenma Okoli
Research Operations Department
Applied Biotech International Nigeria Ltd.
Abuja, Nigeria

L. E. Okoror
Federal University
Ekiti, Nigeria

Gloria E. Okotie
Department of Physiology
University of Ibadan
Ibadan, Nigeria

Grace I. Okwu
Department of Microbiology
Ambrose Alli University
Ekpoma, Nigeria

E. K. Oladipo
Laboratory of Molecular Biology, Immunology and Bioinformatics
Department of Microbiology
Adeleke University
Ede, Nigeria

Oluwaseyi Paul Olaniyan
Nutritional and Toxicological Research Laboratory
Department of Biochemistry Sciences
Osun State University
Osogbo, Nigeria

Akinola Samson Olayinka
Department of Physics
Edo State University Uzairue
Uzairue, Nigeria

Michael Chinedu Olisah
Department of Medical Biotechnology
National Biotechnology Development Agency
Abuja, Nigeria

Julius Kola Oloke
Department of Pure and Applied Biology
Ladoke Akintola University of Technology
Ogbomoso, Nigeria

H. A. Olorunsola
UNESCO Chair on Plant Biotechnology, Plant Science Research Cluster
Department of Biological Science
Covenant University
Ota, Nigeria

T. M. Olotu
Laboratory of Molecular Biology, Immunology and Bioinformatics
Department of Microbiology
Adeleke University
Ede, Nigeria

A. O. Oluduro
Department of Microbiology
Obafemi Awolowo University
Ile-Ife, Nigeria

Olaniyan Tope Olugbemi
Laboratory for Reproductive Biology and Developmental Programming
Department of Physiology
Edo State University Uzairue
Iyamho, Nigeria

Olufemi Idowu Oluranti
Environmental and Applied Physiology Research Unit
Department of Physiology
College of Health Sciences
Bowen University
Iwo, Nigeria

Zipporah Emilomo Omage
Environmental Biotechnology and Sustainability Research Group
Department of Plant Biology and Biotechnology, Faculty of Life Sciences
University of Benin
Benin City, Nigeria

Doris I. Omoigui
Department of Plant Biology and Biotechnology, Faculty of Life Sciences
University of Benin
Benin City, Nigeria

Christie O. Onyia
Biotechnology and Applied Biology Department
Godfrey Okoye University
Enugu State, Nigeria

Ayomide Michael Oshinjo
Cardiometabolic Research Unit, Department of Physiology
College of Health Sciences
Bowen University
Iwo, Nigeria

F. E. Oviasogie
Department of Microbiology
University of Benin
Benin City, Nigeria

O. M. Oyawoye
Laboratory of Molecular Biology, Immunology and Bioinformatics
Department of Microbiology
Adeleke University
Ede, Nigeria

Omotayo Opemipo Oyedara
Department of Microbiology
Osun State University
Osogbo, Nigeria
and
Universidad Autónoma de Nuevo León
San Nicolás de los Garza, Mexico

Oyetunde T. Oyeyemi
Department of Biological Sciences
University of Medical Sciences
Ondo, Nigeria

Wadzani Dauda Palnam
Crop Science Unit
Department of Agronomy
Federal University
Gashua, Nigeria

S. K. Parom
Department of Pharmaceutical Microbiology and Biotechnology, Faculty of Pharmaceutical Sciences
Kaduna State University
Kaduna, Nigeria

Tehreen Riaz
Department of Eastern Medicine
Government College University Faisalabad
Faisalabad, Pakistan

Michael Olugbenga Samuel
Cardiometabolic Research Unit
Department of Physiology
College of Health Sciences
Bowen University
Iwo, Nigeria

Hameed Shah
Department of Chemistry
University of Swabi
Ambar, Khyber Pakhtunkhwa, Pakistan

Mohammad Ali Shariati
K.G. Razumovsky Moscow State University of Technologies and Management (The First Cossack University)
Moscow, Russia

M. Y. Tatfeng
Department of Medical Laboratory Science
Niger Delta University
Amassoma, Nigeria

Benjamin Ewa Ubi
Department of Biotechnology
Ebonyi State University
Abakaliki, Nigeria
and
Arid Land Research Center
Tottori University
Tottori, Japan

Kingsley Eghonghon Ukhurebor
Applied Microbiology, Biotechnology and Nanotechnology Laboratory
Department of Microbiology
Edo University Iyamho
Auchi, Nigeria

Rida Zainab
Department of Eastern Medicine
Government College University Faisalabad
Faisalabad, Pakistan

Part 1

Working Group 04

Molecular Diagnostics and DNA Forensics

Handling Editor: Prof. Ifeoma M. Ezeonu

1 Principles and Techniques for Deoxyribonucleic Acid (DNA) Manipulation

Nwadiuto (Diuto) Esiobu
Florida Atlantic University

Ifeoma M. Ezeonu
University of Nigeria

Francisca Nwaokorie
University of Lagos

CONTENTS

DOI: 10.1201/9781003178903-2

1.1 INTRODUCTION

There are only two types of cells that constitute the three domains of life on planet earth – Prokaryotic and Eukaryotic cells, with prokaryotes constituting two (Archaea and Bacteria) of the three domains. The notion of Kingdoms has gradually faded away in light of the current understanding of the mechanistic molecular details of transcription and translation processes of cells. The domains of life are distinguished among others on the sequence of the ribosomal ribonucleic acid (rRNA) molecule which is often used as the chronometer gene. At any given point in time cells (unlike acellular entities such as viruses, viroids and prions) contain three major macromolecules that in turn control the synthesis and types of other cellular components. These include deoxyribonucleic acid (DNA), RNA and proteins. The DNA is a sophisticated double-stranded helix whose antiparallel strands are held together by hydrogen bonds. It is a polymer of nucleotides with pentose phosphate backbones linked to nitrogenous bases. The DNA in all cells is made up of four nitrogenous bases namely the Adenine and Guanine (the Purines) and Thymine and Cytosine (the Pyrimidines). The backbone of the DNA consists of the deoxy-pentose sugar and the phosphate group. This backbone has little or no role in the genetic function and coding of the DNA. It can be peeled off and replaced without changing the function of the DNA molecule (Esiobu, 2006), confirming that genetic traits are determined by the nitrogenous base sequence. The complementary base pairing of the molecule and the stereochemistry of its bases allow for high fidelity in replication and regulation of gene expression. The RNA is a polymer of ribonucleotides joined by phosphodiester bonds as in the DNA. They are, however, different from the DNA in three ways: they are single stranded, contain uracil base instead of thymine and contain hydroxyl groups in carbon number 2 of the ribose sugar backbone. The "Central Dogma of Life" describes the established flow of genetic information from the DNA to RNA and then to proteins in extant cells. DNA is the blueprint through which all of the traits and aggregate properties of a microbial cells, plants and animals (morphology, physiology, metabolism, yield, disease resistance, etc.) are governed. Expression of the DNA results in various types of RNA which in turn are translated at the ribosome into proteins.

Seeing the critical importance of the DNA in cellular structure and function, this chapter is dedicated to understanding the molecular structure of the macromolecules. It describes structures in detail, replication mechanisms and presents the fundamental techniques used in studying and manipulating the DNA of cells. A thorough understanding of the blueprint of life is indispensable to carrying out any reasonable biotechnological projects – in silico or in wet labs.

1.2 MOLECULAR STRUCTURE OF THE NUCLEIC ACIDS

1.2.1 Chemical Nature of the Nucleic Acids

Once DNA was confirmed as the molecule carrying the genetic information of cells, the next challenge that immediately faced geneticists was determination of its structure.

Two types of nucleic acids are found within cells, DNA and RNA. By the 1940s, a fair amount of information was already available about the nucleic acids. It was known that nucleic acids are large polymers composed of nucleotide subunits. Each nucleotide consists of three composite molecular groups: a central 5-carbon atom sugar, connected to a phosphate group and a single-ringed or double-ringed nitrogenous base (Figure 1.1a).

The sugar moiety differs between DNA and RNA; the exact difference being in carbon-2. In the structure depicted above, the 5-carbon sugar is ribose, with a hydroxyl group at carbon-2. This makes the nucleotide a ribonucleotide and the nucleic acid, RNA. In DNA, on the other hand, the sugar is deoxygenated at carbon-2 (deoxyribose) as shown in Figure 1.1b, making the nucleotide a deoxyribonucleotide and the nucleic acid, DNA.

The nitrogenous base in a nucleotide can be either a purine (double-ringed) or a pyrimidine (single-ringed). Four types of nitrogenous bases are found in DNA: adenine (A), guanine (G), cytosine (C) and thymine (T). Structures of these bases are shown in Figure 1.1c. RNA, on the other hand, contains the bases A, G, C, but T is replaced by another pyrimidine, uracil (U). The structure of uracil is shown in Figure 1.1d.

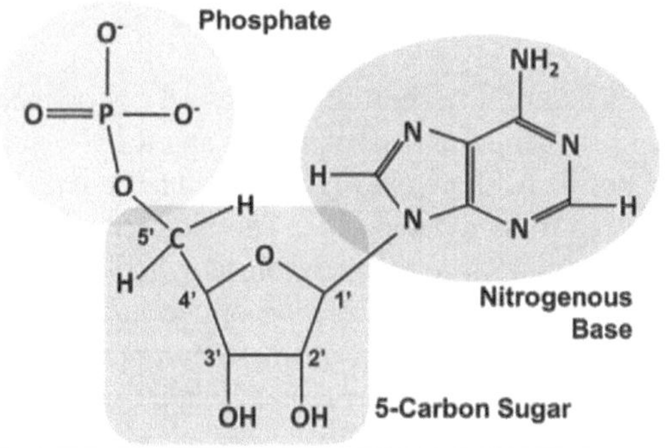

FIGURE 1.1a Molecular structure of a nucleotide.

(Source: Wikipedia.)

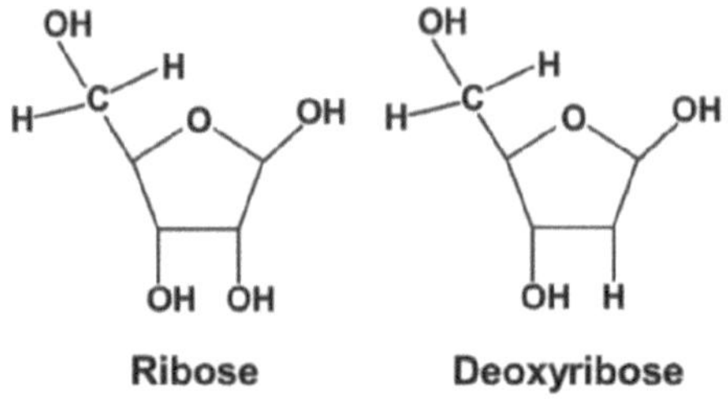

FIGURE 1.1b The chemical structures of the sugar moieties in RNA and DNA.

(Source: Wikipedia.)

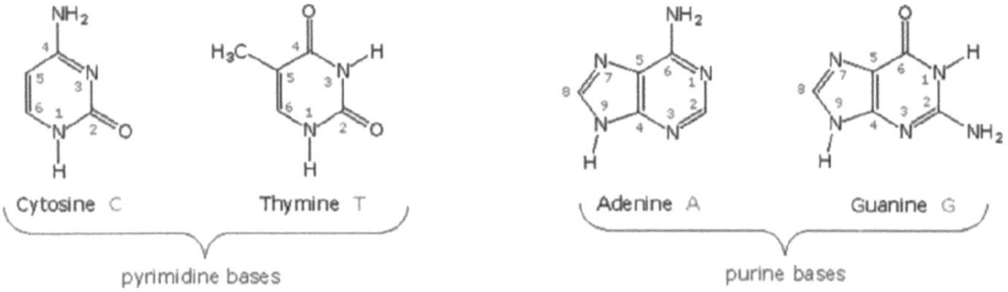

FIGURE 1.1c Structure of DNA bases. Adenine and Guanine are the purine bases, Cytosine and thymine are the pyrimidine bases.

(Source: https://www2.chemistry.msu.edu/faculty/reusch/VirtTxtJml/nucacids.htm.)

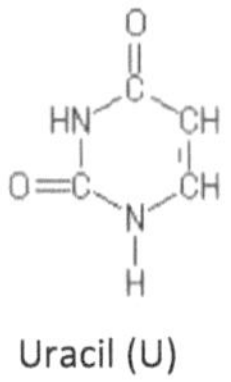

FIGURE 1.1d Structure of the pyrimidine base, Uracil; found only in RNA.

(Source: Wikipedia.)

In the nucleotide structure, the central sugar moiety is connected to the nitrogenous base by a bond between its C-1 carbon atom and the N-1 atom of a pyrimidine or the N-9 atom of a purine as shown in Figure 1.1e. The sugar and base of a nucleotide are together referred to as a nucleoside. The addition of a phosphate makes it a nucleotide. Both DNA and RNA are therefore polynucleotides, because they are polymers of nucleotide subunits. Free unincorporated nucleotides exist as triphosphates, with three phosphate groups, but during incorporation into the nucleic acid chain, only one phosphate group is incorporated, while two phosphates are released as pyrophosphate (PPi).

In DNA and RNA chains, nucleotide subunits are connected into long chains by covalent linkages formed between the phosphate group of one nucleotide and a

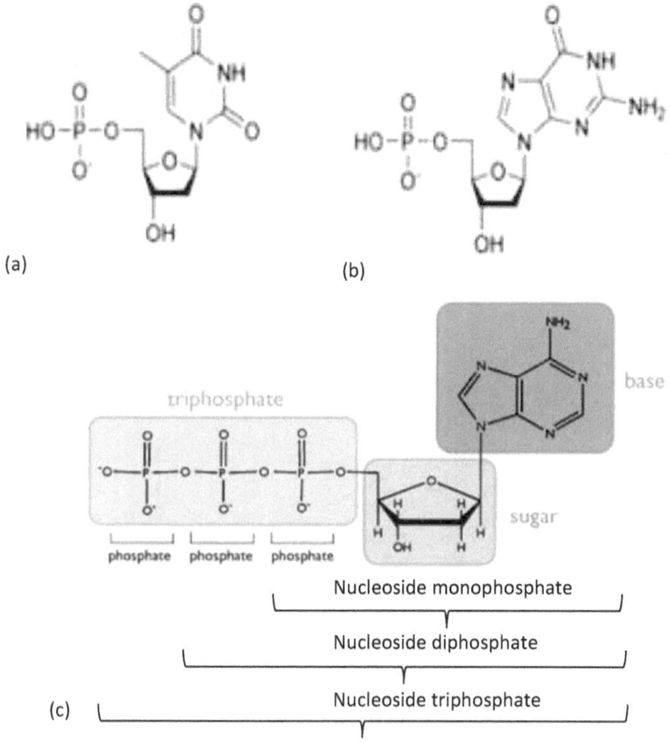

FIGURE 1.1e Basic Nucleotide Structure, showing the base, sugar and phosphate groups. In a deoxyribonucleotide, the C-1 carbon atom of a deoxyribose is bonded to N-1 of a pyrimidine (a) or N-9 of a purine (b). The sugar and base of a nucleotide are together referred to as a nucleoside. The addition of a phosphate makes it a nucleotide (c).

(Source: Wikipedia.)

hydroxyl group on another nucleotide. Such bonds are called *phosphodiester linkages.* In a growing polynucleotide chain, the hydroxyl group (OH) on carbon-3 of the sugar of an existing nucleotide serves as the acceptor molecule, to which the phosphate group on carbon-5 of the incoming nucleotide attaches (Figure 1.1f).

Because the linking of nucleotides is always from the carbon-3 OH group (the 3′ end), the growths of both DNA and RNA are directional, strictly in the 5′ to 3′ direction; the oldest end being the 5′ end, while the youngest end is the 3′ end. This also makes the 3′-OH group an extremely important functional group in a polynucleotide, as without it, the nucleotide chain would cease to grow. This property has been exploited in some DNA technologies, such as Sanger dideoxy or chain terminating sequencing technique.

Although the chemical composition of DNA was already known by the 1940s, an understanding of how its atoms were linked together and the establishment of their three-dimensional arrangements in space was necessary to learn how the information in DNA is transmitted through cell generations (Watson et al., 2013).

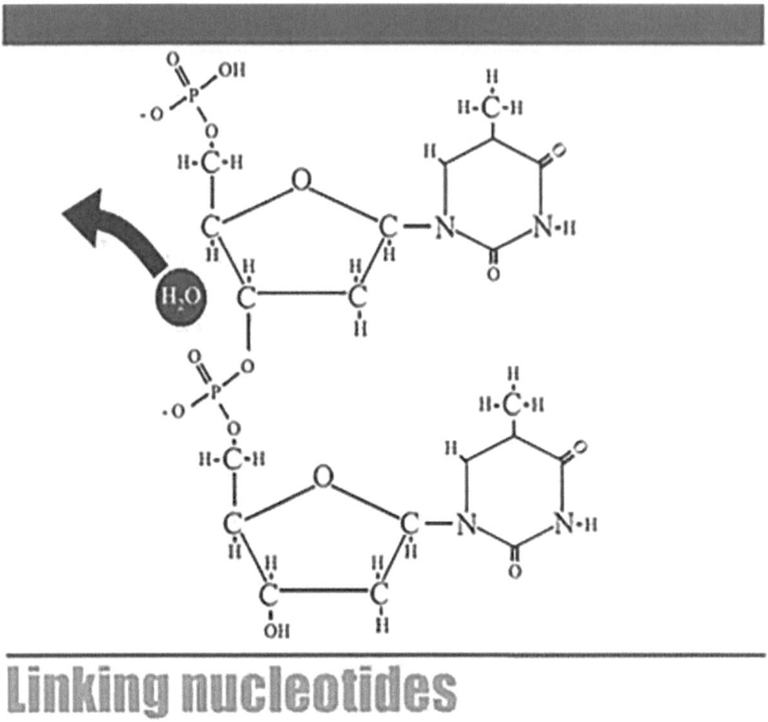

FIGURE 1.1f Formation of a phosphodiester linkage between two nucleotides.

(Source: https://gfycat.com/discover/nitrogenous-base-gifs.)

Erwin Chargaff, a biochemist working in New York, used paper chromatographic techniques to analyze the nucleotide composition of DNA in several types of organisms. By 1949, Chargaff's studies showed that the four DNA nucleotides are not present in equal amounts and that the relative ratios of the nucleotides varied from species to species. The studies also showed that the molar concentration of adenine always equaled that of thymine, while the concentration of guanine always equaled that of cytosine, regardless of the DNA source. This latter finding of A=T and G=C is known as Chargaff's rule. However, the significance of this rule only became clearer after the elucidation of the three-dimensional structure of DNA.

1.2.2 The DNA Double Helix

The first illustrations of DNA structure came from x-ray diffraction patterns produced in 1938 by an Englishman, Astbury. From these, many investigators, such as Pauling and Corey, Fraser and Furberg, tried to formulate models for DNA (Watson and Crick, 1953). Pauling and Corey's model consisted of three intertwined chains and placed the phosphate groups on the inside, while the bases were on the outside.

Watson and Crick did not agree with this model for two major reasons: (1) phosphate groups are negatively charged and placing such groups inside the intertwined chains, in close proximity, would cause the chains to repel each other. On the other hand, with the bases inside, acidic hydrogen atoms from the bases would form hydrogen bonds to hold the chains together. (2) Some of the van der Waals distances appeared to be too small.

Between 1950 and 1952, more photographs of DNA were provided by M.H. F. Wilkins and Rosalind Franklin, and in 1953, Watson and Crick proposed a different DNA model (Figure 1.1g). Much of the data they used came from the unpublished

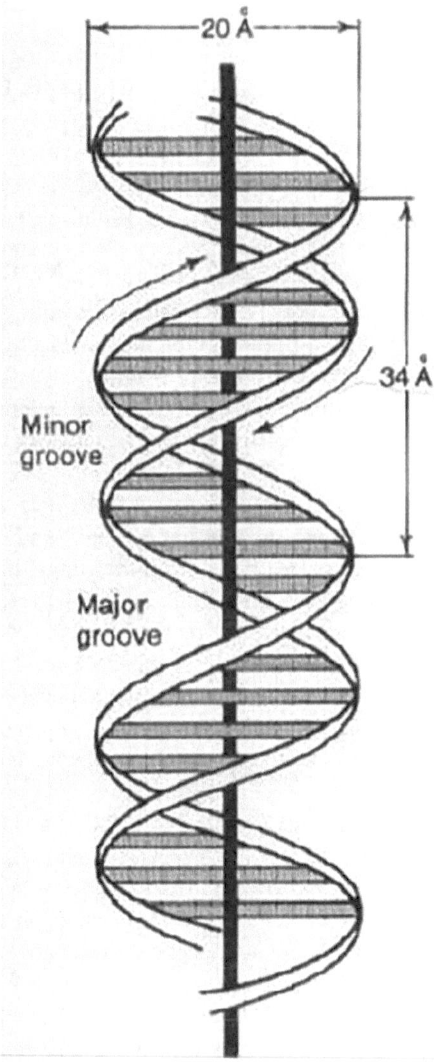

FIGURE 1.1g Watson-Crick Model of DNA.

work of Wilkins and Franklin, whose data were later published in the same issue as the paper by Watson and Crick. Important features of their model were:

1. There are two helical polynucleotide chains coiled around a common axis, with the chains running in opposite directions.
2. The bases are located on the inside of the helix, while the phosphate and sugar are on the outside.
3. The diameter of the helix is 20 A°. Adjacent bases are separated by 3.4 A° along the helix axis and related by a rotation of 36°. The helical structure repeats after 10 residues on each chain.
4. The two chains are held together by hydrogen bonds formed between pairs of bases.
5. Adenine pairs with Thymine by means of two hydrogen bonds, while Cytosine pairs with Guanine by means of three hydrogen bonds.

The most important feature of DNA is the complementarity of its two chains and this complementarity is due to the specific base pairing, A with T and G with C between the bases. It means that wherever a G is found in one strand, C will be found in the other, and wherever a T is found, its complementary strand will have an A. The molar amounts of guanine and cytosine are, therefore, identical in double-stranded DNA from any source; likewise, the molar amounts of adenine and thymine will be the same. In addition to the complementarity of the two strands, the strands are arranged in an antiparallel manner (Figure 1.1h). This means that the individual strands of the DNA molecule are arranged in a "head to toe" arrangement; one strand runs in a 5′ to 3′ direction top to bottom, while its complementary strand runs 5′ to 3′ bottom to top.

Visualized in the simple manner depicted above, DNA may well be imagined as a ladder (though a twisted one) with the alternating sugar and phosphates forming the sides of the ladder while the paired bases form the rungs. In the polynucleotide chain, the sugar and phosphate groups are regular and only the nucleotide bases differ. The sugar-phosphate backbone therefore has no role in the coding function of the DNA. Peeling it off does not alter the product of the DNA molecule (Esiobu, 2006), confirming that the genetic traits are encoded by the sequence of the nitrogenous bases. Therefore, when reference is made to a specific sequence of nucleotides in DNA (or RNA), it actually refers to the order of the purine and pyrimidine bases along the chains.

It may do, for the sake of simplicity, to imagine DNA as a ladder, but this is only so as to better describe the arrangement of its parts. In actual fact, DNA is a double helix. The two glycosidic bonds that attach a base pair to its sugar rings do not lie directly opposite each other. As a result, the two sugar-phosphate backbones of the double helix are not equally spaced along the helical axis, and the grooves that form between the backbones are not of equal size as seen in Figure 1.1g. Moreover, some DNA sequences permit the double helix to twist in the left-handed sense (Z-DNA), as opposed to the right-handed sense originally formulated. And while some DNA molecules are linear, others are circular. There are two well-characterized forms of right-handed DNA, referred to as *A* and *B*. They differ in several ways such as in

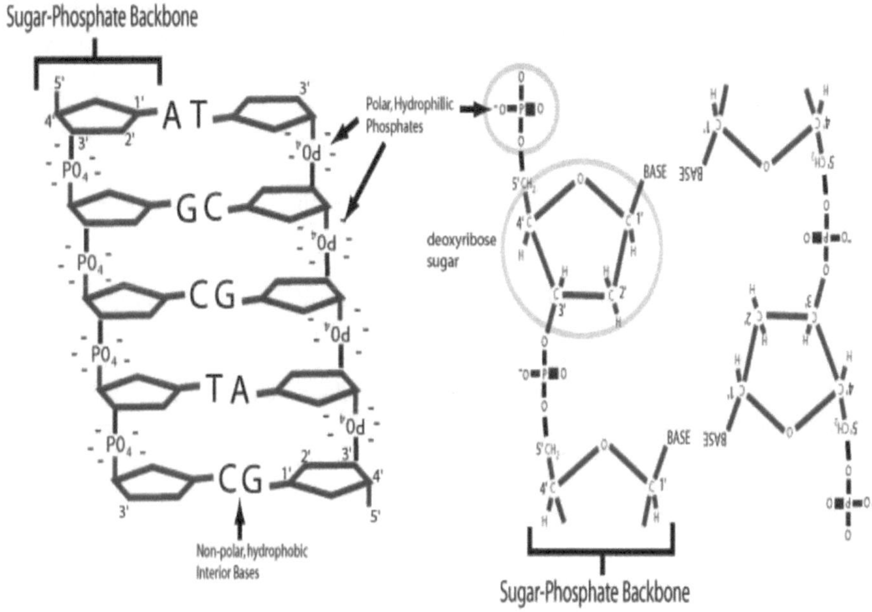

FIGURE 1.1h The structure of DNA showing the complementary and antiparallel arrangement. Note that one chain ends in 5′ hydroxyl group while the other ends in a 3′ hydroxyl.

(Source: https://www.google.com/search?q=antiparallel+arrangement+of+dna&client.)

the distance required to make a complete helical turn (the pitch), in the way their deoxyribose groups are bent, and in the angle of the tilt that the base pairs make with the helical axis. The *B* conformation is the longer, thinner form with the base pairs oriented approximately perpendicular to the helical axis. In solution, DNA usually assumes the *B* form while relatively unhydrated DNA fibers tend to assume the A form (Watson et al., 2013).

1.2.3 DNA Renatures as well as Denatures

When DNA molecules are subjected to high temperatures (95°C–100°C), their hydrogen bonds easily break, causing the complementary strands to separate from each other. This is known as denaturation. Since there are three hydrogen bonds between a G-C base pair and only two bonds between the A-T base pair, higher temperatures are required to separate G-C pairs than A-T pairs. DNA molecules having high concentration of G-C base pairs therefore have higher denaturation or melting temperatures. DNA molecules could also become partially denatured at intermediate temperature or when exposed to alkaline conditions. Under such conditions, regions of the double helix that are A-T rich will melt apart, while the C-G rich regions retain their structure. Denaturation is a reversible process. When heat-denatured DNA solutions are cooled to normal physiological temperatures, the separated strands come back together and the hydrogen bonds are re-formed. This is known as renaturation. This physiological feature of DNA is also exploited in many DNA-based technologies.

1.2.4 Functions of DNA and RNA

DNA is the repository of the genetic information of any cell. This information is stored or encoded in the DNA in the sequence of nucleotides along its chains. To access this information (during DNA replication, transcription into RNA or any other process that requires the information in DNA), the sequence of bases must be "read" in a linear fashion. The details of replication and transcription will be discussed later in the chapter.

RNA, though also a nucleic acid, is a lower-molecular-weight molecule than DNA. It is much more abundant in the cell and is distributed throughout the cell, especially in ribosomes. There are three kinds of RNA found in cells: ribosomal RNA, which is a structural component of ribosomes; transfer RNA (tRNA), which functions in protein synthesis; and messenger RNA (mRNA), which is an intermediate molecule that copies the information in DNA and transmits it to ribosomes where it is decoded into protein.

1.3 DNA REPLICATION AND GENE EXPRESSION IN PROKARYOTES AND EUKARYOTES

1.3.1 DNA Replication

DNA is the chemical basis of heredity. DNA molecules serve as the repository of genetic information in all cellular forms of life and some viruses. The DNA encodes all the information required to make all of the cell's proteins. These proteins perform the life functions of a living organism and determine their features. When a cell reproduces, it passes on its information to the daughter cells. Before a cell can reproduce, it has to make a copy of its DNA. DNA replication therefore is the process by which double-stranded DNA molecules make identical copies of themselves. This process occurs naturally during cell division in prokaryotic and eukaryotic cells, in the cytoplasm and nucleus, respectively. Irrespective of where the DNA replication occurs, however, the basic process is the same in both types of cells with some minor differences (O'Donnell et al., 2013). In both cell types, replication always starts from a site called an Origin and proceeds bidirectionally; one strand moving in the clockwise direction and the other moving counter-clockwise. However, while prokaryotic chromosomes have only one Origin, eukaryotic chromosomes are characterized by thousands of replication Origins. Not all DNA replication is bidirectional, as some bacterial plasmid DNA replicate in a unidirectional mode.

In nature, DNA replication plays an important role in cell proliferation. It occurs to enhance development of any organism during conception. DNA replication helps in the transfer of genetic information from one generation another through the daughter cells or organisms. Similarly, replication helps to grow organism during infant stage, puberty and adulthood. Furthermore, damaged or aged tissues are replaced by this process. DNA replication can also be demonstrated experimentally. This has formed the basis for several discoveries and applications in biotechnology. Our understanding of the mechanisms underlying the in-vivo DNA replication, has fueled many advances in the production of recombinant DNA used for genetic engineering, and in genetic modification. In vitro techniques like polymerase chain reaction (PCR) for example, use synthetic primers (to mimic the role of the primase and RNA primers in vivo) for "replication" of DNA segments. During cloning, the DNA must be replicated within a host to provide a large quantity of identical DNA needed for probes or other research applications. DNA replication takes place continuously and at a faster rate in prokaryotes than in eukaryotes. DNA replication is a timed event that correlates with characteristics and functionalities of gene expression, chromatin state, Guanine and Cytosine (GC) contents, and sub-nuclear structure. In biotechnology, plasmids, which are autonomously replicating circular DNA molecules found in most bacteria, are used extensively in many recombinant DNA projects.

1.3.2 Stages of DNA Replication

1.3.2.1 Initiation

DNA replication begins at specific coding regions within the chromosome, known as "**origins**." The origins are targets for initiator proteins. These proteins are responsible for recognizing a specific DNA sequence within the origin of replication. When activated, they proceed to recruit additional proteins that facilitate the replication process. The association of proteins around the Origin (*OriC*) separates the DNA strands to reveal single-stranded templates. The open region where replication is occurring is known as the **Replication Fork**. The initiation occurs in two steps. In the first step, the **initiator** protein recognizes a specific DNA sequence within the *OriC* and unwinds a short stretch of the DNA double helix. Thereafter, another protein called **helicase** attaches, hydrolyzing Adenosine Triphosphate (ATP)for energy to separate the double stranded molecule (that is, to break the hydrogen bonds holding the complimentary bases of the DNA strands), into two single DNA strands. Molecules of single-strand binding (SSB) protein, encoded by *ssb* gene, bind to the exposed single strands to prevent their coming back together. Each of these strands act as a template for replication of a new DNA strand. At a replication fork, one of the strands is oriented in the 3′ to 5′ direction towards the replication fork. This strand is known as the **leading strand**. The complementary strand is oriented antiparallel in the 5′ to 3′ direction and is called the **lagging strand** (Figure 1.2a).

In the second step, small **RNA primers**, synthesized by a special RNA polymerase called Primase, attach to the template strands to kick off the action of the replicating enzyme, **DNA polymerase III**. DNA polymerase III catalyzes the formation of a phosphodiester bond between the 5′ phosphate group of an incoming nucleotide and the 3′-hydroxyl group of an existing nucleotide, but is unable to start a chain; hence, the need for a primer to provide the 3′-hydroxyl group. When the enzyme is activated, it binds to the unzipped DNA **template** strand and subsequently "walks"

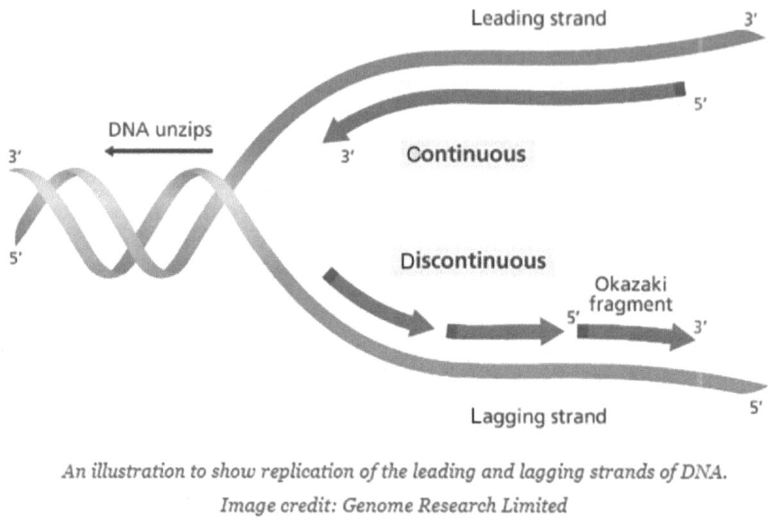

FIGURE 1.2a DNA replication fork.

(Source: https://www.yourgenome.org/facts/what-is-dna-replication.)

through it, adding new complementary nucleotide bases (A, G, C and T), as specified by the template and extending the growing DNA chain in the 5′ to 3′ direction. The replication of the two strands involves the same proteins. However, since the two strands are oriented differently, as described above, and since the replicating enzyme operates in only one particular direction (5′ → 3′), the precise mechanism of replication of the two strands differs slightly.

1.3.2.2 Elongation

Elongation is the extension of the new DNA chain from the primer. DNA polymerase extends the primer by adding free nucleotides triphosphate to the **3′-end direction** of the strand. Replication of the leading strand is simpler. Once the primer has attached at the beginning of the fork, it is a continuous process in which the DNA polymerase proceeds along the template molecule behind the helicase and SSB, incorporating nucleotides, complementary to the template.

The lagging strand is oriented in the 5′ to 3′ direction, which specifies replication in the 3′ to 5′ direction. However, DNA polymerase does not work in this direction. Consequently, replication of this strand takes place by a "backing up" mechanism, in which the DNA Primase synthesizes new RNA primers, internally and sequentially on the template strand, at intervals of approximately 200 nucleotides. Each primer is then extended by DNA polymerase until the enzyme "bumps" into the 5′ RNA end of the previous fragment. In this manner, short stretches of DNA are made discontinuously in a direction opposite that of overall replication. The short fragments, so produced are known as **Okazaki fragments**. In other to allow continuous creation of the RNA primers, the new synthesis is delayed hence the name "**lagging strand**."

1.3.2.3 Termination

The process of expanding the new DNA strands seems continuous. However, it terminates in two ways; either when there is no more DNA template left to replicate especially at the end of the chromosome or when two replication forks meet at the **terminus** called *terC*. On completion of DNA synthesis, it is important to proofread and ascertain that newly synthesized strands are bound and stabilized. In the lagging strands, these activities are performed by two enzymes: **RNAase H** and **DNA Ligase**. The enzyme **RNAase H** catalyzes the removal of the RNA primers present at the beginning of each Okazaki fragment in lagging strands. The new strand is proofread to ensure that there are no mistakes in the new DNA sequence. The **DNA polymerase I** proofreads the newly formed daughter strands and replaces RNA Primer with sections of DNA. In addition to these enzymes, eukaryotic cells produce telomeres to prevent erosion of genes. On the other hand, **DNA Ligase** joins two fragments together to create one complete strand. This enzyme seals up the sequence of DNA into two continuous double strands. Thus, the result of DNA replication is two DNA molecules consisting of one new and one old chain of nucleotides. The daughter DNA molecules produced consist of one progeny and one parental strand. This concept defines why DNA replication is described as **semi-conservative**. That is, one half of the chain is part of the original DNA molecule, while the other half is wholly brand new. When DNA replication is successfully completed, the new DNA automatically winds up into a double helix to give the characteristic feature of DNA (Figure 1.2b).

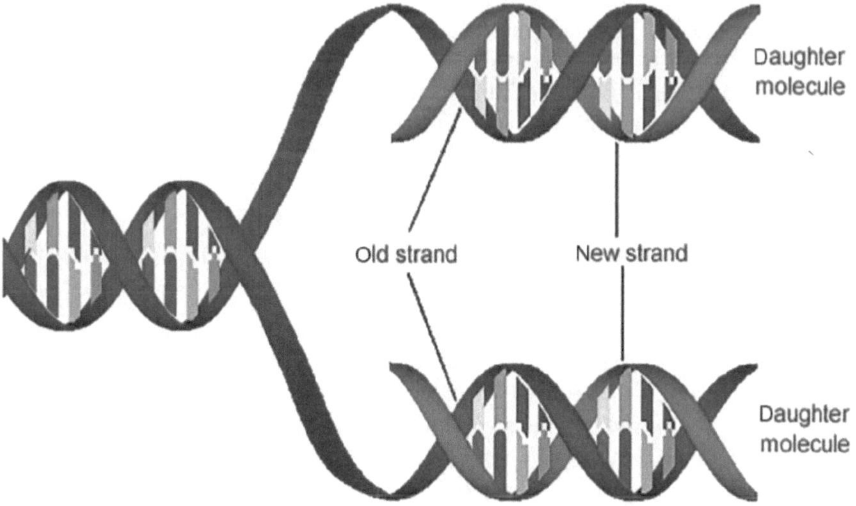

FIGURE 1.2b https://www.eduhk.hk/biotech/eng/classrm/class_gene2.html.

1.3.3 Gene Expression

DNA segments that code for polypeptide products are called **genes**. The genes are the basic units of inheritance. A gene is a segment within a strand of DNA with specific instruction for the production of one specific polypeptide. It is located on a chromosome in its place or locus in the cell nucleus or mitochondria. Genes are sequences of nucleotides in DNA or RNA (for some viruses) that code for a functional molecule. The specific sequence of bases provides instructions on how to make the proteins. Genes therefore code for the production of proteins and proteins dictate cell function. There are, however, other DNA sequences that either have structural purposes or are involved in regulating the expression of genes.

Gene expression is a process by which a gene product is made. Gene products are usually **proteins**. During gene expression, the information from a gene is used in a two-step process to produce a functional gene product. In the first step, the DNA is copied into an RNA molecule. The RNA can be directly functional or can be the intermediate template for a protein that performs a particular function. The latter type is known as messenger RNA (mRNA). The information flow from DNA to RNA is possible because the DNA nucleotide sequence specifies the mRNA sequence, through complementary base-pairing. The normal base-pairing rules apply in RNA synthesis. However, the base thymine is replaced with the base **uracil**. That is, in RNA, uracil is placed instead of thymine. In the second step of gene expression, the information contained in the sequence of bases in mRNA is decoded by the ribosomes, to synthesize a protein product. This shows the flow of information from DNA to RNA and then to protein. This unidirectional flow of genetic information from DNA → mRNA → Protein is known as the "**central dogma of genetics**." The two steps of gene expression are known as **transcription** and **translation**, respectively. By definition therefore, transcription is the production of an mRNA from DNA, while translation is the production of a polypeptide product from mRNA. Each of the steps in gene expression is subject to regulation and as such, a cell is able to control and adjust the amount and types of proteins produced at any particular point in its existence. In prokaryotes, lacking internal compartmentalization, both transcription and translation occur in the cytoplasm, but in eukaryotes, the mRNA is transported out of the cell nucleus and into the cytoplasm, for translation to occur.

1.3.3.1 Transcription

Transcription is catalyzed by an enzyme, RNA polymerase (RNA-P). RNA polymerase synthesizes an mRNA, using four ribonucleoside triphosphates (ATP, CTP, GTP and UTP) as substrate and one DNA strand as a template. It does not require a primer. The RNA-P of prokaryotes has the same subunit structure, $\alpha_2\beta\beta'\sigma$, being composed of four different subunits, one of which (α) is present in two copies. The σ (**sigma**) subunit, also called **sigma factor** or **transcription factor**, has the function of promoter recognition. A **promoter** is a sequence of bases that serves as binding site of RNA-P. Every gene possesses a promoter sequence. Eukaryotes have three types of RNA polymerase: RNA-P I, RNA-P II and RNA-P III, but most of the transcription in the nucleus of the eukaryotic cell is carried out by RNA-P II. Transcription involves three steps in prokaryotes and four steps in eukaryotes:

1. **Initiation** – To initiate transcription, the sigma subunit of RNA-P locates the promoter and RNA-P binds to the promoter. The promoter not only signals where transcription begins, but it also determines which DNA strand will be transcribed, because the sequence of bases on the template strand determines how RNA-P binds to it. The binding causes the hydrogen bonds between the two DNA strands to break, forming an open region (**open complex or transcription bubble**) of single-stranded DNA (Figure 1.2c).
2. **Elongation** – Once RNA-P has bound successfully, polymerization then begins with RNA-P moving down the template DNA strand, synthesizing an mRNA molecule as specified by the sequence of bases on the template DNA. The open complex or bubble moves with it (base pairs of DNA being opened in front and closed behind), and the growing transcript trails behind. Shortly after (when transcription is about 8 bases long), the process reaches maximum speed. The sigma factor having fulfilled its function of promoter recognition is released.
3. **Termination** – In prokaryotes, there are two ways in which transcription is terminated.
   i. **Rho-dependent termination**: Here, a protein called "**Rho factor**" is responsible for disrupting the complex involving the template strand, RNA polymerase and RNA molecule. The Rho protein is an unwinding enzyme that travels along the newly synthesized mRNA, unwinding it from the DNA template. If RNA-P moving ahead of Rho factor pauses in its progress, Rho factor catches up with it and dislodges it from the template, causing transcription termination.
   ii. **Rho-independent termination**: Alternatively, termination may occur after RNA-P has transcribed a G-C rich region of DNA followed by a series of A residues, arranged in a pattern (an inverted repeat) that enables a stem and loop structure to form in the newly made RNA molecule. The stem and loop structure forms within the transcription bubble and causes the process to stall. The subsequent A-U pairs between transcript and template make weak connections that are easily broken, thereby releasing the transcript.

   Termination in eukaryotes is more complicated, involving the addition of adenine nucleotides at the 3′ of the RNA transcript. This process is referred to as **Polyadenylation**.
4. **Processing** – In eukaryotes, the RNA molecule is processed in numerous ways after synthesis: Introns are removed and the exons are spliced together to form a mature mRNA molecule consisting of a single protein-coding sequence.

1.3.3.2 Translation (Protein Synthesis)

The structure, and ultimately function, of a protein is determined by its amino acid sequence. The key objective of protein synthesis, therefore, is the placement of the proper amino acid at the proper place in the polypeptide chain. This is the role of

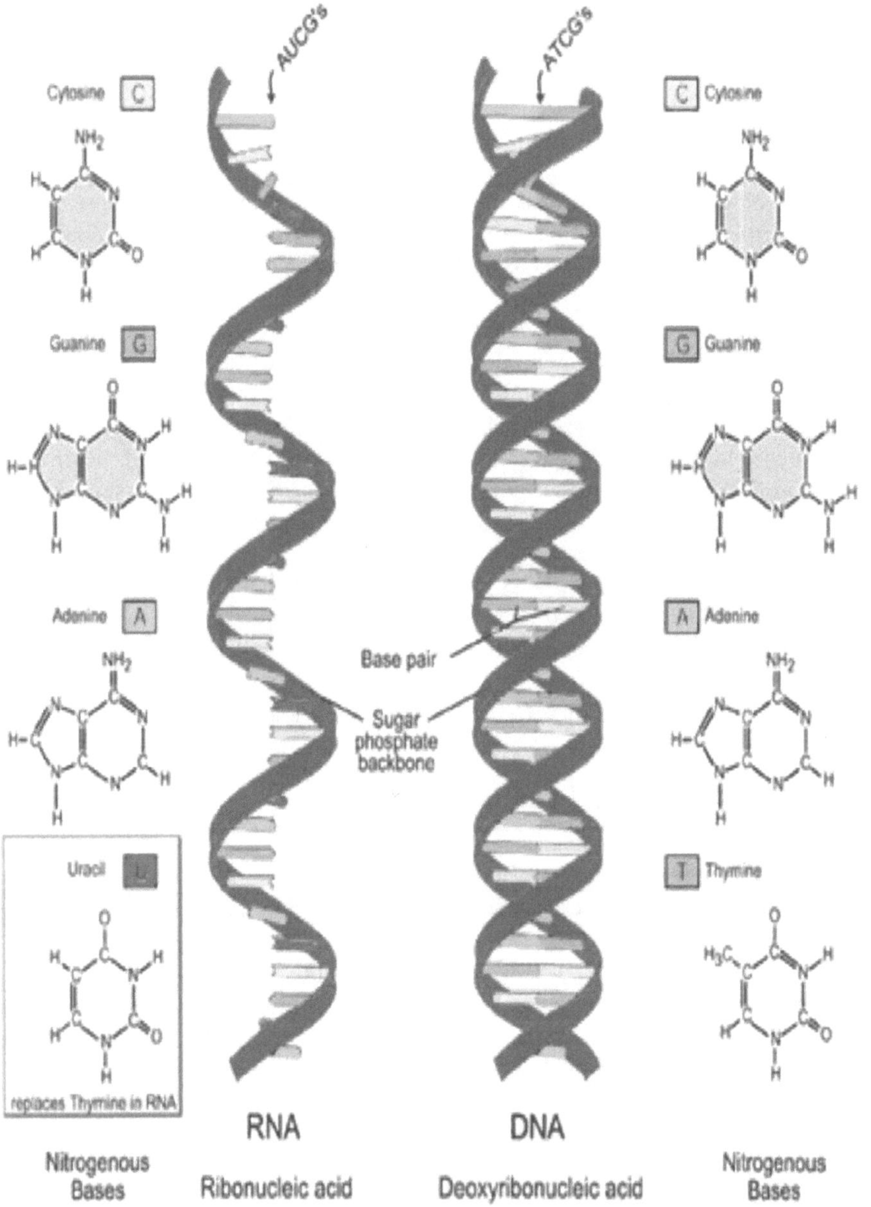

FIGURE 1.2c Transcription process.

(Source: http://www.genome.gov/Pages/Hyperion/DIR/VIP/Glossary/Illustration/rna.shtml.)

the protein-synthesizing machinery of the cell which includes **ribosomes**; tRNAs' several initiation, elongation and termination factors; and of course the transcript (Figure 1.2d). Translation or protein synthesis takes place in four steps: initiation, elongation, termination and post-translation processing.

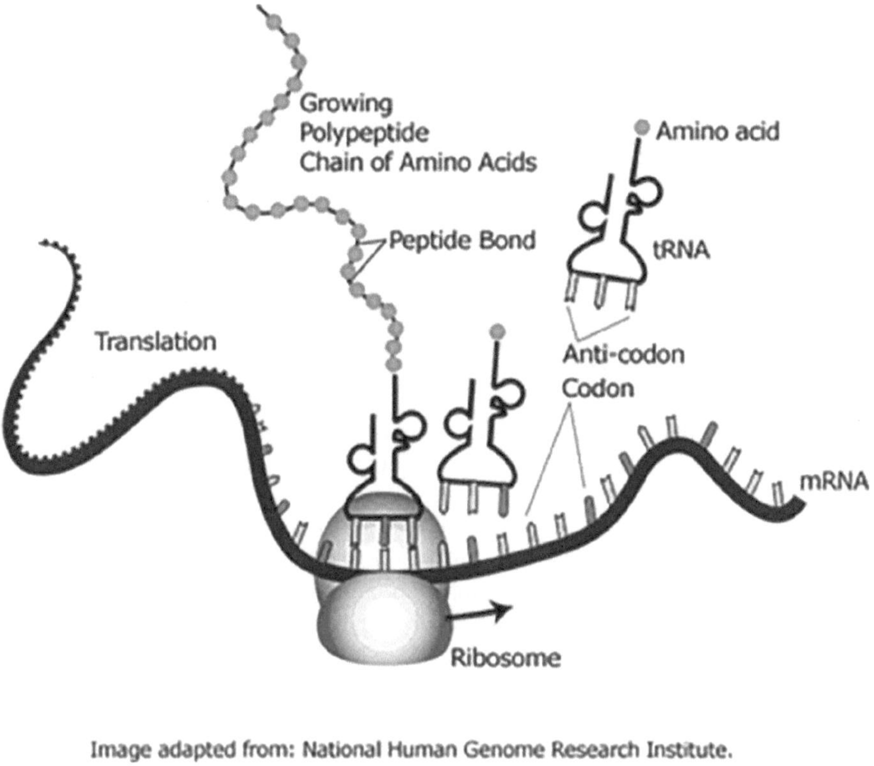

FIGURE 1.2d Translation process.

(Source: Image adapted from National Human Genome Research Institute.)

1. **Initiation**. At this stage, small subunit of the ribosome binds at the 5′ end of the mRNA molecule and moves in a 3′ direction until it meets a start codon (AUG). This is followed by the formation of a "complex" with the large unit of the ribosome complex and an initiation tRNA molecule.
2. **Elongation**. Subsequent codons on the mRNA molecule regulate which tRNA molecule linked to an amino acid would bind to the mRNA. An enzyme peptidyl transferase links the amino acids together using peptide bonds. This continues to produce a chain of amino acids as the ribosome moves along the mRNA molecule.
3. **Termination**. The process of translation is terminated when the ribosomal complex reached one or more stop codons (UAA, UAG and UGA). The ribosomal complex in eukaryotes is larger and more complicated than that in the prokaryotes. Similarly, the processes of transcription and translation are divided in eukaryotes between the nucleus (transcription) and the cytoplasm (translation), which provides more opportunities for the regulation of gene expression.
4. **Post-translation processing of the protein**. Once a peptide or protein has been synthesized and released from the ribosome it often undergoes further

chemical transformation. This post-translational modification may involve the attachment of other moieties such as acyl groups, alkyl groups, phosphates, sulfates, lipids and carbohydrates.

1.3.4 Regulation of Gene Expression

Gene regulation is the process of controlling the rate and manner in which a gene is expressed. A complex set of interactions between genes, RNA molecules, proteins (including transcription factors) and other components of the expression system determine when and where specific genes are activated. It also determines the amount of protein or RNA product produced. The amounts and types of mRNA molecules produced during transcription in a cell usually reflect the metabolic state of that cell.

Depending on the stage at which the expression of a gene is regulated, regulation may be one of four types: (1) Transcriptional regulation – determines whether the gene will be transcribed or not; (2) Post-transcriptional regulation – regulation of the processing, transport and longevity of mRNAs; (3) Translational regulation – determination of whether or not and to what extent an mRNA will be translated; and (4) Post-translational regulation – regulation of protein modification and activity. The commonest type of regulation is transcriptional regulation and within that, the initiation of transcription.

1.3.4.1 Regulation of Gene Expression in Eukaryotes

In eukaryotes, gene expression occurs in both the nucleus and cytoplasm, but since transcription occurs in the nucleus, most of the regulation occurs in the nucleus. All types of regulation described above occur in eukaryotic cells.

1.3.4.2 Regulation of Gene Expression in Prokaryotes

Prokaryotic cells like bacteria and archaea lack internal cellular bodies (organelles, including nucleus), while eukaryotic cells, for example animals, have a "true" nucleus containing their DNA. However, both eukaryotes and prokaryotes contain large RNA/protein structures called **ribosomes**, which produce protein. In prokaryotes, DNA floats freely within the cytoplasm and therefore the process of gene expression (both transcription and translation) occurs within the cytoplasm. In these cells, the processes of transcription and translation occur almost simultaneously. In prokaryotes, synthesis of specific proteins usually continues until the protein is no longer needed. Thus, regulation of gene expression in prokaryotes is primarily at the **transcriptional level**.

1.4 LABORATORY TECHNIQUES FOR DNA EXTRACTION FROM VARIOUS CELL TYPES

1.4.1 DNA Isolation, Purification and Quantification

DNA is found within all living organisms and contains the genetic code responsible for function and different phenotypic features. It is made up of nucleotide subunits, each of which contains a deoxyribose sugar, a phosphate group and a nitrogenous base. The 5′ phosphate group of DNA gives it an overall negative charge which is exploited in DNA extraction/isolation.

1.4.1.1 DNA Extraction Steps

The three main steps of DNA isolation involve (1) cell lysis using either a chemical method using sodium dodecyl sulfate (SDS), a detergent that dissolves the phospholipid bilayer of cells or mechanical (microscopic beads), or heating method to also lyse the cells, and (2) protein denaturation – through the use of proteinase K and sometimes heat to denature enzymes like DNases which can degrade DNA and other cell proteins. Proteins are considered contaminants in DNA extraction and their presence is used to measure the purity of the final DNA sample. (3) DNA precipitation and purification are the final steps which use a salt and cold 90%–100% ethanol to precipitate DNA out of solution. The salt is used to neutralize the negative charge of the phosphate group in DNA and along with the cold ethanol keeps the DNA precipitated. The DNA can be further purified using specialized kits which consists of DNA binding resins or silica membrane and alcohol wash buffers to remove additional impurities (Figure 1.3a). DNA is normally rehydrated in sterile deionized water or Tris-HCL EDTA (TE) buffer at pH 8.

1.4.1.2 Measuring DNA Concentration and Purity

The purity and concentration of the extracted DNA is important for downstream processes of DNA analysis including PCR, restriction enzyme analysis and DNA sequencing. A spectrophotometer or fluorometer is normally used which can measure DNA in small microliter volumes. DNA absorbs optimally at 260 nm wavelength

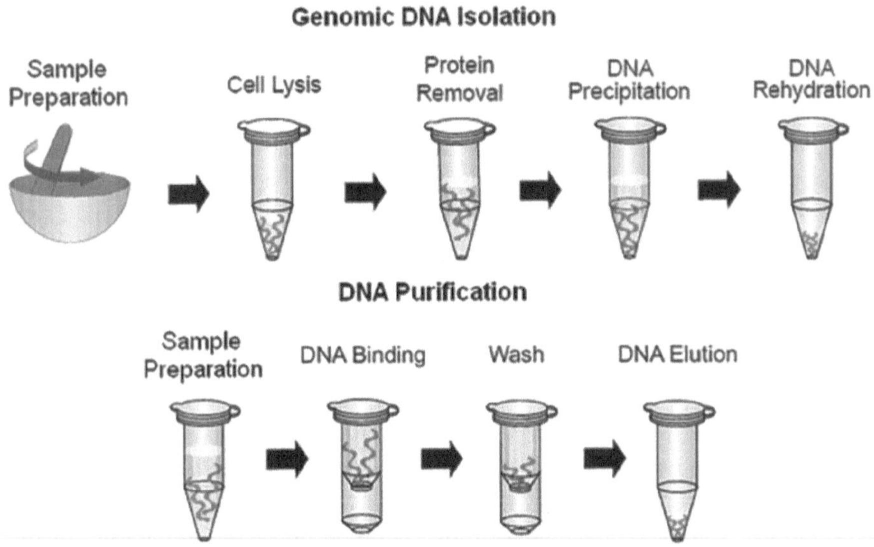

FIGURE 1.3a Major steps involved in DNA isolation and purification.

of light. The standard used to calculate DNA concentration per 1.0 absorbance reading is 50 ng/μl. Proteins however absorb optimally at 280 nm. The ratio of DNA absorbance at 260 nm and proteins at 280 nm is used to calculate the purity of a DNA sample. A value >1.7 is generally considered to be a pure DNA sample.

1.4.2 Agarose Gel Electrophoresis and Polyacrylamide Gel Electrophoresis (PAGE)

The 5' phosphate groups of both DNA and RNA give them an overall negative charge which is exploited in nucleic acid extraction/isolation. When agarose is used as the matrix in gel electrophoresis, DNA and RNA molecules are separated on the basis of their molecular size and charge, as they migrate through the agarose gel pores. The pore size of the gel matrix is determined by the concentration of agarose; as concentration of agarose increases, the pore size will decrease. Therefore, larger molecules will move slower through the gel in comparison with the smaller DNA fragments. As electrical current passes through the gel, the DNA migrates moving from the cathode (–) to the anode (+); the smaller fragments moving faster (Figure 1.3b). Consequently, the smaller fragments migrate faster towards the positive anode. Other factors such as the nature of the DNA – supercoiled, relaxed, linear or circular – also affect their rate of migration through the gel matrix. Fluorescent dyes such as Sybr green and Ethidium bromide (EtBr) are used to visualize DNA bands under UV light. A gel-loading dye added to each sample permits the monitoring of the migration of DNA/RNA by running ahead since it contains an intensely colored dye (e.g. methylene blue) of low molecular weight. The loading dye also contains glycerol to increase the density of the DNA sample and allow it to fall within the wells/holes created within the agarose. Every gel run includes a reference of known molecular weight DNA markers, which helps assess the quality of the electrophoresis and is used to calculate the size of experimental DNA/RNA bands. Agarose gel electrophoresis is used to separate nucleic acid (DNA/RNA) ranging between 50 bp and ~15 kb. The migration rate of fragments is inversely proportional to the log of its molecular weight.

The use of E-Gel techniques and precast agarose gels is a growing trend in well-funded labs because it simplifies the workflow and provides results in real time. E-Gel protocols employ ready-made precast gels containing DNA stain is a self-contained disposable cassette. It does not require buffers, staining or gel boxes. Instead DNA/RNA samples and ladder are loaded on the gel cassette and loaded on an E-Gel Power snap electrophoresis system which generates the electric field and shows the gel migration in real time. This system drastically simplifies workflow but will be of limited application when downstream extraction and purification of a particular fragment is desired.

Agarose gel electrophoresis has numerous applications involving extracted DNA or RNA. It is an important step in the validation of extracted DNA and RNA to assess the purity of the sample denoted by a lack of smearing and presence of distinct bands. Concentration can also be estimated which is proportional to the intensity of the band seen under UV light. Other applications include confirmation of PCR products, restriction enzyme digestion just to name a few.

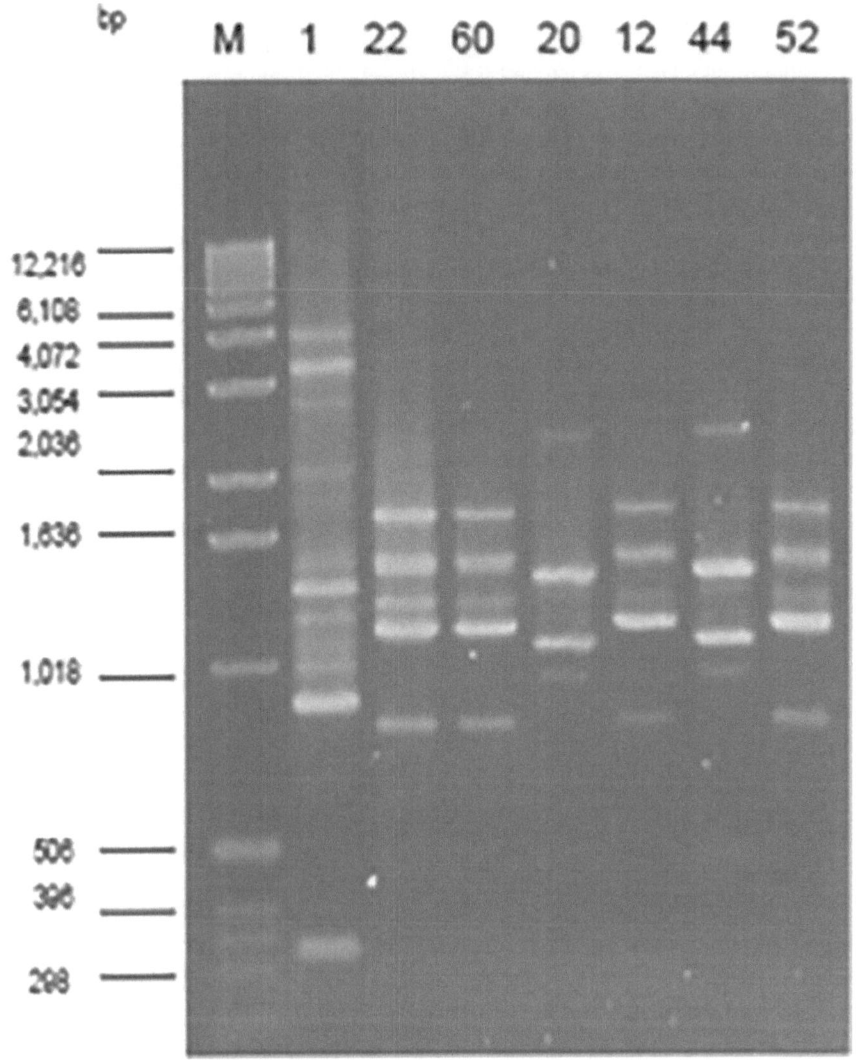

FIGURE 1.3b Gel electrophoresis for AP-PCR.

(Source: Nwaokorie et al., 2012.)

1.4.2.1 Polyacrylamide Gel Electrophoresis Principle

PAGE (polyacrylamide gel electrophoresis) is a type of gel technique often used to separate components of a protein mixture based on their size. Acrylamide is the gel matrix, instead of agarose. It is soluble in water, and readily polymerizes in solution, resulting in the formation of polyacrylamide. To induce the polymerization of acrylamide and bis-acrylamide monomers, ammonium persulfate (APS), which spontaneously decomposes to form free radicals, is added to the gel. TEMED, a free radical stabilizer, is also included to promote polymerization. As in agarose gels, increased concentrations of acrylamide results in decreased pore size after polymerization. The experimenter determines the concentration of the polyacrylamide gel based on the size of the molecules being studied. Like all gel electrophoresis techniques, PAGE is based on the principle that a charged molecule will migrate in an electric field towards an electrode with opposite charge. However, to accurately determine the molecular weight of biological molecules, the mobility of a substance in the gel cannot depend on both charge and size since the varying charges complicate the rate of migration through the matrix! So, during PAGE, the biological samples are pre-treated so that they acquire uniform charge, allowing the electrophoretic mobility to be driven primarily by size. Different protein molecules with differing shapes and sizes are denatured with the aid of an anionic detergent – SDS whose interaction with proteins results in the loss of their secondary, tertiary or quaternary structures. The proteins mixed with SDS are uniformly negatively charged. Loading such samples onto a gel in a vertical electric field (created by the power supply) allows migration towards the anode at a rate determined by a molecular sieving that is based on size (Figure 1.3c). Upon completion of the run, gels are visualized after staining and destaining techniques, with dyes such as Coomasie brilliant blue. The size of a protein can be calculated by comparing its migration distance with that of a known molecular weight ladder.

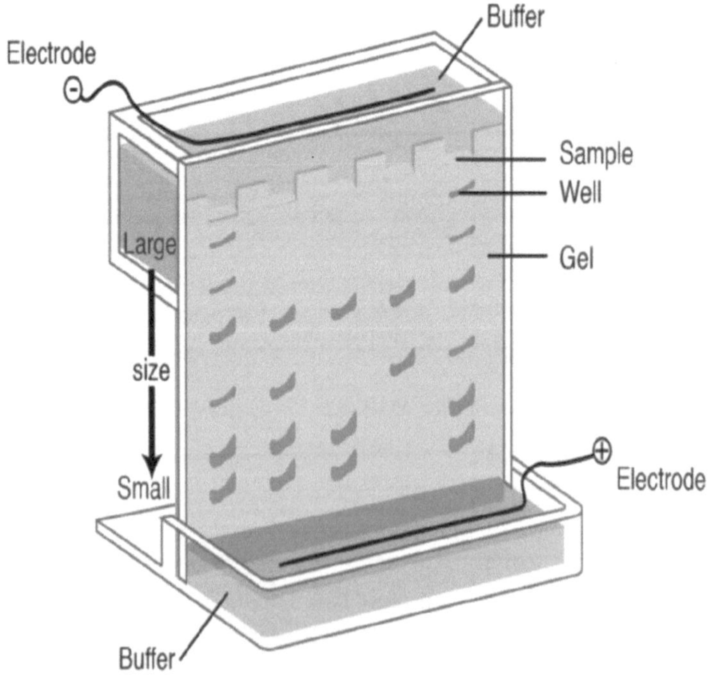

FIGURE 1.3c PAGE apparatus and principle.

1.4.3 Other DNA-Based Molecular Techniques

1.4.3.1 Denatured Gradient Gel Electrophoresis

DGGE (denatured gradient gel electrophoresis) is a type of gel electrophoresis in which constant heat (about 60°C) and an increasing concentration of denaturing chemicals (in a gel matrix) are used to unwind (i.e. denature) DNA molecules. The single strands are then forced through the varying pores of the vertical electrophoresis system in a process, so powerful it can distinguish single nucleotide polymorphisms (DNA that differs only by one nucleotide) in DNA samples. As already described in preceding sections, electrophoresis techniques separate molecules (DNA, RNA and Protein) on the basis of their electrical charge, shape and molecular weight. In DGGE, DNA is made to migrate through the pores of a polyacrylamide gel of varying concentrations. The DNA unwinds (denatures or melts) as it reaches the denaturing reagents, allowing differential sequences to migrate according to the size and base composition of the fragment. DGGE applications include distinguishing between mutated and wild type sequences without prior knowledge of what these sequences are, detection of mutations within closely related organisms and also organismal differences within an ecosystem. Using SNP detections, DGGE can be used in cloning analysis and cancer cell detection. The resulting gel picture bands can be analyzed using software such as Gel Compar II or Bionumerics to provide statistical analysis and UPGMA clustering profiles.

1.4.3.2 Recombinant DNA Technologies

This is one of the most expansive and important fields in biotechnology. It is the driving force behind most discoveries and applications of the DNA technologies, combining other techniques to create new products used in improving lives on earth. Recombinant DNA technology requires several of the following steps:

i. Extraction and purification of the nucleic acid containing the gene of interest (DNA or RNA)

   The details are described elsewhere in this chapter.

ii. Isolation of the gene on interest by restriction enzyme or PCR

   See sections of this chapter on restriction enzyme and PCR.

   Sometimes it is necessary to find a gene without the knowledge of the gene's sequence. In such a situation, a genomic library is constructed by cleaving the genome of an organism into many fragments. Each fragment is then cloned into a vector to make a unique recombinant. The genomic library (preserved as cell clones) is then screened for the gene of interest by either phenotypic rescue (i.e. genetic complementation) or DNA hybridization with oligonucleotide probes. These days it is far easier to utilize the extensive genomic resources on global databases such as the National Center for Biotechnology Information (NCBI) – https://www.ncbi.nlm.nih.gov/ – than to start from scratch to screen for a gene of unknown sequence in a genomic library.

iii. The propagation of many copies of the nucleotide sequence of interest through cloning

There are four main types of cloning vectors: plasmids, viruses, cosmids and artificial chromosomes. A good cloning vector should have an origin of replication, a selectable marker and a multicloning site or polylinker for gene insertion. The most common approach to cloning is to digest both vector and DNA of interest with a same restriction enzyme, incubate both in the presence of DNA ligase, which links the gene and the vector through formation of a phosphodiester bond. In the next step, the recombinant plasmid is introduced into a compatible host. This is done through chemical-mediated transformation of cells or by electroporation (Holmes and Esiobu, 2016). For animal and plant cells, gene guns are sometimes used. The most commonly used hosts are *E. coli* and yeasts – *Saccharomyces cerevisiae*. After transformation, the recombinant clones are then selected using colorimetric or antibiotic resistance markers.

iv. Expression of the genetic traits in a compatible organism

An expression vector is used when it is important to produce large quantities of the desired protein. Such vectors have features like a fast promoter that allow high expression levels of the gene insert. When a eukaryotic gene is expressed in a bacterium, cDNA is used instead of the eukaryotic gene because it lacks introns. In addition, a bacterial leader must be added at the 5′ end of the gene.

v. Purification of desired end products; including but not limited to genes for GMOs, proteins as in insulin, diagnostic probes for disease detection and much more. Purification of recombinant proteins is achieved by fusing the coding region of the protein of interest to multiple histidine residue codons (usually 6) and then using the His-tagged protein to selectively bind and purify the protein of interest. Some expression vectors contain the histidine tags codons.

1.4.3.3 Polymerase Chain Reaction – Types and Applications

The PCR techniques involve in vitro amplification of nucleic acid templates. In 1985, Kary Mullis developed a thermal cycler that automated the process of PCR, allowing an exponential growth of molecular manipulations of DNA in the laboratory. DNA polymerases are used to synthesize new strands of DNA complementary to the original template sequence. The DNA polymerase can only add a nucleotide to preexisting 3′-OH group. This requirement is accomplished by the use of primers, to which it adds the first nucleotide. In addition, well-designed primers are used to delineate the segment of the DNA for in vitro amplification. At the end of the reaction, millions of copies of the specified sequence are generated.

Typically, a basic PCR reaction tube contains a DNA template, primers, nucleotides, DNA polymerase and a buffer.

- DNA template is the experimenter's sample DNA that contains the target sequence. It could be any form of DNA – plasmids DNA, genomic DNA, phage DNA or small amounts of tissue DNA – typically used at low concentrations of 1 pg–1 ng of plasmid or viral templates, 1 ng–1 µg of genomic templates. At the beginning of the reaction, high temperature is applied to the original double-stranded DNA molecule to denature it.

- DNA polymerase is the enzyme that synthesizes new strands of DNA complementary to the target sequence. Several natural and engineered DNA polymerases are used in PCR. The first of these enzymes is the *Taq* DNA polymerase (isolated from *Thermus aquaticus*). The polymerases used in classic PCR are heat tolerant (up to 95°C), allowing for stability during DNA denaturing steps. It performs optimally at 72°C.
- Primers are short pieces of single-stranded DNA (typically 15–25 nucleotides) that bind complementarily to the target sequence, delineating the region of interest and start point for the polymerase to synthesize new DNA. Commercial labs usually synthesize custom or universal oligonucleotide primer sequences on demand by the researchers. The stringency of the Primer annealing is critical for a successful PCR reaction. As a result, the first step to achieving reliable PCR results is to use well-designed primers with high specificity to the DNA region of interest. The forward and reverse primers anneal to the template DNA segment through complementary base pairing, with the 5′ end primer matching the 3′ end of one strand of template DNA segment, and the other primer binds similarly to the second strand as shown in Figure 1.3d.
- Nucleotides (dNTPs or deoxynucleotide triphosphates) – these are essentially the "building blocks" for new DNA strands. During PCR procedures, four types of deoxynucleotide triphosphates (dATP, dCTP, dGTP, dTTP), known as dNTPs, are mixed with the enzyme, primer and template; DNA polymerase picks up these free deoxynucleotides from the reaction mixture and incorporates them at the 3′ end of a primer or growing DNA chain, following base complementarity along the sequence of nucleotides on the template DNA.
- *PCR buffer* – this helps to maintain the right pH and correct balance of ions (especially Magnesium during the reaction cycles especially magnesium (Mg) ions, for enzymes to work). A typical PCR buffer stock solution is provided in a 10× or 5× format, which would need to be diluted to 1× in the PCR reaction.

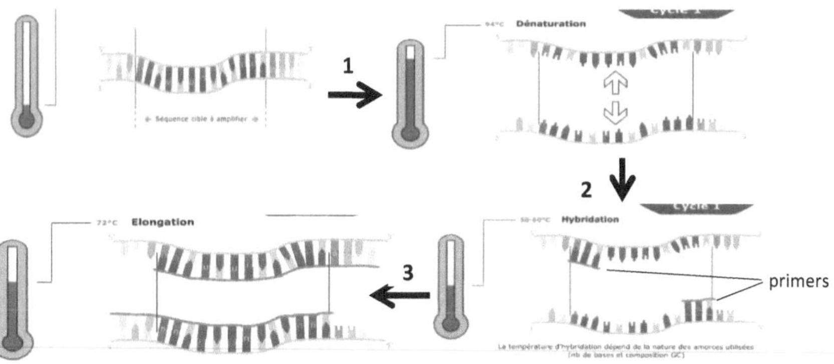

FIGURE 1.3d Sequence of reactions in PCR.

PCR is basically many cycles of three steps:

1. Heating at 94°C–95°C to denature the template DNA (denaturation step)
2. Cooling to 50°C–60°C for primers to anneal to template (annealing step)
3. Heating at 72°C to allow elongation of the new DNA strands (elongation step)

The thermal cycler is programmed to run the PCR reactions which usually consist of a series of 20–40 repeated cycles of the three steps indicated above, with calculated ramps for changes in temperature. Prior to the three repeated steps is a single denaturation temperature step called the initialization step at a high temperature (>90°C) before the cycling starts. The final step is an extended elongation step (70°C–74°C) to allow the polymerases complete the reaction terminals, and then a hold step at 4°C.

The PCR technique has found many useful applications:

- Microbial diagnostics – The use of universal and gene-specific primers allows for the detection of fastidious and resistant pathogens, including those yet to be cultured. Because a specific sequence can be amplified greatly, much less clinical material is needed to make a diagnosis. Furthermore, PCR can be used to detect the presence of resistance genes (Ezeonu et al., 2018), as well as viruses.
- Forensics – The ease and high sensitivity of the PCR technique has helped to drastically enhance forensic investigations using trace amounts of biological material such as semen, blood and hair roots. Probes and or primers for hypervariable regions in the population could be good markers of the DNA source many decades after a crime. PCR can be used to evaluate evidence at the scene of a crime, and help to identify missing persons, and resolve paternity cases.

Basic research – A comparison of sequence homology of conserved genes in different organisms can be made. It is even possible to study extinct organisms using samples of material from bones or museum specimens. The DNA used for PCR amplification can be partially degraded. As long as a few DNA molecules are intact between the two primers, amplification can occur.

The ease and speed associated with the PCR technique has made it the method of choice for amplifying nucleic acids for sequencing. However, one drawback to PCR is that specific information about the target DNA is required to make specific primers. For standard PCR, which uses two primers, sequence information must be obtained for both ends of the target DNA. Variations of the standard PCR technique have been developed. Examples include quantitative PCR (qPCR) and Random Amplified Polymorphic DNA (RAPD).

qPCR – Quantitative PCR, commonly referred to as qPCR or qPCR, is an improvement of the standard PCR, which allows for both detection and quantification of the target DNA. The technique assesses the amplification process of the PCR cycles in real time. The qPCR technique is ideal for monitoring specific DNA material in a mixture of DNA from different sources, for example, viral DNA from human viral infections. The qPCR technique is also the method of choice for monitoring the expression of genes in the human genome (Vira et al., 2016).

Multiplex PCR – Multiplex PCR is a variant of PCR, which allows the simultaneous amplification of many DNA targets in one reaction by applying different primer pairs in the same reaction. For instance, it could be used to detect the presence of several virulence genes from the DNA extracted from the pure culture of a pathogen. It is also used in virology to detect different types of viruses from a sample. The different specific primers used in the reaction are able to specifically find their targets and mark them for amplification.

Nested PCR – Involves two sequential PCR reactions for the same target region. In the first PCR reaction, a primer external to the sequence of interest is used. That is, the reaction targets and amplifies a region including flanking sequences of the DNA of interest. In the second PCR, the product of the first PCR is used as template DNA, in conjunction with one or more primers, with recognition for a sequence within the previous amplicon. In other words, the second amplicon has a size smaller than the first.

RAPD – This technique is a simplified version of PCR requiring only a single primer. The primer size is normally 10–15 nucleotides. The primer can be designed without any genetic information (it is not specific for any gene in particular). The RAPD locus (amplified region) is usually less than 3,000 bp, and a single primer can amplify from 3 to 20 loci within the DNA.

1.4.3.4 Restriction Fragment Length Polymorphism (RFLP) and Ribotyping

Small subunit ribosomal RNA (16S in prokaryotes such as bacteria and archaea, and 18S in eukaryotes such as fungi, algae, animals and plants) can be analyzed without sequencing to distinguish individual isolates or metagenomic DNA as in the oral microbiome studies by Chakraborty et al. (2014). Because restriction enzymes only cut at their recognition sequences, when DNA molecules differ, even in one nucleotide, it is possible to generate bands (restriction fragments) of different sizes when the restriction digest is electrophoresed. There are two methods of using the SSU rRNA to characterize organisms (especially bacteria): RFLP and ribotyping. RFLP requires PCR amplification of the rDNA which encodes the rRNA, followed by digestion with restriction enzymes and then run on a gel. The profile is captured by a Gel-doc and analyzed with software such as the Gel Compare II to distinguish between isolates or populations of organisms. Ribotyping on the other hand involves isolation of total DNA followed by digestion of the DNA with specific restriction enzymes. Sometimes it is necessary to digest the DNA with various enzymes before a useful distinctive restriction profile is obtained. The digested DNA is separated by electrophoresis on an agarose gel and then transferred on to a nylon or nitrocellulose membrane. Thereafter, DNA fragments on the membrane are probed with labeled DNA fragments called probes, complementary to the rDNA gene sequences. After probing, each fragment of bacterial DNA containing a ribosomal RNA gene will be highlighted, creating a fingerprint pattern. The DNA encoding for rRNA is highly conserved among closely related organisms, making ribotyping less discriminatory than other DNA-typing procedures. RFLP and ribotyping are referred to as genomic fingerprinting when used on prokaryotes. Uncultivated organisms identified solely on the basis of their sequence (or other observable characteristics) are called phylotypes.

1.4.3.5 Sequencing – Sanger and Next Gen Applications

One of the most important events that led to the emergence of the genomic revolution was the discovery of how to sequence the DNA. This was first accomplished in 1977 by Alan Maxam and Walter Gilbert (working together) and Frederick Sanger independently. Presently there are many DNA sequencing protocols with multiple variants of each. This section is a very brief introduction to the major types of DNA sequencing.

1.4.3.5.1 The Sanger DNA Sequencing

The Sanger sequencing method uses a DNA template to synthesize new strands whose order of nucleotides are tracked to determine the sequence of the original template molecule. The sequencing reaction contains double-stranded DNA denatured into single strands of the template DNA; oligonucleotide primers (that are complementary to a region of the template to be sequenced), the four deoxyribonucleotide triphosphates (dNTPs – dATP, dCTP, dGTP and dTTP), DNA polymerase and the all-important dideoxyribonucleotide triphosphate (ddNTP). The ddNTPs are different from the dNTPs in the fact that they lack the hydroxyl group of the 3′ carbon in the deoxyribose sugar. During the reaction, DNA synthesis continues as described above until a ddNTP is incorporated to the growing chain. Without the 3′-OH group to connect the 5′-PO_4 of the next dNTP, the extension process is abruptly terminated. This is why the Sanger sequencing technique is called the **chain-termination DNA sequencing** method. In practice, four separate reaction tubes are set up, one for each ddNTP (Tube 1=ddATP, Tube 2=ddGTP, Tube 3=ddCTP and Tube 4=ddTTP) plus all the other sequencing reagents (all four dNTP and primers) and enzymes. As synthesis proceeds the ddNTP is incorporated following complementary base-pairing rules and the chain elongation is terminated! The result is a collection of DNA fragments of varying lengths, each ending in the same ddNTP in each tube. For example, a sequencing reaction containing ddCTP, dATP, dCTP, dGTP, dTTP, the enzymes and oligos will produce fragments ending with C; while ddATP will produce fragments ending with A, and so on. After synthesis the fragments are denatured (usually by heating) and then sequenced. If manual sequencing is used, the fragments of DNA synthesized from radioactively labeled ddNTPs are separated by electrophoresis. Because the rate of migration in inversely related to the log of the molecular weight of the DNA piece, the smaller fragment moves faster through the gel. And because the synthesis always adds a nucleotide to the 3′ OH of the growing strand, the ddNTP at the end of the shortest fragment is designated the 5′ end of the original DNA strand while the largest fragment is the 3′ end. After running the contents of the four tubes in a gel and developing the radioactive labels, the DNA sequence is read sequentially from the smallest to the largest fragment to derive the sequence of the nucleotides. More modern Sanger sequencing techniques are automated and use differently colored fluorescent dye for each of the four ddNTP (green, red, yellow and blue for example). The resulting fragments are separated by capillary electrophoresis. As the fragments exit the bottom of the gel, from the smallest to the largest, a laser beam detects the identity of the ddNTP by its fluorescent color. This is then recorded on a graph called a chromatogram, where the amplitude of each spike represents the fluorescent intensity of that particular fragment. The corresponding DNA sequence is also listed to corresponding peaks of the chromatograms. Sanger's sequencing protocol can accurately read 500–800 bp in a single electrophoretic run. Although it can sequence the longest reads, Sanger's process is expensive and time consuming. It took one decade and over 300 million USD to sequence the first human genome in 2001.

1.4.3.5.2 Next-Generation DNA Sequencing

The next-generation sequencing (NGS) technologies were developed about one decade ago as faster, cheaper and more innovative protocols for DNA reads. They allow for sequencing of DNA and RNA much more quickly and as such revolutionized the study of genomics and molecular biology. Examples of NGS include: Illumina Miseq/Hiseq and 454 Pyrosequencing as well as whole-genome sequencing. All commercial NGS function with the same principle and process. The template DNA is sheared into small pieces and attached to oligos called adapters on both ends. The adapter on one end is fixed to a flow cell surface which is covered with primers. The exposed adapters bend over to anneal to the membrane-anchored primers which initiate the bridge PCR amplification. These result in many copies of same strand bound to flow cell at both ends, which are then sequenced simultaneously. This method is sometimes referred to as the parallel sequence techniques. The need for vectors and cloning is completely eliminated by the NGS.

The most common NGS is the reversible chain-termination sequencing where the DNA is sequenced as each nucleotide is incorporated. The process is also called sequencing by synthesis. After clusters of identical copies of bridge PCR DNA fragments, sequencing reagents are introduced. Like the Sanger protocol, the sequencing by synthesis uses modified fluorescent nucleotide that, when incorporated into the growing strand, stops that reaction because the 3′ OH group is blocked. Unlike the Sanger sequencing, however, ddNTPs are not used. Instead, a small chemical group is added that can be enzymatically removed. In addition, the modified nucleotide does not fluoresce until it is incorporated into the new DNA strand. The automated process in a flow cell involves (1) addition of the DNA template, all four dNTPs, DNA polymerase and all reagents needed for DNA synthesis using the tethered DNA, and (2) incorporation of modified nucleotide as determined by base-pairing rules leading to an emission of its fluorescent tag while it pauses synthesis. During the halt, the identity of the incorporated nucleotide is recorded and (3) synthesis reagents are then flushed out from the flow cell, (4) while the cleavage enzyme reagents are reintroduced to restore the synthesis by cleaving the fluorescent label and the blocking molecule to expose the 3′ OH, and finally (5) the enzyme cocktail is removed and synthesis reagents are added again and cycle repeats. The sequence reads from these NGS are much shorter than that of the Sanger protocol because the fluorescent light must be synchronously detected for sufficient signal to be captured. After 75–150 bases, the synchronicity of incorporation deteriorates.

1.4.3.5.3 Whole-Genome Shotgun Sequencing

The shotgun sequencing technique is widely used in genome sequencing projects. Shotgun sequencing involves shearing large pieces of DNA into smaller fragments that are then sequenced randomly. These random fragments are then realigned and ordered into larger contiguous pieces that are representative of the original large DNA unit. The sequencing method is versatile and can be applied to study different domains of life – bacteria, fungi, protists and other eukaryotes in a sequencing reaction. This method will provide a more comprehensive view of all organisms represented within an ecosystem in one snap shop. It could be used to understand the functional analysis of a given niche when RNA-seq approach is sequenced.

1.5 CONCLUSION AND RECOMMENDATIONS

The blueprint of the cell is a complex polymer. The universal nature of the nitrogenous bases and the genetic codons allows the movement of genes from one organism to another. However, understanding the intricate regulations of gene expression and the many techniques expounded in this chapter will empower the reader unto innovative ventures in the world of cells.

REFERENCES

Chakraborty, S., Persaud, V., Vanegas, S., Gautier, G. and Esiobu, N. (2014). Analysis of the Human Oral Microbiome of Smokers and Non-Smokers Using PCR-RFLP and Ribotyping. *Advances in Microbiology* **4:** 681–691.

Esiobu, N. (2006). Use of Peptide Nucleic Acid Probes for Rapid Detection and Enumeration of Viable Bacteria in Recreational Waters and Beach Sand. In: *Methods in Molecular Biology, vol 345: Diagnostic Bacteriology Protocols*, 2nd Ed. Edited by: L. O'Connor, Humana Press Inc. Totowa, NJ, pp. 131–140.

Ezeonu, I. M., Ononugbo, C. M. and Ike, A. C. (2018). Molecular characterization of *Listeria monocytogenes* isolated from a ready-to-eat fermented milk and cereal product, *Fura-de-Nunu. African Journal of Microbiology* **12:** 448–455.

Holmes, D. and Esiobu, N. (2016). Using Synthetic Biology to Create a Safe, Non-productive Plasmid Surrogate for Ebola Gene Sequences. Presented at the 2016 Florida ASM Annual meeting held at the UMiamiRosentstiel School of Marine and Atmospheric Science on Oct 15–16, 2016. https://twitter.com/flmicrobiology?lang=en.

Nwaokorie, F. O., Coker, A. O., Ogunsola, F. T., Gaetti-Jardim, E. Jr., Avila-Campos, M. J. and Savage, K. O. (2012). AP-PCR and antimicrobial susceptibility patterns of *Fusobacterium nucleatum* associated with chronic periodontitis in Nigerian patients. *British Microbiology Research Journal* **2**(2): 97–107.

O'Donnell, M., Langston, L., Stillman, B. (2013). Principles and Concepts of DNA Replication in Bacteria, Archaea, and Eukarya. Cold Spring Harbor *Perspectives in Biology* **13**:5:a010108.

Vira, H., Bhat, V. and Chavan, P. (2016). Diagnostic Molecular Microbiology and Its Applications: Current and Future Perspectives. *Clinical Microbiology of Infectious Diseases* **1**: 20–31.

Watson, J. D. and Crick, F. H. (1953). Molecular Structure of Nucleic Acids: A Structure for Deoxyribose Nucleic Acid. *Nature* **171**: 737–738.

Watson, J. D., Baker, T. A., Bell, S. P., Gann, A., Levine, M. and Losick, R. (2013). *Molecular Biology of the Gene*, 7th Edn. Pearson, Prentice Hall, NJ.

2 Forensic DNA Profiling and Molecular Identification of Infectious Pathogens

The Nigeria Experience

D. E. Agbonlahor
Edo State University

M. Y. Tatfeng
Niger Delta University

Ifeoma B. Enweani-Nwokelo
Nnamdi Azikiwe University

Ifeoma M. Ezeonu
University of Nigeria

E. A. Brisibe
University of Calabar

Francisca Nwaokorie
University of Lagos

Nwadiuto (Diuto) Esiobu
Florida Atlantic University

A. O. Eremwanarue
Lahor Centre for Molecular Diagnostics

F. E. Oviasogie
University of Benin

DOI: 10.1201/9781003178903-3

Benjamin Ewa Ubi
Ebonyi State University
Tottori University

G. S. George
Niger Delta University

CONTENTS

2.1 INTRODUCTION

DNA typing in forensic has deeply changed the criminal justice system in a number of countries including the USA, the UK, France, and Japan. It has been used in exonerating wrongly convicted individuals; providing evidence of guilt of suspects; and identifying potential perpetrators through DNA databases in cases that otherwise would likely have remained unsolved. Forensic DNA analysis or DNA profiling was first introduced in 1981, and ever since, it has been developed and established into a powerful technique for accurate criminal justice to crack the cases on crime scenes (Fernando and Udesh, 2019).

Forensic science generally focuses on the use of genetic material in criminal justice system to answer queries linked to legal worries including civil and criminal cases. What makes DNA so important is the distinctiveness of this molecule. Every individual person has 99.9% of human DNA sequences; the forensic scientists need only 0.1% of the DNA's unique sequences (Romeika, 2014). With subsequent adornment of DNA analysis methods in crime and civil cases laboratories, even small amounts of blood, buccal swab, saliva, semen, skin cells, or other biological materials can be used to develop investigative leads. These leads may include linking an individual to a crime scene, excluding suspects, persons of interest or consensual partners in the case of sexual assault, obtaining husband's DNA sample for exclusionary purposes, identification of human remains from mass disasters, and establishment of the actual biological Father or Mother of a child in the case of DNA parentage testing. Because of the correctness and fidelity of forensic DNA analysis, this evidence has also become an invaluable tool for an unlimited number of individuals who have been illegally convicted.

On the other hand, the introduction of nucleic acid amplification and detection has resulted in a change from orthodox laboratory strategies that rely on phenotypic determinants or organic chemistry products to molecular methods for the fast identification of a number of infectious pathogens. Molecular methods are increasingly incorporated into the clinical laboratories, particularly for diagnostic purposes involving the detection and characterization of etiologic agents of microbial infections (Nwaokorie et al., 2012). The advantages include fast turn-around time, high sensitivity, and specificity. However, these should be matched by thorough validation and quality control. Molecular detection has mostly returned to the clinical microbiology laboratory in the variety of end-point polymerase chain reaction (PCR) technology such as single run PCR (singleplex) or nested procedures with detection by gel electrophoresis. However, with the advent of automation at the various stages of deoxyribonucleic acid or RNA extraction, amplification, and product detection together with real-time PCR, molecular laboratories will continue to become more economical and cost-effective.

Given the outline indicated above, this chapter is expected to provide an overview of forensic DNA analysis and the clinical applications of molecular methods for the detection of infectious pathogens with emphasis on molecular detection of antimicrobial resistance (AMR) gene testing and pathogens causing infectious bacterial, viral, and fungal infections. It is also expected to highlight some of the advances made in country so far in these areas.

2.2 THE NIGERIA EXPERIENCE

Currently, we are aware of the availability of a number of molecular laboratories in different parts of Nigeria that have facilities for forensic DNA profiling and/or molecular identification of infectious pathogens. Although not exhaustive, some examples include:

i. Lahor Research Laboratories, Benin City (tests for DNA parentage and identification of pathogens).
ii. Lagos State DNA and Forensic Centre, Lagos, Lagos State.
iii. African Center of Excellence for Genomics of Infectious Diseases, Redeemer University, Ede, Osun state (identification and sequencing of pathogens).
iv. Molecular Biology Laboratory at the Department of Medical Laboratory Science, Niger Delta University, Bayelsa State (identification of pathogens).
v. Africa Bioscience laboratory, Ibadan (identification of pathogens).
vi. Institute of Lassa Fever Virus Research and Control Molecular Diagnostic Laboratory, Irrua, Edo State (identification of pathogens).
vii. Molecular Diagnostic Laboratory at the College of Medical Sciences, University of Maiduguri (identification of pathogens).
viii. Biotechnology Molecular Laboratory, NIPRD, Abuja (identification of pathogens).
ix. Applied Biosystems, Abuja (identification of pathogens).
x. Center for Human Virology and Genomics, Nigerian Institute of Medical Research, Yaba, Lagos (identification of pathogens).

xi. Biosolution Technologies, Akure (identification of pathogens).
xii. Molecular Diagnostic Laboratory, Department of Microbiology, University of Nigeria, Nsukka (identification of pathogens).
xiii. Molecular Diagnostics Laboratory of the National Veterinary Research Institute, VOM (identification of pathogens).
xiv. Applied Biotech International Nigeria Ltd, Wuse, Abuja (identification of pathogens).
xv. Molecular Research Foundation for Students and Scientists, Nnamdi Azikiwe University, Awka (identification of pathogens).
xvi. Biotechnology Research Centre, Nnamdi Azikiwe University, Awka (identification of pathogens).
xvii. Molecular Biology Laboratory, Department of Genetics & Biotechnology, University of Calabar, Calabar.

Note: It is worthy to note that the Nigeria Centre for Disease Control (NCDC) recently activated over 40 additional molecular laboratories specifically for the detection of COVID-19 virus during the 2020 pandemic. These molecular diagnostic laboratories are located in the six geopolitical zones of the country.

2.3 DNA PROFILING TECHNIQUES FOR HUMAN IDENTIFICATION

Sample Collection: Numerous factors can have an impact on the molecular weight of the DNA obtained from crime scene samples. A major cause of DNA degradation is the activity of nucleases which are widespread in the environment. While the analyst will do nothing to forestall the damaging action of nucleases before specimen assortment, there are many steps that need to be taken to get rid of their activity once the specimen environment can be controlled.

2.3.1 Crime Scene Preservation and Investigative Techniques

i. Examination of the Crime Scene

Investigators should gather as much information as possible about the scene of the crime. Quick and methodical approach is recommended and mental plan formulated which begins with a walk through the area along the "trail" of the crime.

ii. Documenting the Crime Scene

Video camera is very crucial in a crime scene. It is very essential to obtain still photographs of the scene. A ruler is also needed where relative size is important. Then a crime scene sketch should be made.

iii. Collection and Preservation of Evidence

Collection process begins after the preservation of the crime scene. Majority of evidentiary substances are usually packed in containers of paper origin, and fluid specimens can be shipped in unbreakable sample collection tubes.

Combustible evidentiary materials are commonly kept in sealed containers. Sample should be properly secured and every piece of information recorded. It is noteworthy that evidentiary material(s) should be of major importance to the crime scene examiner. DNA prints are very excellent proofs that a person was actually at any crime scene. Others include bite marks (sexual assaults), broken natural finger nails, blood and body fluids, fire arms and tool marks, shoe prints, tire tracks, fracture matches, and hair with root sheath attached. Proper and adequate documentation and preservation are very essential.

Sampling methods include using forceps, tapes, lifts or vacuuming the items. Biological materials may be sampled by cutting, swabbing, or scraping.

iv. Blood Stain Analysis

Conventional serological analysis include: protein, enzymes, and antigens.

Restriction Fragment Length Polymorphism (RFLP): Direct analysis of certain DNA sequences present in white blood cells

PCR: DNA analysis of certain DNA sequences that have been copied multiple times to a detachable level.

Specimens are maintained cool and dry prior to the commencement of DNA recovery procedures. However, forensic scientists continue to appraise the success of sample assortment strategies, the sincerity of DNA samples, and storage of samples to certify accurate and reliable collection of DNA for any analysis. The cotton swab is a basic and essential tool for collecting DNA proof for forensic analysis (Contents et al., 2008). On the other hand, many factors will have an effect on the recovery of DNA from sample, including the kind of the sample itself.

Examples of samples that can be obtained from the crime scene and/or suspects include blood stains, semen, vaginal fluid, skin/burnt skin/fingerprint, sweat stains, nasal secretions, hairs with root, teeth, bones, nails, tooth-brush, door handle swabs, shoes and clothing, currency (Naira) notes, and cigarette butts, as presented in Figure 2.1 (Bond and Hammond, 2008).

DNA Extraction: DNA extraction is the process by which scientists are able to separate the DNA from other macromolecules in a sample. It is a critical step molecular method, which provides information about the amount of DNA present in an unknown sample. This data can be used successfully to obtain better results preserving the sample for further analysis as shown in Figures 2.2 and 2.3, respectively.

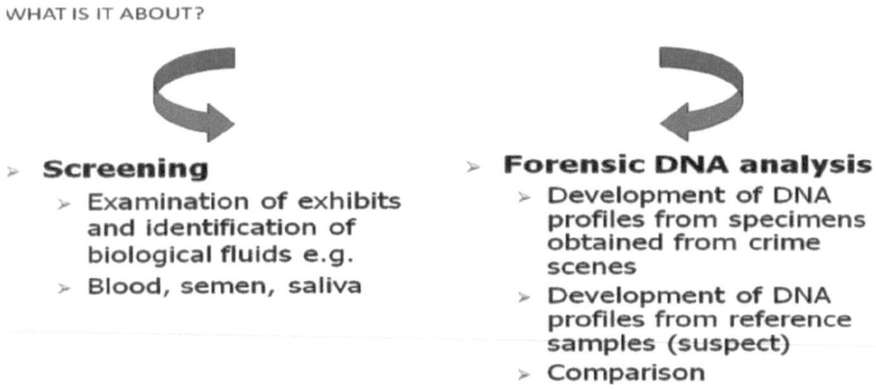

FIGURE 2.1 Illustration of what we do in our local facilities.

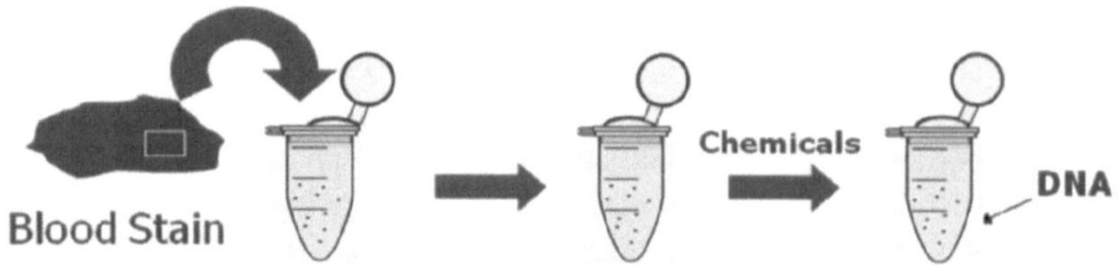

FIGURE 2.2 DNA extraction illustration from blood sample.

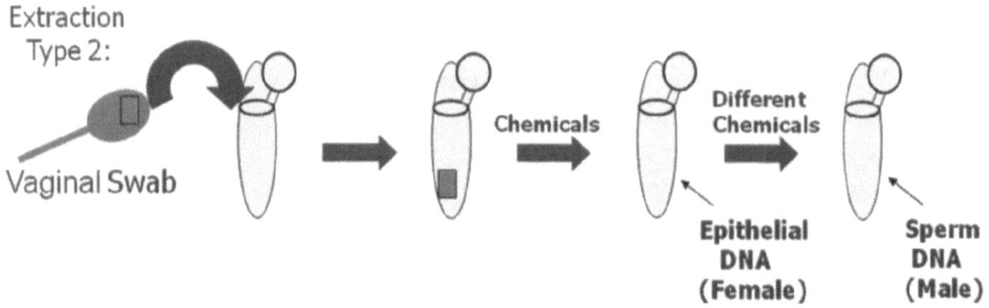

FIGURE 2.3 DNA extraction workflow from vaginal swab sample.

DNA quantification: To determine the average concentration of DNA present in a mixture. PCR inhibitors can greatly affect quantification of DNA as well as amplification process. Several inhibitors have been identified; they include hematin, indigo, melanin, collagen, tannic acid, humic acid, and calcium phosphate (Martins et al., 2015).

2.4 POLYMERASE CHAIN REACTION TECHNIQUE

PCR is a biomolecular technique that was developed in 1983 by Kary Mullis (Saiki et al., 1988), which replicates specific stretches of DNA to create multiple copies. PCR has become so indispensable that it is used in medical and biological research laboratories for a variety of applications including DNA cloning for sequencing, DNA-based phylogeny or functional analysis of genes, the diagnosis of hereditary diseases, the identification of genetic fingerprints (used in forensic sciences), paternity testing, for the analysis of environmental samples, and the detection and diagnosis of infectious diseases (Butler, 2015). In 1993, Mullis was announced a Nobel Prize winner in Chemistry for his work on PCR.

PCR procedures have paved way for DNA amplification and identification of nucleotides in their order of arrangement in individual genes. PCR is a molecular technique employed in the multiplication of certain fragment of DNA, producing many copies by means of DNA enzymes under controlled environment. The technique ensures that a copy of DNA fragment can be multiplied into millions of copies to enhanced detection and visualization.

PCR technology has been applied in various fields of biotechnology including forensics (DNA profiling), identification of infectious diseases, cloning, DNA parentage testing, diagnosis of hereditary diseases, and for the analysis of different environmental samples (Butler, 2015). Repetitive DNA regions located outside the coding regions are used in forensics for DNA profiling. These regions vary from one individual to the other and can be used for identification of one person as well as a group of people, such as a group of family members. PCR can make copies of specific nucleotide sequences from a genome or degraded DNA. PCR uses a small amount of template DNA, two primers that bracket the sequence of interest, nucleotides, and thermostable DNA polymerase to amplify a specific region of DNA, thus creating a large amount of DNA from a very small sample.

2.5 FORENSIC AND PARENTAGE DNA ANALYSIS USING AUTOSOMAL STR PROFILING

In forensic DNA analysis, short tandem repeats (STR) or microsatellites are short sequences of DNA which are repeated consecutively in a number of times and are the most commonly genotyped in order to distinguish between individuals, to incriminate an individual, or to vindicate the innocent. STRs were discovered in the 1980s (Reynolds et al., 1991), and since then, they are the "gold standard" in human identification in forensic investigations. They are made up of one-, two-, three-, four-, five-, or six-nucleotide repeat. An individual can have the same number of repeats one each chromosome pair (homozygous) or different number of repeats on each chromosome pair (heterozygote). Tetra-nucleotide repeats are commonly used for genotyping in forensic DNA analysis (Goodwin et al., 2011). In criminology, STR profiles obtained from biological samples from the scene are compared with other profiles of known suspects identified by police or included in a national forensic DNA database (Jobling and Gill, 2004). Also, the STR profiling is used in the identification of rape perpetrators, DNA paternity/maternity testing, and disaster victim identification (DVI) (Yoshida et al., 2011). Due to the STR profiles, in numerous cases, persons have been excluded in connection with crimes and have been exonerated. A major advantage of the STRs markers is the reality that they can test in a rapid, simple, and simultaneous way more than ten STR loci by multiplexing (Morretti et al., 2001). As a result of this characteristic, it offers an increased degree in the identification of different biological samples. The nomenclature of the STR loci and the allelic variants was established in 1993 by the DNA Commission of the International Society of Forensic Genetics (ISFG). Apart from their usefulness in forensic DNA analysis, STRs became used in medical genetic research, because it has been established that the trinucleotide STR loci is linked with some genetic disorders. The DNA profile refers to the genotype (the number of repeats found in each allele of the analyzed STR marker) of a suspect, victim, or crime scene sample. In the development of STR typing system, in 1997, the Federal Bureau of Investigations (FBI) introduced the database named CODIS (Combined DNA Index System) that included 13 autosomal loci and the amelogenin sex test (Sullivan et al., 1993). These loci are highly polymorphic, localized in non-coding regions which are on different chromosomes. As an improvement in efficiency, the new multiplexes that amplify 20 loci or more in a single reaction (including amelogenin too) have been introduced in the year 2004 (Dixon et al., 2005). In our local facilities, forensic DNA analysis and/or DNA parentage testing is made through multiplex PCR amplification of 16 STRs or more in 310 ABI Genetic Analyzer. Amplified DNA products are analyzed by capillary electrophoresis in which DNA is separated by size and color followed by laser detection of DNA product. Chromatograms are analyzed then using computer software. Since primers with fluorescent tag attaches to DNA and targets region of interest, individuals are assigned to designation for each of the number of loci tested as illustrated in Figures 2.4 and 2.5 (Pitter et al., 2008). The STR-based forensic DNA analysis has been well accepted by population and professionals, as a vital tool in human identifications and in the criminal justice when compared with the traditional serologic markers (Agbonlahor et al., 2013).

For instance, Lahor Research Laboratories in Benin City has routinely carried out forensic DNA analysis involving crime scene investigation and DNA parentage testing as illustrated in Tables 2.1–2.3. The facility may also be used for the identification of individuals in cases of plane crash or motor accidents where victims are burned beyond recognition (see Plate 2.1).

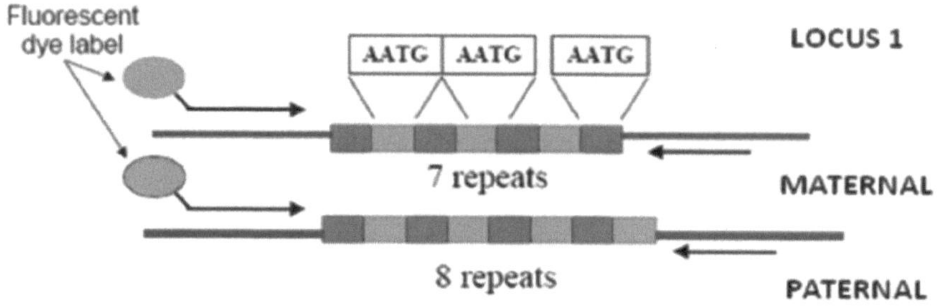

FIGURE 2.4 Illustration of labeled repeats.

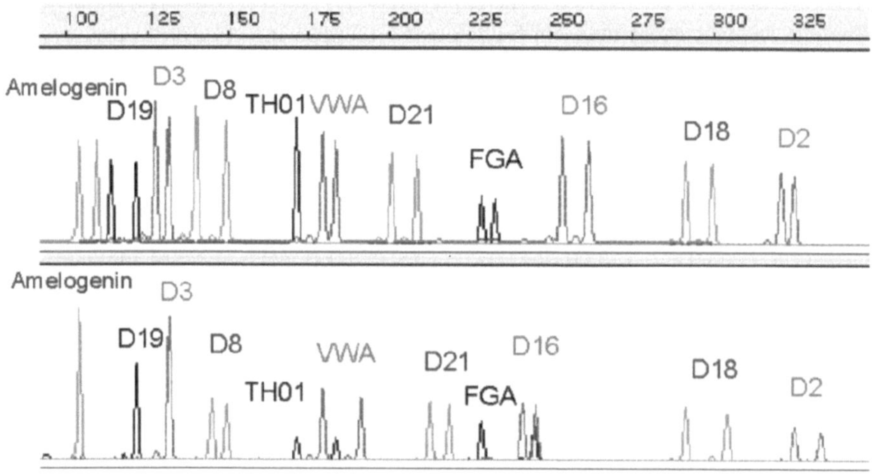

FIGURE 2.5 Illustration of chromatograms after capillary gel electrophoresis.

TABLE 2.1
Comparison of DNA Profiles of Suspect and Crime Scene Sample (Match)

LOCI (STR/Primer)	Suspect 1	Crime Scene Sample
D3S1358	14, 15	14, 15
vWA	15, 16	15, 16
FGA	22, 23	22, 23
Amelogenin	X, Y	X, Y
D8S1179	10, 13	10, 13
D21S11	29, 30	29, 30
D18S51	12, 14	12, 14
D5S818	11, 13	11, 13
D13S317	10, 11	10, 11
D7S820	10, 10	10, 10
D16S539	9, 11	9, 11
THO1	6, 8	6, 8
TPOX	8, 8	8, 8
CSF1PO	12, 13	12, 13

TABLE 2.2
Comparison of DNA Profiles of Suspect and Crime Scene Sample with Mismatch

Loci (STR/Primer)	Suspect 1	Crime Scene Sample
D3S1358	14, 15	14, 15
vWA	**15, 16**	**14, 17**
FGA	22, 23	22, 23
Amelogenin	**X, Y**	**X, X**
D8S1179	10, 13	10, 12
D21S11	29, 30	29, 30
D18S51	12, 14	12, 14
D5S818	**11, 13**	**28, 18**
D13S317	10, 11	10, 13
D7S820	10, 10	9, 10
D16S539	9, 11	9, 11
THO1	6, 8	6, 8
TPOX	**8, 8**	**9, 9**
CSF1PO	12, 13	12, 13

In the above result, a male DNA profile obtained from the crime scene sample was found to match the DNA profile of the suspect.

Table 2.2 shows that there were mismatches in about four different loci (**vWA, Amelogenin, D5S818, TPOX out of the 14 STR tested**) between the suspect and

TABLE 2.3
An Example of DNA Paternity Test Results between an Alleged Father and Two Children

Locus	Child 1	Alleged Father	Child 2
D3S1358	17, 17	16, 16	16, 17
D7S820	9, 10	9, 11	10, 11
vWA	16, 16	16, 16	16, 16
D12S391	17, 21	17, 18	18, 19
FGA	24, 27	21, 24	21, 27
D8S1179	12, 14	12, 14	14, 14
D21S11	29, 30	30, 31	29, 30
D18S51	15, 15	15, 15	15, 15
D5S818	10, 10	12, 13	10, 12
D13S317	12, 12	11, 13	12, 13
D16S539	11, 11	11, 12	11, 11
TH01	7, 8	7, 8	8, 8
TPOX	7, 8	8, 8	8, 9
CSF1PO	10, 11	11, 11	11, 11
D2S1338	19, 22	17, 21	19, 21
D19S433	14, 14	13, 14	13, 14
D22S1045	14, 16	16, 16	14,16
D2S441	11, 11	11, 11	11, 14
D10S1248	12, 13	13, 13	13, 14
D1S1656	13, 18.3	14, 18.3	11, 14
D6S1043	12, 20	18, 20	20, 24

PLATE 2.1 The scene at a plane crash site in Nigeria.

crime scene sample. Therefore, the suspect is excluded as a donor of the DNA profile obtained from the crime scene sample.

From Table 2.3, the DNA profiles for the alleged father and the two children (child 1 and child 2) are clearly showed. There were mismatches between child 1

DNA profile and the alleged father in four different loci indicated in red. On the other hand, there was no mismatch on any of the 21 loci tested between child 2 DNA profile and the alleged father. Therefore, the alleged father is **excluded** from the parentage of child 1 and **cannot be excluded** from the parentage of child 2.

It is important to note that the discrimination power of DNA test is calculated to be at least 99.99%.

2.6 NEED FOR DNA DATABASE IN NIGERIA

Nigeria as a country with a population of about 200 million people is well over due for DNA database profiling of her citizens. This is simply a process where all Nigerians are profiled by either government- or private-owned laboratories which would be certified by the government to do so and kept in a database for subsequent retrieval. This will help to contain criminal activities all over the country as no two individuals have exactly the same DNA profiles, quick identification of unknown individual recovered from war or plane crashes. This DNA database can be established at the local governments, through the states to the federal government as exemplified in Figures 2.6 and 2.7.

2.7 MOLECULAR IDENTIFICATION OF INFECTIOUS PATHOGENS AND RESISTANT GENE MARKERS

Local molecular diagnostic facilities in Nigeria (e.g. at Lahor Research Laboratories, Benin City, African Center of Excellence for Genomics of Infectious Disease, Redeemer's University, Ede, Osun State, Nigerian Institute of Medical Research, University of Nigeria, Nsukka) have over the years engaged in identification of infectious pathogens using different molecular techniques starting from species-specific PCR identification and other meaningful scientific research which has contributed immensely to the body of knowledge in Nigeria and other nations of the world. Malaria parasite has remained a threat to Nigeria and other African nations at large,

FIGURE 2.6 Flow Information in the proposed DNA database in Nigeria.

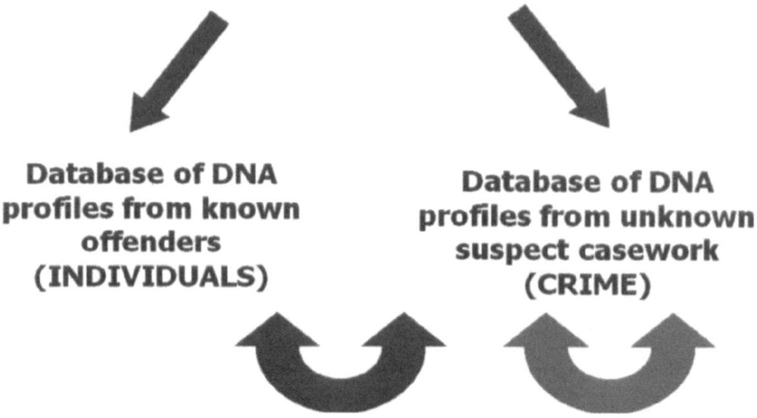

FIGURE 2.7 All DNA profiles are sent to DNA database for future retrieval.

and as such, we have conducted series of research on the subject matter utilizing different molecular techniques in our local facilities which include: Proinflammatory cytokines and hematological indices in HIV seropositive adult (Mirabeau and Agbonlahor, 2008), *Plasmodium falciparum Pfcrt* K76T mutation in people with sickle-cell disease (Tatfeng et al., 2008), cytokines profile of uncomplicated *Plasmodium* malaria infection (Mirabeau and Agbonlahor, 2008), Chloroquine resistance molecular markers in *Plasmodium falciparum* in Edo state (Mirabeau et al., 2009), measurement of some Thl, Th2 cytokines and white cell count in childhood hemoglobinopathies with uncomplicated malaria infection (Tatfeng et al., 2010a,b), T-Helper 4, cytotoxic T-lymphocytes, interleukin-4 and interleukin-10 in HIV seropositive patients with uncomplicated *Plasmodium falciparum* malaria, which has been screened at our facility and documented in Mirabeau et al. (2010).

The scientific world is not unaware of the rising danger of multiple drug-resistant (AMR) bacterial isolates which usually render our routine antibiotic regime ineffective against bacterial infection. Ehiaghe and her colleagues (2017) utilized a number of molecular techniques to detect various antibiotic-resistant genes from isolates of surgical wound infections. The role of plasmid DNA in bacterial resistance cannot be overemphasized, and in line with this fact, we had carried out investigation using our local facilities to ascertain the role of plasmid DNA fluoroquinolone AMR bacteria isolates (Ehiaghe et al., 2013); the role of molecular and phenotypic profiling for identification of bacterial pathogens and detection of resistance markers (Oviasogie et al., 2013; Ezeonu et al., 2018; Agbo et al., 2019); genotyping of *Fusobacterium nucleatum*, an anaerobe associated with oral infections (Nwaokorie et al., 2012); methicillin-resistant *Staphylococcus aureus* isolated from handrails, mobile phones, and automated teller machine (Oviasogie et al., 2014); and plasmid profile of multiple drug-resistant isolates from surgical wound site (Ehiaghe et al., 2016). DNA sequencing technique was recently used in the identification of a novel bacterial isolate named *Cedecea davisae* strain CF26, belonging to a genus which was lately admitted as a member of the family Enterobacteriaceae (Agbonlahor et al., 2018). The authors however inferred that this isolate is rarely incriminated in human infections.

Genotyping and Fluconazole susceptibility of *Candida albicans* strains from patients with vulvovaginal candidiasisis and characterization of *Candida africana* from Jos, Nigeria, had been documented in the study by Nnadi et al. (2012a,b). Molecular characterization of environmental *Cryptococcus neoformans* VNII isolates in Jos, Nigeria, and whole-genome sequencing of uncommon *Cryptococcus neoformans* MLST43 Genotype Isolated in Nigeria were reported by Nnadi et al. (2016, 2019). In this study, the authors reported the whole-genome sequencing of a Nigerian *C. neoformans* VNII, Mat-α, strain with a rare multilocus-sequence-type (MLST) genotype (ST43).

It is also noteworthy to report that virulent gene of *Mycoplasma pneumoniae* from sputum samples of subjects attending NAUTH, Nnewi, Nigeria, was detected by molecular methods by Agbakoba et al. (2018). Osuji et al. (2019) reported a case study of a 24-year-old male who was a repeat blood donor at the University of Abuja Teaching Hospital, Abuja, Nigeria, with viral load of 31,379 copies/ul but negative to hepatitis B surface antigen. Gene sequencing and phylogenetic studies revealed that the isolate belonged to the HBV genotype E with gene sequences similar to an isolate from Sudan. A more advanced next-generation sequencing (NGS) approach was recently adopted for the discovery of rhabdoviruses causing acute febrile illness (Mathew et al., 2015). In this study, the nucleic acid of two novel rhabdoviruses were sequenced and named Ekpoma virus-1 (EKV-1) and Ekpoma virus-2 (EKV-2).

It is also pertinent to note that until 2008, Lassa fever which has plagued Nigeria since 1969, had always been confirmed outside the country. It is gratifying to note that the first in-country confirmation was carried out simultaneously in 2008 at the Institute of Lassa Fever Research and Control Molecular Diagnostic Facility, Irrua, and at the Lahor Centre for Molecular Diagnostics, Benin City (Agbonlahor, personal communication; Akpede, 2008). Agbonlahor et al. (2017) further employed a reverse transcriptase PCR technique at the Lahor Centre for Molecular Diagnostics to identify trapped rodents which harbored Lassa fever virus in three south-south states of Bayelsa, Delta, and Edo.

Arising from the current COVID-19 pandemic, the Federal Government of Nigeria, acting through NCDC, scaled up the number of molecular diagnostic facilities in the country for the confirmation of the virus so as to reduce the pressure on the hitherto few diagnostic centers. It is important to note that the reverse transcriptase PCR technique remains the gold standard for the confirmation of the ravaging Coronavirus.

2.8 CONCLUSION

As at date, significant progress has been made in Nigeria in the development of biotechnology with special reference to forensic DNA profiling and molecular identification of infectious diseases. This has come more through individual efforts, while government has exhibited a lot of reluctance or total indifference in this regard. However, as a result of the ongoing COVID-19 pandemic, the Federal Government of Nigeria, through NCDC, had no option than to activate a number of molecular diagnostic laboratories across the nation in order to increase capacity for viral testing. It is hoped that these Laboratories will be maintained and sustained for testing for any infectious pathogen that could plaque the nation with an outbreak, epidemic or a pandemic. It is also hoped that government will do the needful and show commitment by establishing National Forensic Laboratories across the nation for the long overdue DNA database profiles for her citizens.

REFERENCES

Agbakoba, N.R., Adobe, C.N., Enweani, I.B., Ezeanya, C.C. and Akujiobi, C.N. (2018). Molecular Detection of *Mycoplasma pneumoniae* virulent gene from sputum samples of subjects attending NAUTH, Nnewi. Anambra State. Nigeria. *African Journal of Medicine and Medicinal Sciences* 48(2): 225–231.

Agbo, M. C., Ezeonu, I. M., Odo, M. N., Ononugbo, C. M., Onoduagu, B.O., Eze, C. C., Ezeibe, E. N. and Ozioko, C. A. (2019). Phenotypic and molecular characterization of extended spectrum β-lactamase producing *Pseudomonas aeruginosa* in Nigeria. *African Journal of Biotechnology* 18: 1083–1090.

Agbonlahor, D. E., Erah, A., Agba, I. M., Oviasogie, F. E., Ehiaghe, A. F., Wankasi, M., Eremwanarue, O. A., Ehiaghe, I. J., Ogbu, E. C., Iyen, R. I., Abbey, S., Tatfeng, M. Y. and Uhunmwangho, J. (2017). Prevalence of Lassa virus among rodents trapped in some South South states of Nigeria. *Journal of Vector Borne Diseases*, 54: 146–150.

Agbonlahor, D. E., Ehiaghe, J. I., Eremwanarue, O. A., Ehiaghe, A. F., Oviasogie, F. E., Iyen, R. I., Ogbu, E. C., Ikhiuwu, P. E., Oseji, M. O., Abbey, S. and Tatfeng, M. S. (2018). A primal incrimination of *Cedecea davisae* with post-prostatectomy urinary tract infection in Nigeria. *International Journal of Biological and Chemical Science*, 12(2): 676–688.

Agbonlahor, D. E., Happi, C., Igere, B., Oviasogie, F. E., Abbey, S., Tatfeng, Y. M., Folarin, O. A., Ehiaghe, J. I., Omole, E., and Amengialue, E. (2013). Resolution of disputed parentage, employing traditional serologic markers and DNA short tandem repeats multiplex system in Nigeria. *Journal of Medical Laboratory Science* 22(1): 27–33.

Akpede, G. (2008). A welcome address presented to the honourable minister of health at the opening ceremony at Institute of Lassa fever virus Research and Control Molecular Diagnostic Laboratory, Irrua, Edo State.

Bond, J.W., Hammond, C. (2008). The value of DNA material recovered from crime scenes. *Journal of Forensic Sciences* 53:797–801.

Butler, J. M. (2015). The future of forensic DNA analysis. *Philosophical Transactions of the Royal Society B: Biological Sciences*, 370(1674): 20140252. doi: 10.1098/rstb.2014.0252.

Dixon, L.A., Murray, C. M., Archer, E. J., Dobbins, A. E., Koumi, P., Gill, P. (2005). Validation of a 21-locus autosomal SNP multiplex for forensic identification purposes. *Forensic Science International* 154:62–77.

Ehiaghe, F. A., Ehiaghe, J. I., Agbonlahor, D. E., Oviasogie, F. E., Etikerentse, S. M. O., Nwobu, R. A. U., Akinship, B. O., Ikusemore, A. I., Aladenika, S. T. and Enwa, F. O. (2013). Plasmid profiling and curing analysis of fluoroquinolone multidrug resistant *Pseudomonas aeruginosa* in Benin City, Niergia. *American Journal of Biotechnology and Molecular Biology*, 5(1): 23–32.

Ehiaghe, J. I., Nwobu, R. A. U., Ehiaghe, F. A., Agbakoba, N. R., Agbonlahor, D. E. and Uwabor, C. I. (2016). Plasmid profiling of multidrug resistant bacteria isolates from surgical site infection in Nigeria. *American Journal of Biotechnology and Molecular Biology*, 3: 201–205.

Ehiaghe, J.I., Oviasogie, F.E, Nwobu, R.A.U., Agbonlahor, D.E, Ehiaghe, F.A, Uhunmwangho, J, Eremwanarue, O.A, Uwabor, C.I, Itota, I.R, Ogbu, E.C. (2017). Molecular detection of antibacterial resistance genes in phenotypically characterized multidrug resistant pathogens isolated from surgical wound infections in two Nigerian teaching hospitals. Journal *of Medical Laboratory Science* 26(1): 63–75.

Ezeonu, I. M., Ononugbo, C. M. and Ike, A. C. (2018). Molecular characterization of *Listeria monocytogenes* isolated from a ready-to-eat fermented milk and cereal product, *Fura-de-Nunu. African Journal of Microbiology* 12: 448–455.

Fernando, M. N. and Udesh, N. P. (2019). Review: Recent advances in forensic DNA analysis. *International Research Journal of Natural and Applied Science* 6(6): 18–30.

Goodwin, W., Linacre, A., Hadi, S. (2011). *An Introduction to Forensic Genetics*. Hoboken, NJ: Wiley-Blackwell.

Jobling, M. A., Gill P. (2004). Encoded evidence: DNA in the forensic analysis. *Nature Review Genetics* 5:739–751.

Martins, C., Lima, G., Carvalho, M. R., Cainé, L. and Porto M. J. (2015). DNA quantification by real-time PCR in different forensic samples. *Forensic Science International: Genetics Supplement Series*. Elsevier Ireland Ltd, 5, pp. e545–e546. doi: 10.1016/j.fsigss.2015.09.215.

Mathew, S. H., Kristian, G. A., Onikepe, A. F., Grove, N. J., Odia, I., Ehiane, P. E., Omoniwa, O., Yang, X., Gire, K. S., Winnicki, S., Tariyal, R., Schaffner, F. S., Okokhere, O. P., Okogbenin, S., Akpede, G. O., Asogun, D. A., Agbonlahor, D. E., Walker, P. J., Tesh, B. R., Levin, J. Z., Garry, R. F. Sabeti, P. C. and Happi, C. T. (2015). Discovery of Novel Rhabdoviruses in the Blood of healthy individuals from west Africa. *PLOS Neglect Tropical Diseases*, 1–17.

Morretti, T.R., et al. (2001). Validation of short tandem repeats (STRs) for forensic usage: Performance testing of fluorescent multiplex STR systems and analysis of authentic and simulated forensic samples. *Journal of Forensic Sciences* 46:647–660.

Nnadi, N.E., Ayanbimpe, O.R., Ojogba, G.M., Okolo, M., Enweani, I.B., Scordino, F., and Giuseppe, C. (2012a).Genotyping and Fluconazole susceptibility of *Candida albicans* strains from patients with vulvovaginal candidiasisis in Jos Nigeria. *Asian Pacific Journal of Tropical Disease* 2(1): 48–50.

Nnadi, N.E., Ayanbimpe, O.R., Ojogba, G.M., Okolo, M., Enweani, I.B., Scordino, F. and Giuseppe, C. (2012b). Isolation and characterization of *Candida africana* from Jos, Nigeria. *Medical Mycology* 7: 765–767.

Nnadi, N.E., Enweani, I.B, Cogliati, M., Ayanbimpe, G. M., Okolo, M. O., Kim, E., Sabitu, M. Z., Criseo, G., Romeo, O. and Scordino, F. (2016). Molecular characterization of environmental Cryptococcus neoformans VNII isolates in Jos, Plateau State, Nigeria. *Journal de Mycologie Médicale/Journal of Medical Mycology*, 26(4): 306–311.

Nnadi, N., Giosa, D., Ayanbimpe, G.M., Alessandra, E., Cigliano, R.A., Oheli, C.U., Aguiyi, J.C., Enweani, I.B., Romeo, O. (2019). Whole-genome sequencing of uncommon *Cryptococcus neoformans* MLST43 genotype isolated in Nigeria. *Mycopathologia*, 184(5): 555–557.

Nwaokorie, F.O. (2012). The role of molecular biology in the diagnosis and treatment of infectious diseases. *Nigerian Journal of Clinical and Biomedical Research* 6(1): 19–27.

Nwaokorie, F.O., Coker, A.O., Ogunsola, F.T., Gaetti-Jardim, E. Jr., Avila-Campos, M.J., Savage, K.O. (2012). AP-PCR and antimicrobial susceptibility patterns of *Fusobacterium nucleatum* associated with chronic periodontitis in Nigerian patients. *British Microbiology Research Journal* 2(2): 97–107.

Osuji, A. I., Agbakoba, N. R, Ifeanyichukwu, M.O, Enweani, I, Musa, B.M. and Tatfeng, M. (2019): Occult hepatitis B blood donor certified fit for donation at a tertiary hospital in Nigeria: A case report. *Journal of Infectious Diseases and Diagnosis* 4(2): 1–9.

Oviasogie, F. E., Agbonlahor, D. E., Osagie, E. A. and Igere, B. (2014). Genetics and phenotypic characterisatics of methicillin-resistant *Staphylococcus aureus* isolates from handrails, mobile phones and automated teller machines (ATM) in Benin City, Nigeria. *Journal of Medical Laboratory Science*, 23(2): 51–58.

Pitter, F., Niederstätter, H., Huber, G., Zimmermann, B., Oberacher, H. W. (2008). The next generation of DNA profiling - STR typing by multiplexed PCR - ion-pair RP LC-ESI time-of-flight MS. *Electrophoresis* 29: 4739–4750. 10.1002/elps.200800209.

Reynolds, R., Sensabaugh, G., Blake, E. (1991). Analysis of genetic markers in forensic DNA samples using the polymerase chain reaction. *Analytical Chemistry* 63:2–15.

Romeika, J. M. (2014). Recent advances in forensic DNA analysis. *Journal of Forensic Research* 12(1): 1–13. doi: 10.4172/2157-7145.S12-001.

Saiki, R., Gelfand, D., Stoffel, S., Scharf, S., Higuchi, R., Horn, G., Mullis, K., Erlich, H. (1988). Primer-directed enzymatic amplification of DNA with a thermostable DNA polymerase. *Science* 239 (4839): 487–491. doi:10.1126/science. 239.4839.487. PMID 2448875.

Sullivan, K.M., Mannucci, A., Kimpton, C.P., Gill, P. (1993). A rapid and quantitative DNA sex test: fluorescence-based PCR analysis of X–Y homologous gene amelogenin. *Biotechniques* 15:636–641.

Tatfeng, Y. M. and Agbonlahor, D. E. (2010). Age-related cytokine profile in uncomplicated *Plasmodium malaria* infection. *Colombia Medica* 41(4): 223–227.

Tatfeng, Y. M. and Agbonlahor, D. E. (2008). Proinflammatory cytokines and haematological indices in HIV seropositive adults with uncomplicated *Plasmodium falciparum* malaria. *Tropical Medicine and Health* 36(2): 75–79.

Tatfeng, Y. M., Agbonlahor, D. E. and Abouo, A. M. (2010a). T-Helper 4 and Cytotoxic T-Lymphocites, Interleukin-4 and 10 in HIV seropositive patients with uncomplicated *Plasmodium falciparum* malaria. *Tropical Medicine and Health* 38(4): 139–142.

Tatfeng, Y. M., Agbonlahor, D. E., and Amegor, O. F. (2010b). Measurement of some Th1, Th2 cytokines and white cell count in childhood haemoglobinopathies with uncomplicated malaria infection. *Revista de Cinencias Medicas e Biologicas* 9(1): 13–16.

Tatfeng, Y. M., Agbonlahor, D. E., Olalekan, O. A., Oviasogie, F. and Isibor, J. (2009). Chloroquine resistance molecular markers in *falciparum* malaria in Edo State. *Tropical Medicine and Health* 37(2): 37–41.

Tatfeng, Y. M., Agbonlahor, D. E., Tchounga, K. S., Omolu, P. I., Okodua, C. S. and Adeolu, A. (2008). Chloroquine prophylaxix associated with high prevalence of *Plasmodium falciparum pfcrt* K76T mutation in people with sickle – cell disease in Benin City, Nigeria. *Journal of Vector Borne Diseases* 45: 51–55.

Yoshida, K., Yayama, K., Hatanaka, A., Tamaki, K. (2011). Efficacy of extended kinship analysis utilizing commercial STR kit in establishing personal identification. *Legal Medicine* 13:12–15.

3 DNA Markers in Forensic and Diagnostic Science

M. Y. Tatfeng
Niger Delta University

D. E. Agbonlahor
Edo State University

Ifeoma B. Enweani-Nwokelo
Nnamdi Azikiwe University

Ifeoma M. Ezeonu
University of Nigeria

Francisca Nwaokorie
University of Lagos

E. A. Brisibe
University of Calabar

D. Esiobu
Applied Biotech International Nigeria Ltd.

CONTENTS

DOI: 10.1201/9781003178903-4

3.1 INTRODUCTION

Deoxyribonucleic acid (DNA) is one of the most outstanding molecules in any biological system. Over the years, several attempts by researchers to unravel its structure and functions made it the most researched macro-molecule. It is a very informative and complex molecule that provides life to a cell. Its ability to be transferred from generation to generation has further conferred onto it some unique properties. As the information storage unit of life, understanding DNA roles in living things has tremendously impacted human endeavors. Furthermore, the uniqueness of DNA is in its chemical composition which is identical across biological systems, though there are differences in the sequence arrangement of nucleotides. Two different individuals as well as identical twins will differ at some point in their DNA sequence.

The analysis and understanding of the DNA is known to provide solutions to several of the problems afflicting humanity. These include:

Health issues (diagnosis)
Environmental problems (pollution)
Societal concerns (crime and parentage dispute resolution)
Agricultural development, etc.

3.2 STRUCTURE AND FUNCTION OF DNA

Over six decades ago, the suggested double-helical structure of the genetic material by Watson and Crick offered reliable evidence explaining the transfer of genetic information from generation to generation (Travers and Muskhelishvili, 2015). The sequence complementarity with adenine pairing thymine and guanine pairing cytosine explained the discovery in the past by Avery et al. (1944) that DNA was the transforming principle. The transforming principle is responsible for the transfer of genetic information between different strains of bacteria.

In DNA, the two strands run in opposite direction and are therefore said to be anti-paralleled. Each strand is composed of small set of constituent molecules called nucleotides. Chemically, the nucleotides in the strands of DNA molecules comprise of phosphate group, a base and a backbone made up of a 5′-carbon (Karl, 1992). In association with its derivative, ribonucleic acid (RNA), these two molecules represent the most abundant biological materials in living things.

3.3 ANATOMY OF A NUCLEOTIDE

A nucleotide contains three molecules namely: a sugar, a phosphate and a nitrogenous base. The sugar deoxyribose (from which DNA derives its name and ribose in RNA) and phosphate molecules form the ribbon-like backbone of the DNA strand and are similar in all nucleotides. The third molecule of the nucleotide is the base. These three molecules are linked to one another by a covalent bond, while a phosphodiester bond links all nucleotides together.

There are four different bases defining the four different nucleotides: The pyrimidines are thymine (T) for DNA and uracil (U) for RNA, and cytosine (C), while the purines are adenine (A) and guanine (G).

DNA owes its double-strandedness to base-pair complementarity throughout the molecule. If nitrogenous bases of a given strand are aligned with specific bases on the other strand, the aligned bases hybridize via hydrogen bonds. The specific pairing remains A/T and G/C.

Double hydrogen bonds exist between A and T, whereas three bonds exist between C and G. This renders C-G bonds more rigid than those of A-T. Under "normal" cellular conditions, two DNA strands consisting of complementary bases hybridize and establish a stable double helix. However, the two strands will only hybridize if they are in "antiparallel configuration." This means that the sequence of one strand, when read from the 5′ end to the 3′ end, must be complementary, base for base, to the sequence of the other strand read from 3′ to 5′ (Karl, 1992).

DNA performs only two functions, which are replication and gene expression. Its complex nature has provided researchers with valuable information which have revolutionized the field of forensic, diagnostic and several other specialties in the field of science.

3.4 WHAT ARE GENETIC MARKERS?

Over the years, gene mapping for the identification of organisms was restricted to phenotypic markers. This included genes that encoded easily visible characteristics such as physical features in humans or seed shape in plants. The inadequate number of these types of characteristics in many organisms limited the mapping efforts that could be done. This informed the study of gene markers which could identify genetic characteristics that are not readily observable in organisms (such as protein variations). Consequently, a marker is like an identity card and must possess certain characteristics. These characteristics include the following: a marker should be polymorphic, which means it will exist in different forms so that it is easy to differentiate the chromosomes which carry mutant genes or chromosomes with normal genes in the form of the marker it also carries.

Polymorphism in a marker can be detected at three levels: phenotypic recognition, biochemical markers which represent changes in the gene product such as changes in proteins and amino acids, and variations at the genomic level such as changes in nucleotide sequence. These changes at genetic level could be deletion, duplication, inversion and/or insertion (as molecular or DNA markers).

3.5 DNA OR MOLECULAR MARKERS

The advancement in molecular genetics has laid the foundation for genomics studies. It has opened newer frontiers for the genetic improvement of crops and farm animals. These molecular signatures provide significant and accurate genetic information and better understanding of our biodiversity. A DNA or molecular (genetic) marker is a gene or DNA sequence with a known site on a chromosome and associated with a unique information or trait that can be used to characterize or identify individuals or species. In other terms, DNA markers represent a sequence, which may evolve from a mutation or alteration in the loci that can be noticed. Furthermore, they may be a short DNA sequence in which a single base-pair change (single nucleotide polymorphism, SNP) occurred, or a long one like mini and microsatellites and can be used in DNA-DNA hybridization, polymerase chain reaction (PCR) or restriction mapping experiments for DNA profiling. Recently, great interest towards molecular markers has been on the increase as they reveal polymorphism at the DNA level that is valuable in animal genetics studies (Van and Rodgers, 1996; Gizaw et al., 2007).

Molecular or DNA markers can be categorized as dominant or co-dominant. While dominant markers such as randomly amplified polymorphic DNA (RAPD) allow for analyzing many gene loci at a time, co-dominant markers such as restriction fragment length polymorphisms (RFLPs) and microsatellites or simple sequence repeats (SSR), on the other hand, can analyze one locus at a time. Scientists unfamiliar with these different molecular techniques tend to get lost as each of these tools has its own advantages and disadvantages (Al-Samarai and Abdulkareem, 2015).

Molecular or DNA markers must have certain basic properties namely: (1) a genetic marker must also show specific allelic characters which are very clear, such that it allows different alleles to be recognized easily; (2) they must have a simple inheritance (often co-dominant in terms of their binding patterns), so that heterozygotes can be differentiated from homozygous allelic conditions; (3) a marker must possess copies which are single and exhibit minimum pleiotropic effect; (4) they must not also be expensive and should thus be cost-efficient in terms of development and genotyping; and (5) they must be highly available with "unrestricted use" and suitable to be duplicated/multiplexed, so that the data can be accumulated and shared between laboratories.

3.6 USES OF DNA MARKERS

There are several uses of molecular or DNA markers. They can be deployed in studying the relationship between a genetic disease and its genetic cause (for example, a particular change of a gene that yields a malfunctioning protein). Molecular markers are also employed in genetic evolution to determine genetic distances between individuals or populations. In addition, unique genetic markers (autosomal markers) which are found on mitochondrial or Y chromosomal DNA have provided valuable information on maternal and paternal linkages.

3.7 SINGLE NUCLEOTIDE POLYMORPHISM (SNP)

Landers in 1996 came up with a new molecular marker technology called SNP (Yang et al., 2013). The technology revolved around a DNA sequence variation caused by one nucleotide change at a specific position in the DNA sequence (Yang et al., 2013). This type of polymorphism includes single base transitions, transversions, insertions and deletions (Jarne and Lagoda, 1996), and the least frequent allele should have a distribution of 1% or more. Transitions are the most noticeable (approx. 2/3) among all the SNP mutation types already identified.

SNP markers provide the most popular approaches in genetic studies, in spite of the limitations they provide (simple bi-allelic co-dominant markers) when matched with the highly informative multi-allelic microsatellites. This popularity of the SNPs relied on some preferred properties including their genetic stability, abundance in the genome and amenability to automated analysis. The recent need for highly discriminatory genetic markers for the studies of multifactorial diseases recently opened newer frontiers to SNP analysis (Vignal et al., 2002).

3.8 VARIABLE NUMBER OF TANDEM REPEATS (VNTRs)

VNTRs represent locations in a genetic makeup of an organism where a short nucleotide sequence is organized as a short repetitive sequence called short tandem repeat (STRs). They are found on many chromosomes, and often show differences in length (number of repeats) among individuals.

Microsatellites or SSR loci, otherwise known as VNTRs and simple sequence length polymorphisms (SSLPs), are found throughout the genetic material of most organisms (Varshney et al., 2005). Microsatellite markers are short fragments of DNA in which the same short DNA sequence is repeated several times. They are highly variable across the general population, often conserved from generation to generation and are mostly non-protein coding loci. They vary in length of one to six nucleotides, while minisatellites range from ten to one hundred nucleotides and are classified as mono-, di-, tri-, tetra-, penta- and hexa-nucleotide repeats. The sequences of two, three and four nucleotide repeats are the most reliable choices genetic studies. They are tandemly repeated (usually 5–20 times) in the genome with a minimum repeat length of 12 base pairs (Ellegren, 2004).

3.9 FORENSIC SCIENCE

Forensic science or medical jurisprudence is the usage of scientific data for criminal or civil law or proceedings. In legal matters, forensic science is majorly concerned with the evaluation or assessment of physical and biological evidence settle issues. As a result, forensic science capability is very reliable because it generates information that is more accurate, precise and trustworthy than eyewitness evidences or confessions. Forensic science has specialties from virtually all aspects of modern science, medicine, microbiology, biochemistry, parasitology, engineering, entomology, etc. For example, when the evidence used in court to settle legal issues is medical, the term "Medical Jurisprudence or Forensic Medicine" is most appropriate. So far, legal issues that have over time required medical or scientific evidences include crime, natural disasters and parentage dispute resolution.

Forensic DNA analysis first saw it used in 1984. The technology was developed by Prof. (Sir) Alec Jeffreys, who observed that variations in the genetic code could be of tremendous importance in the identification of humans. DNA profiling in forensic evaluation was first used by Prof. Jeffreys in a double murder mystery in Narborough, England. Following a sexual assault and assassination of a 15-year-old school girl in a psychiatric hospital in 1985, the police was unable to find any suspects but obtained a semen sample from the victim. Furthermore, another 15-year-old girl was equally raped and strangled in the nearby village of Enderby in 1986. Forensic evidence revealed that both killers were of the same blood type. Prof. Jeffreys provided his forensic expertise by analyzing the semen samples. The evidence revealed that there was no match between the samples collected from the victim and the initial suspect who became the first person in the world to be exonerated using forensic DNA evidence; however, he confirmed that the DNA profiles were identical in the semen samples collected from the two murdered victims.

DNA samples were eventually collected from more than 4,000 males in the entire male population and compared to semen samples from the two crime scenes. The criminal was finally arrested in 1987, when his DNA profile was found to be identical to that obtained from the semen samples collected from the raped victims.

In the early 20th century, the science of forensics had become largely developed and established in the field of criminal investigations with DNA markers being used as major tools. And on account of this, DNA databases have been developed in several countries including the national and international databases of the Federal Bureau of Intelligence (FBI) in the United States of America as well as in the European countries (ENFSI: European Network of Forensic Science Institutes). These searchable databases are currently used to match crime scene DNA profiles to existing ones in the database.

3.10 COMBINED DNA INDEX SYSTEM (CODIS) CORE LOCI

CODIS generate investigative insights in cases where biological data recovered from the crime scene tendered as evidence (Table 3.1). Matching of genetic profiles in the Forensic Index can possibly assist in identifying serial offenders and link crime scenes together. From a genetic profile match, police from multiple jurisdictions can coordinate their respective investigations and share the leads they developed independently. The identity of suspected perpetrators can be obtained by investigators from matches made between the Forensic and Offender Indexes. However, qualified DNA experts in the laboratories collaborate in sharing matching profiles in other to confirm the suspects since names and other personally identifiable information are not stored in a single database.

CODIS utilize STR and mitochondrion DNA profiles as well as metadata (such as gender, date of last sighting, age, etc.) to assist in the identification of missing persons, disputed child or crime suspects. CODIS use additional DNA technologies, such a Y-STR and miniSTR. With the synergistic collaboration of legislative bodies

TABLE 3.1
Multiplex Amplification of the CODIS (Combined DN Index System) Core Loci and ESS (European Standard Set)

Markers Used in Human DNA Analysis

TH01	D3S1358	vWA	D21S11	Amelogenin
TPOX	DYS391	D1S1656	D12S391	SE33
D10S1248	D22S1045	D19S433	D8S1179	D2S1338
D2S441	D18S51	FGA	QS2	D5S818
QS1	D16S539	CSF1PO	D13S317	D7S820

Source: www.qiagen.com24plex-qs-kit.

and all components of the criminal justice, community law enforcement, crime laboratories, victims and prosecutors, the future of DNA and CODIS holds even greater promise to solve crime and identify missing persons.

3.11 DIAGNOSTIC SCIENCE AND DNA MARKERS

Identification of pathogenic organisms and detection of many genetic diseases using DNA markers have proven to be specific, fast and very reliable. Several markers have been identified in each taxon for this purpose; they are usually conserved and highly hypervariable.

3.12 16S RIBOSOMAL RNA FOR BACTERIAL IDENTIFICATION

The 16SrRNA gene has one of the most attractive potential in sequence informatics as it provides identification platforms for bacterial isolates up to genus and species level since biochemical profiles less stringent identification. Hence the need for re-characterization of bacterial isolates using the 16SrRNA gene cannot be overemphasized. The application of this marker in bacteriology has led to the renaming of many bacteria. Few studies suggest that 16SrRNA gene sequencing can assist in the identification of bacteria to genus level but less so with regard to species and 1%–14% of the isolates remaining unidentified after sequencing (Mignard and Flandrois, 2006).

Bacterial 16S ribosomal RNA (rRNA) genes contain nine "hypervariable regions" (V1–V9) that provide considerable sequence diversity among different bacteria. Species-specific sequences within a given hypervariable region constitute useful targets for identification and other scientific investigations (Soumitesh et al., 2007).

3.13 INTERNAL TRANSCRIBED SPACER (ITS) FOR FUNGAL IDENTIFICATION

Fungi are as diverse as bioactive compounds and mycotoxins they produce, very ubiquitous in our environment; their species identification is marred by limited biochemicals and sometimes confusing phenotypes. Fungi identification primarily relies on phenotypic and physiological features. Nowadays, there are arrays of molecular methods employed to identify fungal species. One of the methods considered as a new concept to rapidly and accurately identify unknown fungal sample is DNA Barcoding. DNA barcode is a short, highly DNA sequence with approximately 350–700 base pairs in length. It has shown outstanding potentials in the identification of plants, animals and other eukaryotes. ITS region of nuclear DNA (rDNA) is believed to be a highly variable and the most sequenced region used in fungal taxonomy. ITS region is a non-coding and highly diverse region with several taxonomic insights. This marker has a highly discriminatory power; even though as the most used barcode marker for fungi, it still presents some limitations (Fajarningsih, 2016).

3.14 VIRAL GENETIC SIGNATURES

The genetic diversity in viruses is so complex that nucleic acid markers for their identification are highly polymorphic. Whole genome sequencing often provides more reliable information on their identity. However, for clinical applications, strain-specific markers are often recommended, e.g. the spike (S), the envelope (E) gene and other Open Reading Frames (ORF) in nCov-19 virus, V3 and Glycoprotein 120 in Human Immunodeficiency Virus (HIV) and many others.

3.15 NUCLEOTIDE POLYMORPHISMS IN GENETIC DISEASES

Single-gene disease studies in humans have generated many unexpected findings such as the discovery of tri-nucleotide repeat expansions and their association with other forms of neurodegenerative diseases such as Friedreich's ataxia (*FRDA* gene), Huntington's disease (*HTT* gene), fragile X syndrome (*FMR1* gene) and spinocerebellar ataxias.

This line of research has also uncovered "modifier" genes that can modify the severity of phenotypes associated with mutations in the primary disease-associated gene. Cystic fibrosis was one of such disease which was considered a single-gene disease associated with mutations in the cystic fibrosis-associated gene (*CFTR*). However, the initial discovery of the *CFTR* gene preceded the characterization of several additional genes associated with cystic fibrosis and many phenotype-modulating modifier genes associated with mutations in *CFTR* (Guggino and Stanton, 2006). Cystic fibrosis-associated phenotypes due to mutations in the *CFTR* gene are in turn modulated by changes in the following genes: bowel obstruction at birth/meconium ileus, gastrointestinal phenotypes and microbial infections (*NOS1*).

3.16 DETECTION OF GENETIC MARKERS

Over the years, several techniques have been deployed for the analysis and detection of genetic markers. They have exhibited high sensitivity and accuracy. PCR, restriction fragment length polymorphism, microarrays, mass spectrophotometry, sequencing technologies and many others have been deployed for the analysis of genetic markers in forensic and diagnostic sciences.

PCR is an important tool in genetic marker studies. It enables the amplification of a DNA segment of interest through an enzymatic and temperature controlled reaction. Three main steps are involved namely denaturation, annealing and extension. The components of a PCR reaction include the *thermostable enzyme Taq* which derives its name from the *Bacillus thermoaquaticus* from which it is harvested. This enzyme extends the newly synthesized strand by joining the nucleotides to each other. The *deoxynucleotide triphosphates* are the building blocks (deoxycytidine triphosphate, deoxyadenosine triphosphate, deoxyguanosine triphosphates and deoxythymidine triphosphates), the *buffer* maintains the reaction at physiological pH, the cations (magnesium chloride) enhance primer binding and Taq activity, the *primers* also known as oligonuleotides are short synthesized DNA fragment that bracket the DNA segment of interest during amplification, their length ranges between 18 and 30 base pairs with a recommended GC/AT ration of 50:50, and the *template* is the extracted genome housing the DNA fragment of interest. There are several types of PCRs: quantitative or Real-Time PCR, multiplex PCR, singleplex PCR, gradient PCR, reverse transcriptase PCR, touchdown PCR, etc.

3.17 CONCLUSION

Molecular or DNA markers provide and open up new opportunities for evaluating diversity in genetic resources. Its applications in forensic science and diagnostic services have not only improved medico-legal practice but have equally provided platforms for the accurate and timely identification of genetic signatures that can be used for the detection of pathogens as well as predisposing genetic polymorphisms characterizing some non-communicable diseases such as diabetes, hypertension and malignancies.

REFERENCES

Al-Samarai F and Abdulkareem AA (2015). Molecular markers: An introduction and applications. *Eur. J. of Mol. Biotech.* **9**: 118–130.

Alberts B, Johnson A, Lewis J et al., (2002) *Biology of the Cell In: The cell.* 4th edition, New York: Garland Science, p. 304.

Avery OT, Macleod CM & McCarty M (1944) Studies on the chemical nature of the substance inducing transformation of pneumococcal types: Induction of transformation by a deoxyribonucleic acid fraction isolated from *Pneumococcus* type III. *J. Exp. Med.* **79**:137–158.

Ellegren H. (2004): Microsatellites: simple sequences with complex evolution. *Nat. Rev. Genet.* **5**: 435–445.

Fajarningsih ND (2016): Internal transcribed spacer (its) as DNA barcoding to identify fungal species: A review. *Squalen Bull. of Mar. and Fish. Post Harvest and Biotech.* **11** (2): 37–44.

Gizaw S, Van Arendonk JAM, Komen H, Windig JJ, and Hanotte O (2007): Population structure, genetic variation and morphological diversity in indigenous sheep of Ethiopia. *Anim. Genet.* **38**:621–628.

Guggino WB and Stanton BA (2006): New insights into cystic fibrosis: Molecular switches that regulate *CFTR. Nat. Rev. Mol. Cell Biol.* **7**: 426–436.

Jarne P and Lagoda PJL (1996): Microsatellites, from molecules to populations and back. *Trends in Ecol. & Evol.* **11**: 424–429.

Karl D (1992): Understanding DNA and Gene Cloning In: *A Guide for the Curious*, John Wiley and Sons, 2nd edition, p. 240.

Mignard S and Flandrois JP (2006): 16S rRNA sequencing in routine bacterial identification: A 30-month experiment. *J. Microbiol. Methods.* **67**:574–581.

Soumitesh C, Danica H, Michele B, Nancy C and David A (2007): Detailed analysis of 16S ribosomal RNA gene segments for the diagnosis of pathogenic bacteria. *J. Microbiol. Methods* **69**(2): 330–339.

Travers A and Muskhelishvili G (2015): DNA structure and function. *FEBS J.* 23:1–12

Van Wezel IL and Rodgers RJ (1996): Morphological characterization of bovine primordial follicles and their environment in vivo. *Biol. Reprod.* **55**:1003–1011.

Varshney RK, Graner A and Sorrels ME (2005): Genetic microsatellite markers in plants: Features and applications. *Trends Biotechnol.* **23**: 48–55.

Vignal A, Milan D, SanCristobal M and Eggen A (2002): A review on SNP and other types of molecular their use in animal genetics. *Genet Sel. Evol.* **34**: 275–305.

Yang W, Kang X, Yang Q, Lin Y and Fang M (2013): Review on the development of genotyping methods for assessing farm animal diversity. *J. Anim Sci. Bio.* **4**(2): 1–6.

4 Legal Standards Setting in the Use of Forensics/DNA Barcode

Evidence for Wildlife Crime Detection and Prosecution in Nigeria

Christie O. Onyia
Godfrey Okoye University

Obianuju P. Ilo
University of KwaZulu-Natal
National Biotechnology Development Agency

Scott E. Miller
Smithsonian Institution

CONTENTS

DOI: 10.1201/9781003178903-5

4.1 INTRODUCTION

Forensic Science or Criminalistics is defined as the use of scientific methods and procedures to solve a crime. It is the application of Science to civil or criminal laws. A crime is an unlawful act punishable by a State or any designated authority (Farmer, 2008). According to Martin (2003), a criminal offence is an act or public wrong, harmful not only to some individual but also to a community, society, or the state. Before the emergence of standardised forensic practices, in ancient times, criminal investigations and trials depended much on forced confessions and testimony of a witness. Hans Gross was the first scientist to apply scientific methods to crime scenes, leading to the birth of criminalistics. Early in the 20th century, Edmund Locard formulated the "Exchange Principle" which stated, "whenever two objects come into contact with one another, materials are exchanged between them", thereby postulating that "every contact by a criminal, leaves a trace" (Roncacè and Nicosia, 2016). It was not until the late 20th century that Smith and Simpson discovered new forensic Science, while Alec Jeffreys pioneered the use of DNA profiling in forensic Science in 1984. The importance of DNA fingerprinting after that became important in assisting police detective work, as well as in resolving paternity and immigration disputes. In the 21st century and modern society, certain procedures should be followed during investigation and trial. The offender, if found guilty, would be punished according to the existing law. Where the offence is criminal, the investigation should follow legal standards of admissible evidence and criminal procedure. The same standards and procedures should be followed in the prosecution of offenders of Convention on International Trade in Endangered Species of Wild Flora and Fauna (CITES).

Nigeria's network of protected areas includes one biosphere reserve, eight national parks, over 400 forest reserves, 12 strict nature reserves, and 28 game reserves. The total area of land under national parks is about 2.4 million hectares. The game reserves were meant to conserve wildlife as they are important to the ecosystem. However, due to the rapid increase of human population in Nigeria, from about 150 million in 2010 to about 200 million in 2020, there is an increased demand for natural resources, and this poses threats to biological resources. High-level illegal exploitation of these resources endangers their survival and leads to their extinction because of the unsustainable manner in which many species are harvested (Ejidike and Ajayi, 2013). Timber and non-timber forest products (NTFPs) are used for food, medicines, oil, resin, tannin, household equipment, fuelwood, furniture, and building materials. They are therefore being used and trafficked for income generation. Animal species are not left out. Species prioritised for conservation include Pangolins (*Phataginus* spp.), chimpanzee (*Pan troglodytes*), lowland gorilla (*Gorilla gorilla*), ostrich (*Strutio camelopedalus*), black rhinoceros (*Diceros biocornis*), giraffe (*Giraffa camelopardalis*), pygmy hippopotamus (*Choeropsis liberiensis*), and water chevrotain (*Hyemoschus aquaticus*). There is evidence that some of these species have since become extinct and there is a need for more species to receive special attention. Some of these species have been included in CITES lists I, II, and III depending on their level of endangerment (CITES, 2020) and the International Union of Conservation of Nature (IUCN, 2020).

Nigeria is progressively becoming a transit route for wildlife dealers. For example, Omifolaji et al. (2020) stated that there were seizures of 57 pangolin occurrences, comprising an estimated 478,010 pangolins from 2012 to 2019 in Nigeria, while CITES (2012) reported a seizure of 50 elephant tusks in Nigeria meant to be delivered to Thailand. Border inspectors and wildlife officials across nations are constantly monitoring endangered species that are being killed and trafficked in violation of national laws and international treaties. Sometimes many intercepted objects or samples are easily identified, but very often their taxonomic identification may not go beyond family or subfamily level (Sweeney et al., 2011), making room for some samples to go unconfiscated. Another limitation in using a morphology-based taxonomic identification tool to assign specimen is the issue of subjective interpretation of the morphological characteristics of the specimen (Ko et al., 2013). Many of the samples are concealed or trafficked at a stage that expert taxonomists are not able to identify them (McCullough et al., 2006). For example, certain mammals might have been poached and their tusks and skins, tanned or dyed, and exported for commercial purposes. Endangered plants might have been ground or processed into powder and trafficked for its medicinal use. The integration of DNA barcoding offers a solution to the challenges posed by morphology-based identifications for the fight against illegal trade on CITES species. DNA barcoding is a standardised molecular technique with several uses and applications for the identification of all life stages of animals (including immature stages) and processed products of plants and animals (Hebert et al., 2003; Wilson and Schiff, 2010; Frewin et al., 2013).

Google in 2012, through Global Impact Award, provided three million dollars ($3 million) administered by the Consortium for Barcode of Life (CBOL) hosted in Smithsonian Institution to check illegal trafficking and poaching of CITES-listed species. Nigeria was one of the six partner countries selected globally to set legal standards, among other agenda items, to demonstrate and adopt the use of DNA barcode evidence in the investigation, prosecution, and conviction of violators of the CITES regulatory framework.

The National Planning Committee (NPC) was set up and mandated to define the agenda and roadmap for implementation of the project, which was to last for 2 years (Appendices I and II). The Legal Standard Workshop was thereafter held for legal practitioners and enforcement agencies in Nigeria to determine the admissibility and effectiveness of DNA barcode data in the legal systems. The workshop brought together scientists and researchers, enforcement agencies, prosecutors and magistrates, and policymakers. The key players included: Federal Ministries of Science and Technology (FMST) and Environment (FMEnv) – CITES Unit; National Environmental Standards and Regulations Enforcement Agency (NESREA); Ministry of Justice; Sheda Science and Technology Complex (SHESTCO);

University of Lagos; University of Maiduguri; University of Port Harcourt; Obafemi Awolowo University – Natural History Museum, Ile-Ife; Raw Materials Research and Development Council; Nigeria Institute of Trypanosomiasis and Onchocerciasis Research, Kaduna; Customs; Immigration; Forestry Research Institute of Nigeria (FRIN), Ibadan; Gashaka Gumti National Park Taraba; Kanji Lake National Park, New Bussa, Niger State; Interpol, Abuja; Nigerian Television Authority (NTA) International; Radio Nigeria; Trust international. An important item on the Legal standards Workshop was the procedures for collecting, documenting, handling, processing, analysing, and presenting barcode voucher specimens and data.

4.2 POLICY, REGULATIONS, AND ENFORCEMENT ISSUES

The Federal Ministry of Environment has the mandate for policy issues that relate to environment and biodiversity and hosts the national Secretariat on CITES matter, while the NESREA has the responsibility of setting standards and regulations for the protection and conservation of Nigeria's biodiversity and natural resources in general, including CITES matters (NESREA, 2011). Their responsibility also includes coordination and liaison with relevant national and international stakeholders on matters of enforcement of environmental standards, regulations, rules, laws, and guidelines.

According to NESREA's regulatory guideline, it is an offence:

i. to import, export, re-export, or introduce from the sea, or attempt to import, export, re-export or introduce from the sea, any specimen of species listed in CITES Appendices I, II, & III and the Schedules to the Endangered Species Act without a valid permit or certificate;
ii. for any person to have in his or her possession or under his or her control, or to offer or expose for sale or display to the public, any specimen of the species listed in Appendices I, II, & III of the Convention or the Schedules to the Act;
iii. to make or attempt to make either oral or written false or misleading statements in connection with an application for a permit or certificate or registration;
iv. to obstruct or otherwise hinder an officer in the performance of his or her duties;
v. for any unauthorised person to alter, deface, or erase a mark used by the Management Authority to individually and permanently identify specimens.

The violator of CITES laws, if apprehended, whether at home or abroad attracts one or more of the following penalties: forfeiture of the specimen; arrest of the offender; heavy fine – not exceeding five million Naira (N5 million) for the individual; and not exceeding twenty million Naira (N20 million) for corporate bodies; imprisonment; or both fine and imprisonment – depending on the gravity of the offence. According to Omifolaji et al. (2020), lack of strict punishments may not deter people from engaging in the crime, as the delinquents can effortlessly pay the fine and return to the trade.

4.2.1 Monitoring of CITES-Related Matters

Monitoring and enforcement of CITES matters are conducted through the **Customs' Single Window Platform**, involving the following Stakeholders: Nigerian Customs Service; Nigerian Police Force (INTERPOL); National Drug Law Enforcement Agency (NDLEA); Agricultural Quarantine Services; and the Enforcement Authority – NESREA. Nigerian law provides that exhibit that is in processed form is forfeited to the enforcement agents, the exhibit that is in life form is returned to National Park Service, while non-living exhibits are safely stored in NESREA's facilities (Appendix I).

4.2.2 Procedure for Enforcement and Prosecutions

i. The apprehension of a CITES offender by any of the stakeholders, and subsequent transfer of the case to NESREA.
ii. Institution of criminal action against the suspect at the Federal High Court, which is the arm of the government, vested with jurisdiction for trial of such offences.
iii. Seizure of the CITES specimens for further investigation while the suspect is remanded in prison custody pending trial.
iv. Pre-trial in-house meeting of prosecuting counsel to review the facts of the case and strategies to be adopted in the prosecution – the various factors necessary for the successful prosecution of cases are considered, including appropriateness of the charge filed and whether there is a need for forensic or scientific tool in the successful prosecution of the case.
v. The punishment provided under the National Environmental Protection of Endangered Species in International Trade Regulations 2011 is forfeiture, imprisonment, and/or fine.

4.2.3 Previous CITES Offenders Trial Cases

A fiat was signed in 2008 in favour of the National Parks Services concerning CITES offence. Nigeria had many cases that warranted seizures of CITES specimens. Many of the seized items were either recognisable, live, and processed specimens such as ivory products (worked/semi-worked), whole tusk, cat, and reptile skin including crocodile, stuffed trophies, pangolin product, live turtles, and parrots. A number of the offenders had been tried, convicted, and prosecuted. Based on the presentation made by Barr. Alabo Wakoma, Esq., the Legal Adviser to NESREA at the National Workshop on Legal Standards on Barcode of Wildlife Project in Nigeria held in Abuja on 28 August 2013, it was reported that in 2012, 404 arrests were made in which 90 offenders were compounded and 34 jailed, while from January to July 2013, 70 arrests were made, 29 compounded and 52 jailed. All these prosecutions were based on the admittance of guilt; none of these cases had necessitated the need for scientific/forensic investigation on the seized CITES specimens. With the advancement in technology, culprits are getting tougher and technologically smarter, as well as advancing technologically in their practice of committing crime and the ability to evade conviction and prosecution in the absence of scientific proof of evidence. There is, therefore, a need for the adoption of scientific/forensic proof for conviction as "confessional statements made under threat or duress" are not admissible in the Nigerian court. In criminal trials, forensic Science is usually applied in contentious areas. Expert opinion is therefore needed where the expert can furnish the court with scientific or other information of a technical nature that is likely to be outside the experience and knowledge of the court.

Two main considerations permissible for the admission of forensic or scientific evidence in Nigeria include:

i. The technique must be relevant to the fact in the issue
ii. The evidence must be presented in court by an expert.

A person is usually accepted as an expert if the person is specially trained and skilled in the field in which the evidence is given or in occasional cases, as decided by the judge based on his discretion.

4.3 ADMISSIBILITY OF DNA BARCODE IN CRIME DETECTION

The first adopted step towards setting standards for the use of DNA Barcode in crime detection by the CBOL was to identify the global area of application. The agreed-upon area was monitoring and control of illegal trade and poaching of wildlife species listed under CITES levels I and II, including endangered species of national importance.

4.3.1 Development of Standard Operating Procedures (SOPs)

Setting legal standards for the use of DNA Barcode to track the illegal movement of endangered species in whatever form requires "step-by-step" procedures to ensure authenticity and legal integrity of the reference samples developed by the Barcode of Wildlife Project Nigeria team. Maintenance of the chain of custody of the samples/specimens and issuance of a unique identifier are essential requirements.

4.3.1.1 SOPs for Mammals/Vertebrates in Nigeria

The SOP for mammal/vertebrate collection was prepared by Mammals Working Group, comprising Staff from National Parks Services, National Biotechnology Development Agency, Abuja, Obafemi Awolowo University, Ile-Ife. The vouchers and tissue biopsies of most mammals and other vertebrates were obtained from protected areas, National parks, open markets, and animal orphanages in Nigeria. The sample kits for animal specimens included Personal Protective Equipment (PPE), spreadsheet data form, immobilising (dart) gun, Global Position System (GPS), camera, surgical kit for the removal of tissues, 95% ethanol, and bar-coded vial. Blood tissues were collected from living animals that were released after collection. The edges of the ears or tail were punctured, and a capillary tube was fixed for suction. Forty microlitres of blood were placed on FTA cards and allowed to dry before placing the cards in bar-coded vials. Information such as the collector's name, coordinates of the location, date and time, and unique ID were recorded. Voucher specimens were processed for animals collected live for conversion into vouchers. Some of the animals were sacrificed and preserved as either wet specimens or stuffed animals. Ethanol was used for the preservation of wet specimens, while taxidermy was used on stuffed animals. Details related to the collection event for each specimen, such as field number, GPS coordinates, date, and name of collector, were recorded in a field datasheet. The tissue biopsies were put in FluidX tubes containing 90% ethanol. Information such as date, time, and mode of transportation, kind of preservatives used were recorded. On return to the museum, final taxonomic designation and confidence level for each of the taxonomic identification were recorded according to CBOL's Strawman proposal. A catalogue number was assigned. To reduce the denaturation of DNA, samples were protected from sunlight and preserved in a cooler with ice packs until conveyed to the laboratory. The samples were transported to the laboratory as soon as possible and unprocessed tissues were stored at –20°C freezer.

4.3.1.2 SOPs for Plants

The SOPs for plant specimen collection were prepared by the Plant Working Group made up of scientists from FRIN, Ibadan, University of Lagos, University of Port Harcourt, and National Environmental Standards, Regulations and Enforcement Agency. The tools and materials for plant specimen collection were: PPE, GPS, secateurs, plant presses, camera, absorbent paper, envelopes/vials, rope, and tags. Images of the plants showing their taxonomic features were taken. The minimum number and size collected from each plant were in replicates and sufficient to fill a herbarium sheet, which measures 42 by 27 cm. Each plant was labelled with a unique number and collectors' details. Other information such as location, habitat, description, and collection date were also recorded. On return to the herbarium at FRIN, steps such as pressing, drying, and registration were carried out for storage. Plant tissues were kept in ambient temperature and moisture conditions.

4.3.2 Field Information Management Systems (FIMS)

To ensure reliability, interoperability, and value of biodiversity data, the critical information requirements include the description, the location, and at what time the sample was captured or collected (Deck et al., 2012). The Barcode of Wildlife Project-Field Information Management System (BWP-FIMS) is a database that stores information relating to fieldwork such as specimen records, events collection, identification, and images. The specimen metadata made up of many components, such as the name of the species, country code, and unique identifier, were inputted in a BWP-created spreadsheet. The advantages of BWP-FIMS over the traditional spreadsheet are that it helps to validate the field data to ensure that they are in the right format and that all required field data are up to standard, to keep data safe and accessible, as well as to avoid loss of data. Another advantage of the BWP-FIMS is that there is no mandatory set outline; each project determines its list of terms. Therefore, the project was built around the Darwin Core Terms set. This is a common standard that includes a wordlist of terms anticipated to expedite the sharing of information about biological diversity by specifying identifiers, tags, and meanings (Darwin Core Task Group, 2009). Once the spreadsheet was generated, it was validated to ensure it is of GenBank data standard and uploaded.

4.3.3 Laboratory Information Management System (LIMS)

The objectives of LIMS include standards-setting and procedures for sample collection, documentation, handling, processing, analysing, and presenting barcode voucher specimens and sequenced data. The essence of adopting the system was to minimise manual documentation as well as improve the efficiency and accuracy of data. LIMS software when integrated with a properly calibrated instrument and operated by skilfully trained staff helps laboratories to automate the collection of test data, as well as enabling them to meet regulatory requirements. LIMS has been usefully employed in different sectors, such as in food safety testing, water quality, and treatment plants, pharmaceutical production and testing, and tracking the movement of regulated species and products (Lyal and Miller, 2020). LIMS software program is used in outlining all molecular procedures conducted with each sample and linked to field specimen metadata (FIMS). It detects and re-runs unsuccessful reactions and systematic errors, thereby improving efficiency and quality control. The BWP-NIG laboratory Management data are available at https:// software.mooreabiocode.org/.

4.3.4 Sequenced Data

Out of the 98 plant species listed by the BWP-Plant Working group, 53 plant species were collected and identified by a certified taxonomist and sampled in triplicates, in line with the project's SOP for plant collection and processing. The plants were curated and the vouchers were deposited in the herbaria hosted in the Forestry Research Institute of Nigeria. The sampled animal tissues (including tissues from parrots, love birds, monkeys, civet cats, python, baboon, serrated tortoise and porcupine, elephants, lions, gorilla) were transported to the International Institute for Tropical Agriculture (IITA) for proper storage in –20°C freezer. Laboratory processing was delayed due to the sudden outbreak of the Ebola virus in the West African sub-region in the middle of the project. The steps involved in LIMS include the generation of 96 well extraction plates, Polymerase Chain Reaction (PCR) plates, cycle sequencing plates, and cherry-picking. Three plugins, namely the Biocode plugin, Biocode GenBank Submission plugin, and MySQL connector, are required for the Biocode LIMS. The MySQL connector joins the FIMS and LIMS database, while the Biocode GenBank Submission plugin automatically allows the submission of finalised contigs to make the DNA sequences accessible to the public at the end of the workflow. The BWP-NIG used the Biocode LIMS which incorporates Geneious to the laboratory workflow; from DNA extraction, PCR, sequencing to consensus assembly. Geneious is a suite of cross-platform bioinformatics software applications with functionality such as sequence alignment, Chromatogram assembly, and accessing databases.

4.4 ACTION PLAN

The adopted project's goals/action plan of the Barcode of Wildlife Project-Nigerian (BWP-NIG) team was systematically implemented in two phases (Appendix II), and the highlight of the level of completion and achievement are hereby discussed:

i. **FIMS** – The list of endangered species under CITES Appendix I and II, including species of national interest, was created. The list contained 201 species that were actively traded, of which 98 were plant species; while 103 were animal species. Specimen metadata, consisting of several elements such as a unique identifier, name of the species, country code, were inputted in a Microsoft Excel spreadsheet (Table 4.1) as there was no internet connectivity in the field/forest. The Institutional code given for the samples is National Biotechnology Development Agency because it is registered as an institutional collection under the Global Registry of Biodiversity Repositories (GRBio). The information was thereafter validated and uploaded to Biocode FIMS. To ensure a successful upload, the data were queried. The Biocode FIMS used for the project is available at https://biscicol.org.

ii. **Resources Surveillance** – Surveillance of taxonomic and forensic laboratory resources was carried out nation-wide. FRIN, Ibadan, was identified as the taxonomic laboratory for plant species, while the Natural History Museum – Obafemi Awolowo University, Ile-Ife, was designated the taxonomic lab for animal species.

iii. Samples Standardisation – Standardisation of sampling and analytical methods were successfully handled by the various mandated institutions, listed above.

iv. Development of SOPs – All participating institutions, including legal practitioners – lawyers, magistrates; regulators; wildlife enforcement officers; scientists; and policymakers were present at the Legal Standards Workshop organised by CBOL under the effective leadership of Dr. David Schindel (the Executive Secretary of CBOL). The main objective of the workshop was to share experiences in the development of SOPs in their areas of jurisdiction. The outcome of the workshop guided the subsequent implementation of the project. Field sampling and data curation were carried out by the herbaria and museum alongside expert taxonomists.

v. **Training of Technicians and Scientists** – Technicians and curators were trained in the Forestry Research Institute of Nigeria, while scientists/molecular biologists and enforcement officers were trained at the University of Lagos and National Biotechnology Development Agency, Abuja.

vi. **Sample Processing** – DNA extraction, PCR, and sequencing of samples were carried out at IITA, Ibadan. Some of the primers used for the vertebrates were: VR1-ti (66.3°C), VF1-ti (67.2°C), HCO2198 (55.3°C), LCO1490 (50.5°C), VF1d – ti (67.4°C), VR1d – ti (66.9°C), VRLi – ti (68.2°C), VFLi – ti (68.6°C), Colbird F1 (59.3°C), Colbird R1 (60.4°C), Bird F1 (59.8°C), Bird R1 (56.9°C), RepCO1F (51.2°C), RepCO1R (55.8°C), RepTBCF (52.6°C),

TABLE 4.1
Specimen Metadata with Unique Identifier

Processing Lab	Sequencing Lab	Extraction Plate ID	Extraction Barcode	Extraction Well	Tissue Barcode	Tissue Rack	Tissue Position	Tissue Type
NABDA	IITA	NABDA_ IOP_001	FR06483050	A01	FR05335775	NABDA_ IOP_001	A01	SKIN SCRAPING
NABDA	IITA	NABDA_ IOP_002	FR06483051	A02	FR05335777	NABDA_ IOP_001	A02	BLOOD
NABDA	IITA	NABDA_ IOP_003	FR06483052	A03	FR05335776	NABDA_ IOP_001	A03	BLOOD
NABDA	IITA	NABDA_ IOP_004	FR06483053	A04	FR05335774	NABDA_ IOP_001	A04	BLOOD
NABDA	IITA	NABDA_ IOP_005	FR06483054	A05	FR05335780	NABDA_ IOP_001	A05	SKIN SCRAPING
NABDA	IITA	NABDA_ IOP_006	FR06483055	A06	FR05335783	NABDA_ IOP_001	A06	BLOOD
NABDA	IITA	NABDA_ IOP_007	FR06483056	A07	FR05335782	NABDA_ IOP_001	A07	BLOOD
NABDA	IITA	NABDA_ IOP_008	FR06483057	A08	FR05335781	NABDA_ IOP_001	A08	BLOOD
NABDA	IITA	NABDA_ IOP_009	FR06483058	A09	FR05335786	NABDA_ IOP_001	B01	SKIN SCRAPING
UNILAG	IITA	NABDA_ IOP_095	FR06483144	H11	FR04499594	UNILAG_ OTO_002	F05	Leaf
UNILAG	IITA	NABDA_ IOP_096	FR06483145	H12	FR04499595	UNILAG_ OTO_002	F06	Leaf
UNILAG	IITA	NABDA_ IOP_097	FR06483146	A01	FR04499596	UNILAG_ OTO_002	F07	Leaf
UNILAG	IITA	NABDA_ IOP_098	FR06483147	A02	FR04499597	UNILAG_ OTO_002	F08	Leaf
UNILAG	IITA	NABDA_ IOP_099	FR06483148	A03	FR04499540	UNILAG_ OTO_002	G01	Leaf
UNILAG	IITA	NABDA_ IOP_100	FR06483149	A04	FR04499541	UNILAG_ OTO_002	G02	Leaf
UNILAG	IITA	NABDA_ IOP_101	FR06483150	A05	FR04499542	UNILAG_ OTO_002	G03	Leaf
UNILAG	IITA	NABDA_ IOP_102	FR06483151	A06	FR04499543	UNILAG_ OTO_002	G04	Leaf
UNILAG	IITA	NABDA_ IOP_103	FR06483152	A07	FR04499544	UNILAG_ OTO_002	G05	Leaf

Preservative	Voucher ID	Institution Code	Collection Code	Catalog Number	Chain of Custody	Year Collected	Decimal Latitude	Decimal Longitude
EDTA	NABDA_MAMMALS_1000	NABDA	MAMMALS	1000	Yes	2014	9.026222	7.432111
EDTA	NABDA_MAMMALS_1001	NABDA	MAMMALS	1001	Yes	2014	9.026222	7.432111
EDTA	NABDA_MAMMALS_1002	NABDA	MAMMALS	1002	Yes	2014	9.026222	7.432111
EDTA	NABDA_MAMMALS_1003	NABDA	MAMMALS	1003	Yes	2014	9.026222	7.432111
EDTA	NABDA_MAMMALS_1004	NABDA	MAMMALS	1004	Yes	2014	9.026222	7.432111
EDTA	NABDA_MAMMALS_1005	NABDA	MAMMALS	1005	Yes	2014	9.026222	7.432111
EDTA	NABDA_MAMMALS_1006	NABDA	MAMMALS	1006	Yes	2014	9.026222	7.432111
EDTA	NABDA_MAMMALS_1007	NABDA	MAMMALS	1007	Yes	2014	9.026222	7.432111
EDTA	NABDA_MAMMALS_1008	NABDA	MAMMALS	1008	Yes	2014	9.026222	7.432111
Silica gel	NABDA_Magnoliidae_2045	NABDA	Magnoliidae	2045	Yes	2014	6.345167	5.355389
Silica gel	NABDA_Eudicots_2046	NABDA	Eudicots	2046	Yes	2014	6.349417	5.343083
Silica gel	NABDA_Eudicots_2047	NABDA	Eudicots	2047	Yes	2014	7.443056	3.897667
Silica gel	NABDA_Magnoliopsida_2048	NABDA	Magnoliopsida	2048	Yes	2014	7.442778	3.897139
Silica gel	NABDA_Magnoliids_2049	NABDA	Magnoliids	2049	Yes	2014	7.392528	3.862139
Silica gel	NABDA_Eudicots_2050	NABDA	Eudicots	2050	Yes	2014	7.392639	3.8625
Silica gel	NABDA_Eudicots_2051	NABDA	Eudicots	2051	Yes	2014	7.391694	3.85775
Silica gel	NABDA_Eudicots_2052	NABDA	Eudicots	2052	Yes	2014	7.3925	3.858056
Silica gel	NABDA_Eudicots_2053	NABDA	Eudicots	2053	Yes	2014	7.39175	3.863028

and RepTBCR (58.8°C). For DNA Barcoding of animal species, Cytochrome c oxidase subunit 1 (CO1) was found generally effective in differentiating species. MATK and rbcl primers were used for the plant species; however, MATK had a higher resolution level. The laboratory management data are available at http://software.mooreabiocode.org/.

vii. **Custody Transfer** – The Customs Single Window Platform which has been in existence for custody transfer of confiscated items was adopted.

viii. **Forensic Laboratory** – There was no forensic lab in Nigeria that could identify the carcasses of dead animals or parts and processed samples. The only Police forensic laboratory was in Lagos, but the police representative at the Legal Standards Meeting could not attest to its functionality. Therefore, NESREA that has regulatory and enforcement responsibility for the protection of CITES-listed species was designated as the host institution. NESREA's environmental and air quality laboratory in Lagos was identified as facility to be re-designed and re-modelled as a wildlife forensic laboratory. The architectural design that would meet the chain of custody requirement was furnished by CBOL; the re-furbishing of the laboratory was still on-going at the time the project grant elapsed at the end of the 2-year target period.

4.5 CONCLUSION

DNA barcodes permit experts to empirically identify endangered species that may have been impaired or industrially processed. However, for any forensic evidence, or in this case, DNA barcode, to be used as proof of evidence in crime detection and conviction, chain of custody protocols and regulations (to eliminate possible sample contamination) for managing evidence until it is presented in court must be put in place. This requires the creation of a laboratory that could conduct DNA barcoding under the chain of custody rules and in compliance with the forensic standard by the Nigerian government. The regulation should be **gazetted** so that the DNA barcode could be a recognised database that could be referenced, indicating how any sample caught with anybody can be handled. Other requirements for acceptance of forensic evidence include certification of personnel and accreditation of the laboratory facility. Organisations involved in the chain of custody should include scientists and technicians, drawn from accredited institutions/organisation or bodies. The enforcers and courtroom officials should be well trained in the handling of specimens. Lastly, since the use of DNA Barcode evidence in wildlife crime prosecution is still new to some members of the judiciary, there is a need for awareness creation and training for lawyers and judges. The Customs Single Window Platform which has been in existence for custody transfer of confiscated items should be expanded to include scientists in specialised disciplines, to facilitate case dismissal or prosecution.

CONFLICT OF INTEREST

The authors declare no conflict of interest.

ACKNOWLEDGEMENTS

We thank Google Global Impact Awards for making this work possible. The authors would like to acknowledge the following institutions/organisations and individuals for their various contributions in the successful implementation of the project to the level it got to:

i. Dr. David Schindel, for effective coordination of the global Barcode of Wildlife Project
ii. CBOL – the initiator and sponsor of the project
iii. Prof. Ogbe Solomon, the former Director-General of the National Biotechnology Development Agency (NABDA) and the Chairman of NPC for directing the project and intellectual/material contribution
iv. NABDA – the co-sponsor and provider of human and material resources
v. NESREA for the provision of human resources and laboratory space
vi. Dr. Mrs. Ngeri Benebo, the then Director-General of NESREA and her team of Directors and Scientific Officers for their cooperation in the provision of useful materials that form part of this presentation
vii. IITA, Ibadan, Nigeria, for sample storage and processing
viii. Prof. Ogundipe and his team of Scientists at the University of Lagos, for the training of Scientists.

Others include Federal Ministry of Environment and her parastatals, National Parks Services, Forestry Research Institute of Nigeria, Natural History Museum – Obafemi Awolowo University (OAU), Ile-Ife.

APPENDIX 4.A: SOME OF THE SEIZED ILLEGALLY TRAFFICKED SAMPLES ARCHIVED IN NESREA

APPENDIX 4.B: PROJECT'S ACTION PLAN – PHASES 1 AND 2

CBOL-BWP-NIG'S ACTION PLAN PHASE 1	CBOL-BWP-NIG'S ACTION PLAN PHASE 2
i. Prioritisation of species	i. Custody transfer
ii. Identification of taxonomic and forensic laboratory resources	ii. Design, remodelling and refurbishing of forensic laboratory
iii. Standardisation of sampling and analytical methods	iii. Library construction
iv. Development of Standard Operating Procedures (SOPs) –Legal Standards Workshop	iv. Adoption of DNA barcode for crime investigation and prosecution
v. Field sampling and data curation	
vi. Training of technicians/scientists and enforcement officers	
vii. Tissue sampling and sample processing – generating sequences data	

APPENDIX 4.C: ROADMAP FOR BWP-NIG PROJECT

NPC Formed	Library Construction	Academic Labs	Enforcement Agencies	Forensic Labs
Phase 1: Planning	• Prioritise species • Identify taxonomic /lab resources • Assess training needs	• Assess capabilities • Strategy for in-country processing vs export • Assess training needs	• Prioritise species • Hold planning workshop • Strategy for in-country processing vs export	• Identify forensic lab • Design specimen flow • Assess capabilities • Assess training needs
Phase 2: Training	• Technicians trained • SOPs developed /tested	• Technicians trained • SOPs developed	• Officials trained • SOPs developed	• Technicians hired/trained • SOPs developed/tested
Phase 3: Testing	• Additional training and proficiency testing if needed; technical support systems finalised			
	Library construction		Pilot testing	
	• Tissue sampling • Data curation	• Sample processing • Data sharing	• Field sampling • Custody transfer	• Sample processing • Data sharing
Phase 4: Implementation	Library completion		Full Implementation	
	• Tissue sampling • Data curation	• Sample processing • Data sharing	• Investigations • Custody transfer	• Sample processing • Evidence preparation

REFERENCES

CITES (Convention on International Trade in Endangered Species) (2012). International Consortium on Combating Wildlife Crime, SC62 Doc. 14.7, Official document to the 62nd meeting of the Standing Committee, Geneva, 23–27 July. Available on https://cites.org/sites/default/files/eng/com/sc/62/E62-14-07.pdf [Accessed on 23rd November, 2020].

CITES (2020). Appendices I, II, and III. Available on https://cites.org/sites/default/files/eng/app/2020/E-Appendices-2020-08-28.pdf [Accessed on 23rd November, 2020].

Darwin Core Task Group (2009). Darwin Core (Kampmeier G, review manager). Biodiversity Information Standards (TDWG). Retrieved from http://www.tdwg.org/standards/450.

Deck J., Gross J., Stones-Havas S., Davies N., Shapley R., Meyer C. (2012). Field Information Management Systems for DNA Barcoding. In: Kress W., Erickson D. (eds) *DNA Barcodes. Methods in Molecular Biology (Methods and Protocols)*, vol 858. Humana Press, Totowa, NJ. https://doi.org/10.1007/978-1-61779-591-6_12.

Ejidike, B.N. and Ajayi, S. R. (2013). Trends in wildlife conservation practices in Nigeria. *International Journal of Biodiversity and Conservation* 5(4), pp. 185–190. https://doi.org/10.5897/IJBC11.127.

Farmer, L. (2008). Crime, Definitions of. In: Cane and Conaghan (editors), *The New Oxford Companion to Law*, Oxford University Press. ISBN 978-0-19-929054-3, p. 263 (Google Books).

Frewin, A., Scott-Dupree, C., and Hanner, R. (2013). DNA barcoding for plant protection: applications and summary of available data for arthropod pests. *CAB Review* 8(18): 1–13. https://doi.org/10.1079/ PAVSNNR20138018.

Hebert, P. D. N., Cywinska, A., Ball, S. L., and deWaard, J. R. (2003). Biological identifications through DNA barcodes. *Proceedings of the Royal Society B* 270: 313–321. https://doi.org/10.1098/rspb.2002.2218 PMID: 12614582.

International Union for Conservation of Nature (IUCN) (2020). Regional Action plan for the conservation of Western chimpanzees (Pan troglodytes versus) 2020–2030. Available on https://www.iucn.org/content/regional-action-plan-conservation-western-chimpanzees-pan-troglodytes-verus-2020-2030 [Accessed on 23rd November, 2020].

Ko, H. L., Wang, Y. T., Chiu, T. S., Lee, M. A., Leu, M. Y., and Chang, K. Z. (2013). Evaluating the accuracy of morphological identification of larval fishes by applying DNA barcoding. *Plos One* 2013: 8. https://doi.org/10.1371/ journal.pone.0053451 PMID: 23382845.

Lyal, C. H. C. and Miller, S. E. (2020). Capacity of United States federal government and its partners to rapidly and accurately report the identity (taxonomy) of non-native organisms intercepted in early detection programs. *Biol Invasions* 22: 101–127. https://doi.org/10.1007/s10530-019-02147-x.

Martin, E. A. (2003). *Oxford Dictionary of Law*, 7th edn. Oxford: Oxford University Press. ISBN 978-0-19-860756-4.

McCullough, D. G., Work, T. T., Cavey, J. F., Liebhold, A. M., and Marshall, D. (2006). Interceptions of nonindigenous plant pests at US ports of entry and border crossings over a 17-year period. *Biol Invasions*. 8: 611–630. https://doi.org/10.1007/s10530-005-1798-4.

NESREA (2011). S. 1. No. 16. Retrieved from https://www.nesrea.gov.ng/wp-content/uploads/2020/02/Protection_of_Endangered_Species_in_International_Trade_Regulation_2011.pdf. Accessed: 12/06/2020.

Omifolaji, J. K., Ikyaagba, E. T., Jimoh, S. O., Ibrahim, A. S., Ahmad, S. and Luan X. (2020). The emergence of Nigeria as a staging ground in the illegal pangolin exportation to South East Asia. *Forensic Science International: Reports*. 2:100138. https://doi.org/10.1016/j.fsir.2020.100138.

Roncacè, S.I. and Nicosia, U.I. (2016). ...Every contact leaves a trace... Available from: https://www.researchgate.net/publication/305640858_Every_contact_leaves_a_trace_Locard_1920 [accessed May 18, 2020].

Sweeney, B. W., Battle, J. M., Jackson, J. K., and Dapkey, T. (2011). Can DNA barcodes of stream macroinvertebrates improve descriptions of community structure and water quality? *Journal of the North American Benthological Society* 30(1): 195–216. https://doi.org/10.1899/10-016.1.

Wilson, A. D., and Schiff, N. M. (2010). Identification of *Sirex noctilio* and Native North American Woodwasp Larvae using DNA Barcode. *Journal of Entomology* 7: 60–79. https://doi.org/10.3923/je.2010.60.79.

Part 2

Working Group 07

Medical and Pharmaceutical Biotechnology

Handling Editor: Prof. Hajiya Mairo Inuwa

5 Harnessing Indigenous Bioresources for Development of Diagnostics for Detection and Confirmation of Infectious Agents in Nigeria

E. O. Ekundayo
Michael Okpara University of Agriculture

J. C. Igwe
Kaduna State University

Ifeoma M. Ezeonu
University of Nigeria

CONTENTS

DOI: 10.1201/9781003178903-7

5.1 INTRODUCTION

Infectious diseases are caused by pathogenic bacteria, fungi, parasites and viruses in humans and animals. Infectious diseases are otherwise called communicable diseases because they can spread from one person to another, either directly or indirectly, with possibility of causing epidemics and pandemics. Infectious diseases in animals which can be spread to humans are called zoonotic diseases. In the past and as in the present, infectious agents have always posed serious threat to public health and human well-being. Infectious agents like human immunodeficiency virus (HIV) causing the acquired immunodeficiency syndrome (AIDS) and the corona viruses (CoV) responsible for Severe Acute Respiratory Syndrome (SARS) such as COVID-19 are the causes of global devastating pandemics for which cures have not been found (WHO, 2020). Apart from the pandemics which are often associated with high morbidity and mortality, other infectious diseases like malaria, tuberculosis (TB), viral hepatitis, drug-resistant bacterial infections, newly emerging viral infections like Ebola, Lassa fever and Zika viruses are posing significant threats to health in different parts of the world. Infectious diseases are responsible for over 17 million deaths annually (WHO, 2020).

The diseases caused by infectious agents come in different forms and present symptoms and signs that may be similar. Most pathogens can cause a wide spectrum of clinical syndromes in humans and animals. Conversely, a single clinical syndrome may result from infection with any one of several pathogens. Diagnostic laboratory tests are used to detect, identify and confirm the causative agents of infectious diseases. The reagents, techniques, devices and instruments used to achieve these are collectively called diagnostics. Laboratory detection and confirmation of infectious agents causing any infectious diseases involves the examination of specimens to detect, isolate and identify the causative agents or their products or components. Diagnostics are crucial to the successful diagnosis of diseases, detection and confirmation of infectious agents and epidemiological surveillance of emerging infectious diseases.

5.2 THE NEEDS FOR DIAGNOSTICS FOR INFECTIOUS DISEASES IN NIGERIA

Nigeria has the highest burden of infectious diseases in sub-Saharan Africa and is amongst the countries with highest burden of HIV/AIDS, malaria and tuberculosis (TB) in the world (WHO, 2018). Diagnostics for infectious diseases are some of the most critical needs in Nigeria. Nigeria depends on importation for over 99% of her medical devices needs including diagnostics for infectious diseases (USCS, 2020). In 2018, total cost of medical devices was \$134.7 million. Imports value of medical devices in Q1 of 2019 was \$24 million (NBS, 2020). Therefore, total lack of capacity in manufacturing of diagnostics in Nigeria is a huge loss of foreign exchange. Lack of local R&D in infectious diseases diagnostics affects the scientific and economic development of the country. It has negative impact on the quality of healthcare. The right diagnostics needed for effective healthcare are often not readily available. Connected with this is the poor quality of imported diagnostics due to differences in the properties of the local strains or due to importation of low-quality diagnostic tests with inadequate quality assurance. The imported diagnostics may not be optimized for sensitivity and specificity for strains of microorganisms found in the country. Some test kits or diagnostic devices manufactured abroad do not work as well in developing countries due to differences in the antigenic and genetic properties of the indigenous strains. Thereby, the quality of results obtained from the use of these test kits is substandard. These are some of the compelling reasons for the need to harness the bioresources available in the country for the development and production of diagnostics for infectious diseases in Nigeria.

The diagnostics commonly imported for diagnosis of infectious agents include stains for microscopy, culture media for cultivation of bacteria and fungi, immunodiagnostic kits and increasingly, molecular diagnostics. Diagnostic tests are based on components derived from the infectious agents. Therefore, infectious agents themselves represent huge bioresources from which raw materials can be developed for the production of diagnostics for the detection and confirmation of specific agents responsible for a particular infection. The raw materials for production of diagnostics for infectious diseases are available in the country. Ingredients for composition of culture media are derived from easily available materials like meat offal from abattoir and slaughter houses as well as agricultural waste products from cassava and rice processing. These can be converted to raw materials as ingredients in the manufacture of culture media. Nigeria is endowed with abundance of bioresources for productions of diagnostics; what is lacking is the strategy and necessary policies to turn these natural assets into valuable products to meet the critical needs of the nation in the area of diagnostics.

For researchers, biotechnologists and entrepreneurs, diagnostics for infectious diseases remain an unexplored world of abundant opportunities. For the national economy, investment and promotion of research and development (R&D) and entrepreneurship in diagnostics, will save the country billions in foreign exchange annually, create employment for teeming multitudes of Nigerian young graduates in biological and biomedical sciences and contribute to the realization of better control of infectious diseases in the country. The utilization of the resources provided by the many infectious agents plaguing the country, for development of diagnostics, is one area of turning adversity to an advantage for the country.

Obtaining suitable specimens and isolates of infectious agents are significant challenges for many diagnostic companies in developed countries. However, these are things that are available in abundance in the country but the value is little recognized. As part of strategy to utilize these bioresources, biorepositories of specimens and culture collections of infectious agents should be established. These will become valuable sources of materials for R&D of diagnostics. They can also serve as sources of reasonable amount of revenues. Currently, many Nigerian researchers have to import organisms from abroad as control strains in their researches. The various infectious agents in the country need to be isolated, characterized and preserved. From these characterized agents, researchers can identify and produce antigens for development of serological tests. The antigens can also be employed in the development of monoclonal and polyclonal antibodies that are needed in the development of immunodiagnostic tests. Furthermore, molecular characterization of these agents will provide the knowledge of the unique genetic characteristics that can be used to develop molecular diagnostics. Our aim in this chapter is to describe the various components of the infectious diseases' diagnostics and the principles of various diagnostic methods and technologies with emphasis on the methods and process of their development and production. The overall objective is to stimulate interest in the R&D of diagnostics for infectious diseases using indigenous bioresources available within our reach.

5.3 OVERVIEW OF DIAGNOSTICS FOR INFECTIOUS DISEASES

Diagnostics for infectious diseases are products of basic and applied research on the infectious agents or the pathophysiology of the diseases they cause. Many diagnostics in use for infectious diseases today originated from researches conducted in academic laboratories while some are the products of R&D of some diagnostic companies. Modern biotechnology has revolutionized the methods of development and production of diagnostics. Biotechnology has also tremendously expanded the possibilities of types of diagnostics and the speed of diagnosing diseases. Commercial diagnostics may be packaged into various forms and formats as kits or components that are used together to carry out a particular technique or procedure. Diagnostics are based on methods, techniques and technologies. The diagnostic methods used in the diagnosis of infectious disease (as detailed in Table 5.1) fall into four broad groups, namely:

1. Morphological and microscopic methods
2. Culture-based diagnostic methods
3. Serological and immunodiagnostic methods
4. Molecular diagnostic methods

TABLE 5.1
Technology and Some Diagnostics for Infectious Diseases

Method/Technology	Examples of Diagnostics	Components	Uses
Morphological and microscopy method **Stains**	Gram stain	Crystal violet, neutral red/ Safranin, Lugol's iodine, alcohol-acetone decolourizer	Separation of bacteria into Gram positive or Negative groups
	Ziehl-Neelsen stain	Carbol fuchsin, Malachite green, alcohol decolourizer	Detection of acid-fast tubercle bacilli

(Continued)

TABLE 5.1 (*Continued*)
Technology and Some Diagnostics for Infectious Diseases

Method/Technology	Examples of Diagnostics	Components	Uses
	Giemsa stain	Giemsa stain solution in methanol	Detection of malarial parasites in blood smear
	Field stain	Field Stain A & B aqueous solution	Detection of malarial parasites in thick and thin blood films
	Leishman stain	Leishman stain solution	Detection of malarial parasite and blood cell morphology
Culture technique **Culture media**	Nutrient agar MacConkey Mannitol salt Agar Mueller-Hinton Etc.	Peptones, meat extracts, yeast extracts and other substances required for growth	Cultivation of bacteria and fungi for isolation, characterization and identification and antibiotic susceptibility testing
Immunodiagnostics	**Serological tests** in latex beads or lateral flow or ELISA formats for detection of antibodies in serum or detection of antigens for HIV, malaria, typhoid, viral hepatitis, etc.	Antigens or antibodies immobilized on membrane in lateral flow tests or in polystyrene wells in ELISA	Detection of antibodies or antigen to infectious agents
Molecular diagnostics	Hybridization and PCR-based tests for detection of viruses, bacteria, etc.	DNA and RNA extraction kits, Primers, Probes, DNA Ladders, enzymes, e.g., Taq Polymerase	Molecular detection and confirmation of infectious agents in specimen, e.g., HIV, Hepatitis virus, Lassa, Ebola, Corona virus (COVID-19)
Antibiotic susceptibility disks	Antibiotic disks for essential antibiotics	Disks of common antibiotics	Antibiotic susceptibility testing of bacterial isolates

5.4 MORPHOLOGICAL AND MICROSCOPIC METHODS

Morphological and microscopic methods are the oldest methods of detection and confirmation of infectious agents of diseases. The methods are commonly used for the detection of bacteria, fungi and parasites in stained smear of specimens using the microscope to examine the morphology and staining reactions. The microscopic techniques have developed over the past two centuries from the simple microscope of Anton van Leuwenhoek to the sophisticated electron microscope and various computer-aided imaging systems of the 21st century (Conn, 1930; Madson, 2001,

Sandle, 2004). However, the light microscopy remains the widely used technique in developing countries. In this technique, specimens are prepared in various ways to make the visualization of the morphology of microorganisms possible. Staining was developed in the 18th century as an important advancement in the detection and confirmation of microorganisms as causative agents of diseases (Sandle, 2004). The early bacteriologists experimented with coloured materials in attempt to differentiate the microorganisms from diseased tissues they were studying. The most successful of these attempts resulted in the development of the Gram and Ziehl-Neelsen staining techniques (Madson, 2001). Ziehl-Neelsen staining is used for the detection of acid-fast bacilli for diagnosis of tuberculosis (Madson, 2001). Gram Staining remains one of the fundamental techniques in detection and identification of bacteria from culture (Sandle, 2004). Romanowsky stains like Giemsa, Field and Leishman stains are widely used in microscopic diagnosis of malaria (Satpathi et al., 2014).

Staining employs certain biological dyes to colour microorganisms or the background for the purpose of easy visualization under the bright field microscope. The point of interest to us is that the dyes were used for other purposes before the bacteriologists of the 19th century began to use them to stain biological materials for pathological work. The stains used at the beginning of the development of staining techniques were made from natural dyes derived from biological sources like plants and animal tissues. Dye making was an ancient art until methods for manufacturing of synthetic dyes were developed.

Dyes and stains remain important items of international trades. Nigeria like most of the developing countries import the dyes and stains needed for diagnostic applications. The stains commonly used include crystal violet, Safranin, neutral red, malachite green, Romanowsky stains like Giemsa, Field and Leishman stains. Large amounts of these reagents and stains are used in medical laboratories in hospitals and clinics and in laboratories in academic institutions. Local capability in their production is highly limited except for preparation of stain solutions from imported stain powders. Although the methods for the manufacture of the stain powders and dyes have been developed for over two centuries, Nigeria still imports dyes and stains to meet most of her needs.

Stains are made from dyes. Dye is a very important indigenous industry in Nigeria. Most of the dyes used in the traditional textile industry in Nigeria are derived from plants. There are many plants in Nigeria which have been used for generations as sources of dyes in local textile and dyeing industries (Okonkwo et al., 2019). The traditional practitioners have developed expertise in the art of making dyes. Several studies have reported the extraction and characterization of natural dyes from plants in Nigeria, including "dogoyaro" plant, bark of whistling pine and root of African peach (Osabohien et al., 2009, 2014), guinea corn leaves (Nwonye and Ezema, 2017) and *Rothmannia whitfieldii*, *Baphia nitida* and *Whitfieldia lateritia* (Nnorom et al., 2018; Okonkwo et al., 2019). These plants represent potential sources of bioresources for production of biological stains for use in diagnostic applications. However, little research has been done on the uses of these dyes in biological staining. It is important to remind ourselves that the dyes and stains that are used today were just like the local dyes until scientists like Robert Koch and Gram developed their staining techniques using the locally available dyes. Research on natural dyes from plants which could serve as the basis for developing staining techniques for diagnostic need is needed. This is one of the areas in which researchers can harness these important indigenous bioresources to meet the needs of the country.

5.5 CULTURE-BASED METHODS

Culture techniques are the most widely used methods for the detection and identification of bacteria and fungi causing infections (Sandle, 2011). The development of techniques to culture disease-causing agents outside the body of the host represented a major milestone in the development of diagnostic microbiology (Merck Microbiology Manual, 2010). The ability to culture infectious agents remains one of the most important aspects in R&D of diagnostics. The culture of microorganisms enables the colonies of pure growth to be isolated for identification. This also makes it possible to conduct further research on the isolated agent. Culture media are used for the cultivation of microorganisms outside the body of the host. Culture media are essential diagnostics for growth and identification of bacteria and fungi. Culture media consist of preparations containing nutrients and supplements that support the recovery and growth of microorganisms in vitro. In diagnostic microbiology, culture media are designed in various ways to allow for selective, differential and diagnostic cultivation of pathogenic bacteria and fungi.

5.5.1 Brief History of the Development of the Culture Media

In line with our objective of stimulating interest in the R&D of diagnostics for infectious diseases, a brief history of the development of the culture media will be helpful. The history of the development of culture technique and culture media is an interesting one. From the desire to disprove the theory of the spontaneous generation, to the establishment of the germ theory of disease or to discredit the alleged miracle of red blood like stains on the Eucharistic bread, early bacteriologists sought for ways to cultivate microorganisms on artificial media (Merck Microbiology Manual, 2010). Sandle (2011) has given an inspiring historical account of the pioneering works of the early bacteriologists of the 19th century in the development of the culture media. Drive for scientific inquiry, severe necessity during the world wars and the developments in the meat industry combined to shape the direction of the development of the microbiological culture media. The first successful attempt of direct relevance to diagnostic interest was made by Louis Pasteur in 1860 (Sandle, 2011). He grew a culture of bacteria in a medium made of yeast ash, sugar and ammonium salts. Robert Koch obtained better results using broths made from fresh beef, serum or meat extracts for liquid culture. Robert Koch is credited with making significant contribution to the development of solid culture media. He grew bacteria on solid surfaces of substrates like cut potato slices. There are several shortcomings of this method. He perceived that it would be better to solidify his liquid medium by incorporating a suitable gelling agent. He experimented with egg albumin, serum and gelatin. Although he got improved results with gelatin, it was not completely satisfactory as a gelling agent in culture media. It was in his laboratory that agar-agar, which was used as a gelling agent in preparation of fruit jams, was first introduced as a gelling agent in culture media. Thus, first attempts at producing culture media were based on materials that were readily available. As the growth requirements of bacteria and fungi became better understood, culture media with more precise compositions were developed.

The development of culture technique and microbiological culture media has spanned more than 150 years. The methods for the manufacturing of ingredients and culture media have long been established and manufacturing of culture media is a large global industry. The development of culture media has reached a level of sophistication in the 21st century that one would almost think that there is nothing left to be developed as far as the manufacture of culture media is concerned. However, manufacture of culture media remains almost an exclusive preserve of big diagnostic companies in Europe and America. Logically, companies that started manufacturing culture media with more than 100 years of experience continue to dominate the manufacture and supply of culture media. Most of the developing countries appear to be content with importation of culture media from the big diagnostic companies in the developed countries. Nigeria, like many developing countries, depends on importation for their needs for culture media. However, the dwindling economic resources and the desire for scientific development and self-reliance are the factors driving the needs to develop strategies to build local capacity for manufacture of these important products required for diagnostic services and research.

The methods for manufacture of culture media in the 21st century appears to be essentially similar to the ones developed by pioneering manufacturers of the past except for mass production made possible by industrialization and automation. The same essential ingredients that were used in the past are still the same today. The raw materials used have not changed significantly. The major active substance in culture media remains peptones. Today peptones are still largely manufactured from meat, milk and vegetables. These are bioresources that are readily available. Indigenous bioresources that can be utilized for production of culture media and the components are available in Nigeria. In recent times, the rising costs of importation of culture media are making the scientists of the developing countries to look inward (Akpor et al., 2019). R&D on available indigenous resources for manufacture of culture media will be a right step in the right direction for the country. Continued dependence on importation for basic diagnostic needs is not sustainable and is detrimental to the scientific, economic and healthcare development of the country.

5.5.2 Methods for Production of Microbiological Culture Media

The high demand for culture media by the increasing number of microbiology laboratories in the latter part of the 18th century and the difficulty of in-house production of the culture media necessitated the development of methods for commercial production of base materials and culture media. By the 1890s, all the characteristics of culture media familiar to the microbiologists of today, like clear broths, solid media in Petri dishes and wide use of peptones and agar, have been developed (Sandle, 2011). Peptones were originally produced for pharmaceutical uses by the meat industry as enzymatic digest by companies like Liebig Extract of Meat Company (LEMCO) established in 1865 and for microbiological uses by Digestive Ferments Company (Difco) originally founded in 1895 as Ray Chemical and the Baltimore Biological

Laboratory founded in 1935 which are now part of Beckton, Dickson and Company (Difco & BBL Manual, 2009). These companies remain well-known brands and some of the largest manufacturers of culture media. The development of these companies illustrates what can happen when scientists take initiative to harness resources at their disposal for scientific and economic developments to meet perceived needs. Capability to manufacture culture media from readily available resources will reduce the problems of scarcity and high cost associated with importation.

The manufacture of culture medium has undergone tremendous development over the past 150 years. However, the basic requirements and sources of raw materials have not changed much. The culture media used in the bacteriology laboratory can be divided into three main types: (1) natural or empirical culture media, (2) synthetic or defined culture media, and (3) living media (Sandle, 2011). The earliest culture media produced by Koch and the pioneer researchers and developers of culture media were natural or empirical culture media. Such media are still in common use today. In such media, the requirements for the growth of bacteria are provided by inclusion of natural substances such as meat extracts or infusions, milk, urine, blood or derivative of blood like serum and plasma and vegetable juices. In modern empirical media, peptone is used as the major ingredient. Peptones are obtained from enzymatic digestion or acid hydrolysis of protein. In culture media for cultivation of fungi and yeasts, extracts of vegetables like carrots, lettuce and fruits including tomatoes and oranges are utilized. Many excellent manuals on the detailed procedure for the production of peptones have been published including the various editions of the Merck Microbiology Manual, the Difco & BBL Manual of Microbiological Culture Media.

Peptones have been produced from some indigenous bioresources like soya bean (*Glycine max* L) and African locust bean (*Parkia biglobosa*) by Nigerian Scientists (Uzeh et al., 2006). Akpor et al. (2019) evaluated the chicken feather hydrolysate as alternative source of peptone for microbial cultivation. Lasekan et al. (2013) evaluated chicken by products as sources of useful biological resources for manufacture of culture media ingredients. Others have evaluated cassava starch as possible alternative gelling agents to agar which is expensive and not manufactured locally (Maliro and Lameck, 2004; Kwoseh et al., 2012). These represent preliminary works but interesting developments that should be continued.

5.5.3 Common Constituents of Culture Media

Microorganisms require nutrients and other substances which include carbon, nitrogen, inorganic phosphate and sulphur, trace elements, vitamins and water for growth. A typical microbiological culture medium (as shown in Table 5.2) must contain available sources of these nutrients and substances. Traditionally, these growth requirements were supplied through sources such as meat infusion, beef or yeast extracts. However, it was discovered that addition of peptones improved the performance of culture media. Peptones are digests of proteins which provide readily available nitrogen, carbon and essential amino acids that enhance rapid growth of microorganisms. Different types of peptones have been produced to meet the different requirements of different organisms. Some fastidious bacteria may also require additional growth factors to grow in artificial medium. Culture media formulations are based on the ability

TABLE 5.2
Common Constituents of a Typical Culture Media

Constituents	Sources
Amino-nitrogen	Peptone, protein hydrolysate, infusions and extracts of meat, yeast and vegetables
Growth factors	Blood, serum, yeast extract or vitamins, NAD
Energy sources	Sugar, alcohols and carbohydrates
Buffer salts	Phosphates, acetates and nitrates, commonly as ammonium salts
Mineral salts and metal ions	Phosphate, sulphate, magnesium, calcium, iron
Selective agents	Chemicals, antimicrobials and dyes
Indictor dyes (Differential agents)	Phenol red, neutral red
Gelling agents	Agar, gelatin, alginate, silica gel
Chromophore	Various chemicals

of bacteria to utilize the components of the media. For diagnostic purposes, media also contains substances like reagents, dyes or antibiotics that make them selective for certain groups of bacteria. Others contain indicators that make them differential.

The nitrogen and vitamins required in culture media are usually obtained from peptone, protein hydrolysates, infusions and extracts of meat, plant materials like soya beans and cow pea. These ingredients can be obtained from indigenous bioresources like meat offal and entrails from meat industry. Milk and milk residues like whey from dairy operations can be used to produce casein from which peptones can be obtained. Soya beans and many other wild beans in Nigeria can be exploited to obtain peptones. Nigeria is a major producer of cassava. Cassava can be an important source of carbohydrates as an energy source in culture media. Incorporation of local ingredients into the manufacture of culture media will reduce the dependence on importation of culture media in the country and lower the cost. A large number of different types of culture media have been formulated. There are some very common ones that are routinely used in diagnostic bacteriology. The following examples (as shown in Table 5.3) illustrate the typical compositions of common culture media.

TABLE 5.3
Typical Compositions of Common Culture Media

Culture Medium	Composition(g/L)
Peptone Water Medium	
Meat peptone	10.0
NaCl	5.0
Nutrient Agar	
Peptone	5.0
Beef extract	1.5
Yeast extract	1.5
NaCl	5.0

(*Continued*)

TABLE 5.3 (*Continued*)
Typical Compositions of Common Culture Media

Culture Medium	Composition(g/L)
Agar	15
MacConkey Agar	
Peptone	1.5
Tryptone	1.5
Gelatin peptone	17
Lactose	10
Bile salts mixture	1.5
Neutral red	0.03
With NaCl[a]	5.0
Crystal violet	0.001
Mueller-Hinton Broth/Agar	
Meat extract	2.0
Casein peptone, acid digest	17.5
Soluble starch	1.5
Agar	13.0
Broth, no agar	
Mannitol Salt Agar	
Casein peptone	5.0
Meat peptone	5.0
Meat extract	1.0
Mannitol	10
NaCl	75
Agar	15
Sabouraud Dextrose Agar	
Casein peptone	5.0
Meat peptone	5.0
Glucose	20.0
Tryptone Soy Agar	
Casein peptone	15.0
Soya peptone	5.0
NaCl	5.0
Agar	15.0
Cysteine Lactose Electrolyte	
Deficient agar (CLED)	4.0
Gelatin peptone	4.0
Casein peptone	3.0
Meat extract	0.128
L-Cysteine	10.0
Lactose	15
Agar	0.02

[a] Some formulations are without NaCl and crystal violet. Casein/Meat peptone (50/50 3.0 g) may replace peptone and tryptone

(*Continued*)

TABLE 5.3 (*Continued*)
Typical Compositions of Common Culture Media

Culture Medium	Composition(g/L)
Bromothymol blue	0.10
Andrade indicator	
Simons Citrate Agar	
$(NH_4)_2PO_4$	1.0
K_2HPO_4	1.0
NaCl	5.0
$MgSO_4{\cdot}7H_2O$	0.2
Agar	13
$Na_3C_6H_5O_7{\cdot}2H_2O$	2.0
Bromothymol blue	0.08
Thioglycolate Broth	
Yeast extract	5.0
Casein peptone	15.0
Glucose	5.5
Sodium thioglycolate	0.5
NaCl	2.5
L-Cysteine	0.5
Triple Sugar Iron Agar	
Casein peptone	14.0
Yeast extract	3.0
Meat extract	3.0
Glucose	1.0
Lactose	10.0
Sucrose	10.0
NaCl	5.0
$Na_2SO_2O_3{\cdot}5H_2O$ (Sodium thiosulphate)	0.3
Ammonium ferric citrate	12.5
Agar	12.5

Source: Adapted from WHO (1995) Production of Basic Diagnostic Laboratory Reagents.

5.6 IMMUNODIAGNOSTICS AND IMMUNOLOGICAL TESTS

Antigen tests can be done for early detection of an infectious diseases or when it is not possible to use any other method (Cheesbrough, 2006). These methods involve the use of specific antibody or antisera or labelled antibody. Nowadays, these tests are formatted into techniques like the Enzyme-Linked Immunosorbent Assay (ELISA) or Immunochromatographic Test (ICT) or Rapid Diagnostic Tests (RDT). Many of these tests are now common. The drawback on the use of these tests includes the high cost of importation and lack of specificity sometimes for the local strains of the infectious agents. This underscores the need to develop more accurate tests from local strains. These require the mastery of techniques to isolate antigens and produce monoclonal or polyclonal antibodies. These techniques are described later in this chapter.

PRINCIPLES OF AND TYPES OF IMMUNODIAGNOSTIC TESTS:

Direct slide agglutination:

Antiserum + bacterial cells => Bacteria agglutinated

This is the principle used in the Widal test for *Salmonella* infection. *Salmonella* infections are common in Nigeria. Large quantity of Widal test kits is imported into Nigeria, annually. Apart from the intrinsic low sensitivity and specificity of the test, the kits imported into Nigeria have additional problem of being produced from strains that may be different from the local strains in Nigeria. This will further compromise the accuracy of the test kits for use in Nigeria. Since typhoid is endemic in Nigeria, the logical approach is to develop more specific tests using the local isolates from Nigeria.

Latex agglutination

Latex particles are coated with specific antibody. The specimen containing microbial antigens is mixed with the latex reagents, resulting in agglutination of the latex particles.

Antibody latex reagent + Antigens in specimen => Latex particles agglutination

Direct Immunofluorescence

Fluorochrome antibody reagent + Antigen (Pathogen) => Pathogen Fluoresces

Enzyme Immunoassays (EIA) to detect antigen

Antibody against the antigen to be detected is fixed to the well of a microtitration plate or membrane of an individual test device such as a plastic plate.

1. **Fixed antibody + Antigen in specimen => Antigen binds to antibody**
2. **Enzyme conjugated antibody added => Binds to antigen–antibody complex**
3. **Chromogenic substrate added =>Colour produced**

Immunochromatographic (IC) or Lateral Flow techniques to detect antigens.

These are the popular RDT. They are most useful in developing countries. They are used for testing for HIV, hepatitis, syphilis, chlamydial infection, malaria and others. Specific antibody or antigen is immobilized on a membrane. By capillary action, the antigen in the specimen first meets specific antibody conjugated to colloidal gold particles and the antigen binds to the antibody. The antigen–antibody colloidal gold complex migrates up the strip where it is captured by a line of specific antibody, producing a pink line in the test result window or line.

1. **Antigen in the serum + Antibody colloidal gold conjugate => Antigen binds to antibody**
2. **Antigen–antibody colloidal gold meets antibody line on strip => Complex captured (pink line produced)**

Serological tests: Testing serum for antibodies
In these tests, antigen is used to detect and measure antibody in a person's serum.

These are commonly used for pathogens when the antigen is not present in routine specimens or if present is not easily isolated and identified by other available techniques. They are used most commonly for viral infections such as dengue, HIV and bacterial infections like brucellosis, rickettsia, syphilis and leptospirosis.

5.7 MOLECULAR METHODS

Rapid advances in Molecular Biology have led to new insights into the genetic and structural features of a considerable number of human pathogens. These discoveries have led to the development of innovative molecular diagnostics techniques (Reischl, 1996). Molecular methods have revolutionized the diagnostics of infectious diseases due to the increasing availability of sequence data of many infectious agents (Speers, 2006). Molecular diagnostic techniques explore the physical and biochemical properties of specific biomolecules for the development of sensitive and specific molecular diagnostics. Most of the molecular-based diagnostic tests are based on various nucleic acid–based technologies such as nucleic acid analysis without amplification and nucleic acid amplification techniques (NAAT) such as the Polymerase Chain Reaction (PCR) and other emerging amplification methods (Sarkinfada et al. (2014).

Nucleic acid–based molecular diagnostics started with the discovery of the duplex structure of DNA and its complementarity rules by Watson and Crick in 1953. Nucleic acid–based molecular diagnostics are based on four major techniques, namely, enzymatic DNA restriction, nucleic acid hybridization, DNA amplification by PCR and DNA sequencing (Caliendo et al, 2013; Pomerville, 2017). The most outstanding biotechnological advancement in DNA technology and methodology as far as molecular diagnostics are concerned are the development of PCR and rapid sequencing methods which have made available sequence data of many pathogens (Procop, 2007).

5.8 HARNESSING INDIGENOUS BIORESOURCES FOR DEVELOPMENT OF DIAGNOSTICS

Diagnostics for infectious diseases fall into the following categories

I. Preparations for detection of infectious agents in the morphological and microscopic methods.

These are mainly stains prepared from natural or synthetic dyes. Stains commonly used for preparation of stains for diagnostic purposes include crystal violet, neutral red, malachite green, safranine, Romanowsky stains like Field stain, Giemsa stain and Leishman stains. Dyes are also used as indicators in culture media. Biological stains have been in use for diagnosis of infectious diseases since the 1800s and have undergone tremendous development (Madson, 2001).

Natural dyes are derived from biological sources like plants, insects and animals. Plants have provided the major sources of natural dyes. Several plants are used in the traditional dyeing industry in different parts of Nigeria. These dyes are used for dyeing fabric in the traditional textile industry and for body decorations. They include plants such as uri (*Rothmannia whitfieldii*), uhie (*Baphia nitida*) among the Igbo of the South Eastern Nigeria, laali (*Lawsonia inermis*) and elu (*Philenoptera cyanescens*) among the Yoruba in the South West and lalle (*Lawsonia inermis*) used among Hausas and other tribes of the Northern Nigeria. R&D on these natural dyes can lead to development of biological dyes which can reduce the dependence on importation of stains and introduce new innovations in staining techniques.

II. Culture Media.

A very large number of culture media have been formulated for different purposes in the cultivation of microorganism. Culture media are mainly produced by big manufacturing companies like Merck, Oxoid, Difco and BBL of the Becton, Dickson and Company (BD) group. However, smaller companies in India and China now manufacture culture media which are usually cheaper than those manufactured by the leading manufacturers. Culture media is not currently manufactured at the commercial level in Nigeria. The utilization is lower than expected because of non-availability and cost. Many diagnostic procedures in the clinical laboratories cannot be done. Likewise, microbiological experiments in schools cannot be conducted because of lack of culture media. One approach to solving some of these challenges is to manufacture culture media locally using bioresources that are available within easy reach. The constituents of culture media have been outlined in Table 5.3.

The most important components of culture media are peptones and other sources of amino-nitrogen and carbohydrates. These can be manufactured from bioresources that can be sourced locally. Availability of these ingredients from local sources will facilitate the manufacture of culture media in Nigeria. Peptones are usually prepared from meat, milk and plant materials like soya bean and peas. Carbohydrates like starch and sugars can be manufactured from food processing wastes from cassava, rice and grains processing factories. Yeasts can be obtained from local sources like palm and from breweries for production of yeast extracts.

Methods for preparation of peptones and other ingredients of culture media have been described in the publication of the World Health Organization, Mediterranean Region (Production of Basic Laboratory Diagnostic Reagents). The equipment required for laboratory level production is minimal. For commercial production, industrial scale manufacturing infrastructure will be required. The methods for productions of key ingredients of culture media are described briefly as outlined in the WHO (1995) publication (Production of Basic Laboratory Diagnostic Reagents).

Production of Meat Peptones

Meat peptones are produced from muscular tissue or offal like waste parts and entrails and gelatin. Muscular tissues and offal for production of meat peptones can be fresh or frozen or dried. Gelatin extracted from collagen by boiling. Collagen is the fibrous protein found in connective tissues, bone and cartilage. Local sources of these materials include abattoir and meat industry. Peptones have also been prepared from chicken feathers and wastes (Lasekan et al., 2013; Akpor et al., 2019). A variety of proteolytic enzymes or proteases are used to achieve the enzymatic hydrolysis of animal protein. They include pepsin, trypsin, carboxypeptidase A and B and elastase. A common source of pepsin is stomach of pigs or other animals. Trypsin along with chymotrypsin, carboxypeptidase A and B and elastase can be obtained from the pancreas of various animals.

Production of Meat Extracts

Meat extract is an essential component of culture media. It is commonly produced as an enzymatic digest of beef muscle. It is prepared by putting lean minced meat like beef heart or fat-free meat in pan or pot with desired amount of water. Depending on the amount, for example about 500 g of meat can be processed in 1 L of water. The meat is kept in a refrigerator at 4°C for about 14 hours. The meat is then brought to boiling point and left to simmer for 15 minutes with constant stirring with a glass rod. The boiled meat is then filtered through a wet paper filter to remove fat. Water is added to make it up to 1 L to replace water lost during boiling. The extract can be freeze-dried to obtain dehydrated extract or the water can be evaporated.

Production of Casein peptones

Casein peptones are obtained from cow milk. Casein and whey are the major protein components of milk with casein making up to 80% and whey constituting the remaining 20%. After the removal of cream or fat from bovine milk, casein can be precipitated as insoluble portion by treatment with hydrochloric acid or sulphuric acid. The casein recovered is called acid casein. It is insoluble in water. To make it soluble in water, acid casein is usually dissolved in a suitable hydroxide such as sodium hydroxide (NaOH). The resulting sodium caseinate is used as the basis for hydrolysed caseins. Sodium caseinate consists of 87%–90% protein. Casein peptones are very important in culture media because they are very nutritive as they contain all of the amino acids and are rich in the essential ones (Merck Microbiology Manual, 2010).

Casein peptones are manufactured by either acid or enzymatic hydrolysis. Although many acids can be used, hydrochloric acid is the most commonly used. Acid hydrolysis leads to complete breakdown of casein into amino acids and other simple chemicals. On the other hand, enzymatic hydrolysis is less drastic. Proteases from pancreas are used. Pancreatic digests produce peptones that are rich in peptides.

Soy Peptones

Soy peptones are enzymatic digests of soya bean flour. To eliminate the several protease inhibitors in soya bean, it is first necessary to heat or toast the defatted beans. The toasted beans are then milled into flour. Soya bean flour is rich in high quality protein, carbohydrates, calcium and vitamin B complex. The enzymes used in digestion of soya flour are usually from animal-free sources. They include papain from pawpaw and proteases from microorganisms. Uzeh et al. (2006) described the production of peptones from soya beans (*Glycine max* L.) and African locust beans (*Parkia biglobosa*), which are indigenous bioresources.

Yeast Extracts

Yeast extracts are important component in some formulation of culture media. Various pharmacopoeia described yeast extract as water-soluble, peptone-like derivative of yeast cells (*Saccharomyces*). Yeast extract supplies not only vitamins in culture media but also proteins, carbohydrates and some micronutrients. Yeasts can be readily isolated from indigenous sources like palm. They can also be obtained as a by-product of the brewing industry (Brewer's yeast). Baker's yeast (*Saccharomyces cerevisiae*) is a food grade strain that can be obtained from food processing industry. The manufacture of Baker's yeast is a well-established procedure. The yeast cells are usually grown in molasses-based medium as a fed- batch type of fermentation that lasts for between 12 and 20 hours. The yeast extract is obtained as autolysate. Cell hydrolysis is accomplished by the endogenous enzymes of the yeast cells. Autolysis is initiated by either a controlled temperature shock or by an osmotic shock which causes the yeast cells to swell and explode. After autolysis, soluble material is separated from insoluble by means of centrifugation and several filtration steps. The final filtration product can be concentrated and spray-dried to obtain the product.

III. Immunodiagnostic Reagents

Immunodiagnostic reagents consist of antigens and antibodies. Antigens are extracted from infectious agents. Antibodies can be produced by immunizing laboratory animals like rabbit, goat and sheep with antigens derived from infectious agents. Antibodies can also be produced using the hybridoma technology for the production of monoclonal antibodies. Infectious agents like bacteria, viruses, fungi and parasites are huge bioresources for production of diagnostics. They possess antigens and genetic resources that are used for production of diagnostics.

IV. Molecular Diagnostics

Molecular diagnostics are based on the genetic properties of infectious agents and the proteins that the genes code for (Begum and Begum, 2011). The infectious agents are bioresources from which production of molecular diagnostics can be realized. To develop molecular diagnostics, the infectious agents must be identified and characterized. Isolation and characterization of infectious agents require basic microbiological work on the agents of diseases. After isolation, molecular characteristics of the agents have to be determined. This involves the extraction of the nucleic acid of the infectious agents. The DNA sequences of the agents need to be determined. Increasing availability of sequence data of many infectious agents is available in the public domain. However, there is need to sequence the genes and genomes of local strains of infectious agents in Nigeria.

Methods for molecular characterization of bacteria, fungi and viruses have been developed (Speers, 2006). The details of methods for molecular characterization and sequencing of genes of microorganisms are beyond the scope of this text. However, there are many excellent published standard manuals and procedures (WHO, 2017). The important thing to note here is that DNA extraction kits can be manufactured locally. Where it is not possible, the sequencing of isolates of infectious agents can be accomplished with collaboration with advanced laboratories. Based on the sequence data, primers and probes can be developed for PCR-based and other types of molecular diagnostic methods. An overview of developmental pathway for diagnostics for infectious diseases is shown in Figure 5.1.

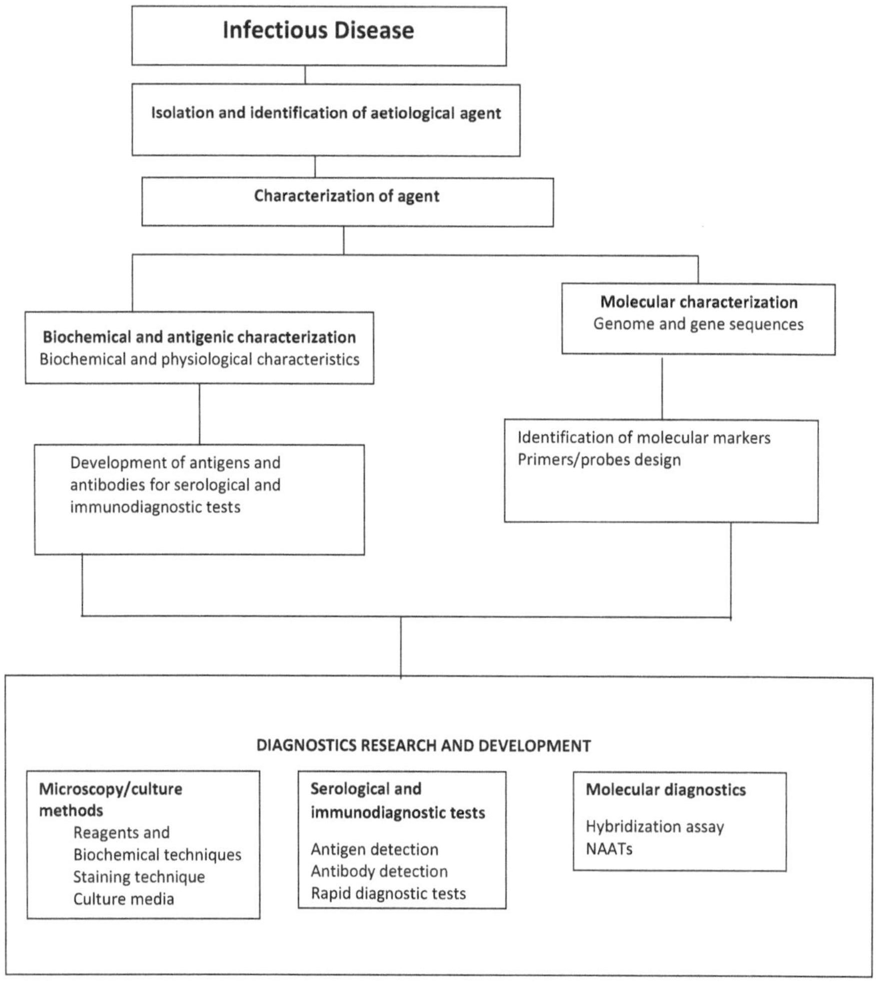

FIGURE 5.1 Overview of developmental pathway for diagnostics for infectious diseases.

5.9 STRATEGY FOR HARNESSING THE INDIGENOUS BIORESOURCES FOR PRODUCTION OF DIAGNOSTICS FOR INFECTIOUS DISEASE

In order to harness the indigenous bioresources of the country for development of diagnostics, certain critical issues need to be addressed. Probably the most important is the issue of priority. The total dependence on importation and near zero level of capacity in local manufacture of diagnostics must become a national concern. Apart from the primary effect on the healthcare sector, it has enormous effect on the scientific and economic development of the nation. The nation losses billions in foreign exchange annually and jobs that a vibrant diagnostic manufacturing industry should create are lost.

There is a need to address educational and capacity development in the area of R&D of diagnostics. The current approach to training of students with little or no practical experience in biotechnology needs to be addressed. For long, the universities have become a place where students are trained into dependence. Textbooks and lecturers and even the appraisal system encourage patronage of foreign scientific products at the expense of local capacity and self-reliance. Science and technology students are not trained to innovate and develop technologies but to use the technologies developed by others. Consciously or unconsciously, we pass across the message to the students that manufacture of scientific items and products are beyond us or are not feasible in our setting. For instance, instead of teaching the final year students the process and methods of producing their culture media, as part of their research training, we often encourage them to obtain expensive brands of culture media in the bid to satisfy our claim to be using internationally recognized brands of diagnostics and scientific products in our research. While this is not completely wrong, it will be better to include the production of similar products from indigenous resources in comparison with the known brands. In this way, the capacity of the graduates to produce important scientific and diagnostic consumables will be achieved. Our students must be encouraged to harness the indigenous resources to manufacture products that are usually imported. This approach will inspire the students and graduates to know that they can also innovate and develop products that contribute to the realization of economic and scientific development of the nation.

The current situation of total dependence on importation for research and diagnostic applications calls for urgent action. As a matter of urgency, the government in association with professional bodies like the Biotechnology Society of Nigeria, the Nigeria Society for Microbiology and other stakeholders must pursue a policy direction that make self-reliance in diagnostics, particularly for infectious diseases a national priority. The state of unpreparedness in which the COVID-19 pandemic met the country that there was a total lack of diagnostics should not be allowed to continue. Harnessing indigenous bioresources for diagnostics through well-organized R&D will have a multiplier effect on the educational system, the research and scientific development as well as the economic benefits in terms of conservation of foreign exchange and job creation.

Special mechanisms of funding diagnostics R&D with particular emphasis on the utilization of indigenous resources must now be vigorously pursued. There is a need to organize and coordinate the various research activities on diagnostics of specific infectious diseases. Relevant research institutes with mandate for technological, healthcare and scientific development need to partner with non-governmental organizations like the professional bodies to identify priority diagnostic needs for development. The National Biotechnology Development Agency (NABDA), the Nigerian Medical Research Institute (NIMR), the National Centre for Disease Control (NCDC), the Nigerian Institute for Pharmaceutical Research (NIPRD) and other relevant agencies of the Federal Government need to work with Biotechnology Society of Nigeria (BSN) to create innovative ways through dedicated funding mechanisms for R&D in diagnostics of infectious diseases. The Tertiary Education Trust Fund (TETFund) has an important role to play in this strategy.

We propose the creation of a Centre of Excellence for Diagnostics for Infectious Diseases. Such a centre will focus on R&D with the aim of developing specific products for commercial production. Such a centre should also work closely with commercial entities or create start-up companies that will take up the commercialization of the products of the centre. This way, the research outcomes will not end up on the shelf like many researches of the past. The initiative of the Central Bank of Nigeria (CBN) to provide special funding for healthcare-related business ventures as part of national responsive to COVID-19 is a step in the right direction. Researchers and scientists must be able to work out modalities to establish start-up biotechnology companies. Innovative business models practiced in developed countries need to be applied to ensure the viability of biotechnology companies (Kurtzman, 2005; Caliendo et al., 2013). The government provide policy direction and create enabling environments for those companies to prosper. Diagnostic as part of the larger biotechnology enterprise must be given utmost priority. Biotechnology has potential to greatly create technological advancement and prosperity for the nation.

5.10 CONCLUSION

Nigeria faces acute shortage of diagnostics for infectious diseases. Bioresources including the infectious agents are in abundance in the country. What is needed is to build capacity in diagnostics R&D to harness the indigenous bioresources for the development of diagnostics for infectious diseases. New policy direction and innovative mechanisms of funding diagnostics R&D must become the priority of the Federal Government and stakeholders.

REFERENCES

Akpor, O. B., Odesola, D.E., Thomas, R.E and Oluba, O. M. (2019) Chicken feather hydrolysate as alternative peptone. *Source for Microbial Cultivation F1000 Research* 7:1918.

Begum, A. and Begum, S. (2011) Role of molecular biology as diagnostic tools of infectious disease-an update. *Bangladesh Journal of Medical Microbiology* 05(02):27–29.

Caliendo, A.M., Gilbert, D N., Ginocchio, C. C., Hanson, K. E., May, L., Quinn, T.C., Tenover, F C., Alland, D., Blaschke, A. J., Bonomo, R. A., Carrol, K. C., Ferraro, M. J., Hirschhorn, L. R., Joseph, W. P., Karcme, T., MacIntyre, A. T., Reller, L. B., Jackson, A. F and Infectious Diseases Society of America (IDSA) (2013) Better tests, better care: Improved diagnostics for infectious diseases. *Clinical Infectious Diseases*: 57(S3): S139–S170.

Cheesbrough, M. (2006) *District Laboratory Practice in Tropical Countries Part 2*, 2nd edition, Cambridge University Press, 16–18.

Conn, H.J. (1930) The history of staining: The development of bacteriological staining methods. *Stain Technology* 5(2): 39–48.

Difco & BBL (2009) *Manual of Microbiological Culture Media*, 2nd edition, BD Diagnostics-Diagnostic Systems, Sparks, MD, https://2016.export.gov/industry/health/eg-main-1302019-asp#p10-70.

Kurtzman, G. (2005) A business Model for a New Generation of Diagnostics Companies. Biotechnology Healthcare. October 2005. Available online (www.bioadvance.com).

Kwosrh, C.K., Asomani-Darko, M. and Adubofour, K. (2012) Cassava starch-agar blend as alternative gelling agent for mycological culture media. *Botswana Journal of Agriculture and Applied Science* 8(1).

Lasekan, A., Bakare, A.F and Hashim, D (2013) Potential of chicken by-products as sources of useful biological resources. *Waste Management* 33(3): 552–565.

Madson, B.M (2001) Application of stains in clinical microbiology. *Biotechnic & Histochemistry* 76(3): 119–125.

Maliro, M. F. A and Lameck, G (2004) Potential of Cassava flour as a gelling agent in media for plant tissue cultures. *African Journal of Biotechnology* 3(4): 244–247.

Merck Microbiology Manual (2010) 12th edition, Darmstadt, Germany: Merck Group, 8–28.

NBS (2020) Nigerian Bureau of Statistics: Nigerian trade Statistics. www.Nigerianstat.gov.ng/tradestat (accessed 28/05/2020).

Nnorom, O.O., Iheone, A.E., Obidiegwu, M.U. and Iheaturu, N.C. (2018) Extraction and Application of natural dyes from *Rothmanniawhitfieldii*(uri) and *Baphianitida* (uhie). *FUTO Journal Series (FUTOJNLS)* 4(1): 188–198.

Nwonye, U. N and Ezema, P. N (2017). Extraction and utilization of natural dye extract from guinea corn leaf. *International Journal of Development strategies in Humanities, Management and Social Sciences* 7(1): 40–51.

Okonkwo, S. N., Ohanuzue, C.B.C., Onuegbu, G.C., Obasi, H.C and Nnorom, O.O (2019) Extraction of Natural Dyes from Whitfieldialateritia plant and its Application on Cotton fabric. *Journal ofTextile Science & Engineering* 9: 392 doi:10.4172/2165-8064.1000392.

Osabohien, E (2009). Extraction and Utilization of natural dyestuffs from dogonyaro plant. *Advanced Natural and Applied Science Research* 7:187–190.

Osabohien, E and Otutu, J.O (2014) Extraction and Utilization of natural dyestuffs from the bark of whistling pine and the root of African peach. *British Journal ofApplied Science and Technology* 4(20): 2921–2930.

Pomerville, J (2017) *Fundamentals of Microbiology*. Jones and Bartlett Learning, 248–249. ISBN 978-1-284-10095-2.

Procop, G.W (2007) Molecular diagnostics for detection and characterization of microbial pathogens. *Clinical Infectious Diseases* 45: S99–S111.

Reischl, U (1996) Application of molecular biology-based methods to the diagnosis of Infectious Diseases. *Frontiers in Biosciences* 1:e72–e77.

Sandle, T (2004) Gram's stain: History and explanation of the fundamental technique of determinative bacteriology. *IST Science and Technology Journal* 54:3–4.

Sandle, T (2011) History and development of microbiological culture media. *The Journal (Institute of Science and Technology)* 2011:10–14.

Sarkinfada, F., Auwal, I. K and Manu, A. (2014) Applications of molecular diagnostic techniques for infectious diseases. *Bayero Journal of Pure and Applied Sciences* 7(1): 37–45.

Satpathi, S., Mohanty, A.K., Satpathi, P., Mishra, S.K. and Behera, P.K. (2014) Comparing Leishman and Giemsa staining for the assessment of peripheral blood smear preparations in a malaria-endemic region in India. *Malaria Journal* 13:512.

Speers, D. J. (2006) Clinical applications of molecular biology for infectious diseases. *Clinical Biochemical Review* 27:39–51.

United States Commercial Service Nigeria USCS (2020) American Embassy Trade and Commerce Section, https://www.buyusa.gov (accessed 22/05/2020).

Uzeh, R.E., Akinola, S.O and Olatope, O. A (2006) Production of peptone from soya beans (*Glycine max* L merr) and African locust beans (*Parkiabiglobosa*). *African Journal of Biotechnology* 5(18): 1684–1686.

WHO (1995) Production of basic diagnostics laboratory reagents. WHO Regional Publications, Eastern Mediterranean Series 11.

WHO/European Observatory on Health Systems and Polices (2016) Ensuring innovation in diagnostics for bacterial infection: Implications for policy. European Observatory on Health Systems and Policies, WHO Regional Office for Europe, UN City Copenhagen, Denmark.

WHO (2017) WHO Manual for the preparation of secondary reference materials for in vitro diagnostic assays designed for infectious diseases nucleic acid or antigen detection: calibration to WHO International Standards. WHO Expert Committee on Biological Standardization, 67th report, WHO Technical Report Series No 1004, Annex 6. Geneva.

WHO (2018) Global tuberculosis report 2018. World Health Organization https: apps.who.int/ins/handle/10665/274453.Licence:CC BY-NC-SA 3.0 IGO (accessed 25/05/2020).

WHO (2019) Guidance on control materials for antigen detecting malaria RDTs: Tools for preparation and validation. Geneva. World Health Organization. Licence CC BY-NC-SA3.0 1GO.

WHO Health Statistics (2020) Monitoring health for the SDGs, sustainable development goals. Geneva: World Health Organization. Licence: CC BY -NC-SA3.0 IGO.

6 Antibiotic Resistance
Global Trends, Impact and Mitigation

Ifeoma M. Ezeonu
University of Nigeria

I. R. Iroha
Ebonyi State University

Nwadiuto (Diuto) Esiobu
Florida Atlantic University

CONTENTS

6.1 WHAT ARE ANTIBIOTICS/ANTIMICROBIALS?

Originally, an antibiotic was defined as a low molecular weight substance produced by a microorganism that kills other microorganisms at low concentration (Waksman, 1947), while an antimicrobial was defined as any natural, semi-synthetic or synthetic substance that kills or inhibits the growth of microorganisms but causes little or no

DOI: 10.1201/9781003178903-8

damage to the host (AMRLS, 2020). According to those definitions, synthetic drugs such as quinolones and sulphonamides, which are currently used for treatment of bacterial infections, would not be called antibiotics and neither would methicillin and amoxicillin, which are semi-synthetic. Currently, however, antibiotics have been redefined to include semi-synthetic and synthetic antimicrobial variants, particularly those used against bacteria. Antibiotics are mainly active against bacteria but may have activity against some parasites. They are, however, not active against viruses. Antimicrobials, on the other hand, are agents that are used against microorganisms in general, including non-bacterial pathogens, such as fungi, protozoa and viruses. Therefore, all antibiotics are antimicrobials, but not all antimicrobials are antibiotics.

A historical review of the advent of antibiotics and antimicrobials reveals that from the launching of the germ theory of disease, which proposed that diseases are caused by microorganisms, many studies were undertaken by scientists to find effective means of killing these microbes. Many substances were developed and evaluated, but ultimately rejected for clinical use, due to either instability of the compounds or toxicity, until 1910, when the German physician, Paul Ehrlich, developed the antimicrobial, Salvarsan. Salvarsan was used for treatment of syphilis until the revolutionary discovery of the first antibiotic, penicillin, by Sir Alexander Fleming, in 1928. Penicillin replaced Salvarsan because it was safer, more effective and more readily available. By 1940, penicillin was being produced commercially. Soon after the discovery of penicillin, other antibiotics followed.

Antibiotics function by interfering with one of the processes vital to the survival of invading bacteria, such as synthesis of the cell wall, RNA synthesis, DNA synthesis or protein synthesis. Different types of antibiotics have different sites of action on the bacterial cell structure and metabolism, as shown in Table 6.1.

6.2 DEVELOPMENT AND SPREAD OF ANTIBIOTIC RESISTANCE

Antibiotic resistance is the ability of bacteria to overpower the antagonistic effects of antibiotics to which they were earlier sensitive, resulting in their survival despite exposure to standard doses of the antibiotic (Sharma et al., 2018). While it is now common knowledge that the first natural antibiotic, Penicillin, was discovered in 1928, by Sir Alexander Flemming, what is perhaps not so common knowledge is that antibiotic resistance is also not a new phenomenon. The first penicillin-resistant bacteria were identified in 1967, not long after penicillin was introduced into the market (AMRLS, 2020). In fact, it is now established that antibiotic resistance predates the discovery of penicillin.

As earlier mentioned, the antimicrobial compound, Salvarsan, was developed in 1910, but by the 1920s, resistance to Salvarsan began to emerge and by 1924, 4 years before the discovery of penicillin, a paper was published to that effect by S. Silberstein in 1924 (Stekel, 2018). More evidence of the ancient origin of antibiotic resistance was provided by a defining study conducted on the Canadian permafrost, isolated caves and human specimens preserved for hundreds of years. In the revolutionary study, modern molecular biology techniques of nucleic acid sequencing, metagenomic analysis and bioinformatics were used to analyse samples collected from the frozen sediments. The analyses revealed the presence of genes encoding resistance to β-lactam,

TABLE 6.1
Different Types of Antibiotics and Their Modes of Action

Target	Mode of Action	Examples of Antibiotic/ Antimicrobial
Cell wall synthesis	These antibiotics selectively interfere with synthesis of peptidoglycan, an important component of the bacterial cell wall. Such antibiotics are very effective, since the host cells do not possess cell walls.	Penicillins, Cephalosporins, Bacitracin, Vancomycin
Cell membrane function	The cell membrane functions to regulate the movement of substances in and out of the cell. Therefore, any disruption or damage to its structure may cause the leakage of important solutes required for the cell's survival. However, because the membrane structure is similar in both eukaryotic and prokaryotic cells, these types of antibiotics may also be toxic for the host.	Polymixin B, Colisitn
Protein synthesis	Protein synthesis is one of the key processes in all bacterial cells. Some antibacterial agents target bacterial protein synthesis by binding to either the 30S or 50S subunits of the ribosomes, thereby disrupting the normal flow of protein synthesis. This ultimately leads to the death of the organism or the inhibition of its growth and multiplication.	Aminoglycosides, Macrolides, Lincosamides, Streptogramins, Tetracyclines, Chloramphenicol
Nucleic acid synthesis	Some antibiotics inhibit DNA and RNA synthesis by binding to some of the components involved in the process of DNA or RNA synthesis. Inhibition of DNA or RNA synthesis interferes with survival of the cell.	Quinolones, Metronidazole, Rifampin
Other metabolic processes	Other metabolic processes, such as biosynthesis of important precursors, may also be targeted by antibiotics. For example, both sulphonamides and trimethoprim disrupt the folic acid pathway, which is a necessary step for bacteria to produce precursors important for DNA synthesis.	Sulphonamides, Trimethoprim

tetracycline and glycopeptide antibiotics in metagenome samples of 30,000-year-old permafrost (D'Costa et al., 2011). Since then, other studies, using metagenomic analyses, have also shown the presence of resistance genes in preserved human skeletons of at least 5,000 years old (Petrova et al., 2009; Perron et al., 2015; Santiago-Rodriguez et al., 2015). Studies are still ongoing as some scientists attempt to define the age of antibiotic resistance, using modern molecular biology techniques on older fossils and finding more and more evidence of functional resistance genes and their transmissibility in these ancient specimens (Perry et al., 2020). It would not be surprising to find eventually that antibiotic resistance is as old as bacteria themselves.

To understand how resistance develops, it is important to understand one or two facts about the genetics and biochemistry of bacterial cells:

- Bacteria possess a great diversity of capabilities, encoded in their chromosomes, and easily call up these capabilities to adapt to changing environments. For instance, a particular bacterial species may possess two or more pathways for achieving a particular metabolic need, and therefore when one pathway is targeted by a potential inhibitory agent, the organism may simply switch to another pathway, thus surviving the attack. This essentially constitutes resistance to that inhibitory substance.

- Mutations occur naturally in living cells, including bacterial cells. A mutation could occur in a bacterial cell at an antibiotic target site, rendering the antibiotic ineffective against the cell (resistance). Many known types of resistance have been reported to be results of spontaneous or even point mutations, including Rifampicin-resistance in *Mycobacterium tuberculosis* (Davies and Davies 2010; Pang et al., 2013), some extended spectrum β-lactamase (ESBL)-mediated resistance in enterobacteria (Rubstova et al., 2010) and resistance to quinolones among others (Davies and Davies 2010).
- In addition to their chromosomal genes, some capabilities are encoded by genes which may be located on extrachromosomal genetic elements such as plasmids, integrons and transposons. These extrachromosomal elements, also sometimes referred to as mobile genetic elements (MGEs), are transferrable between strains, thus increasing the options available for a bacterial cell to survive an adverse environment.

  Many types of resistance are due to the acquisition of genes of unknown origin on plasmids; for example, most resistance to β-lactam antibiotics, tetracyclines, sulphonamides, aminoglycosides and chloramphenicol (Davies and Davies, 2010).

The mutation of existing genes or acquisition of new genes, on MGEs, from the environment and other species or strains could lead to the development of antibiotic resistance through any of four mechanisms: antibiotic inactivation or its modification; alteration in the antibiotic target site; modification in the metabolic pathways to overcome antibiotic effect; and by minimizing entry and/or promoting active efflux of the antibiotics. Resistance developed through the mutation of an existing gene is referred to as vertical resistance, while that resulting from acquisition of external genes is known as horizontally acquired resistance. Once one or two cells in a population become resistant, the rest happens by natural selection. For instance, if one or two cells at an infection site are resistant, application of an antibiotic eliminates the susceptible cells, leaving the resistant cells to proliferate. Over time, the resistant strains could then predominate. Antibiotic resistance is therefore a natural phenomenon, but is further facilitated by anthropogenic activities such as use of antibiotics in animal production, improper disposal of antibiotic residues and waste water, poor hygiene and misuse of antibiotics in human infections, which enhance the horizontal transfer and spread of resistance genes (Wellington et al., 2013; Pehrsson et al., 2016). According to Woodford et al. (2011), bacteria are well-known potent originators for the dissemination of antimicrobial-resistant genetic apparatus and are competent to offer secured platforms for the upholding and transmission of genes accountable for antimicrobial resistance through MGEs.

The widespread usage of antibiotics in food-animal production as growth-promotion and infection-control agents, for example, allows resistant strains of microbes to proliferate within the intestinal systems of these animals. Subsequent contamination of the environment with animal wastes containing resistant bacteria, in turn, becomes a major route through which human populations become infected by such resistant or multidrug-resistant bacteria. Once humans acquire antibiotic-resistant bacteria, either through ingestion of food animals or products contaminated with resistant bacteria or misuse of antibiotics by the humans, spread of the resistant strains continues through poor hygiene and unsanitary environmental conditions. Environmental microorganisms including Gram-negative bacteria such as *Escherichia coli*, *Klebsiella* species and *Pseudomonas aeruginosa* represent the most relevant reservoir of resistance to antibiotics and other antimicrobial agents in the community; and this is due in part to their ability to acquire and incorporate into their genome antibiotic resistance genes from their environment. The resistant strains are also spread between the community and hospital environments as humans move to and from these environments.

6.3 GLOBAL TRENDS AND IMPACT OF ANTIBIOTIC RESISTANCE, WITH EMPHASIS ON DEVELOPING COUNTRIES, PARTICULARLY NIGERIA

Antimicrobial resistance is a serious global problem that has been recognized as a threat to the world's sustainability and development efforts, having direct impact on a number of the United Nations' Sustainable Development Goals (SGDs), as documented by Jasovsky et al. (2016). The World Health Organization (WHO) estimates that globally about 700,000 persons die, every year, from infections due to antibiotic-resistant bacteria. However, this number could be an underestimation, given the poor surveillance and reporting in many parts of the world, particularly the developing nations of the world; multidrug-resistant and extremely drug-resistant tuberculosis (TB) alone are reported to be responsible for nearly 200,000 deaths, each year (WHO, 2018).

Antimicrobial resistance is a global problem that is currently threatening the progresses made since the monumental discovery of antibiotics in the early 1920s, especially in developing countries like Nigeria. The antimicrobial resistance problem in the developing world is something like a vicious cycle. Firstly, the developing countries of the world have the greatest burden of infectious disease and therefore have more need of antibiotics. Then, fortunately or unfortunately, expansion of resources and technological advances has made it possible for these developing countries to obtain larger quantities of antimicrobials for management of their infectious disease burden. However, these developing countries lack adequate control of the use of antimicrobials, with the attendant problems of misuse and abuse of antimicrobials, which in turn promote antimicrobial resistance (AMR), thereby making management of the high infectious disease burden difficult; a situation which completes the cycle and returns health care in these countries back to square one.

Reports from different parts of Africa have shown that 70%–90% of isolates of pathogens such as *Staphylococcus aureus*, *Pseudomonas aeruginosa*, *Streptococcus pneumoniae*, *Escherichia coli*, *Shigella* and *Salmonella* are exhibiting multiple drug resistance to antibiotics such as ampicillin, amoxicillin, streptomycin, tetracycline, chloramphenicol and other broad-spectrum antibiotics (Kimang'a, 2012). The combination of high rates of resistance and lack of affordable and effective treatment implies lower success rates for treatment of infectious diseases. In developing countries, AMR has direct impact on health in a number of ways:

1. *Maternal and child health*: Antibiotics play a central role in the provision of maternal and child care, especially in low- and middle-income countries (LMICs) like Nigeria, which have the highest maternal mortality rate in the world. In LMICs, diarrhoea accounts for a significant number of deaths of children under 5 years old. Therefore, increasing antibiotic resistance would mean ineffective treatment of these diarrhoeal diseases and increase in child mortality. In Nigeria, diarrhoea accounts for more than 16% of all child deaths, and studies have shown increasing resistance among both Gram-positive and Gram-negative bacterial isolates from these infections (FMAEH, 2017). AMR is therefore a serious threat for LMICs.
2. *Communicable diseases*: Diseases such as TB, pneumonia, gonorrhoea, neonatal sepsis, blood infections, wound infections, etc. which hitherto have been managed with antibiotics are becoming increasingly difficult to manage due to antibiotic resistance. This leads to higher morbidity and mortality rates.
3. *Non-communicable diseases*: AMR also indirectly impacts non-communicable diseases (NCDs). For instance, patients undergoing surgery, transplants or cancer treatments often need to be administered antibiotics for infection control. AMR therefore poses a serious threat to such individuals.

In Nigeria, the most common bacterial infections are community-acquired diarrhoeal diseases, respiratory, urinary tract and invasive infections. Studies across Nigeria have shown that there is a high rate of resistance in organisms causing these infections (FMAEH, 2017). For diarrhoea-causing organisms, there is reported widespread antimicrobial resistance among enteric *Escherichia coli* in Nigeria particularly to penicillins, aminoglycosides, cephalosporins, chloramphenicol, tetracyclines and cotrimoxazole. Similar resistance patterns have also been documented for *Shigella* and non-typhoidal *Salmonella*. High rates of resistance have also been documented for organisms associated with urinary tract infections (UTI) to all the drugs commonly prescribed for UTI. It is reported that most of these organisms demonstrated 100% resistance to ampicillin and cotrimoxazole which have long been used as first line drugs in the treatment of UTI (FMAEH, 2017).

Most problems of antibiotic resistance occur in the hospital environment where antibiotics are usually used heavily for the treatment and management of some infectious diseases, but this trend is not always the case – as some other cases of antibiotic resistance have been reported in the non-hospital environment, especially in abattoirs, poultry farms, livestock farms or in agricultural practices, where antibiotics especially those meant for human medicine are used irrationally for promoting animal growth. In most of the cases where antibiotics are used for non-clinical purposes, such as in the use of antibiotics to promote animal growth, antibiotics are usually used in sub-inhibitory concentrations that only kills highly susceptible bacteria and allows other strains to go on thriving and thus become resistant to the antibiotic under usage.

The problem of antibiotic resistance is worsened in developing countries mainly due to the lack of control of use of antibiotics, as antibiotics can still be obtained or procured over-the-counter (OTC) even without a doctor's prescription. To further compound the matter, many hospitals in developing nations, inclusive of Nigeria, still lack state-of-the-art facilities and equipment required for the prompt detection, reporting and surveillance of antibiotic resistance cases both from the hospital and non-hospital environment. This scenario has contributed to the sustained emergence and spread of antibiotic resistance in such economies.

Newer resistance mechanisms are emerging globally, threatening the effective treatment of common infectious diseases. Gram-negative bacteria from environmental isolates including *E. coli*, *Klebsiella* species and *P. aeruginosa* harbour some novel antibiotic and multidrug resistance genes such as ESBLs, MBLs (metallo-beta lactamases), AmpC and plasmid-mediated colistin resistance genes that give them the exceptional ability to resist the antimicrobial onslaught of potent antibiotics directed towards them. Antibiotic resistance is fast becoming such a significant public health problem that there are bacteria for which most antibiotics no longer work. The effective monitoring of the development and spread of antimicrobial resistance in zoonotic pathogens especially those that produce high-profile antibiotic degrading enzymes like MBLs and AmpC beta-lactamases are critical to the containment of any disease outbreak due to these microbes. Antibiotic resistance (especially those mediated by ESBLs, MBLs and AmpC β-lactamases) is a global health problem that is posing a menace to our ability to effectively treat and manage infections caused by organisms producing these enzymes. Studies conducted in different parts of Nigeria indicate that the country is already recording a high prevalence of multidrug-resistant and extensively drug-resistant bacterial strains (Iroha et al., 2009, 2019; Agbo et al., 2019, 2020).

While the problem of antimicrobial resistance may be receiving serious attention in the developed economies as a result of increased surveillance, detection and reporting measures on antibiotic resistance put in place in such countries, the menace of antimicrobial resistance still rages on in the developing economies – where no official or coordinated surveillance, detection or reporting measures are put in place or backed up by legislation to assuage their negative effects. The incidence and problem of antibiotic resistance is still on the increase in the developing parts of the world due to multiple factors:

- **Overuse and misuse of antibiotics in humans and animals**

  The contribution by users of antibiotics is especially worthy of mention for Africa and Nigeria in particular, as this is a major factor contributing to antibiotic resistance in this region. It has been reported that in Africa, out of about 90.1% of individuals that seek medical care outside the home, approximately 36.2% receive antibiotics, and of all those who receive antibiotics, 31.7% do so without prescriptions, while about 26.4% obtain antibiotics from informal vendors (Vialle-Valentin et al., 2012). This prevalent practice of self-prescription is one that must be given a serious consideration in the fight against AMR. Furthermore, the use of antibiotics in animal production exerts a selection pressure in favour of resistant strains (or genes), increasing the reservoir of resistant strains.

- **Inadequate quality assurance**

  There are often reports of sales of sub-standard and poor-quality drugs in the markets of many developing countries. It is estimated that as much as 50% of drugs sold in some African countries are counterfeit or sub-standard. When consumed, these sub-standard antibiotics produce sub-inhibitory concentrations in vivo, thereby enhancing the selection of resistant strains.
- **Poor development and enforcement of national policies guiding antimicrobial use and disposal**

  Unlike the developed countries of the world, many developing countries are yet to formulate and implement AMR action plans. In addition, there is poor enforcement of existing policies relating to production and distribution of sub-standard drugs, which increase the selection and spread of resistant bacteria. Furthermore, improper disposal and treatment of hospital and industrial wastes often make it possible for antibiotics and their residues to find their way into the environment, where they select for resistant bacteria.
- **Absence of national policies governing surveillance, detection and reporting of antibiotic resistance**

  Many developing countries lack national strategic plans governing the detection, surveillance and reporting of antibiotic resistance. There are no policies regulating laboratory practices as regards the handling and disposal of resistant isolates. Often, research laboratories dispose resistant isolates carelessly, thus discharging such strains or their resistance genes back into the environment.

In the developed nations such as in Europe, America and parts of Asia, cases of antibiotic resistance have been sustainably managed and even contained effectively. European countries, especially those of western Europe, have been able to develop comprehensive healthcare systems, with integrated and strict monitoring of antibiotic use, prescription and regulation, all coordinated nationally. Using this strategy, it has been possible for these countries to slow the rate of antimicrobial resistance. Similar measures could be applied in other parts of the world to reduce antimicrobial resistance, globally. The African nations especially, Nigeria, can draw from these strategies to develop their own approaches that will help to strengthen the continent's health institutions and systems to effectively control the spread of antibiotic resistance in the region.

6.4 DETECTION METHODS FOR ANTIBIOTIC RESISTANCE AND THE NEED FOR NATIONAL SURVEILLANCE

The detection methods for antimicrobial resistance are usually two-fold:

- **Genotypic detection methods**
- **Phenotypic detection methods**

In the genotypic detection method, the genetic elements or genes responsible for the resistance nature in the target pathogen are investigated using molecular biology techniques including polymerase chain reaction (PCR), multilocus sequencing technique (MLST), and whole genome sequencing (WGS), among others. The genotypic detection methods are important because they help to determine the genetic elements or genes that are responsible for resistance, and this is important because it is the genes responsible for resistance that are usually transmissible from resistant pathogens to other susceptible organisms via genetic transfer mechanisms such as transformation, conjugation and transduction. The genotypic detection methods are usually proceeded or carried out after doing a preliminary antimicrobial susceptibility testing – which is usually captured in the phenotypic detection methods.

In the phenotypic techniques, conventional antibiotics and other standard microbiology techniques including determination of MIC (minimum inhibitory concentration), MBC (minimum bactericidal concentration) and inhibition zone diameter (IZD) determination are employed. These methods are usually less cumbersome and less expensive than the genotypic detection methods. Unlike the genotypic detection methods where specialized training and equipment are needed, the phenotypic methods can be carried out without much training or equipment.

The type of antibiotic detection technique that an institute or nation or hospital laboratory decides to use is usually affected by so many factors such as availability of resources, manpower, equipment and technical know-how. However, it is always best to employ both the genotypic and phenotypic detection methods in order to properly contain and assuage the spread of antibiotic-resistant pathogens in any environment.

The routine antimicrobial susceptibility testing (AST) carried out in Nigeria's hospitals is not capable of detecting some unique and silent emerging and spreading multidrug-resistant phenotypes. Some of these unique resistance phenotypes include extended spectrum beta lactamase, metallo-beta lactamase, AmpC enzymes, methicillin-resistance, vancomycin resistance and mupirocin resistance, among others. Without a modification of the routine AST techniques in our hospital laboratories, some of these important, unique and multidrug resistance phenotypes can be easily missed during testing in the laboratory. What is more worrisome is that most hospital laboratories in Nigeria, both private and government-owned hospitals, do not follow and use the stipulated guidelines on AST and other studies related to detecting, surveillance and reporting of multidrug-resistant bacteria from clinical samples. The **Clinical Laboratory Standard Institute (CLSI)** is the body saddled with the responsibility of producing guidelines and antibiotic breakpoints that will help guide AST experiments in the hospital laboratories across the globe. But it is quite unfortunate that most hospital laboratories in Nigeria do not follow the CLSI guideline in their detection, surveillance and reporting of antibiotic resistance cases in their different establishments.

Currently, Nigeria does not have a National Action Plan for the surveillance, detection and reporting of antimicrobial resistance. This has compounded the problem of antimicrobial resistance in Nigeria because many novel resistance phenotypes have emerged and spread across the hospitals and outside the hospital community without any form of containment, and these phenomena have negatively impacted patient's prognosis (even without the health authorities knowing) because majority of antimicrobial therapy has been blindly administered and/or wrongly administered to patients.

The time for Nigeria to set up a national surveillance and monitoring units across the geopolitical zones of the country is now. Such units currently do not exist anywhere in the nation. The national surveillance unit should be given constitutional backing and sustainable financial empowerment to carry out their activities – as is applicable in the developed economies where the problem of antimicrobial resistance has been sufficiently contained. The tasks of such a unit shall include among other things:

- Surveillance of the emergence of antibiotic-resistant phenotypes, especially novel resistance phenotypes in both the hospital and non-hospital environment.
- Reporting and monitoring of antibiotic resistance cases in hospitals and non-hospital environment.
- Monitoring and stopping the use of antibiotics in non-clinical purposes such as in the propagation of farm animals.
- Development and monitoring of AST procedures in hospital laboratories and other private laboratories.
- Creation of awareness on the danger of self-medication and the use of antibiotics without doctor's prescription.
- Monitoring and controlling the use of biocides in the oil and other-related industries.

6.5 STRATEGIES FOR MITIGATION, INCLUDING NOVEL AND PROPOSED NEW STRATEGIES

Antimicrobial/antibiotic resistance is a natural phenomenon, which is aggravated by human activities, and being a natural phenomenon, which has been shown to even predate the discovery of antibiotics, it is safe to assume that no matter the antimicrobial/antibiotic developed, it is only a question of time before resistant strains of microorganisms emerge. The fight against AMR should therefore focus not on stopping the occurrence of resistance but on stopping or reducing the selection and spread of resistant strains. It has become an unquestionable fact that the global problem of AMR has resulted over time from the errors of 'commission' (doing things that should not be done) and errors of 'omission' (leaving what should be done undone) of humans. Therefore, solving the problem of AMR should also involve a concerted effort, globally, to retrace our steps and reverse our actions, through improved antibiotic stewardship.

The adoption of good antimicrobial stewardship should be institutionalized by all healthcare systems and governments of various nations if the war against AMR is ever to be won. Since the rate of development of novel and potent antimicrobials can never match the development of resistance, it is important that the global healthcare systems and governments come up with strategies that will help them to tackle antibiotic resistance at their respective nations, continents and borders. Nigeria and other African countries must key into this and begin now to draw from some international best practices in order to play her role in the effective containment of antibiotic resistance in the country and its continent. This, of necessity, means that all groups that have hitherto contributed to the AMR crises must reverse their acts of commission or omission, as shown in the examples below.

6.5.1 Users of Antibiotics

Users of antibiotics mainly contribute to antibiotic resistance through the practice of self-mediation and the attendant consequences. Therefore, antibiotic users can, as individuals, help to prevent and control the spread of resistance by using antibiotics only when prescribed by a physician and at doses and manners prescribed by the physician.

Other groups can also participate in mitigating antibiotic resistance in various ways.

6.5.1.1 Government and Regulatory Agencies

Government and regulatory agencies can:

- Develop and implement a national plan to monitor the use or misuse of antibiotics.
- Put in place strict and reliable surveillance measures to monitor the occurrence of drug-resistant bacteria in the community especially from abattoirs and poultry sources.
- Strengthen the health systems in their nations in order to provide state-of-the-art equipment and research facilities that will help to effectively detect, monitor and report antibiotic resistance cases.
- Strengthen policies, programmes and implementation of infection prevention and control measures. The establishment of an antibiotic resistance research institute or committee is also paramount because such an institute or committee will help to coordinate the government's efforts in the fight against antibiotic resistance – so that all data regarding antibiotic resistance both from the community and hospital environments can be properly coordinated and harnessed to develop policies and programmes that will further help to effectively control and prevent antibiotic resistance.
- Regulate and promote the appropriate use and disposal of quality medicines. Government should ensure that the purchase of antibiotics OTC is prohibited and prevented from happening because such a practice allows people to use antibiotics irrationally and even at sub-inhibitory concentrations that allows antibiotic-resistant bacteria to emerge and spread.
- Make information available on the impact of antibiotic resistance. Public awareness and education on the judicious use of antibiotics should be carried out regularly in both the community and hospital environment to assuage the spread of resistant microbes.

6.5.1.2 Health Professionals

Health professionals can help reduce antibiotic resistance by:

- More prudent prescription of antibiotics.
- Prescription based on proper diagnostics and sensitivity testing.
- Educating patients on the need to adherence to prescription.
- Following up on patients and reporting resistant infections.

6.5.1.3 Healthcare Industry

To prevent and control the spread of antibiotic resistance, the health industry can:

- Invest in research and development of new antibiotics, vaccines, diagnostics and other tools.

- Set up antimicrobial resistance surveillance and monitoring units across the nation's healthcare systems in order to keep the emergence and spread of drug-resistant pathogens at bay.

6.5.1.4 Agriculture Sector

To prevent and control the spread of antibiotic resistance, the agriculture sector can:

- Only give antibiotics to animals under veterinary supervision.
- Not use antibiotics for growth promotion or to prevent diseases in healthy animals, since these activities allow drug-resistant bacteria to emerge and spread in the community.
- Vaccinate animals to reduce the need for antibiotics and use alternatives to antibiotics when available.
- Promote and apply good practices at all steps of production and processing of foods from animal and plant sources.
- Improve biosecurity on farms and prevent infections through improved hygiene and animal welfare.

In addition to the measures listed above, which are mostly directed at correcting the mis-actions of human groups, new and emerging strategies are being considered for mitigating the AMR crises.

6.5.1.5 Non-Antimicrobial Alternatives

The Federal Ministry of Agriculture, Environment and Health advocates that concerted efforts should be made for the provision and use of alternatives to antimicrobials, in Nigeria, that will be effective against pathogenic microorganisms, without promoting resistance. Some of the alternatives suggested include vaccines, probiotics, immune-boosting trace elements/vitamins and bioactive phytochemicals (FMAEH, 2017).

6.6 THE MICROBIOME APPROACHES

The important role of the human microbiome in health and immunity has become a major area of exploration for the mitigation of a variety of medical problems, including antibiotic resistance. Two basic strategies are considered:

- Specific targeting of resistant pathogens – This involves the development of vaccines or narrow spectrum antimicrobial agents against pathogenic and/or drug-resistant strains and species.
- Direct manipulation of the microbiota – This includes delivery of individual, small collectives or whole-scale communities of strains (probioitcs) or of nutrients (prebiotics) or other growth factors to strengthen or restore beneficial functions of the microbiota or exclude invasive and/or antibiotic-resistant strains and species. This approach has proved promising in recent studies (Ayukekbong et al., 2017; Ezeonu et al., 2020).

Both strategies could lead to: a reduction in the use of broad-spectrum antibiotics and hence a reduction in the selection of antibiotic-resistant organisms; the preservation of beneficial microbiota; and elimination or exclusion of antibiotic-resistant organisms.

6.7 ALTERNATIVE ROUTES OF DELIVERY OF ANTIMICROBIALS

It has been suggested that oral ingestion of antibiotics may be a contributing factor to the rise in antibiotic resistance, by increasing the reservoir of resistant gut bacteria. Therefore, using alternative delivery routes, such as skin patches and transdermal administration, would slow the trend.

6.8 CONCLUSIONS AND RECOMMENDATIONS

Antibiotic resistance is the unresponsiveness of a pathogenic microorganism to the onslaught of antimicrobials or antibiotics that are supposed to cause its death and it is one of the biggest threats to global health, food security, and development today. Any use of antibiotics is capable of causing resistance. Therefore, overuse or misuse of antibiotics, such as in the Nigerian situation, where there is a prevalence of self-medication and antibiotics can easily be obtained OTC, exacerbates the problem of antibiotic resistance. This is also further compounded by the uncontrolled use of antibiotics and other antimicrobial agents for non-clinical purposes such as in animal husbandry, poultry production and in other agricultural and veterinary practices. Antibiotic-resistant microbes can spread to humans from human-to-human contact or animal-to-human contacts as a result of these activities, and it can also spread from raw food and from the environment into the human population through the consumption of food animals or animal products containing antibiotic residues. Since human practices are largely responsible for the spread of antibiotic resistance, humans must also put up a concerted effort to reverse the trend by correcting those practices that lead to the emergence and spread of resistance. Nigeria, as a nation, must also put in place measures that will help to monitor, detect and report antimicrobial resistance from both the hospital and non-hospital environment, through the institution of a National Surveillance Network on antimicrobial resistance. Delaying such an important network means leaving the lives and wellbeing of Nigerians to the mercy of the resistant microbes.

REFERENCES

Agbo, M.C., Ezeonu, I. M., Ike, A. C. and Ugwu, C. C. (2020). Multidrug-resistance patterns and detection of *pstS* gene in clinical isolates of *Pseudomonas aeruginosa* from Nsukka, southeast Nigeria. *Asian J Pharm Clin Res* 13: 115–119.

Agbo, M.C., Ezeonu, I. M., Odo, M. I., Ononugbo, C. M., Onoduagu, B. O., Eze, C. C., Ezeibe, E.N. and Ozioko, C. A. (2019). Phenotypic and molecular characterization of extended spectrum β-lactamase producing Pseudomonas aeruginosa in Nigeria. *Afr J Biotech* 18: 1083–1090.

Antimicrobial Resistance Learning Site. (2020). https://amrls.umn.edu/pharmacology. Retrieved 31/3/2020.

Ayukekbong, J. A., Ntemgwa, M. and Atabe, A. N. (2017). The threat of antimicrobial resistance in developing countries: causes and control strategies. *Antimicrob Resist Inf Cont* 6:47 doi 10.1186/s13756-017-0208-x.

D'Costa, V. M., King, C. E., Kalan, L., Morar, M., Sung, W. W., Schwarz, C., Froese, D., Zazula, G., Calmels, F., Debruyne, R., Golding, G. B., Poinar, H. N. and Wright, G. D. (2011). Antibiotic resistance is ancient. *Nature* 477: 457–461.

Davies, J. and Davies, D. (2010). Origins and evolution of antibiotic resistance. *Microbiol Mol Biol Rev* 74: 417–433.

Ezeonu, I. M., Louis, G. O., Eze, C. A., Chah, K. F. and Ugwu, N. E. (2020). Direct instillation of *Lactobacillus plantarum* into rat bladder inhibits urinary tract infection (UTI) and protects the integrity of urinary organs. *Microbiol Res Int* 8: 1–9.

Federal Ministries of Agriculture, Environment and Health. (2017). Antimicrobial Use and Resistance in Nigeria: Situation Analysis and Recommendations.

Iroha, I.R., Oji, A.E., Afiukwa, T.N., Nwuzo, A.C and Ejikeugwu, P.C (2009). Extended Spectrum Beta Lactamase (ESBL) mediated resistance to antibiotics among *Klebsiella pneumonia* in enugu metropolis. *Macedonian J Med Sci* 2(3):196–199.

Iroha, I.R., Onyia, U., Moses, I.B., Ejikeugwu, C.P., Nwakaeze, A.E., Ugbo, E.N. (2019). Prevalence and antibiotic resistance profiles of extended spectrum β-lactamase-producing *Escherichia coli* among paediatric patients with urinary tract infection in St. Patricks' Hospital, Mile Four, Abakaliki, Ebonyi State, Nigeria. *Afr. J. Clin. Exper. Microbiol* 20(4): 332–336.

Jacoby, G. A. and Low, K. B. (1980). Genetics of antimicrobial resistance. In: Stallones, R. A. et al. (Eds). *The Effect on Human Health of Subtherapeutic Use of Antimicrobials in Animal Feeds.* National Academies Press, Washington (DC), pp. 92–129.

Jasovsky, D., Littmann, J., Zorzet, A. and Cars, O. (2016). Antimicrobial resistance – a threat to the World's sustainable development. Development dialogue paper no.16. www.daghammarkjold.se.

Kimang'a, A. N. (2012). A situational analysis of antimicrobial drug resistance in Africa: Are we losing the battle? *Ethiop J Health Sci* 22:135–143.

Pang, Y., Lu, J., Wang, Y., Song, Y., Wang, S. and Zhao, Y. (2013). Study of the rifampin monoresistance mechanism in *Mycobacterium tuberculosis. Antimicrob Agents Chemother* 57: 893–900.

Pehrsson, E. C., Tsukayama, P., Patel, S., Mejia-Bautista, M., Sosa-Soto, G., Navarrete, K. M., Calderon, M., Cabrera, L., Hoyos-Arango, W., Bertoli, M. T., Berg, D. E., Gilman, R. H. and Dantas, G. (2016). Interconnected microbiomes and resistomes in low-income human habitats. *Nature* 533:212–216.

Perron, G. G., Whyte, L., Turnbaugh, P. J., Goordial, J., Hange, W. P., Dantas, G. and Desai, M. M. (2015). Functional characterization of bacteria isolated from ancient arctic soil exposes diverse resistance mechanisms to modern antibiotics. *PLoS One* 10: e0069533.

Perry, J., Waglechner, N. and Wright, G. (2020). The prehistory of antibiotic resistance. Cold Spring Harbor Perspectives in Medicine, http://perspectivesinmedicine.cshlp.org Retrieved on 29/2/2020.

Petrova, M., Gorlenko, Z. and Mindlin, S. (2009). Molecular structure and translocation of a multiple antibiotic resistance region of a *Psychrobacterpsychrophilus* permafrost strain. *FEMS Microbiol Lett* 296:190–197.

Rubstova, M. Y., Ulyashova, M. M., Bachmann, T. T., Schmid, R. D. and Egorv, A. M. (2010). Multiparametric determination of genes and their point mutations for identification of Beta-lactamases. *Biochemistry* 75: 1628–1649.

Santiago-Rogriguez, T. M., Fornaciari, G., Luciani, S., Dowd, S. E., Toranzos, G. A., Marota, I. and Cano R. J. (2015). Gut microbiome of an 11th century AD pre-Columbian Andean mummy. *PLoS One* 10: e0138135.

Sharma, C., Rokana, N., Chandra, M., Singh, B. P., Gulhane, R. D., Gill, J. P. S., Ray, P., Puniya, A. K. and Panwar, H. (2018). Antimicrobial resistance: Its surveillance, impact, and alternative management strategies in dairy animals. *Frontiers Vet Sci* 4:237.

Stekel, D. (2018). Antibiotic resistance pre-dates penicillin. *Nature* 562: 192.

Vialle-Valentin, C. E., Lecates, R. F., Zhang, F., Desta, A. T. and Ross-Degnan, D. (2012). Predictors of antibiotic use in African communities: Evidence from medicines household surveys in five countries. *Trop Med Int Health* 17:211–222.

Waksman, S. A. (1947). What is an antibiotic or an antibiotic substance? *Mycologia* 39: 565–569. PMID: 20264541.

Wellington, E. M. H., Boxall, A. B., Cross, P., Feil, E. J., Gaze, W. H., Hawkey, P. M., Johnson-Rollings, A. S., Jones, D. L., Lee, N. M., Otten, W., Thomas, C. M. and Williams, A.P. (2013). The role of the natural environment in the emergence of antibiotic resistance in Gram negative bacteria. *Lancet Infect Dis* 13:155–165.

Woodford, N., Turton, J. F., Livermore, D. M. (2011). Multi resistant Gram-negative bacteria: the role of high-risk clones in the dissemination of antibiotic resistance. *FEMS Microbiol Rev* 35: 736–755.

World Health Organization (2018). Antimicrobial resistance. http://www.who.int/news.

7 Safeguarding Nigeria Territory from the Use of Antibiotic Resistant Microbes in Bioterrorism

J. C. Igwe
Kaduna State University

E. O. Ekundayo
Michael Okpara University of Agriculture

Ifeoma M. Ezeonu
University of Nigeria

S. K. Parom
Kaduna State University

CONTENTS

7.1 INTRODUCTION

Antibiotic resistant microorganisms are agents, which are of significant public health importance, as they contribute largely to mortality and morbidity in both hospital and community. If such agents are infectious or communicable, and are isolated among significant population could aggravate panic of global concern.

DOI: 10.1201/9781003178903-9

The emergence of infectious diseases does trigger a complex of sociopolitical, economical, environmental, ecological and microbiological dynamics. The outbreak of severe acute respiratory syndrome-related coronavirus (SARS) in December 2019, which is a species of coronavirus that infects humans, bats and certain other mammals had spread to over 213 countries and territories around the world and two international conveyances with a global cases of over 10, million and 5,000 deaths within 6 months. This outbreak has no exception to neither white nor black, rich nor poor, geographic nor political boundaries, often arriving silently and lethally. In 1916, the Russian Armed Forces arrested a German agent who grew a culture of *Burkholderia mallei* (a contagious zoonotic infectious disease that occurs primarily in horses, mules and donkey). Germany and its allies infected French cavalry horses and many of Russia's mules and horses on the Eastern Front. These actions hindered artillery and troop movements, as well as supply convoys. In 1972, police in Chicago arrested two college students, Allen Schwander and Stephen Pera, who had planned to poison the city's water supply with typhoid and other bacteria. Schwander had founded a terrorist group, "R.I.S.E.", while Pera collected and grew cultures from the hospital where he worked (Seth, 2000). In 1984, the cases of Oregon caused by the followers of Bhagwan Shree Rajneesh, where 11 restaurants were infected with *Salmonella typhimurium*, which resulted in the deaths of over 751 persons with food poisoning (US Institute of Medicine, 2003). The outbreaks of anthrax in the United States in September and October 2001, which led to the death of five persons through poster letters delivery to media houses and congress members, were apparently classified as a bioterrorist act (CNN, 2001). These classes of microorganisms constitute grave threat to the health of humans when infected. The overlook of this form of biomass especially among intellectuals could liquidate the territorial integrity of any Nation. The accumulative effect is majorly overwhelming and could lead to increased economic and sociopolitical collapses. These overwhelming effects had prompted multiple definitions of bioterrorism, either from government agencies such as the Centers for Disease Control (CDC) or as found in federal and state laws. Irrespective of the sources, they generally include some or all of the following factors: The intentional use or threat of use of any biological agent to cause a limitation in the supply of the quality of food, air or water supply, which in turn will affect human, animal, plant or other living organism, with the goal of influencing government conduct, policy or intimidating/coercing a civilian population (Ainscoug, 2002).

7.2 NIGERIAN SOCIETY AND POSSIBLE MICROBIAL THREATS

The outbreak of COVID-19 pandemic has sprouted a dynamics in global view of infectious disease with an indelible mark on most sectors of life. This unprecedented emergence of microorganisms which could serve as bioagent especially the antibiotics resistant ones has once again called for reevaluation of microbial threat and its possible use by state actors, extremist and terrorist to achieve/drive their goals. It is important to recall that in 2013, Muhammad Nazeef Yunus, an Assistant Lecturer with the Department of Arabic and Islamic Studies at Kogi State University, Nigeria, and his cohorts were arrested and paraded for collaborating/aiding the activities of Boko Haram Terrorist in North East Nigeria. This academia might in some cases be biological scientist, who are knowledgeable in the field of microbial engineering and could work to suit the interest of this group. Globally, academia are known as reservoir of transferable knowledge, as they serve as epitome/architectural map for any societal development. If left unchecked could serve as channels by extremist sympathizers to breed bioagents that could be used against the society. Most other reports have highlighted the impact and contributions of religious and traditional leaders, serving personnel's, indigenes in abetting the activities of this extremist. This in the advent of COVID-19 should bring to mind the possibility of exploring bioagents in achieving common interest. Other significant factors that could play much role in the use of bioagents as weapon is the negligence of Nigeria Port Health Authority, ineffective activity of the Chemical and Biological Convention Unit, Nigeria Immigration Services and Nigeria Population Census in gathering of data. The proliferation of private biotechnology laboratories, emerging in different suburbs/states, and National Center for Disease Control (NCDC) approval of these labs for use as COVID-19 sample collection centers without proper provision for guidelines on decontamination and bio-banking of samples could serve as time bomb in future to come. If any of these samples are misappropriated/stolen, it could be reused as inoculum for infection of the populace to compel the government to approve their goal. Some of the considerable factors which might trigger spontaneous malignant effect in Nigeria and elicit infectious diseases proliferation when introduced include urbanization, deforestation, changes in land use and climate, population growth, poverty, malnutrition, political instability and terrorism. It is therefore important that Nigeria government respond to microbial threats from detection to prevention and control in such a manner that will involve a multidisciplinary effort from all sectors of the public health, clinical medicine, natural sciences and veterinary medicine communities. This set of multidisciplinary should form a committee to whose recommendations, will be considered criteria for immediate action, provided it is justified solely on the basis of humanitarian need.

It is important to also note that with advances in biotechnology, isolation, engineering and reintegration of more virulent diseases such as HIV, tuberculosis and malaria could contribute to destabilization of the country and damaging our social and political infrastructures. It will also affect the country's per capita gross domestic product (GDP), resulting in significant decrease in investment in public health and worsening of the epidemic. For example, the resurgence of multidrug resistant tuberculosis has devastated many countries, particularly among developing countries. The available statistics showed that Nigeria has 318 cases per 100,000 populations in 2019, which is the highest TB burden in Africa. The high rising figures observed in Nigeria TB cases are attributed to antimicrobial resistance (AMR), which has become a major barrier to treatment of tuberculosis and malaria worldwide. AMR has also been reported to threaten the effectiveness of antiretroviral therapy in persons with AIDS and has made treatment of common bacterial infections more difficult in the United States and elsewhere.

7.3 BIOAGENTS

The CDC had identified the six most dangerous pathogens (Table 7.1) that could be used in a bioterrorist attack. Most interestingly, these bioagents can be isolated from the environment. Although, these agents have been isolated and known drugs

TABLE 7.1
Six Most Dangerous Bioagent Known to Mankind

Name	Description	Symptoms	Treatment
Anthrax (excluding cutaneous)	Transmission by inhaling; kills 85% of those infected often within 1–3 days	Fever and fatigue; progresses to chest pain, cough, rapid decline	Antibiotics (cipro) before symptoms appear; vaccine not widely available
Smallpox	Physical contact with infected fluids or objects or inhalation of droplets; fatal in 30% of unvaccinated patients	Fever, aches, vomiting; rash develops into pustules	No treatment, but vaccine within 4 days after exposure may mitigate
Pneumonic plague	Airborne; almost 100% fatality rate if untreated	Fever, headache, bloody cough; progresses to respiratory failure and death	Antibiotics within 48 hours of exposure
Viral hemorrhagic fevers	Viruses spread by mosquitoes, rodents and ticks; Ebola death rate up to 90%, Dengue rate 1%	Some variance include fever, aches, exhaustion, internal bleeding	Antiviral treatments vary for the specific form
Botulism	Transmission by inhaling, could be aerosolized; too few cases to know fatality rate	Toxin blocks nerve signals and muscle movement; paralysis; inability to swallow	Ventilator; antitoxin given quickly may stop progression
Tularemia	Inhaling or contact with contaminated substances; could be aerosolized; overall mortality low but untreated severe cases from 30% to 60% fatalities	Fever, headaches, chills, infection of eyes, skin, mucosal tissue	Antibiotics usually successful in naturally occurring cases

Source: CDC (2018).

identified to have activities against them. The possibility of them been reengineered into more profiled antibiotic resistant bio-warfare are inevitable in the advent of interest. It is therefore important that countries with terrorist activities need to undertake preventive measures for protection of their potential use. Table 7.2 shows the historical perspective of biological weapon use and bioterrorism.

7.4 BIOSURVEILLANCE

Biosurveillance, which is the science of real-time disease outbreak detection of both natural and man-made epidemics (bioterrorism), is important for the survival of every independent Nation (Chery, 2011). Nigeria like every other developing countries has significant health risk factors that trigger mortality and morbidity but undocumented due to lack of surveillance such as low education level, low levels of access to safe drinking water and sanitation, lack of access to family planning services, increased poverty index and pollution, poor infrastructure, high proportion of people with

TABLE 7.2
Historical Biological Weapon Use and Bioterrorism

Year	Incidence
5th c. BC	Assyrians poisoned enemy wells with rye ergot (*Claviceps purpurea*), a fungus containing mycotoxins
590 BC	Athenians poisoned enemy water supplies with hellebore, an herb purgative, during the Siege of Krissa
3rd c. BC	Persian, Greek and Roman Literature describe the use of dead animals being used to contaminate enemy water supplies
184 BC	Carthaginian General Hannibal ordered his sailors to hurl clay pots filled with poisonous snakes onto the decks of enemy ships during a naval battle. Hannibal won the battle
1155	Emperor Barbarossa poisons wells with decomposing human bodies
1346	Tartur army catapulted deceased bodies of plague victims over city walls during the siege of Caffa
1495	Spanish sell wine mixed with the blood of lepers to their enemies
1763	British distribute variola virus contaminated blankets to Native Americans resulting in a smallpox outbreak
1797	Napoleon floods fields around Mantua to promote malaria
1915–1918	Germans attempt to infect Allied horses with anthrax and glanders
1932–1945	Japanese operate Unit 731 in Manchuria conducting experiments that included infecting prisoners with a variety of lethal pathogens
1942	British test anthrax bombs on Gruinard Island off the coast of Scotland
1950–1969	The United States and USSR grow offensive biological weapons programs
1969	US President Nixon ends the US offensive biological weapons program
1972	The United States and USSR sign the Biological Weapons Convention agreeing an end to offensive programs
1978	Assassination of Bulgarian exile, Georgi Markov in London with an injected ricin pellet
1979	Accidental anthrax release from a secret Soviet facility in Sverdlovsk kills 66
1984	In the Dalles, Oregon, the Rajneesh cult contaminated local salad bars with *Salmonella* sickening more than 750 people
1990	Japanese Aum Shinrikyo cult unsuccessfully attempt botulinum toxin releases in Tokyo
1991	US troops receive anthrax vaccinations
1991	After the first Gulf War, UN inspectors begin inspections of biological weapons capability in Iraq. Iraqi government officials confirm they had researched the use of anthrax and botulism
1993	Aum Shinrikyo cult unsuccessfully attempts a second botulinum toxin attack on the wedding of the Crown Prince. Later the same month they unsuccessfully attempted to release anthrax from a Tokyo high rise
2001	Anthrax contaminated letters mailed to US Senate offices and media outlets sickening 22 and killing 5
2004	Ricin sent to US Senate Majority Leader Bill Frist's office

Source: Hunter (2018).

tropical and infectious diseases, high number of road traffic accident, climate vulnerability, corruption and bad governance (Hidayat et al., 2004). These accruing factors influenced the unavailability of real-time outbreak disease surveillance system creation in most health institutions. This surveillance system, which would have been expected to collect data from clinics, non-health-care sources, laboratory and over-the-counter drug sales for the purpose of bioterrorism signal detection, at the earliest possible moment is not in place in Nigeria (Nasir et al., 2015). This, we hope, could yield an organized National Retail Data Monitoring system for all possible diseases. Data which potentially could assist in early detection of a bioterrorism event include many categories of information. Health-related data such as that from hospital computer systems, drinking water system, clinical laboratories, electronic health record system, physiologic monitors, school attendance recording, medical examiner record-keeping systems, herdsmen data, emergency call center computers and veterinary medical record systems. The system to be developed in Nigeria should be designed in such a way that infected persons and the epidemiology of the disease could be traced. Visa vice the Department of Chemical weapon convention under the Secretary General to the Federation should be designated as an independent Agency to ensure proper coordination of these activities. The Agency might use very small electronic chips or Fiber-optics tubes containing nerve cells or antibodies coupled to light-radiating molecules as detector in public places to monitor broad spectrum of microbial toxins. As observed in some studies, avalanche of Ultraviolet photodiodes provides higher reliability and robustness required to detect anthrax and other bioterrorism agents in the air. The fabrication methods and device characteristics were described at the 50th Electronic Materials Conference in Santa Barbara on June 25, 2008 (Russell and Steve, 2008). Details of the photodiodes were also published in the February 14, 2008, issue of the journal *Electronics Letters* and the November 2007 issue of the journal *IEEE Photonics Technology Letters.* As observed in the United States, the United States Department of Defense conducts global biosurveillance through several programs, including the Global Emerging Infections Surveillance and Response System. This should be replicated in Nigeria through her Agency (Nikoleli et al., 2016).

7.5 THE INFLUX OF BIOLOGICAL SUBSTANCE INTO THE COUNTRY WITHOUT PROPER SCRUTINY AND EVALUATION

In most scientific communities the use of typed culture microorganism in research is highly recommended, and this practice standardizes the findings as it makes it comparable with international standard. Although this practice is good, the means of importing this biological organisms in most studies are not legal and do not follow the standard procedures. The different categories of the bioagents are shown in Table 7.3. Some multidrug resistant bacteria commonly isolated from clinics that could be of concern/engineered as bioagents include:

a. **Methicillin Resistance *Staphylococcus aureus* Typed Culture:** Methicillin resistant *Staphylococcus aureus* (MRSA) is a strain of *Staphylococcus aureus* (pathogen) with a great impact on global human health and is associated with endocarditis, blood stream infections, pneumonia, skin and soft tissue infections and necrotizing fasciitis. This form of bacteria confers resistance to most betalactams, e.g. amoxicillin, penicillin, oxacillin and methicillin, and is majorly multidrug resistant. This resistance trait do become common and spread worldwide due to the diffusion of epidemic clones. Globally, MRSA has become one of the most frequently isolated nosocomial, community-acquired and livestock pathogens. This has made the bacteria to gain importance in clinical microbiology. Importation of this form of bacteria without following standard procedure could contaminate our reserved and revered biodiversity, eliciting a spread within the community and contributing to the proliferation of diseases with multidrug bacteria.

b. **Multidrug resistance *Pseudomonas aeruginosa:*** *Pseudomonas aeruginosa* is an important pathogen frequently implicated in healthcare-associated infections (HAIs), particularly in critically ill or immuno-compromised patients. It is a versatile pathogen with the ability to cause diverse infection types. Data from the US National Nosocomial Infections Surveillance System from 1986 to 2003 reported *P. aeruginosa* as the second most common cause of pneumonia (18.1%), the third most common cause of urinary tract infection (16.3%) and the eighth most frequently isolated pathogen from the bloodstream (3.4%). While the overall proportion of infections caused by *P. aeruginosa* has remained controllable, the proportion of resistance in this bacterial is alarming. The rate of resistance of this bacteria to imipenem, quinolones and third-generation cephalosporin has over the years increased by 15%, 9% and 20%, respectively. Similarly, a national surveillance study of intensive care unit (ICU) patients in Nigeria had reported a significant increase in multidrug resistance in this bacteria to imipenem, ceftazidime, ciprofloxacin and tobramycin (Makanjuola et al., 2018). This form of bacteria mediates resistance to commonly prescribed antibiotic by several mechanisms including multidrug efflux systems, enzyme production, outer membrane protein (porin) loss and target mutations. Inappropriate empirical therapy has been associated with increased mortality in *P. aeruginosa* infections. Other reports have also proven that worse clinical outcomes may be associated with multidrug resistant infections linked to *P. aeruginosa* owing to limited effective antimicrobial options for this bacterial.

c. **Enterohemorrhagic *Escherichia coli* O157:H7 (*E. coli* O157:H7):** This is a major food-borne pathogen causing severe disease in humans worldwide. Healthy cattle are a reservoir of *E. coli* O157:H7, and bovine food products and fresh produce contaminated with bovine waste are the most common sources for disease outbreaks. After ingestion of *E. coli* O157:H7, the bacteria bind to the intestinal mucosa and begin releasing Shiga toxin that causes hemorrhagic colitis (HC) and the life-threatening sequelae hemolytic uremic syndrome (HUS) in humans. The toxin, in turn, disrupts protein synthesis in the epithelial cells lining intestinal mucosa, leading to cell death, sloughing of the mucosa and eventual bloody diarrhea (Gani et al., 1991). Other virulence factors of the bacteria products of the pathogenicity island are the locus of enterocyte effacement and products of the F-like plasmid pO157. According to the CDC, this bacteria causes an annual

TABLE 7.3
Categories of the Bioagents, Their Characteristics and Examples/Sources

Category	Characteristics	Examples/Sources
A	- High-priority agents include organisms that pose a risk to national security because they - Can be easily disseminated or transmitted from person to person; - Result in high mortality rates and have the potential for major public health impact; - Might cause public panic and social disruption; and - Require special action for public health preparedness.	- Anthrax (*Bacillus anthracis*), - Plague (*Yersinia pestis*), -Botulism (*Clostridium botulinum*), - Smallpox (*Variola major*), -Tularemia (*Francisella tularensis*), - Viral hemorrhagic fevers, e.g. a. Filoviruses (Ebola, Marburg), b. Arenaviruses (Lassa, Machupo)
B	- Second highest priority agents - Are moderately easy to disseminate; - Result in moderate morbidity rates and low mortality rates; and - Require specific enhanced disease surveillance.	- Brucellosis (*Brucella* species), - Epsilon toxin of *Clostridium perfringens* - Food poison-associated bacteria: *Salmonella* spp., *Escherichia coli* O157:H7, *Staphylococci* spp., *Shigella* spp. - Water safety threats: *Vibrio Cholerae*, *Cryptosporidium parvum* - Glanders (*Burkholderia mallei*) - Melioidosis (*Burkholderia pseudomallei*) - Psittacosis (*Chlamydia psittaci*) - Q fever (*Coxiella burnetii*) - Ricin toxin [*Ricinus communis* (castor beans)] - Staphylococcal enterotoxin B - Typhus fever (*Rickettsia prowazekii*) - Viral encephalitis (alphaviruses, such as eastern equine encephalitis, Venezuelan equine encephalitis and western equine encephalitis)
C	Third highest priority agents include emerging pathogens that could be engineered for mass dissemination in the future because of – availability; ease of production and dissemination; and potential for high morbidity and mortality rates and major health impact.	Emerging infectious diseases such as Nipah virus and hantavirus

Source: CDCP /NCEZID (2021).

73,000 illnesses, 2,200 hospitalizations, and 60 deaths in the United States and also influences more than 405 million dollars cost, including lost productivity, medical care and premature deaths. In Africa, burdens associated with *E. coli* O15:H7 has been attributed to 3,890 cases of HUS and 230 deaths with an incidence rate of 1.4 cases per 100,000 person-years.

It is important to note that these microorganisms, which could be engineered and used as bioagents, can be isolated from zoonotic sources and our environment. As a result, an attack on human populations with such an agent might elicit significant health risk to animal populations in the target area, including pets, which could act as channel to the human population. However, animal species in most cases of bioterrorism tends to be more susceptible and serves as channels of disease dissemination and proliferation. This might be attributed to the fact that diseases caused by the agent do have shorter incubation period or life circle or vulnerable to continuous contact than the human population. Hence, simultaneous appearance of signs and symptoms of disease in animals should be well monitored by veterinarian as this could serve as route to rapid identification of biological warfare producing nonspecific effects in humans (Gregory and Waag, 1997). Interesting to note is that when bioagents are released, sometimes they persist in the environment (such as soil, water or air) and as such could trigger sporadic illness in animals and deaths. Active surveillance for this form of outbreak might indicate the persistence of a biological threat.

From data available in different publication sites online, various research findings in the field of biological sciences with root from Nigeria have established the analysis of molecular association of gene in multidrug resistant microorganism and their expression/activities in human. The pathogenicity and virulence nature of these microorganism as shown in bioagents a, b and c above could result in significant mortality and morbidity. Critical observation have shown that the mode of importation, bio-banking and decontamination of biological products/organism among researchers and postgraduate students are not standard, and proper documentation of research activities in most Lab/Universities is not commensurate with international standard; hence, there could be easy misuse of biologicals as bioagents within our community.

The agencies that involve in the importation of biological substance should as a matter of urgency gear up in their model of evaluations especially at the port of entry, as the influx of biological material for the purpose of research and biotechnology activity could serve as source of biological weapon in the hands of terrorist. This biological material may serve as weapon which may have graft impact on the populace if not curtailed. More so, NUC could send circulars to all Universities on the use/acquiring of biological materials. Nigeria government through her agencies should as a matter of urgency begin a robust communication, coordination and collaboration with private organizations and foundations towards mitigating the outcry of the COVID-19 pandemic, as it opens a rethinking gap in various minds. These efforts could be in the form of financial investments and technical assistance, operational research, enhanced surveillance and efforts to share both knowledge and best public health practices across national boundaries.

Nigeria could also adopt a rebuttal-skilled training system which hopes to strengthen the country capacity in microbiology and epidemiology through the

NCDC and enhance the capacity of national public health laboratories, supporting field epidemiologist and improving the capacity to detect and respond to disease outbreaks. The overall objective of the program will be to strengthen diagnostic and surveillance capabilities at all levels and also increase diagnostic capabilities for communicable diseases; the development of appropriate core public health administrative practices; the development of rapid, sustainable national and international laboratory communications networks; the development of rapid, efficient and safe means for shipment of diagnostic materials and laboratory specimens; and the establishment of appropriate quality control principles and practices. This, we believe, will go a long way in checkmating and mitigating the dilemma which might arise due to use of antimicrobial resistant microorganism in bioterrorism.

Other recommendations by the World Health Organization are development of capacity from hospitals, industries and academia in essential laboratory diagnostic practices and techniques, development of programs in infectious disease research and training, biosafety, basic disease epidemiology, data collection and management, statistical analysis and personnel management and administration.

7.6 HOW ANTIBIOTIC RESISTANCE MICROORGANISMS DEVELOP

Resistance to antibiotic develops as a result of the following factors: mutational change(s) in the genome of the organism, selective pressure, gene transfer, inappropriate use, inadequate diagnosis, increased use of antibiotics in animal husbandry. All these basic natural factors reduce the affinity and effectiveness of drugs, chemicals or other agents designed to cure or prevent bacteria proliferation in an environment. In the presence of such agent (antibiotics) the bacteria survive and reproduce.

7.7 POSSIBLE IMPACT OF THE USE OF ANTIBIOTIC RESISTANCE MICROORGANISM IN BIOTERRORISM

i. Effect on Health Fund: It is a well-known fact that the outbreak of COVID-19 had created a big lacuna in the focus of national health fund from other diseases to mainly COVID-19. This will evidently aggravate high death record in other forms of diseases peculiar to Nigeria such as malaria, diarrhea, cholera, diabetics, cardiovascular diseases, etc. Although the outpouring of global solidarity and support that could be sparked by this shared challenge will be a phenomenal, the strategic preparedness and response plan of most health agencies would be deficit. This could influence the outlined funding need of other sectors to be eroded for critical response efforts. It is important to note that as the associated effects evolve, funding needs are likely to increase and the country recession worsens. This will inadvertently lead to loss of jobs, poverty, fear, deaths and economic crisis especially among private sectors and those in remote villages. In view of the lack of funds, the country health sector might be limited in their response plan, which includes among others: putting in place activities to track and understand the spread of the microorganism; ensuring patients get the care they need; buying and shipping of essential supplies for frontline workers; producing evidence-based guidelines and advice and making sure health workers and responders get the information and training to detect and treat affected patients; producing guidance for the general public and for particular groups (immuno-compromised persons) on measures to take to prevent the spread, and also prevent themselves and others; accelerating efforts to develop better therapeutic outcome, tests and treatments.

ii. Loss of Confidence in Health Sector: In Nigeria, the health sector is already overloaded due to the low ratio of health workers to patients especially in primary healthcare centers. An outbreak of infectious diseases will further showcase the glaring attention required in the already overburdened public health systems and the challenges faced in recruiting, deploying, retaining and protecting sufficient well-trained, supported and motivated health workers. There is a great gap in manpower for sustainable investment in health systems including the health workforce, and for decent working conditions, training and equipment especially in relation to personnel protective equipment and occupational safety. While the healthcare workers were on strike in some states in Nigeria, there were significant breakdown in manpower, health and increased morbidity and mortality. Further observed was the importance of deploying social dialogue in building resilient health systems, as it has critical role both in crisis response and in building a future that is prepared for health emergencies.

iii. Wide Health Inequality: The use of antibiotic resistance microorganism as bioterrorism tool in Nigeria will further showcase the significant disparity in healthcare delivery, as people living in the urban tend to have more medical facility and delivery than those in the rural area. Most people in the rural area tend to die of other diseases and negligence due to transfer of competent healthcare professionals to the centers designated for the pandemic, creating a gap and gutter in healthcare delivery. This inequality will also increase death rate among people living in the poorest parts of the country, making it to out-double those in affluent areas. This is further mirrored by the Nigeria economic landslide situation where 60.9% of her population live on less than $1.90 per day. The likely economic recession that may accrue due to the use of bioagent will further worsen the inequalities.

iv. Other Diseases: Various report have acknowledged that in the United States, there were 20 times more deaths per week from COVID-19 than from the flu in the deadliest week of an average influenza season. In the event of such outbreak in Nigeria with low healthcare facilities and unequal manpower, there would be a gruesome mortality and morbidity increase. Most interesting is that the fatality record of COVID-19 is more deadly than the SARS, H1N1, MERS and Ebola epidemics combined. Although due to lack of records, death fatality rate from other diseases outweighed SARS in Nigeria. This low mortality record might be attributed to weather, total microbiome and adaptive immune system. The review of the peculiarity of Africa further showed that when indigenous antibiotics resistant microorganisms are used as bioagent, high negative impact and economic collapse might elicit in the absence of immediate intervention (Ayukekbong et al., 2017). According to the Global Burden of Disease, an annual study by the Institute for Health Metrics and Evaluation (IHME) that tallies the lives lost to 282 illnesses and injuries across 195 countries and territories, diabetes was responsible for 20,000 deaths a week, on average, in 2017 (the latest year for which data are available). Road accidents killed 25,000 a week; lung and tracheal cancers killed 36,000; while lower respiratory infections, such as pneumonia and bronchitis, killed another 49,000. During one particularly deadly 7-day period in April, the coronavirus had killed more than 50,000. This has made SARS-2 the biggest killer of 2020 as its global toll exceeds that of breast cancer or malaria.

7.8 CONCLUSION

As part of post-COVID-19 reshuffling of thought and strategies, the Nigeria government through its National Intelligence Council should produce a National Intelligence Estimate, which will examine the most lethal diseases globally and by region; develop alternative scenarios about their future course; examine national and international capacities to deal with them; and assess their national and global social, economic, political and security impact. The Estimate should also assess the infectious disease threat from international sources to Nigeria, Boko haram, possible state actors, military personnel overseas and any country to which Nigeria has significant equities and treaty.

REFERENCES

Ainscough J.M. (2002). Next Generation Bio-weapons: The Technology of Genetic Engineering Applied to Biowarfare and Bioterrorism. https://fas.org/irp/threat/cbw/nextgen.pdf.

Arya S.C. (2004). Antibiotics prescription in hospitalized patients at a Chinese university hospital. *Journal of Infectious Diseases* 48: 117–118.

Ayukekbong, J. A., Ntemgwa, M. and Atabe, A. N. (2017). The threat of antimicrobial resistance in developing countries: Causes and control strategies. *Antimicrobial Resistance & Infection Control*, 6:47.

Centers for Disease Control (CDC) (2018). Emergency Preparedness and Response: Bioterrorism Agents/Diseases. https://emergency.cdc.gov/agent/agentlist-category.asp.

Centers for Disease Control and Prevention (CDCP) and National Center for Emerging and Zoonotic Infectious Diseases (NCEZID) (2021). Bioterrorism Agents/Diseases. https://www.cdc.gov/ncezid/.

Chery P. (2011). Global Nature of Terrorism Drives Biosurveillance. Archived January 12, 2013, at the Wayback Machine American Forces Press Service, 27 October 2011.

CNN (2001). U.S. officials declare researcher is anthrax killer. https://edition.cnn.com/2008/CRIME/08/06/anthrax.case/index.html.

Gani, L., Arif, H., Widjaja, S. K., Adi, R., Prasadja, H., Tampubolon, L.H., Lukito E., Jauri R. (1991). Physician's prescribing practice for treatment of acute diarrhoea in young children in Jakarta. *Journal Diarrhoeal Diseases Research* 9: 194–199.

Gregory, B. and Waag, D. (1997). Military Medicine: Medical aspects of biological warfare (PDF), Office of the Surgeon General, Department of the Army, Library of Congress 97–22242, archived from the original (PDF) on 2009-06-12, retrieved 2009-05-22.

Hidayat B., Thabrany H., Dong H., Sauerborn R. (2004). The effects of mandatory health insurance on equity in access to outpatient care in Indonesia. *Health Policy Plan* 19: 322–335.

Hunter N.D. (2018). The Law of Emergencies (2nd Edition): Public Health and Disaster Management, 245–271.

Institute of Medicine (US) (2003). *Informing the Future: Critical Issues in Health: Second Edition*. Washington, DC: National Academies Press. Human Security and Bioterrorism. Available from: https://www.ncbi.nlm.nih.gov/books/NBK216161.

Makanjuola, O. B., Fayemiwo, S. A., Okesola, A. O., Gbaja, A., Ogunleye, V. A., Kehinde, A. O., and Bakare, R. A. (2018). Pattern of multi-drug resistant bacteria associated with intensive care unit infections in Ibadan, Nigeria. *Annals of Ibadan Postgraduate Medicine* 16(2): 162–169.

Nasir, I. A., Babyo, A., Emeribe, A. U. and Sani, N. O. (2015). Surveillance for antibiotic resistance in Nigeria: Challenges and possible solutions. *Trends in Medical Research* 10*:* 106–113.

Nikoleli G.P., Karapetis S., Bratakou S., Nikolelis D.P, Tzamtzis N., Psychoyios V.N., Psaroudakis N. (2016). Biosensors for security and bioterrorism: Definitions, history, types of agents, new trends and applications. In: Nikolelis D., Nikoleli G.P. (eds) *Biosensors for Security and Bioterrorism Applications. Advanced Sciences and Technologies for Security Applications*. Springer, Cham. https://doi.org/10.1007/978-3-319-28926-7_1.

Russell D. and Steve W.C. (2008). Avalanche Photodiodes Target Bioterrorism Agents News wise, Retrieved on June 25, 2008. https://phys.org/news/2008-06-avalanche-photodiodes-bioterrorism-agents.html.

Seth, C. W. (2000). *"R.I.S.E." in Toxic Terror*. Assessing Terrorist Use of Chemical and Biological Weapons, 55–69.

World Health Organization (2019). How to reduce the spread of antibiotic resistance. http://www.euro.who.int/en/health-topics/disease-prevention/antimicrobial-resistance/news/news/2012/11/antibiotic-resistance-a-growing-threat/how-to-reduce-the-spread-of-antibiotic-resistance.

8 Deep Learning and Economic Prospects in Medical and Pharmaceutical Biotechnology

Charles Oluwaseun Adetunji,
Kingsley Eghonghon Ukhurebor,
and Olugbemi Tope Olaniyan
Edo State University Uzairue

Juliana Bunmi Adetunji
Osun State University

Gloria E. Okotie
University of Ibadan

Julius Kola Oloke
Ladoke Akintola University of Technology

CONTENTS

DOI: 10.1201/9781003178903-10

8.1 INTRODUCTION

It has been established that chemoinformatics play a significant role in many industrial sectors, most especially those that utilize numerous computing techniques for their various activities (Ekins et al., 2010). The application of artificial intelligence (AI) software, especially those that are built on the application of machine learning for effective forecast, is most necessary (Rost et al., 2016). This implies that most of the software could help us in healthcare sector–associated research for adequate processing of necessary information.

In the context of pharmaceutical and biomedical research, it has been established specifically that chemoinformatics through the application of machine learning techniques could play a crucial role towards the delivery of adequate health care and in some other sectors (Adetunji and Anani, 2021a,b,c; Adetunji et al., 2021; Dwivedi et al., 2021; Olaniyan and Adetunji, 2021, Mishra et al., 2020; Thangadurai et al., 2021; Anani et al., 2020; Islam et al., 2021). Typical examples of such techniques include k-Nearest Neighbours, support vector machines, Decision Trees, and Naïve Bayesian (Mitchell et al., 2014). All these aforementioned techniques have allowed rapid enhancement in the utilization of our datasets which has developed towards big data platform (Zhu et al., 2014; Clark et al., 2015; Ekins et al., 2014a,b; Nwankwo et al., 2020). These techniques have been shown to have some other greater applications, but they all have their advantages and disadvantages which can be utilized for continuous data, binary classification and multiple classes (Mitchell et al., 2014). The utilization of various computational techniques as well as machine learning algorithms which have a greater application in resolving challenges has been discovered to follow the development of datasets (Ekins et al., 2007; Nwankwo et al., 2020).

In term of pharmaceutical datasets, some techniques such as quantitative structure activity relationships and pharmacophores constitute some of the major techniques that are normally applied. However, in recent times, an increase in the level of biological data that are generated as a result of a very high content screens and throughput screening has necessitated the application of numerous tools for their several purposes most especially those that deal with bigger data (Ekins et al., 2014a).

The high level of enhancement in more complex non-linear challenges could be resolved through the power of computer processing. This is possible because of the availability of comparative economical computing resources. Also, it has been observed that most of these outcomes derived through the application of machine leaning models could be utilized most especially on mobile phones (Clark et al., 2014; Ekins et al., 2015). Furthermore, there has been increase in the level of the application of deep learning which was developed based on several years in knowledge gained through the application of artificial neural network research, which has been established to possess several merits most especially in the learning from language and image (LeCun et al., 2015; Baskin et al., 2016). This may then represent a novel next-generation era of pharmaceutical and chemoinformatics research which leverage on mining the dissimilar big data accruing from more refined algorithms such as deep learning.

Therefore, this chapter intends to provide detailed information on the application of deep learning and economic prospects in Medical and Pharmaceutical Biotechnology. Special highlights are provided on some of the applications of these recent techniques most especially in the area of biomedical sectors.

8.2 PROCESS ENGINEERING IN BIOTECHNOLOGY PROJECTS

Guoyao and Fuller (2019) revealed that biotechnology-based breeding is the needed breakthrough to meet the increasing response for genetic traits and demand of quality pork protein. They grouped gene cloning by identical gene or genetic engineering to generate modified microorganisms or animals. It is known that cloning conserves the genetic trait or breed, while recombinant DNA technology for bioengineering uses multiple sources of genes. Genome editing utilizes insertion, deletion, and silencing to produce genetically modified products. Many tools of gene editing are presently deployed for biotechnological processes such as zinc finger nuclease, clustered regularly interspaced short palindromic repeats-associated nuclease-9 and transcription activator-like effector nuclease targeting cell, bacteriophages or plasmid via transfection like electroporation, lipid-based ligands, microinjection, and nucleofection. The authors revealed that the use of genetically modified pigs has been able to produce bacterial phytase, growth hormone, *C. elegans* fatty acid desaturases, fungal carbohydrases, uncoupling protein-1; lack of myostatin, CD163 or α-1,3-galactosyltransferase; and concluded that biotechnology provides opportunity to improve efficiency of swine generation plus develop a substitute to antibiotics in the nearest future. Fang et al. (2016) revealed that with the advancement in molecular biology tools, more development is being seen in DNA-based technologies to provide and promote quality and safe crop breeding products, increase agricultural output, and equally protecting the eco-environment. Hence, their role in modern day agro-science sector is becoming more important. The authors in their study described the utilization of biotechnological skills and bioengineering to improve agricultural yield such as molecular markers, genetic engineering, and genome editing.

Gupta and Savaliya (2012) discussed the importance of biotechnology to livestock improvement such as the quality of meat, milk, productivity, physico-chemical properties, nutritional qualities, and livestock economy. The authors have also predicted the application of biotechnology in high resilient farm animals, improved quality products, heightened production of food preservatives, and robust waste product utilization. They noted that the utilization is also possible in quality assurance schemes in reducing bad products deleterious to public safety. Wang et al. (2009) revealed that biotechnology plays a central role in the detection of species-species and genus-transferred microsatellites, or simple sequence repeat biomarkers. Studies have shown that DNA are spread throughout the eukaryotic cells in the mitochondrial or nucleus and chloroplasts. The utilization of microsatellite provides answers to the study of replication, evolution, repair, mutation or recombination. Hence, it has become a powerful tool in insect, animal, and plant gene study like genetic mapping, diversity studies, or assisted-marker selection.

Mebratu et al. (2014) revealed that genetic engineering can be described as a process involved in the manipulation of genetic components of cells like RNA and/or DNA in order to improving yield, changing or modifying products. It involves in vitro or in vivo techniques, gene therapy or generation of new strains from microorganisms for industrial or pharmaceutical utilization. The authors revealed that the utilization of genetic engineering for animal production with resistance to pathogens, production of vaccines, and increasing yield in agriculture has grown exponentially. Though many concerns have been raised like alteration of natural process of genetic equilibrium, cost and ethical concerns, the utilization of this cutting-edge technology holds a promising future for livestock production.

8.3 APPLICATIONS OF GENETIC ENGINEERING

8.3.1 Disease Resistance

In fighting pathogens, biotechnological tools are utilized such as in genetic selection of traits for livestock production with disease or pathogenic resistance. Also, genetic engineering can be used to introduce disease-resistant genes (additive) into new products or depleting or deleting (knockout) a susceptible gene or locus (Alison and Davis, 2009).

8.3.2 Increasing Meat and Milk Production

The application of genetic engineering for enhancement of meat or milk production has become a value-added innovative opportunity in the agricultural sector. For instance, a 10%–20% change in casein in milk generated from transgenic cows could enhance proteolysis, thus promoting faster ripening of cheese (Scott and Mattew, 2011).

8.3.3 Improved Hair and Fibre

In livestock production, fibre, wool, or hair for fabric and yarn production is a growing area of interest for many biotechnologists. Manipulation in terms of quality, refinement, crimp, and length of hair, wool, and fibre from livestock has been exploited. Studies have shown that lack of amino acid cysteine is a major barrier in wool production industries among livestock; hence, biotechnological tools are deployed to transfer cysteine enhancing genes from bacterial to livestock genome (Scott and Mattew, 2011).

8.3.4 Improving Growth Rates and Carcass Component

Exploiting genetically engineered livestock and poultry has opened up novel insights into mechanisms stimulating growth rate through the application of transgenic tools, and it is now possible to stimulate growth factors, growth modulators, and growth factor receptors (Lai et al., 2006). The authors also revealed that carcass can be improved through selective breeding.

8.3.5 Enhancing Reproductive Activities and Fecundity

Several genes of importance on reproduction have been identified so far such as the Boroola fecundity genes and that of oestrogen receptor. Through genetic manipulation, increased litter size and increased ovulation rate have been made possible (Rothschild et al., 1994).

8.3.6 Vaccine Production

Most of the traditional vaccines are used to inactivate microorganisms, kill bacteria, or immobilize toxic organisms. This may generate adverse effects on the immune system by rendering it redundant; therefore, newer technologies are emanating as alternative solutions to the classical ones. Bacteria or viruses can be used as vectors or carriers of vaccine or act as living genetic modified vaccines. Also, synthetic polypeptides extracted from cloned genes or disease pathogens as recombinant vaccines are also being developed to induce immune response. Most subunit vaccines are given with an adjuvant mainly because they cannot replicate in the host, for instance virus-like particles which are immunogenic. Genetic vaccines are made up of spherical pieces of DNA referred to as plasmids made up of a promoter gene and foreign gene for protein expression in the host. Recombinant plasmids are purified foreign naked gene from bacteria given to a host directly via intramuscular means. Genetic vaccines can stimulate any part of immune compartment to bring about the desired change. The importance of DNA vaccines is the unique sequence referred to as CpG motifs having ability to act as adjuvant activating the immune system to promote the efficiency of the vaccine (Jackwood et al., 2008).

8.3.7 Plant-Based Vaccines

Studies have shown that plant vaccines are protein units as host biological bioreactors. Antigen is known to produce the vaccine so as to induce immune reactions when taken orally. Studies have demonstrated that soybean, potatoes, and corn can be inserted with foreign genes and fed to livestock to build immunity against the same pathogen. The advantage of edible plant vaccines is mainly due to the fact that many animals can be treated at a go without the need of personnel. Also, many agricultural products can be effectively stored without spoiling before distribution (Streatfield, 2005).

8.3.8 Gene Therapy

Gene therapy can be described as a way of attacking diseases from its origin by repairing mutated gene and involves the utilization of DNA molecule as therapeutic agent to cure diseases. Researchers use DNA made up of drug in the form of therapeutic protein packaged as a vector delivered into the cell to correct or treat infectious diseases via modification of cellular machinery. This therapeutic approach is commonly used for inheritable disorders or infectious diseases (Sheridan, 2011).

8.4 PROSPECTS OF GENETIC ENGINEERING IN DEVELOPING COUNTRIES

8.4.1 Recombinant DNA Technology

Ethiopians have adopted the utilization of biotechnology in most of their agricultural production even though they are at the beginning of their research and development in agro-science biotechnology. In livestock production, biotechnology is utilized for most processes such as artificial insemination, vaccine production, molecular genetic analysis, and molecular diagnostics. Also, government is setting up agricultural biotechnology research centres and building capacities in the educational sector, though there are few setbacks due to technical issues, lack of public trust, and regulatory policies. Across relevant agencies of government, efforts are geared towards establishing a policy statement and guidelines that will affect national and international standards in the area of cell and tissue culture, genetic engineering, and molecular genetics (Adane, 2009).

8.4.2 Socio-economic Impacts

In most developing countries, most of the livestock owners are peasant farmer with very limited resources and funding to support and commercially develop a large-scale biotechnology research and capacity that will transform the agro-science industry. Hence, government must set up a platform that will enable biotechnology to thrive and develop considerable commercial and socio-economic realities of most livestock operators. In livestock production, many desirable traits and biodiversity can be identified and selected through molecular markers for gene improvement and disseminate using various skills in assisted-reproductive technologies for overall economic returns (Madan, 2005).

8.4.3 Impacts on Human and Animal Health

Another vital area where biotechnology is being used to improve and bring about development is in the health sector. Through many applications of genetic engineering and biotechnology on animals for experiment, major breakthroughs in cancer, organ transplant, and infertility have been made possible. This is mainly due to the similarity in the genome of some animals and humans; hence, translational experiment is made possible. Also, xenotransplantation procedure in human should be possible with the advancement in genetic engineering in transferring organ from one species to another (Chad, 2006).

8.5 LIMITATIONS OF GENETIC ENGINEERING

8.5.1 Environmental Impacts

It has been discovered that plants could express high resistance to beneficial insects or microorganisms thereby reducing pollination, climatic condition, and unintended non-target agents as well as disruptions in the biodiversity and ecosystem (Sears et al., 2001). Several ecological problems that could be generated from these techniques have been identified such as accidental production of a more virulent strain of a species leading to epidemic.

8.5.2 Health Hazards

Serious consideration is also placed on the health consequences of genetically engineered products and organisms; thus, subsequent instability in phenotypic and genetic expressions has been revealed to be a big health issue. Also, the physiological consequences of genetically modified protein or DNA entering into the gastrointestinal system may induce the pathogenesis of chronic disease conditions in humans or animals (Tereje and Jack, 2007).

8.5.3 Religious, Cultural, and Ethical Issues

Recent development in biotechnology and its application in medicine, crime detection, drug delivery, industries, agriculture, and environment have posed a question of how this technology should be well managed in order not to cause serious problems in our society with implications on our cultural value, religion, and humanity. Many concerns have been raised about the ethical issues of many innovations in biotechnology such as species to species transfer of genes thereby playing God and disrupting the natural order. Many opinions have been put forward that life should not be seen as a biochemical marker to be altered through the lens of biotechnology (Conrad and Harold, 2009).

8.5.4 Economic Constraints

The establishment of biotechnology companies is capital intensive with huge budgetary requirements which is yet to be seen in most developing countries. Financing and sustainability can only be attained through seed money, free accessibility to loan, and venture capital (Nyerhovwo, 2004).

8.5.5 Higher Level of Antibiotic Resistance

The issue of consuming antibiotic-resistant genes in food has also generated a lot of concerns as it portends a danger of reducing the effectiveness of antibiotics in humans or animals. Many bacteria with ability to pick up resistant genes can also become resistant to the antibiotics and cause serious issues by making treatment more difficult in pathogenic bacteria (Prakash, 2002).

8.6 ARTIFICIAL INTELLIGENCE (AI) AND DEEP LEARNING APPLICATIONS

Matheny et al. (2019) revealed that the introduction of AI in health sector as a technique provides tremendous opportunity for improving clinical skills and outcomes and impacts positively on the populace to reduce cost. The authors also pointed out that caution must be followed in implementing and driving the health sector completely with AI due to user disillusionment plus technology-driven disparities. Many areas of generating knowledge of AI have been highlighted by the authors such as clinicians and patients, clinical implementers, policy makers and regulators, and model developers, just to mention a few. Also, moving forward, there must be regulatory and legal landscape designed for healthcare application, human rights consideration, and adoption and maintenance.

Carbonell et al. (2019) revealed that the inability of man to determine the attitude or behaviour of system biology from alteration of gene to phenotypic changes impedes the advancement of biomedical science and bioengineering. Recently, a novel technique was introduced with the capacity to determine biological behaviour referred to as machine learning. Some of the characteristic features are ability to use synthetic biology to generate data with increased reproducibility. Hence, it was suggested that increased investment and funding in the area of machine learning and synthetic biology will enable algorithms to drive the prediction processes.

Zhavoronkov and Mamoshina (2019) revealed in their study that the first novel prediction of senescence was done utilizing deep learning approach developed through AI. The authors showed that deep senescence clock can be useful to a broad variety of biomedical fields such as spanning target identification, pharmaceutical industry, drug discovery, artificial patient data generation plus data economics. Deep learning biomarkers have generated tremendous attention among biomedical scientists as a predictive tool for many biological processes such as disease condition, multi tissue methylation, microbiome, gene expression, imaging, and mortality.

Minssen et al. (2020) revealed that in the healthcare system, companies are utilizing AI in developing novel machine learning algorithms to guide clinical decisions. The authors further showed the challenges posed by machine learning such as regulation and evaluating medical machine learning established medical devices to ensure their effectiveness and safety, and international perspectives and regulations. Uddin et al. (2019) showed that in precision medicine, the goal is to optimize and design the pathway for prognosis, analysis, beneficial intervention by utilizing biological datasets of genes offering personalized treatment to individuals. The authors revealed that with AI algorithms and high-performance computer, predictions can be made for cardiometabolic disorders and cancer detection, from biological and multidimensional data. Even though slow progress has been made so far from AI on neurosciences and neurodegenerative diseases, much prospect still lies ahead with gene characterization and genomic variation.

Bach et al. (2019) demonstrated that AI-based technologies are growing rapidly with a myriad of applications in health care and medicine, even though till today reports on AI are lacking in the area of workflow, productivity, research landscape, and other important topics. The authors evaluated the global development and growth of scientific advancement of AI in medicine from the years 1977 to 2018. From their findings, it was discovered that AI has rigorously impacted prognosis and diagnosis, rehabilitation and regeneration, robot-assisted surgery, precision medicine, data science, and policy and decision making. It was concluded that three clinical areas are majorly impacted such as material, policies, plus practices and suggested that it could be used to constrict the disparity in the healthcare sector between the developed and developing countries. Support and technology transfer from industrialized countries are majorly the source of advancement for developing countries.

Fröhlich et al.'s (2018) Precision, personalized, stratified medicine or P4 is known as a medical method in which patients are categorized based on their disease risk, subtype, treatment, or prognosis response utilizing particular diagnostic tests. The essence of this is to make medical decisions on individual patient's behavioural and molecular markers which are dependent on machine learning and data. The authors revealed that in the last few decades, a lot of renewed interest on AI was observed with the current low impact due to inadequate performance and inability to predict and interpret clinical trials. Thus, they suggested that there is need for collaborations between health professionals such as data scientists, computational physiologist, physicians, health regulators, patient advocates, and health insurance organizations.

AI has become the emerging science and technology with increasing developmental processes emanating from automobile industries to medical science and manufacturing. Today, AI is used in boosting crop and plant yields and quality, improving weather forecasting, cancer detection, improving industrial growth and productivity, and prediction of an epidemic (LeCun et al., 2015). The application of AI in biomedical science include food science, neuroscience, regenerative medicine, medical imaging, bioinformatics, public health, biological engineering, physiological parameter monitoring, biomechanics, nutrition, drug discovery, neurorobotics, medical informatics, etc. Most models in deep learning are instigated by enhancing the number of layers in a neural network (Giles, 2018). AI is applicable in the following areas:

Augmented reality: This can deliver an interactive understanding of a real-world environment, where elements from the real world are "increased" by computer-generated sensory knowledge and covered with the natural environment.

Biometrics: This area of science uses physiological characteristics like fingerprint, face, iris or vascular pattern, and behavioural traits or characteristics like speech or gait to recognize individuals. This is a combination of computer sight with extensive knowledge in human physiology and neuroscience.

Character recognition: This subsumes image recognition popularly called optical recognition or the process of using printed, reading-typed, handwritten text converted to machine codes.

Computer sight: This system can be described as the machine special sense similar to the biological sense organ dealing with how computers sense or see digital images and characters.

Distributed AI: This is designed to solve large data in a complex learning or decision-making processes of bringing together individual data and offer temporal solution via communication nodes of different individual networking agents.

Video and image segmentation: Analysis of images in a fragmented form through labelling to simplify the picture representation.

Information extraction: This system deals with the retrieval of organized information from disorganized sources.

Natural language processing: The utilization of computer to interpret and communicate human language through the use of algorithms can be written or verbal forms.

Object tracking: The process of analysis and identification of a moving object over a period of time in a video.

Predictive analytics: The procedure of making forecasts about events or indefinite future applying dissimilar statistical package to evaluate historical or current facts.

Robotics: The ability of a machine to perform work without supervision, control, or monitoring using hardware and AI techniques.

Scene understanding: This is the process of analysing and interpreting 3D structure of a scene in relation to its spatial, layout, semantic association between objects and functions.

Semantics: The instant disambiguation or recognition of subjects and ideas in crude form, image, video, or text, and utilization of logic for further identification of novel facts or associations.

Speech processing, recognition, synthesis: Systems linked with analysis of speech signals, recognition, and language processing synthesis, in spoken words and converting into text for production of speech (Bergstein, 2017).

8.7 APPLICATION OF ARTIFICIAL INTELLIGENCE IN VARIOUS FIELDS

Banking and finance: Machine learning was previously incorporated into several features of financial systems: from internet banking, machine money withdrawal, assets management, risks assessment, and forex trading. Different algorithms to make business decision proficient and fast, fraud and insecurity activity.

Business: In the area of advertising, marketing, following trends, decision making, and products endorsement utilizing different algorithms of AI in fast real time.

Publication and documentation: In the last few decades, AI has improved data extraction, conversion, plus translation of documents and structuring. Improved protection of vital documents powered by document management systems for security and customer's data protection.

Manufacturing industry: Reduced cost of labour and production efficiency through artificial algorithms deployed for quality control, maintenance, reduced products complexity, plus decision making.

Medical sciences: Diagnostic systems, detection of diseases, drug delivery, and regeneration are the novel applications of machine learning.

Security: Cyber-security and surveillance, military requirements such as logistics, intelligence, planning, decision making, surveillance, and monitoring.

Telecommunications: Documentation, tracking, service optimization network performance, and customer service improvement.

Transportation: Transportation is seen as one of the biggest sectors of economic importance in the world. Reduced transport emission, cost, safety and traffic management autonomous vehicles, cargo ships, and automated package delivery (WIPO, 1991).

8.8 CONCLUSION AND FUTURE RECOMMENDATION TO KNOWLEDGE

This chapter has provided information on the application of deep learning techniques as well as their economic prospects in Medical and Pharmaceutical Biotechnology. There is a need for further research involving life sciences with their utilization in the discovery of drugs which are still limited. The application of emerging fields of molecular informatics could help in the provision of necessary basic information on protuberant deep learning techniques, as well as serve as a motivation that could be utilized for diverse application most especially for effective discovery in computer-assisted drug discovery and design. Specific information was provided on some other techniques on the Decision Trees, support vector machines, Naïve Bayesian, and k-Nearest Neighbours. Moreover, highlights on the process of genetic engineering in biotechnology projects and its socio-economic impact should be harnessed especially in developing countries. There is a need to also develop more robust framework that involves the application of convolutional networks, deep neural networks, and restricted Boltzmann machine networks.

REFERENCES

Adetunji, C.O., Anani, O.A., and Panpatte, D. (2021) Mechanism of actions involved in sustainable ecorestoration of petroleum hydrocarbons polluted soil by the beneficial microorganism. In: Panpatte, D.G., and Jhala, Y.K. (eds.), *Microbial Rejuvenation of Polluted Environment. Microorganisms for Sustainability*, vol 26. Springer, Singapore. https://doi.org/10.1007/978-981-15-7455-9_8

Adetunji, C.O., and Anani, O.A. (2021a) Recent advances in the application of genetically engineered microorganisms for microbial rejuvenation of contaminated environment. In: Adetunji C.O., Panpatte D.G., Jhala Y.K. (eds.), *Microbial Rejuvenation of Polluted Environment. Microorganisms for Sustainability*, vol 27. Springer, Singapore. https://doi.org/10.1007/978-981-15-7459-7_14

Adetunji C.O., and Anani O.A. (2021b) Utilization of microbial biofilm for the biotransformation and bioremediation of heavily polluted environment. In: Panpatte, D.G., and Jhala, Y.K. (eds.), *Microbial Rejuvenation of Polluted Environment. Microorganisms for Sustainability*, vol 25. Springer, Singapore. https://doi.org/10.1007/978-981-15-7447-4_9

Anani O.A., and Adetunji C.O. (2021c) Bioremediation of polythene and plastics using beneficial microorganisms. In: Adetunji C.O., Panpatte D.G., and Jhala Y.K. (eds.), *Microbial Rejuvenation of Polluted Environment. Microorganisms for Sustainability*, vol 27. Springer, Singapore. https://doi.org/10.1007/978-981-15-7459-7_13

Anani, O.A., Mishra R.R., Mishra P., Enuneku A.A., Anani G.A., and Adetunji C.O. (2020) Effects of toxicant from pesticides on food security: Current developments. In: Mishra P., Mishra R.R., Adetunji C.O. (eds.), *Innovations in Food Technology*. Springer, Singapore. https://doi.org/10.1007/978-981-15-6121-4_22

Adane, A. (2009). Agricultural biotechnology research and development in Ethiopia. *African Journal of Biotechnology*, 8: 71–97.

Alison, V. E. and Davis, U. C. (2009). *Potential Effects of Biotechnology on Animal Health and Well-Being. Applications of Animal Biotechnology in Animal Health*, Department of Animal Science, University of California, California, pp. 1–5.

Bach, X. T., Nghiem, S., Sahin, O., Manh Vu, T., Giang, H. H., Giang, T. V., Hai, Q. P., Hoa, T. D., Latkin, C. A., Tam, W., Ho, C. S. H., Roger C. M. (2019). Modeling research topics for artificial intelligence applications in medicine: Latent Dirichlet allocation application study. *Journal of Medical Internet Research* 21(11):e15511. doi: 10.2196/15511.

Baskin, I. I., Winkler, D., and Tetko, I. V. (2016). A rennaissance of neural networks in drug discovery. *Expert Opinion on Drug Discovery* 11:785–795.

Bergstein, B. (ed.) (2017). The artificial intelligence issue. *MIT Technology Review* 120(6).

Carbonell, T. R. and Héctor, G. M. (2019) Opportunities at the intersection of synthetic biology, machine learning, and automation Pablo. *ACS Synthetic Biology* 8: 1474–1477.

Chad, W. (2006). Economics and ethics in the genetic engineering of animals. *Harvard Journal of Law and Technology* 19: 415–424.

Clark, A. M., Ekins, S. (2015). Open Source Bayesian Models: 2. Mining A "big dataset" to create and validate models with ChEMBL. *Journal of Chemical Information and Modelling* 55:1246–1260.

Clark, A. M., Sarker, M., Ekins, S. (2014). New target predictions and visualization tools incorporating open source molecular fingerprints for TB Mobile 2.0. *Journal of Cheminformatics* 6 (1):38.

Conrad, G. and Harold, C. (2009). *Acceptable Genes. Religious, Traditional and Genetically Modified Foods.* State University of New York Press, Albany, pp. 135–140.

Dwivedi, N., Dwivedi, S., and Adetunji, C. O. (2021) Efficacy of microorganisms in the removal of toxic materials from industrial effluents. In: Adetunji C.O., Panpatte D.G., Jhala Y.K. (eds.), *Microbial Rejuvenation of Polluted Environment. Microorganisms for Sustainability*, vol 27. Springer, Singapore. https://doi.org/10.1007/978-981-15-7459-7_15

Ekins, S., Clark, A. M., Wright, S. H. (2015) Making transporter models for drug-drug interaction prediction mobile. *Drug Metabolism and Disposition* 43:1642–1645.

Ekins, S., Clark, A. M., Swamidass, S. J., Litterman, N., Williams, A. J. (2014a) Bigger data, collaborative tools and the future of predictive drug discovery. *Journal of Computer-Aided Molecular Design* 28(10):997–1008.

Ekins, S., Gupta, R. R., Gifford, E., Bunin, B. A., Waller, C. L. (2010). Chemical space: missing pieces in cheminformatics. *Pharmaceutical Research* 27(10): 2035–2039.

Ekins, S., Ecker, G. F., Chiba, P., Swaan, P.W. (2007) Future directions for drug transporter modelling. *Xenobiotica* 37(10):1152–1170.

Ekins, S., Freundlich, J. S. and Reynolds, R. C. (2014b) Are bigger data sets better for machine learning? Fusing single-point and dual-event dose response data for mycobacterium tuberculosis. *Journal of Chemical Information and Modelling* 54:2157–2165.

Fang, J., Zhu, X., Wang, C. and Shangguan, L. (2016) Applications of DNA technologies in agriculture. *Current Genomics* 17: 379–386.

Fröhlich, H., Balling, R., Beerenwinkel, N., Kohlbacher, O., Kumar, S., Lengauer, T., Maathuis, M. H., Moreau, Y., Murphy, S. A., Przytycka, T. M., Rebhan, M., Röst, H., Schuppert, A., Schwab, M., Spang, R., Stekhoven, D., Sun, J., Weber, A., Ziemek, D. and Zupan, B. (2018). From hype to reality: data science enabling personalized medicine. *BMC Medicine* 16:150 https://doi.org/10.1186/s12916-018-1122-7.

Giles, M. (2018). The GAN father: The man who's given machines the gift of imagination. *MIT Technology Review* 121(2): 49–53.

Guoyao, W. U. and Fuller, W. B. (2019). Application of new biotechnologies for improvements in swine nutrition and pork production. *Journal of Animal Science and Biotechnology* 10: 28. https://doi.org/10.1186/s40104-019-0337-6.

Gupta, S., and Savaliya, C. V. (2012). Application of biotechnology to improve livestock products. *Veterinary World* 5(10): 634–638. doi: 10.5455/vetworld.2012.634-638.

Islam, S., Thangadurai, D., Adetunji, C. O., Nwankwo, W., Kadiri, O., Makinde, S., Michael, O. S., Anani, O. A., and Adetunji, J. B. (2021) Nanomaterials and nanocoatings for alternative antimicrobial therapy. In: Kharissova O.V., Martínez L.M.T., Kharisov B.I. (eds.), *Handbook of Nanomaterials and Nanocomposites for Energy and Environmental Applications*. Springer, Cham. https://doi.org/10.1007/978-3-030-11155-7_3-1

Jackwood, M. W., Leslie, H., Kapil, S. and Sanjay, K. (2008). Vaccine development using recombinant DNA technology. *Council for Agricultural Science and Technology* 38.

Lai, L., Kang, J., Li, R., Wang, J., Witt, W., Yong, H., Hao, Y., Wax, D., Murphy, C., Rieke, A., Samuel, M., Linville, M., Korte, S., Evans, R., Starzl, T., Prather, R. and Dai, Y. (2006). Generation of cloned transgenic pigs rich in omega-3 fatty acids. *Nature Biotechnology* 24:435–436.

LeCun, Y., Y. Bengio and G. Hinton (2015). Deep learning. *Nature* 521: 436–444.

Madan, M.L. (2005). Animal biotechnology: Applications and economic implications in developing countries. *Revue Scientifiqueet Technique Office International Des Epizooties*, 24:127–139.

Matherny, M., Israni, S. T., Ahmed, M. and Whicher, D. (2019). *Artificial Intelligence in Health Care: The Hope, the Hype, the Promise, the Peril.* NAM Special Publication. Washington, DC, National Academy of Sciences.

Mebratu, A., Habtamu, B. and Mekonnen, G. (2014). Genetic engineering in animal production: Applications and prospects. *Biochemistry and Biotechnology Research* 2(2): 12–22.

Minssen, T., Gerke, S., Aboy, M., Price, N. and Cohen, G. (2020). Regulatory responses to medical machine learning. *Journal of Law and the Biosciences* 1–18. doi:10.1093/jlb/lsaa002.

Mitchell, J. B. (2014). Machine learning methods in chemoinformatics. *Wiley Interdisciplinary Reviews: Computational and Molecular Science* 4(5):468–481.

Mishra, P., Mishra, R. R., & Adetunji, C. O. (Eds.). (2020). *Innovations in Food Technology: Current Perspectives and Future Goals.* Springer Nature Singapore Pte Ltd. 2020. Springer, Singapore. pp. 143-162. 1st Edition. DOI: 10.1007/978-981-15-6121-4

Nwankwo, W., Olayinka, A.S., Ukhurebor, K.E. (2020). Nanoinformatics: Why Design of Projects on Nanomedicine Development and Clinical Applications may Fail? *Proceeding of the 2020 International Conference in Mathematics, Computer Engineering and Computer Science (ICMCECS)*, Lagos, Nigeria, IEEE Xplore, 1–7.

Nyerhovwo, J. T., (2004). Fostering biotechnology entrepreneurship in developing countries. *African Journal of Biotechnology* 3:291–300.

Olaniyan, O.T., and Adetunji, C.O. (2021) Biochemical Role of Beneficial Microorganisms: An Overview on Recent Development in Environmental and Agro Science. In: Adetunji C.O., Panpatte D.G., Jhala Y.K. (eds.), *Microbial Rejuvenation of Polluted Environment. Microorganisms for Sustainability*, vol 27. Springer, Singapore. https://doi.org/10.1007/978-981-15-7459-7_2

Prakash, C. S. (2002). Look Mom! No Antibiotic Resistance Marker Genes! Available at http://www.biotechinfo.net/LOOK_MOM.html.

Rost, B., Radivojac, P., and Bromberg, Y. (2016). Protein function in precision medicine: Deep understanding with machine learning. *FEBS Letters* 590(15):2327–2341.

Rothschild, M., Jacobson, C., Vaske, D., Tuggle, C., Short, T., Sasaki, S., Eckardt, G. and McLaren, D, (1994). A major gene for litter size in pigs. *Proceedings of the Fifth World Congress Genetics Applied to Livestock Production* 21: 225–228.

Scott, G. and Mattew, B. (2011). *Genetically Engineered Animals and Public Health.* Compelling Benefits for Health Care, Nutrition, the Environment, and Animal Welfare, pp. 22–30.

Sears, M. K., Hellmich, R. L., Stanley-Horn, D. E., Oberhauser, K. S., Pleasants, J. M., Mattila H. R., Siegfried, B. D. and Dively, G. P. (2001). Impact of Bacillus thuringiensis corn pollen on monarch butterfly populations: A risk assessment. *Proceedings of the National Academy of Science* 98: 11937–11942.

Sheridan, C. (2011). Gene therapy finds its niche. *Nature Biotechnology* 29: 121–128.

Streatfield, S. J. (2005). Plant-based vaccines for animal health. *Revue Scientifiqueet Technique-Office International Des Epizooties* 24:189–199.

Tereje, T., Jack, H. (2007). Genetic engineering and omitted health research: Still no answers to ageing questions. *TWN Biotechnology and Biosafety Series* 7:1–14.

Thangadurai, D., Dabire, S. S., Sangeetha, J., Said Al-Tawaha, A. R. M., Adetunji, C. O., Islam, S., Shettar, A. K., David, M., Hospet, R., Adetunji, J. B. (2021) Greener composites from plant fibers: Preparation, structure, and properties. In: Kharissova O.V., Martínez L.M.T., and Kharisov B.I. (eds.), *Handbook of Nanomaterials and Nanocomposites for Energy and Environmental Applications*. Springer, Cham. https://doi.org/10.1007/978-3-030-11155-7_21-1

Uddin, M., Wang, Y., and Woodbury-Smith, M. (2019). Artificial intelligence for precision medicine in neurodevelopmental disorders. *npj Digital Medicine* 2:112. https://doi.org/10.1038/s41746-019-0191-0.

Wang, M. L., Barkley, N. A. and Jenkins, T. M. (2009). Microsatellite markers in plants and insects. Part I: Applications of biotechnology. Genes, genome, and genomics. *Global Science Books* 3(1): 54–67.

WIPO (1991). Conference Proceedings of the WIPO Worldwide Symposium on the Intellectual Property Aspects of Artificial Intelligence, Stanford University, Stanford CA, March 25–27.

Zhavoronkov, A. and Mamoshina, P. (2019) Special issue: Rise of machines in medicine science & society deep aging clocks: The emergence of AI-based biomarkers of aging and longevity. *Trends in Pharmacological Sciences* 40(8): 546–549.

Zhu, H., Zhang, J., Kim, M.T., Boison, A., Sedykh, A., Moran, K. (2014) Big data in chemical toxicity research: The use of high-throughput screening assays to identify potential toxicants. *Chemical Research in Toxicology* 27(10):1643–1651.

9 Health and Nutritional Benefits of Numerous Bioactive Components Derived from Fish and Fish Products

Juliana Bunmi Adetunji
Osun State University

Charles Oluwaseun Adetunji
Edo State University Uzairue

Michael Olugbenga Samuel
Bowen University

Genevieve Dable-Tupas
Davao Medical School Foundation, Inc. (DMSFI)

Chukwuebuka Egbuna
Chukwuemeka Odumegwu Ojukwu University
University of Port Harcourt

CONTENTS

DOI: 10.1201/9781003178903-11

9.1 INTRODUCTION

Advances in understanding human dietetics and nutrition, food biotechnology, and food technology have led to an elevation in food product enhancement in recent times. It has been forecasted that there will be an increase in growth of population globally in the year 2030. Increase in population growth would also correspond to a rise in demand for food and other basic necessities, but lifestyle diseases could worsen as well (Adetunji and Anani, 2021a,b,c; Adetunji et al., 2021; Dwivedi et al., 2021; Olaniyan and Adetunji, 2021, Mishra et al., 2020; Thangadurai et al., 2021; Anani et al., 2020; Islam et al., 2021). With the understanding that prevention is better than cure, more and more people tend to look for and consume foods with health benefits. This is where functional foods can be beneficial to mankind. Most people now agree that it is better to eat functional foods that could boost their immunity and increase their resistance to various diseases. This has resulted into improved acceptance of functional foods that could enhance and promote human health (Hasler 2002; de Jong et al., 2003). Even government and policy makers have now recognized the significance of functional foods in supporting public health through joint effort of various sectors from food, pharmaceutical, and various national research centers of excellence dealing with fisheries and agriculture (Department of Agriculture, Fisheries and Forestry, 2003).

Seafood has been highlighted as a depository of numerous bioactive and functional components that are necessary for the sustainability of human health. Moreover, many of these sea foods and their by-products are richly endowed with bioactive and nutraceuticals that can be obtained or extracted and added to different types of food so as to increase the functionality of the food as well as their benefits to human health (Alasalvar and Taylor, 2002; Shahidi, 2003).

Fish has been recognized as the only hope and rich food for the poor people which contains an array of essential components including minerals, proteins, vitamins, and fats (especially omega-3 PUFAs). Some of the health benefits that could be derived from regular intake of sea foods as a typical example of inherently functional and as functional agents include prevention of high blood pressure, vitamin and mineral deficiencies, arthritis, blood clotting problems, inflammatory diseases, cancer, and promotion of cardiovascular health (Gormley, 2006).

In view of the aforementioned, this chapter highlights some prominent biomolecules available in fish and fish products that have been used for the prevention and management of chronic ailments, with emphasis on in vivo and in vitro biological actions of bioactive components including formulation of stable products from fish.

9.2 PEPTIDES FROM FISH AND FISH PRODUCTS

Bioactive peptides obtained from the hydrolysis of proteins are recently gaining a lot of recognition. Some of these peptides are derived from fish species and normally absorbed within the villi present in the small intestines particularly the intestinal enterocytes which are later absorbed into the bloodstream and subsequently produce significant biological effects on human body function and condition. The discovery of peptides obtained from fish has now opened an opportunity for the development of biotechnological products with various bioactive properties. Some of these peptides have shown many significant biological activities including antihypertensive, antioxidative, and antimicrobial effects among others (Kris-Etherton et al., 2004 and Ibañez et al., 2012). Fish protein hydrolysates derived from fish species like mackerel have been shown to display antioxidant action (Je et al., 2007). Lysine and methionine have also been highlighted as the two major amino acids available in large quantities in fish protein (Hamed et al., 2015).

Increase in blood pressure or hypertension has been recognized as a major menace responsible for cardiovascular illnesses. The application of functional foods and natural health products produced by the nutraceutical industries can perform a noteworthy function in the management of cardiovascular sicknesses. This could be linked to the fact that these functional foods can serve as one of the cheapest alternative therapy that could be utilized for the treatment of cardiovascular diseases (Muir, 2005). Fish has been shown to possess some antihypertensive peptides regarded as angiotensin I-converting enzyme inhibitors (Gormley, 2006). Most of these bioactive constitutes are enclosed in the hydrolysates obtained from fish wastes.

De Leo et al. (2009) observed that angiotensin I-converting enzyme inhibitors obtained from fish products possess the capability to reduce blood pressure. This could be linked to the fact that inhibition of angiotensin I-converting enzyme could reduce the vaso-constricting influence of Angiotensin II and enhance the vasodilatory effect of bradykinins (Gormley, 2006). Je et al. (2008) isolated novel bioactive peptides through the aid of enzymatic hydrolysis of big-eye tuna dark muscle. The bioactive peptides were later purified through the help of mass spectrometry and the most active peptide was characterized as H-Leu-Asn-Leu-Pro-Thr-Ala-Val-Tyr-Met-Val-Thr-OH having a molecular weight of MW 1222Da. The isolated peptide portends the capability to remove any available cellular free radicals and improved the capability of ter-butyl-hydroperoxide-induced cytotoxicity while it also prevented PUFA peroxidation.

Vercruysse et al. (2005) reported the effect of angiotensin converting enzyme (ACE) inhibitors obtained from enzymatic hydrolysates of fish muscle and claimed that these peptides obtained from fish muscle could serve as functional foods. Berge (2005) also highlighted the significance of enzyme-treated fish protein hydrolysates as potential anti-atherogenic agents. Furthermore, it has been observed that Polyl Endopeptide (PEP) inhibitors could inhibit the effect of the enzyme prolyl endopeptidase. The availability of a large amount of prolyl endopeptidase could result in a disturbance in memory and cognition, neuro-degeneration, and some neurological disorders like autism, depression, and schizophrenia (Husain and Nemeroff, 1990). It has also been observed that high rate of PEP activity could be linked to Alzheimer's patients having psychiatric disorders. Therefore, the existence of PEP inhibitors obtained from trout flesh, cod, and salmon could form permanent and sustainable functional food that could be used for the maintenance of the health status of humans (Sorensen, et al., 2004).

9.3 VITAMINS FROM FISH AND FISH PRODUCTS

Vitamins are among the essential components that the human body needs for normal physiological function. Fish oil is richly endowed with numerous vitamins which include vitamins A, D, and E. A typical source of vitamin A is liver oils mostly from cod liver oils and halibut. Sardine fish has about 500 IU of vitamin D, 4,500 IU of vitamin A, and an average of 125 μg/g of oil. It has been noted that there is high bioavailability of vitamin A in small fish types, although species like salmon, herring, lake trout, and mackerel also contain vitamin D in their tissues. It has been observed that people who ingest at least 40% fish products in their diet might not be more susceptible to diseases when compared to people who consume little or no amounts of fish or fish products. Also, one of the dietary sources of vitamin D is fatty fish. Some of the deficiency of vitamin D causes osteomalacia in a mature human being while it causes rickets in little progeny. The major bioactive vitamins present in fish are vitamins A and D (Anon, 2017).

9.4 LIPIDS FROM FISH AND FISH PRODUCTS

Polyunsaturated fatty acids (PUFA) derived from fish have been revealed to possess some significant bioactive components that could promote the health status of a human being when consumed (Bahri et al., 2002). It has been validated that PUFAs that originate from fish could be used in the management of blood clotting, cardiovascular disease, high blood pressure (Morris et al., 1993), enhanced cognitive purpose (Kalmijn et al., 2004), and as anticancer agents (Wolk et al., 2006; English et al., 2004). Castro-Gonzalez et al. (2007) evaluated ω-3 PUFA content available in 18 Mexican marine fish and their application as a typical example of functional food. Results obtained showed that Picuda and Sargo showed high values of ω-3 PUFA which ranged from 200 to 300 mg/100 g. These two types of fish have been suggested to be among the essential functional fish species. Huang et al. (2009) also validated the effect of ω-3-PUFAs obtained from fish oil and fish as a significant functional food necessary for signal-transduction pathways and for the regulation of membrane lipid composition.

Tan et al. (2012) established that eicosapentaenoic acid (EPA)/ docosahexaenoic acid (DHA) rich diet could decrease the incidence of brain aging. This was carried out using 1,500 dementia-free patients with an average age of 67 years. Each patient underwent a brain scan and evaluation for body mass, EPA/DHA available in their red blood cells, and mental functioning. The group was separated into four depending on the amount of EPA/DHA. Results showed that those in the bottom quartile had reduced brain volumes and also observed to have reduction in visual memory evaluation and executive capabilities, such as difficulty in deciphering and multi-tasking compared to those in the upper three quartiles. Virtanen et al. (2013) affirmed that the amount of ω-3 PUFAs circulating in the blood is a typical example of a biomarker as a result of fish consumption that could prevent certain levels of subclinical brain dysfunction normally observed in adult human beings.

9.5 MINERALS FROM FISH AND FISH PRODUCTS

Fish contains minerals and numerous trace elements. One example of these essential minerals includes calcium (Ca). The presence of adequate amount of Ca has been observed to be important for the development of bone during pregnancy and lactation. Deficiency of Ca might lead to the development of diseases (Gormley, 2012; Iji and Kadam, 2013; Nguyen et al., 2011).

9.6 IN VITRO AND IN VIVO BIOLOGICAL ASSAYS USING FISH AND FISH PRODUCTS

9.6.1 Anticancer Activity

Najafian and Babji (2012) evaluated the effect of omega-3 fatty acids among mice induced with lung cancer in an in vivo assay. The following factors were evaluated such as lung metastasis, tumor growth, and body weight. Soya beans were used as control during the experiment. The result obtained shows that fish oil prevented and decreased the metastatic load. Furthermore, the anticancer effect obtained from fish oil enriched with omega-3 PUFAs (EPA and DHA) could be linked to the fact that these bioactive constituents could inhibit the growth of these cancer cells. Also, the inclusion of EPA in the diet has led to the reduction in size and number of rectal polyps among affected individuals with familial adenomatous polyposis. It was also noted that familial adenomatous polyposis was repressed to a level that could be comparable to celecoxib synthetic drug (Gormley, 2012). Picot et al. (2006) evaluated the antiproliferative influence of fish protein hydrolysates against human breast cancer cell lines. Results showed that the protein hydrolysates have the capability to inhibit the spread of these cancer cells. The bioactive component comprises of free amino acids and peptides which have different molecular weights with 7 kDa along with small sodium chloride and small quantities of lipids.

9.6.2 Antibacterial Activity

Lirio et al. (2019) assessed the antimicrobial influence of mucus obtained from three freshwater fishes (*Clarias batrachus*, *Oreochromis niloticus*, *Channa striata*) against human and fish pathogens. This was performed using microbroth plate dilution and minimum inhibitory concentration by agar-overlay diffusion respectively. The result obtained indicates that all the fish mucus extracts exhibit antimicrobial effects against all the tested bacteria which were *Enterococcus faecalis*, *Aeromonas hydrophila*, *Klebsiella pneumonia*, *Serratia marcescens*, *Pseudomonas aeruginosa*, and *Escherichia coli*. Moreover, the mucus obtained from catfish exhibited an inhibitory effect against Gram-positive *Micrococcus luteus* and *Staphylococcus aureus* when evaluated at the smallest concentration (1:4 dilutions). The study shows that the antimicrobial activity obtained from freshwater fish mucus could be utilized against some medically imperative pathogens.

Simplice et al. (2018) evaluated the antibacterial effect of oils obtained from two species of fish namely: *Hepsetus odoe* and *Chrysichthys nigrodigitatus* against food poisoning bacteria. The chemical constituents of the essential oil were determined using gas chromatography coupled to flame ionization detector, while the antibacterial effect was assessed using broth microdilution techniques. The antimicrobial action of the oil acquired from these fishes might be linked to the presence of docosahexaenoic acid (C22:6), eicosapentaenoic acid (C20:5), ω-6 family; arachidonic acid (C20:4), and linolenic acid (C18:3). Furthermore, *C. nigrodigitatus* oil acquired by maceration and pressing techniques indicated MIC values which varied from 32 to 64 mg/ml, while the oil obtained from *H. odoe* by pressing techniques showed MIC values which varied between 8 and 64 mg/ml respectively. This research indicated that the oil obtained from *C. nigrodigitatus* and *H. odoe* contain some antimicrobial compounds (Omega-3 and Omega-6 fatty acids) which might be the active components exhibiting antimicrobial activity against food poisoning bacteria.

Warren et al. (2018) composed a widespread report on the antimicrobial efficiency of omega-3 PUFAs when evaluated against microbial pathogens. The authors highlighted the need to screen for a more potent antimicrobial agent that might be available in the omega-3 PUFAs which could serve as a potential replacement to synthetic drugs used for the treatment of various microorganisms and opportunistic pathogens which are responsible for microbial infections. Moreover, many of such microorganisms have developed significant resistance to numerous antibiotics. Therefore, there is a need to isolate and identify many phytochemical dietary compounds that possess antimicrobial activities. The significance of linoleic acid and its by-products which are docosahexaenoic acid and eicosapentaenoic acid, respectively, rests on their possible use in the inhibition and management of human infections and illnesses particularly affecting the cardiovascular system, the eyes, and in the development of the brain and growth of human beings. The authors suggested that there is a need to establish the modes of action through which these antimicrobial agents from omega-3 PUFAs exhibit their effectiveness against the tested pathogens as well as their level of toxicity and route of administration.

Mil Homens et al. (2016) studied the antimicrobial influence of fish oil-based formulas containing omega-3 fatty acids containing triacylglycerols, ethyl esters, and free fatty acids against *Pseudomonas aeruginosa* PAO1 and *Burkholderia cenocepacia* K56-2. These pathogenic microorganisms were selected because they have demonstrated multiple resistances to numerous antibiotics. The result obtained showed that fatty acids containing docosahexaenoic acid and eicosapentaenoic acid exhibited the highest microbial activity against all the tested pathogens with a MIC value of 187% v/v. Another study was implemented in order to ascertain the efficacy of fish oil containing omega3 fatty acids as a prophylactic and therapeutic agent as well as their effectiveness against *Galleria mellonella* caterpillar model of infection. The management of diseased larvae with a single dose after 7 h of post-infection improved the persistence of larvae, which might be linked to the presence of free fatty acid. Their investigation established the potential of fatty acid present in the fish oil as an effective antimicrobial agent against *Burkholderia* spp. and *P. aeruginosa*.

Rybin et al. (2000) evaluated the antimicrobial effectiveness of fish oil obtained from different sources. The influence of different factors such as the content of oil oxidation products during preparation and the level of unsaturation of the initial oil were assessed. The level of antimicrobial activity from the preparation was carried out using oil. The result revealed that 33.95% of PUFAs obtained from *Sardinops melanostica* (Sardine) showed the highest antimicrobial activity. The antimicrobial activity was carried out in water-soluble oil oxidation products.

9.6.3 Antioxidant Activity

Antioxidants are compounds that portend the capability to nullify the effect of free radicals present in the human body. This implies that antioxidants have the capability to defend the human body against sickness and oxidative stress (Jun et al., 2004) and affirms that an antioxidant could be used in the management of cancer, diabetes, and inflammatory disorders. Reactive nitrogen species and reactive oxygen species are naturally formed in the body all through the process of cellular metabolism and have the capability to adversely affect the cell metabolism in the human body which might lead to the initiation of several diseases. These changes can trigger the body to produce its own antioxidants in an effort to nullify the effect of oxidants; however, endogenous antioxidants may not be enough to do the job. Therefore, there might be a need for the introduction of exogenous antioxidants from functional foods like eggs, fish, and other proteinaceous substances in order to assist the body to nullify the effect of these substances (Aluko, 2012). A typical example of such is fish protein hydrosalates obtained from fish species such as mackerel, Alaskan Pollock, tuna, and yellowfin sole. These have been confirmed to possess significant amounts of antioxidant agents (Jospeh et al., 2011). Majority of these peptides or hydrosalates have been obtained from fish derived from bone, muscles, skin, scales, and viscera.

Fish hydrolysates or peptides can potentially protect the human body against reactive nitrogen and oxygen species which have the capability to attack membrane lipids, DNA, and proteins resulting in the development of degenerative diseases like cancer, obesity, dementia, diabetes, etc. (Ryan et al., 2011).

There is a need to intensify effort on the isolation of natural antioxidants as potential replacement to synthetic antioxidants which has numerous disadvantages associated with synthetic antioxidants. Raghavan et al. (2008) validate these potentials as follows: fish protein hydrolysates/peptides obtained from mackerel exhibited antimicrobial activity in vitro. This also corroborates the observation of Jospeh et al. (2011) who discovered that peptides obtained from hydrolysis of the muscles of mackerel in the presence of protease enzyme prevented the auto-oxidation of linoleic acid. The authors also affirmed the antioxidant efficacy of fish protein obtained from tilapia.

Mozaffarian et al. (2011) also validated the effectiveness of low molecular weight peptides hydrolysates obtained from the muscles of salmon fish portends the capability to enhance the rate of hydrolysis of salmon fish muscle, which in turn amplified the free radical scavenging proficiency of the hydrolysate obtained from salmon fish muscle. This could be linked to the capability of tuna fish protein hydrolysate to demonstrate its potential to avoid hydroxyl radical-induced DNA impairment. Moreover, another interesting known fact is that DNA damage has been shown to be responsible for the development of many diseases. Therefore, consistent consumption of fish would go a long way in the prevention and/or in the treatment of many human diseases and sicknesses (Aluko, 2012).

9.7 CONCLUSION AND FUTURE OUTLOOK

This chapter provided detailed information on some important biomolecules available in fish and fish products that have been used for the prevention and management of chronic ailments, with emphasis on in vivo and in vitro biological actions of bioactive components, including formulation of stable products from fish. Therefore, there is a need to also build a library that will contain all the useful information on the biological active constituents that are present in Nigerian local fishes. There is a need to perform structural elucidation of the active components available in all these fish and fish products using nuclear magnetic resonance (NMR), Liquid chromatography mass spectrometry (LCMS), Fourier transform Infrared spectroscopy (FTIR), thin layer chromatography (TLC), etc. Moreover, the application of genetic engineering and structural biology will play a crucial role toward the development of more relevant biological active constituents that might possess high therapeutic properties and release of essential nutrients and minerals that might help in fighting against food insecurity and malnutrition. Moreover, the application of metabolomics, genomics, and proteomics will also give a better insight on how to improve the amount of bioactive components in fishes most especially those that have enormous healing properties. In conclusion, this study has provided information on the numerous biological active components contained in fish and fish products that could be used for the prevention and management of diseases and could also serve as functional foods in replacement of synthetics drugs.

REFERENCES

Adetunji, C.O., Anani, O.A., and Panpatte, D. (2021) Mechanism of actions involved in sustainable ecorestoration of petroleum hydrocarbons polluted soil by the beneficial microorganism. In: Panpatte, D.G., and Jhala, Y.K. (eds.), *Microbial Rejuvenation of Polluted Environment. Microorganisms for Sustainability*, vol 26. Springer, Singapore. https://doi.org/10.1007/978-981-15-7455-9_8

Adetunji, C.O., and Anani, O.A. (2021a) Recent advances in the application of genetically engineered microorganisms for microbial rejuvenation of contaminated environment. In: Adetunji C.O., Panpatte D.G., Jhala Y.K. (eds.), *Microbial Rejuvenation of Polluted Environment. Microorganisms for Sustainability*, vol 27. Springer, Singapore. https://doi.org/10.1007/978-981-15-7459-7_14

Adetunji C.O., and Anani O.A. (2021b) Utilization of microbial biofilm for the biotransformation and bioremediation of heavily polluted environment. In: Panpatte, D.G., and Jhala, Y.K. (eds.), *Microbial Rejuvenation of Polluted Environment. Microorganisms for Sustainability*, vol 25. Springer, Singapore. https://doi.org/10.1007/978-981-15-7447-4_9

Anani O.A., and Adetunji C.O. (2021c) Bioremediation of polythene and plastics using beneficial microorganisms. In: Adetunji C.O., Panpatte D.G., and Jhala Y.K. (eds.), *Microbial Rejuvenation of Polluted Environment. Microorganisms for Sustainability*, vol 27. Springer, Singapore. https://doi.org/10.1007/978-981-15-7459-7_13

Anani, O.A., Mishra R.R., Mishra P., Enuneku A.A., Anani G.A., and Adetunji C.O. (2020) Effects of toxicant from pesticides on food security: Current developments. In: Mishra P., Mishra R.R., Adetunji C.O. (eds.), *Innovations in Food Technology*. Springer, Singapore. https://doi.org/10.1007/978-981-15-6121-4_22

Alasalvar, C. and Taylor, Y. (eds). (2002). *Seafoods – Quality Technology and Nutraceutical Applications*. Springer-Verlag, Heidelberg, Germany, 508 p.

Aluko, R. E. (2012). Functional Foods and Nutraceuticals. Springer. 121–126.

Anon (2017). https://www.delamaris.si/healthy-diet/that-fish-are-an-important-source-of-vitamins-and-minerals.

Bahri, N., Gusko, A., Hamm, M., Kasper, H., Klor, H.U., Neuberger, D. and Singer, P. (2002). Significance and recommended dietary intake of long-chain omega-3 fatty acids- A consensus statement of the omega-3 working group. *Ernahrungs-Umschau* 49(3), 94–98.

Berge, R. (2005). Fish protein hydrolysates. Patent Number WO 2005002605 A1.

Castro-Gonzalez, M.I., Ojeda, V.A., Montano, B.S., Ledesma, C.E. and Perez-Gil, R.F. (2007). n-3 fatty acid evaluation in eighteen Mexican marine fishes as functional food. *Archivos Latino Americanos de Nutricion* 57(1): 85–93.

de Jong, N., Ocke, M. C., Branderhorst, H. A. and Friele, R. (2003). Demographic and lifestyle characteristics of functional food consumers and dietary supplement users. *British Journal of Nutrition* 89:273–81.

De-Leo, F., Panarese, S., Gallerani, R. and Ceci, L.R. (2009). Angiotensin converting enzyme (ACE) inhibitory peptides. *Current Pharmaceutical Design* 15(31), 3622–3643.

Department of Agriculture, Fisheries and Forestry (2003). The National Food Industry Strategy. http://www.nfis.com.au. Accessed 24 April 2021.

Dwivedi, N., Dwivedi, S., and Adetunji, C. O. (2021) Efficacy of microorganisms in the removal of toxic materials from industrial effluents. In: Adetunji C.O., Panpatte D.G., Jhala Y.K. (eds.), *Microbial Rejuvenation of Polluted Environment. Microorganisms for Sustainability*, vol 27. Springer, Singapore. https://doi.org/10.1007/978-981-15-7459-7_15

English, D.R., MacInnis, R.J., Hodge, A.M., Hopper, J.L., Haydon, A.M. and Giles, J.G. (2004). Red meat, chicken and fish consumption and risk of colorectal cancer. *Cancer Epidemiology Biomarkers and Prevention* 13(9), 1509–1514.

Gormley, T.R. (2006). Fish as a functional food. *Food Science and Technology* 20(3), 25–28.

Gormley R. (2012). Fish for health and life. *Sea Health UCD* 1–2.

Hamed, I., Özogul, F., and Özogul, Y. (2015). Regenstein JM. Marine bioactive compounds and their health benefits: A review. *Comprehensive Reviews in Food Science and Food Safety*. 14(4), 446–465.

Hasler, C. (2002). Functional foods: Benefits, concerns and challenges: A position paper from the American Council on Science and Health. *Journal of Nutrition* 132, 3772–3781.

Huang, T., Sinclair, A.J., Shen, L.R., Yang, B. and Li, D. (2009). Comparative effects of tuna oil and salmon oil on liver lipid metabolism and fatty acid concentration in rats. *Journal of Food Lipids* 16(4), 436–451.

Husain, M. M. and Nemeroff, C. B. (1990). Neuro-degeneration and disturbance in memory cognition. *Journal of the American Geriatric Society* 38 (8), 918–925.

Ibañez, E., Herrero, M., Mendiola, J. A., Castro-Puyana, M. (2012). *Extraction and Characterization of Bioactive Compounds with Health Benefits from Marine Resources: Macro and Micro Algae, Cyanobacteria, and Invertebrates. Marine Bioactive Compounds*. Springer, 55–98.

Islam, S., Thangadurai, D., Adetunji, C. O., Nwankwo, W., Kadiri, O., Makinde, S., Michael, O. S., Anani, O. A., and Adetunji, J. B. (2021) Nanomaterials and nanocoatings for alternative antimicrobial therapy. In: Kharissova O.V., Martínez L.M.T., Kharisov B.I. (eds.), *Handbook of Nanomaterials and Nanocomposites for Energy and Environmental Applications*. Springer, Cham. https://doi.org/10.1007/978-3-030-11155-7_3-1

Iji, P. and Kadam, M. (2013). *Prebiotic Properties of Algae and Algae-Supplemented Products. Functional Ingredients from Algae for Foods and Nutraceuticals Cambridge, UK.* Woodhead Publishing, 768.

Je, J. Y., Qian, Z. J., Byun, H. G., Kim, S. K. (2007). Purification and characterization of an antioxidant peptide obtained from tuna backbone protein by enzymatic hydrolysis. *Process Biochemistry* 42(5):840–846.

Je, J.Y., Qian, Z.J., Lee, S.H., Byun, H.G., and Kim, S.K. (2008). Purification and antioxidant properties of Bigeye tuna (*Thunnus obesus*) dark muscle peptide on free radical-mediated oxidative systems. *Journal of Medicinal Food* 11(4), 629–637.

Jun, S. Y., Park, P. J., Jung, W. K., Kim, S. K. (2004). Purification and characterization of an antioxidative peptide from enzymatic hydrolysate of yellowfin sole (*Limanda aspera*) frame protein. *European Food Research and Technology* 219(1), 20–26.

Kalmijn, S., van Boxtel, M.P.J., Ocke, M., Verschuren, D., Kromhout, D. and Launer, L.J. (2004). Dietary intake of fatty acids and fish in relation to cognitive performance at middle age. *Neurology* 2(2), 275–280.

Kris-Etherton, P., Lefevre, M., Beecher, G., Gross, M., Keen, C. L., Etherton, T. (2004). Bioactive compounds in nutrition and health research methodologies for establishing biological function: The antioxidant and anti-inflammatory effects of flavonoids on atherosclerosis. *Annual Review of Nutrition* 24, 511–538.

Lirio, G. A. C., De Leon, J. A. A., Villafuerte, A. G. (2019). Antimicrobial activity of epidermal mucus from top aquaculture fish species against medically-important pathogens. *Walailak Journal of Science and Technology*. 16 (5), 311–340.

Mishra, P., Mishra, R. R., and Adetunji, C. O. (Eds.). (2020). *Innovations in Food Technology: Current Perspectives and Future Goals* (1st Ed.) Springer; https://doi.org/10.1007/978-981-15-6121-4.

Mil-Homens, D., Ferreira-Dias, S., Fialho, A. M. (2016). Fish oils against *Burkholderia* and *Pseudomonas aeruginosa*: *in vitro* efficacy and their therapeutic and prophylactic effects on infected *Galleria mellonella* larvae. *Journal of Applied Microbiology* 120 (6), 1509–1519.

Morris, M.C., Sacks, F. and Rosner, B. (1993). Does fish oil lower blood pressure? A meta-analysis of controlled trials. *Circulation* 88(2), 523–533.

Mozaffarian, D., Wu, J. H. (2011). Omega-3 fatty acids and cardiovascular disease: Effects on risk factors, molecular pathways, and clinical events. *Journal of the American College of Cardiology* 58(20), 2047–2067.

Muir, A.D. (2005). Natural peptides in blood pressure control. *Agro Food Industry Hi-Technology* 16(5), 15–17.

Najafian, L. and Babji, A. S. (2012). A review of fish derived antioxidant and antimicrobial peptides: Their production, assessment, and applications. *Peptides* 33(1), 178–185.

Nguyen, M. H. T., Jung, W. K. and Kim, S. K. (2011). Marine algae possess therapeutic potential for Camineralization via osteoblastic differentiation. *Advances in Food and Nutrition Research* 64, 429–441.

Olaniyan, O.T., and Adetunji, C.O. (2021) Biochemical role of beneficial microorganisms: An overview on recent development in environmental and agro science. In: Adetunji, C.O., Panpatte, D.G., and Jhala, Y.K. (eds) *Microbial Rejuvenation of Polluted Environment. Microorganisms for Sustainability*, vol 27. Springer, Singapore. https://doi.org/10.1007/978-981-15-7459-7_2

Picot, L., Bordenave, S., Didelot, S., Fruitier-Arnaudin, I., Sannier, F., Thorkelsson, G., et al. (2006). Antiproliferative activity of fish protein hydrolysates on human breast cancer cell lines. *Process Biochemistry* 41(5), 1217–1222.

Raghavan, S., Kristinsson, H. G. and Leeuwenburgh, C. (2008). Radical scavenging and reducing ability of tilapia (*Oreochromis niloticus*) protein hydrolysates. *Journal of Agricultural and Food Chemistry* 56, 10359–10367.

Ryan, J. T., Ross, R. P., Bolton, D., Fitzgerald, G. F., Stanton C. (2011). Bioactive peptides from muscle sources: meat and fish. *Nutrients* 3(9), 765–791.

Rybin, V.G., Shul'gina, L.V., Kuklev, D.V., Rybin, V. G., Byval'tseva, T. M., Blinov, Y. G., Davletshina, T. A., Akulin, V. N. (2000). Activity of a new antimicrobial preparation from fish oils from various sources: Effects of different factors. *Applied Biochemistry and Microbiology* 36, 425. https://doi.org/10.1007/BF02738056.

Shahidi, F. (2003). Nutraceuticals and bioactives from seafood products. In: Bechtel, P. J. (ed.) *Proceedings of 2nd International Seafood By-Products Conference*. Alaska Sea Grant College Programme, Fairbanks, Alaska, 247–263.

Simplice, M. R., Macaire, W. H., Hervé, N. N. F., Fabrice, T. D., Justin, D. D. J., François, T., Jules-Roger, K. (2018). Chemical composition and antibacterial activity of oils from *Chrysicthys nigrodigitatus* and *Hepsetus odoe*, two freshwater fishes from Yabassi, Cameroon. *Lipids in Health and Disease* 17, 45.

Sørensen, R., Kildal, E., Stepaniak, L., Pripp, A.H. and Sørhaug, T. (2004). Screening for peptides from fish and cheese inhibitory to prolyl endopeptidase. *Nahrung/Food* 48(1), 53–56.

Tan, W.S., et al. (2012). Red blood cell omega-3 fatty acid levels and markers of accelerated brain ageing. *Neurology* 78(9), 658–664.

Thangadurai, D., Dabire, S. S., Sangeetha, J., Said Al-Tawaha, A. R. M., Adetunji, C. O., Islam, S., Shettar, A. K., David, M., Hospet, R., and Adetunji, J. B. (2021) Greener composites from plant fibers: Preparation, structure, and properties. In: Kharissova O.V., Martínez L.M.T., Kharisov B.I. (eds.), *Handbook of Nanomaterials and Nanocomposites for Energy and Environmental Applications.* Springer, Cham. https://doi.org/10.1007/978-3-030-11155-7_21-1

Vercruysse, L., van Camp, J. and Smagghe, G. (2005) ACE inhibitory peptides from enzymatic hydrolysates of animal muscle protein: A review. *Journal of Agricultural and Food Chemistry* 53(21), 2244–2245.

Virtanen, J.K., Siscovik, D.S. and Lemaitre, R.N. (2013). Circulating omega-3 polyunsaturated fatty acids and subclinical brain abnormalities on MRI in older adults: The Cardiovascular Health Study. *Journal of the American Heart Association* doi 10.1161/JAHA.113.000305.

Warren, C., Thomson, P. J., Xue-fang, G., Wen-dong, W., Min, L., Miza, S. V., Arshad, A. P., Min-tao, Z. (2018). Effectiveness of omega-3 polyunsaturated fatty acids against microbial pathogens. *Journal of Zhejiang University Science B* 19(4), 253–262. doi: 10.1631/jzus. B1700063.

Wolk, A., Larsson, S. C., Johansson, J. E., Ekman, P. (2006). Long-term fatty fish consumption and renal cell wall carcinoma incidence in a population-based prospective cohort of women. *The Journal of the American Medical Association* 296, 1371–1376.

10 Repositioning Antiviral Drugs as a Rapid and Cost-Effective Approach to Discover Treatment against SARS-CoV-2 Infection

Omotayo Opemipo Oyedara
Universidad Autónoma de Nuevo León
Osun State University

Folasade Muibat Adeyemi
Osun State University

Charles Oluwaseun Adetunji
Edo State University Uzairue

Temidayo Oluyomi Elufisan
Instituto Politécnico Nacional, Centro de Biotecnología Genómica

CONTENTS

DOI: 10.1201/9781003178903-12

10.1 INTRODUCTION

Despite the achievement recorded in the recent discovery of vaccines against COVID-19, there are still ongoing efforts to find drugs for the treatment and management of this deadly disease which has claimed over 2.23 million lives, with over 103 million cases reported globally. Of major public health concern and threat to the vaccine success story are reports about new SARS-CoV-2 variants, which are more deadly with a faster rate of transmission than the parent viral strain. Three of such variants, including the Brazil (P.1) variants, the South Africa B.1.351, and the UK B.1.1.7 which are characterized with large number of mutations, have been reported by the Centre for Disease Control (CDC) (https://www.cdc.gov/coronavirus/2019ncov/transmission/variant.html). With this new development, efforts need to be intensified in drug discovery to stem the menace of COVID-19, which has been ravaging the globe from the time it was discovered in the Province of Wuhan, China.

In the conventional drug development process, a new drug produced for COVID-19 treatment will have to pass through three important stages (Figure 10.1). The first stage is the drug discovery and preclinical trial, which involve the study of disease mechanism and identification of disease-specific target for the drug. This is followed by obtaining drug-like compounds from different sources, which may include plants, animals, microorganisms, or even the environment (Adetunji and Anani, 2021a,b,c; Adetunji et al., 2021; Dwivedi et al., 2021; Olaniyan and Adetunji, 2021, Mishra et al., 2020; Thangadurai et al., 2021; Anani et al., 2020; Islam et al., 2021). The in vitro assays on the potency of the drug-like compounds against the disease causative

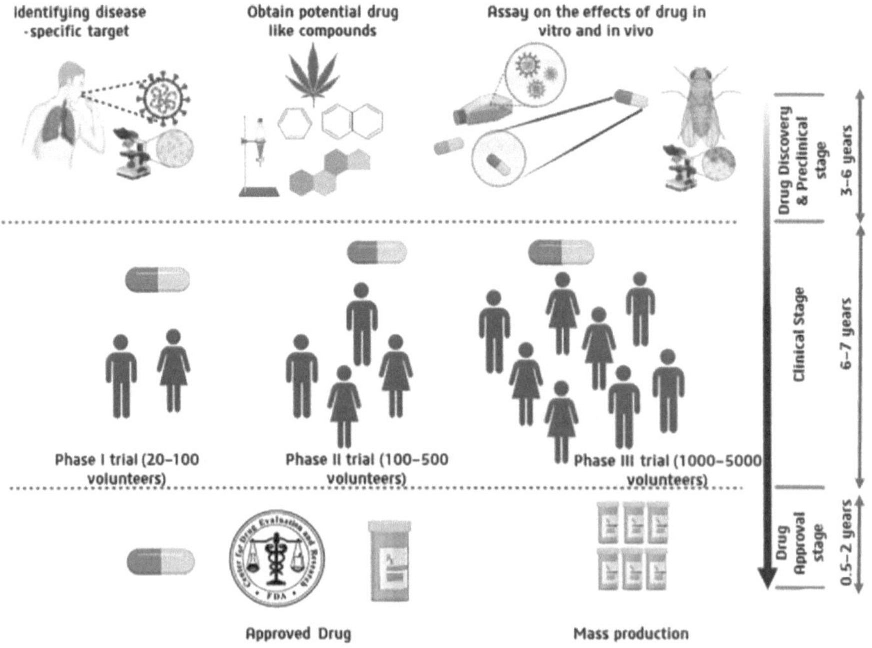

FIGURE 10.1 The three important stages of drug discovery. (Created with Biorender.com).

agent then follows before best drug candidates are tested in vivo using animal models. Drug candidates observed to pass the preclinical stage will then go through the second stage of clinical trial stage, which is also divided into three phases; phase I trial, which involves testing the safety and determining the dosage of drug in about 20–100 volunteers, phase II trial where the potency and short-term side effects of the drug are tested among 100–500 volunteers, and lastly the phase III clinical trial where the drug effects are confirmed and the side effects monitored. The third stage involves the United State Food and Drug Administration (FDA) giving approval for the drug and the post-market surveillance, where the efficacy, risks, and safety of the drug is monitored. The challenge of following through the stringent process of conventional drug discovery is the long period it takes for drugs to be made available in the market because it may take an average of 10–15 years (Parvathaneni et al. 2019) before one novel drug is clinically approved for use. Moreover, the cost of bringing a successful drug to market has been estimated to be US $2–3 billion dollars, hence making it a very expensive venture. However, for COVID-19, a rapid approach of drug discovery is needed to find suitable drugs for its treatment. One of the approaches is drug repurposing which is also known as drug repositioning (Pushpakom et al. 2018).

10.2 DRUG REPURPOSING

The application of a drug with established therapeutic indications in the treatment of a different diseased condition is known as drug repurposing. In other words, a new clinical application is found for a drug or molecule already approved to treat a particular disease. Drugs that can be repositioned can also include drugs that fail clinical trial or withdrawn from the market for safety reasons (Jourdan et al. 2020). Drug repositioning has advantages over the conventional drug discovery development process because it shortens the time required for the availability of drug in the market and also reduces the cost of drug production. These are possible because the safety, pharmaceutical formulation, and distribution channel for repurposed drugs might have already been established. However, in drug repositioning, some important stages of drug development cannot be ruled out, and these include the determination of drug activity against the novel diseased condition, establishment of repurposed drug efficacy, determination of the possible adverse effects in group of volunteers that present the targeted diseased condition, drug pharmacokinetics (Absorption, distribution, metabolism, and excretion), and dosage.

The advent of drug repurposing could be linked to the accidental discovery of new clinical applications for drugs (Figure 10.2). The side effects observed for a particular drug can be the basis for its novel application in the treatment or management of another disease. For instance, reprofiling thalidomide for the treatment of multiple myeloma stemmed from adverse effects observed when it was originally used as a sedative drug (Würth et al. 2015). Another example is aspirin which was originally marketed as an analgesic but repositioned as a platelet agglutination inhibitor in cardiovascular conditions. Moreover, the application of aspirin in oncology has been foreseen because of its potential to prevent colorectal cancer (Jourdan et al. 2020). Thus, phenotypic screening of already existing drugs through in vitro and in vivo assay can also lead to the discovery of drugs that can be repurposed for new indications.

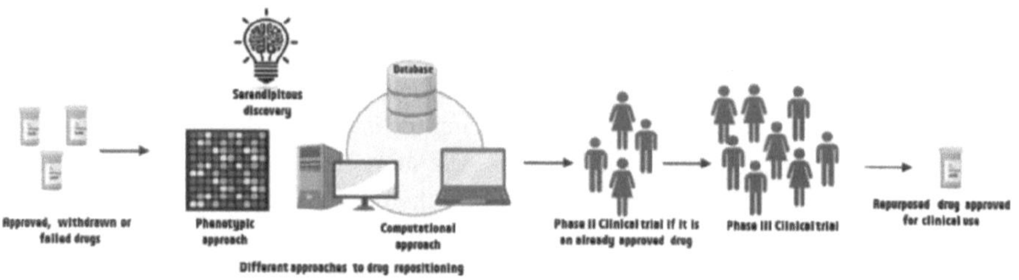

FIGURE 10.2 The steps and approaches involved in drug repurposing. (Created with Biorender.com).

The advancement in technology paved way for the quick identification of new applications for approved drugs with the aid of high-throughput computational approaches. The application of computational techniques in drug repurposing is possible with the availability of fast and cheap sequencing technologies and databases that houses wide arrays of information that can be exploited in understanding the mechanism of diseases and available drugs or compounds for drug discovery and repositioning. Examples of such database include GenBank, PubMed, KEGG, protein data bank (PDB), DrugBank, and ChemBank. Moreover, with the development of softwares, bioinformatics tools, and machine learning, drug or molecule interactions and kinetics can be accurately predicted through data mining from databases for the purpose of repositioning. Application of the simple and quick approaches (serendipitous or accidental discovery, phenotypic assay, and computation methods) to drug discovery have led to an increased interest among scientists and stake holders in the pharmaceutical industries in the exploration of novel clinical applications for existing drugs.

10.3 REPURPOSING ANTIVIRAL DRUGS AS INTERVENTIONS FOR COVID-19

Owing to the pressing need to quickly find drug to stem the spread of COVID-19, drug repurposing presents one of the ways to address this global menace. With drug repurposing, approved drugs can be studied and made available for COVID-19 treatment within the shortest possible time. There are several ongoing researches related to providing interventions to COVID-19, and more than 4,000 studies related to such interventions are listed in the ClinicalTrials.gov database (https://clinicaltrials.gov/). Some of the repurposed drugs listed in the database include approved and experimental or investigational drugs with different pharmacological properties.

Currently, several approved or experimental antiviral drugs are at different stages of clinical studies to check their effectivity and safety in COVID-19 treatment and management (Table 10.1). These antiviral drugs used different mechanisms of action to exert their antiviral effects against viruses such as human immunodeficiency virus (HIV), influenza virus, and Ebola virus, among others. Repurposed drugs can be classified based on their antiviral mechanism of action (Figure 10.3).

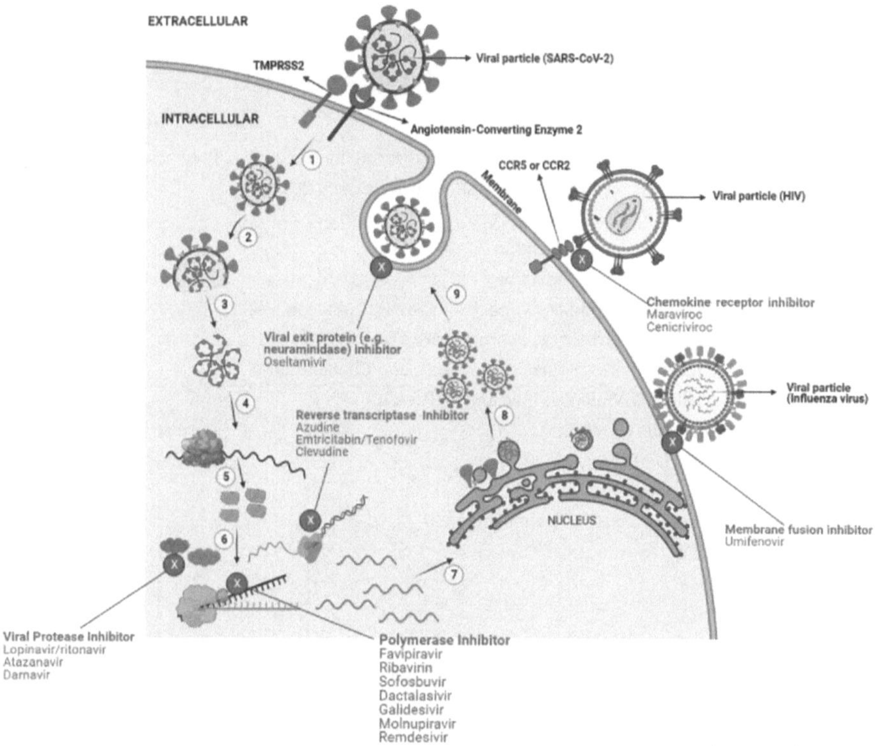

FIGURE 10.3 Replication of SARS-Co-V-2 (1). Binding of SARS-CoV-2 virus to Angiotensin-Converting Enzyme 2 (ACES 2) receptor (2). Entry of virus into the host cell by endocytosis (3). Viral RNA release (4). Viral RNA translation to produce early proteins including proteases (5). Proteases aid the formation of replicase-transcriptase complex (helicase and polymerase) for transcription and replication of viral RNA (7). Translation to make viral structural proteins (nucleocapsid, spike, membrane, envelope and accessory proteins) (8). Assembly of viral particles in the Golgi vesiscles (9). Release of mature virion via exocytosis. (Created with Biorender.com).

10.3.1 Polymerase Inhibitors

Favipiravir was recently approved in Japan under the brand name Avigan. It is active against different strains of influenza virus, including antiviral-resistant strains. It functions by interfering with the activity of the viral RNA polymerase (Furuta et al. 2005). In a multicentre randomized study, administration of favipiravir was suggested to reduce the length of hospital stay and the need to oxygenate COVID-19 patients (Dabbous et al. 2020). In another multicentre phase III clinical study, Udwadia et al. (2021) reported that clinical signs and symptoms in COVID-19 patients may be reduced with early administration of oral favipiravir. High dosage of favipiravir has been suggested to be required to achieve significant antiviral effects in COVID-19 patients (Dauby et al. 2021). However, some studies have linked the application of favipiravir to liver injury (Yamazaki et al. 2021), nephrotoxicity

TABLE 10.1
Antiviral Drug Repurposed for the COVID-19 Treatment

Drug	Indications	Mechanism
Favipiravir	Influenza virus, Ebola virus, Lassa fever, Rabies	Polymerase inhibitor
Ribavirin	Paediatric respiratory syncytial virus infections (RSV), hepatitis C, and Lassa fever	
Sofosbuvir	Hepatitis C virus infection	
Daclatasvir	Hepatitis C virus infection	
Galidesivir	Ebola virus, Marburg virus, bunyaviruses, arenaviruses, paramyxoviruses, and coronaviruses	
Molnupiravir	Influenza virus, human coronavirus, and Ebola virus	
Remdesivir	MERS-CoV, Ebola virus, and SARS-CoV	
Azvudine	Hepatitis C, HIV	Reverse transcriptase inhibitors
Emtricitabin/ tenofovir	HIV-1 infection	
Clevudine	Chronic hepatitis B virus infection	
Lopinavir/ritonavir	HIV	Protease inhibitor
Atazanavir	HIV	
Darunavir	HIV	
Maraviroc	CCR5-tropic HIV infection	Chemokine receptor inhibitor
Cenicriviroc	CCR5 and CCR2-tropic HIV infection	
Umifenovir	Anti-influenza	Virus-mediated fusion inhibitor
Oseltamivir	Anti-influenza	Influenza virus neuraminidase inhibitor
Inosine pranobex (Isoprinosine)	Herpes simplex virus, varicella, human papillomavirus, cytomegalovirus, and Epstein–Barr virus	Immunomodulatory and antiviral

(Nasa et al. 2021), and pancreatitis (Khan et al. 2021) in COVID-19 patients. Favipiravir was also reported to cure two COVID-19 patients effectively in combination with corticosteroid and other interventions such as extracorporeal membrane oxygenation and mechanical ventilation (Shinoda et al. 2021).

Ribavirin was first approved for the treatment of paediatric respiratory syncytial virus infections (RSV). It was a treatment option during the coronavirus outbreaks in China, North America, and Middle East. However, studies to proof the efficacy of ribavirin against SARS-CoV and MERS-CoV (Middle East Severe Respiratory Coronavirus) were inconclusive. It inhibited SARS-CoV in vitro but without any antiviral effects in vivo or improvement in the health conditions of SARS-CoV patients (Lee et al. 2004; Barnard et al. 2006). Similarly, Tong et al. (2020) reported the inability of intravenously administered ribavirin to reduce mortality rates in COVID-19 patients, while Li et al. (2021) reported that a combination of ribavirin and interferon-α therapy should be avoided in COVID-19 patients since it does not improve clinical conditions of patients.

Sofosbuvir, an inhibitor of polymerase in hepatitis C virus, has been suggested as a promising drug for the treatment of COVID-19 (Sali et al. 2020). Furthermore, combination therapy using sofosbuvir and daclatasvir was reported to possibly increase the efficacy of sofosbuvir (Sadeghi et al. 2020; Simmons et al. 2021). From molecular docking analysis reports, galidesivir bind strongly to the catalytic site of SARS-CoV-2 RNA polymerase (Ataei and Hosseinjani 2020; Elfiky, 2020), hence making it a good candidate drug for COVID-19 treatment.

Molnupiravir is an investigational prodrug of the ribonucleoside analogue ß-D-N4-hydroxycytidine (NHC) with inhibitory activity against influenza virus, chikungunya virus, respiratory syncytial virus, Venezuelan equine encephalitis virus (VEEV), hepatitis C virus, norovirus, influenza A and B viruses, Ebola virus, and human coronaviruses. It is undergoing a phase II clinical assessment to evaluate its suitability for SARS-CoV-2 infection treatment in terms of safety and efficacy (NCT number: NCT04405570).

Remdesivir has antiviral activity both in vivo and in vitro against Middle East Respiratory Syndrome Coronavirus, SARS-CoV-1, and SARS-CoV-2 (Beigel et al. 2020; Wang et al. 2020). It was the first antiviral drug approved by the FDA in May 2020 for the treatment of COVID-19 after studies showed that it can reduce the recovery time in critically ill patients. Remdesivir is a controversial COVID-19 drug because studies later showed that it does not reduce recovery rate in critically ill patients, rather effective in patients with mild COVID-19 illness. Moreover, remdesivir can only reduce recovery time and not mortality rate. Rahimi et al. (2021) also reported the possibility of remdesivir causing kidney and liver failure in critically ill patients. However, remdesivir in combination with lopinavir/ritonavir is undergoing phase IV clinical trial, with NCT number: NCT04738045.

10.3.2 Reverse Transcriptase Inhibitors

Azvudine is a nucleoside-based antivirus earlier developed to treat HIV infection but now undergoing clinical trial to be repurposed to treat patients with mild COVID-19. It was recommended as a safe drug for COVID-19 without adverse effect (Ren et al. 2020; Yu and Chang, 2020). In a recent study, HIV patients receiving tenofovir disoproxil fumarate/emtricitabine are reported to a lower risk of COVID-19 (Amo et al. 2020). Clevudine is an inhibitor of reverse transcriptase indicated in the treatment of chronic hepatitis B virus infection. Phase II clinical study of clevudine in patients diagnosed with moderate COVID-19 is currently ongoing to evaluate its safety and efficacy (NCT number: NCT04347915).

10.3.3 Protease Inhibitors

Lopinavir and ritonavir are anti-HIV drugs marketed with the trade name Kaletra. Studies have shown that the use of lopinavir/ritonavir reduces acute respiratory challenges in patients suffering from COVID-19 (Horby et al. 2020). Moreover, the combination therapy of Interferon Beta-1a (IFN-β1-a) and lopinavir/ritonavir was reported to reduce the need for mechanical aeration in COVID-19 patients (Baghaei et al. 2021). Molecular docking and in vitro analysis revealed atazanavir and darunavir as promising drugs to combat SARS-CoV-2 (de Sousa Silva et al. 2020; Fintelman-Rodrigues et al. 2020). On the contrary, De Meyer et al. (2020)

reported that darunavir lack antiviral activity against SARS-CoV-2, while Riva et al. (2020) reported the inability of darunavir to prevent HIV patients from SARS-CoV-2 infection. A combination therapy involving the use of atazanavir (antiviral), Traneurocin (NA-831, a drug for Alzheimer's disease), and Dexamethasone (a glucocorticoid) is undergoing phase III clinical trial (NCT04452565). In another clinical study, the efficacy and safety of nitazoxanide plus atazanavir/ritonavir is being evaluated to treat moderate to severe COVID-19 (NCT04459286).

10.3.4 Chemokine Receptor Inhibitor

Maraviroc and cenicriviroc are both anti-HIV drug that binds to the chemokine receptor expressed on human membrane thereby preventing the interaction between the receptor and viral glycoprotein. While maraviroc is limited to treating C-C chemokine receptor 5 (CCR5) tropic HIV1 infection in adults, cenicriviroc is indicated for both CCR5 and CCR2 tropic HIV infection. The ability of maraviroc to inhibit the main protease (M^{pro}) of SARS-CoV-2 in silico (Shamsi et al. 2020) and cenicriviroc to inhibit its replication in vitro (Okamoto et al. 2020) makes them a drug candidate for the treatment of COVID-19.

10.3.5 Inhibition of Virus-Membrane Fusion and Release of Viral Particles

Oseltamivir is an anti-influenza virus that inhibits the virus neuraminidase and thus affecting the release of viral particles. It was positioned as a good drug candidate against COVID-19 based on its ability to bind stably with SARS-CoV-2 main protease in combination with lopinavir and ritonavir, as revealed by an in silico study (Muralidharan et al. 2020). Moreover, combination of antibacterial and oseltamivir therapy, and early administration of oseltamivir was reported to reduce fever in COVID-19 patients without hypoxia (Chiba, 2020). It also hastened the recovery of COVID-19 patients with illnesses similar to influenza (Coenen et al. 2020). In an in vitro study, the inability of oseltamivir to exert inhibitory effects on SARS-CoV-2 or improve patients' signs and symptoms was reported (Tan et al. 2020). Notwithstanding, oseltamivir is undergoing several clinical studies for use in COVID-19 treatment. Umifenovir blocks the entry of virus into the cell by preventing fusion of the virus with target membrane. It is a promising drug undergoing clinical trial for the treatment of COVID-19 disease.

10.3.6 Immunomodulatory Antiviral Drugs

Inosine pranobex is an immunomodulatory antiviral agent that stimulates the production of cytokines and lymphocytes and also enhances natural killer cell cytotoxicity to destroy aetiological agents of herpes simplex virus, varicella, and human papillomavirus. It was reported to reduce case-mortality rates among elderly infected with SARS-CoV-2 (Beran et al. 2020). It is undergoing phase III clinical trials in combination with Levamisole (NCT number: NCT04383717).

10.4 CONCLUSION

Rapid discovery of antiviral therapy against SARS-CoV-2 is one of the ways to stem the menace of COVID-19. Repurposing antiviral drugs with known indications is a good option to rapidly develop drugs for the treatment of COVID-19.

REFERENCES

Ataei, M. and Hosseinjani, H., 2020. Molecular mechanisms of Galidesivir as a potential antiviral treatment for COVID-19. *Journal of Pharmaceutical Care*, 8(3), pp.150–151.

Adetunji, C.O., Anani, O.A., and Panpatte, D. (2021) Mechanism of actions involved in sustainable ecorestoration of petroleum hydrocarbons polluted soil by the beneficial microorganism. In: Panpatte, D.G., and Jhala, Y.K. (eds.), *Microbial Rejuvenation of Polluted Environment. Microorganisms for Sustainability*, vol 26. Springer, Singapore. https://doi.org/10.1007/978-981-15-7455-9_8

Adetunji, C.O., and Anani, O.A. (2021a) Recent advances in the application of genetically engineered microorganisms for microbial rejuvenation of contaminated environment. In: Adetunji C.O., Panpatte D.G., Jhala Y.K. (eds.), *Microbial Rejuvenation of Polluted Environment. Microorganisms for Sustainability*, vol 27. Springer, Singapore. https://doi.org/10.1007/978-981-15-7459-7_14

Adetunji C.O., and Anani O.A. (2021b) Utilization of microbial biofilm for the biotransformation and bioremediation of heavily polluted environment. In: Panpatte, D.G., and Jhala, Y.K. (eds.), *Microbial Rejuvenation of Polluted Environment. Microorganisms for Sustainability*, vol 25. Springer, Singapore. https://doi.org/10.1007/978-981-15-7447-4_9

Adetunji C.O., and Anani O.A. (2021c) Bioremediation of polythene and plastics using beneficial microorganisms. In: Adetunji C.O., Panpatte D.G., and Jhala Y.K. (eds.), *Microbial Rejuvenation of Polluted Environment. Microorganisms for Sustainability*, vol 27. Springer, Singapore. https://doi.org/10.1007/978-981-15-7459-7_13

Anani, O.A., Mishra R.R., Mishra P., Enuneku A.A., Anani G.A., and Adetunji C.O. (2020) Effects of toxicant from pesticides on food security: Current developments. In: Mishra P., Mishra R.R., Adetunji C.O. (eds.), *Innovations in Food Technology*. Springer, Singapore. https://doi.org/10.1007/978-981-15-6121-4_22

Baghaei, P., Dastan, F., Marjani, M., Moniri, A., Abtahian, Z., Ghadimi, S., Valizadeh, M., Heshmatnia, J., Mirenayat, M.S., Abedini, A. and Kiani, A., 2021. Combination therapy of IFNβ1 with lopinavir–ritonavir, increases oxygenation, survival and discharging of sever COVID-19 infected inpatients. *International Immunopharmacology*, 92, p.107329.

Barnard, D.L., Day, C.W., Bailey, K., Heiner, M., Montgomery, R., Lauridsen, L., Chan, P.K. and Sidwell, R.W., 2006. Evaluation of immunomodulators, interferons and known in vitro SARS-coV inhibitors for inhibition of SARS-coV replication in BALB/c mice. *Antiviral Chemistry and Chemotherapy*, 17(5), pp.275–284.

Beigel, J.H., Tomashek, K.M., Dodd, L.E., Mehta, A.K., Zingman, B.S., Kalil, A.C., Hohmann, E., Chu, H.Y., Luetkemeyer, A., Kline, S. and de Castilla, D.L., 2020. Remdesivir for the treatment of Covid-19—preliminary report. *The New England Journal of Medicine*.

Beran, J., Špajdel, M., Katzerová, V., Holoušová, A., Malyš, J., Finger Rousková, J. and Slíva, J., 2020. Inosine Pranobex significantly decreased the case-fatality rate among PCR positive elderly with SARS-CoV-2 at three nursing homes in the Czech Republic. *Pathogens*, 9(12), p.1055.

Chiba, S., 2020. Effect of early oseltamivir on outpatients without hypoxia with suspected COVID-19. *Wiener klinische Wochenschrift*, pp.1–6.

Coenen, S., van Der Velden, A.W., Cianci, D., Goossens, H., Bongard, E., Saville, B.R., Gobat, N., de Paor, M., Ieven, M., Verheij, T.J. and Butler, C.C., 2020. Oseltamivir for coronavirus illness: Post-hoc exploratory analysis of an open-label, pragmatic, randomised controlled trial in European primary care from 2016 to 2018. *British Journal of General Practice*, 70(696), pp.e444–e449.

Dabbous, H.M., Abd-Elsalam, S., El-Sayed, M.H., Sherief, A.F., Ebeid, F.F., Abd El Ghafar, M.S., Soliman, S., Elbahnasawy, M., Badawi, R. and Tageldin, M.A., 2021. Efficacy of favipiravir in COVID-19 treatment: A multi-center randomized study. *Archives of Virology*, pp. 1–6.

Dauby, N., Van Praet, S., Vanhomwegen, C., Veliziotis, I., Konopnicki, D. and Roman, A., 2021. Tolerability of favipiravir therapy in critically ill patients with COVID-19: A report of four cases. *Journal of Medical Virology*, 93(2), pp. 689–691.

Del Amo, J., Polo, R., Moreno, S., Díaz, A., Martínez, E., Arribas, J.R., Jarrín, I. and Hernán, M.A., 2020. Incidence and severity of COVID-19 in HIV-positive persons receiving antiretroviral therapy: A cohort study. *Annals of internal medicine*, 173(7), pp.536–541.

de Sousa Silva, J.D., da Costa Leite, S., Da Silva, M.T.S., Meirelles, L.M.A. and Andrade, A.W.L., 2020. In silico evaluation of the inhibitory effect of antiretrovirals Atazanavir and Darunavir on the main protease of SARS-CoV-2: Docking studies and molecular dynamics. *Research, Society and Development*, 9(8), pp.e826986562–e826986562.

De Meyer, S., Bojkova, D., Cinatl, J., Van Damme, E., Buyck, C., Van Loock, M., Woodfall, B. and Ciesek, S., 2020. Lack of antiviral activity of darunavir against SARS-CoV-2. *International Journal of Infectious Diseases*, 97, pp.7–10.

Dwivedi, N., Dwivedi, S., and Adetunji, C. O. (2021) Efficacy of microorganisms in the removal of toxic materials from industrial effluents. In: Adetunji C.O., Panpatte D.G., Jhala Y.K. (eds.), *Microbial Rejuvenation of Polluted Environment. Microorganisms for Sustainability*, vol 27. Springer, Singapore. https://doi.org/10.1007/978-981-15-7459-7_15

Elfiky, A.A., 2020. Ribavirin, Remdesivir, Sofosbuvir, Galidesivir, and Tenofovir against SARS-CoV-2 RNA dependent RNA polymerase (RdRp): A molecular docking study. *Life Sciences*, 253, p.117592.

Fintelman-Rodrigues, N., Sacramento, C.Q., Lima, C.R., da Silva, F.S., Ferreira, A.C., Mattos, M., de Freitas, C.S., Soares, V.C., Dias, S.D.S.G., Temerozo, J.R. and Miranda, M.D., 2020. Atazanavir, alone or in combination with ritonavir, inhibits SARS-CoV-2 replication and proinflammatory cytokine production. *Antimicrobial agents and chemotherapy,* 64(10).

Furuta, Y., Takahashi, K., Kuno-Maekawa, M., Sangawa, H., Uehara, S., Kozaki, K., Nomura, N., Egawa, H. and Shiraki, K., 2005. Mechanism of action of T-705 against influenza virus. *Antimicrobial Agents and Chemotherapy*, 49(3), pp. 981–986.

Horby, P.W., Mafham, M., Bell, J.L., Linsell, L., Staplin, N., Emberson, J., Palfreeman, A., Raw, J., Elmahi, E., Prudon, B. and Green, C., 2020. Lopinavir–ritonavir in patients admitted to hospital with COVID-19 (RECOVERY): A randomised, controlled, open-label, platform trial. *The Lancet*, 396(10259), pp.1345–1352.

Jourdan, J.P., Bureau, R., Rochais, C. and Dallemagne, P., 2020. Drug repositioning: A brief overview. *Journal of Pharmacy and Pharmacology*, 72(9), pp. 1145–1151.

Khan, A.E.A., Hoque, M.M., Mallik, M.U., Basu, K.C., Hasan, M.M., Mahmud, I., Adnan, M.M.A. and Rahman, M.M., 2021. Favipiravir induced acute pancreatitis in a COVID-19 patient. *Journal of Medicine*, 22(1), pp. 69–71.

Lee, N., Chan, K.A., Hui, D.S., Ng, E.K., Wu, A., Chiu, R.W., Wong, V.W., Chan, P.K., Wong, K.T., Wong, E. and Cockram, C.S., 2004. Effects of early corticosteroid treatment on plasma SARS-associated Coronavirus RNA concentrations in adult patients. *Journal of Clinical Virology*, 31(4), pp.304–309.

Li, H., Xiong, N., Li, C., Gong, Y., Liu, L., Yang, H., Tan, X., Jiang, N., Zong, Q., Wang, J. and Lu, Z., 2021. Efficacy of ribavirin and interferon-α therapy for hospitalized patients with COVID-19: A multicenter, retrospective cohort study. *International Journal of Infectious Diseases*, 104, pp.641-648.

Muralidharan, N., Sakthivel, R., Velmurugan, D. and Gromiha, M.M., 2020. Computational studies of drug repurposing and synergism of lopinavir, oseltamivir and ritonavir binding with SARS-CoV-2 protease against COVID-19. *Journal of Biomolecular Structure and Dynamics,* pp.1–6.

Mishra, P., Mishra, R. R., and Adetunji, C. O. (Eds.). (2020). *Innovations in Food Technology: Current Perspectives and Future Goals.* Springer Nature Singapore Pte Ltd. 2020. Springer, Singapore. pp. 143-162. 1st Edition.

Nasa, P., Shrivastava, P., Kulkarni, A., Vijayan, L. and Singh, A., 2021. Favipiravir induced nephrotoxicity in two patients of COVID-19. *Journal of The Association of Physicians of India*, 69, p. 88.

Olaniyan, O.T., and Adetunji, C.O. (2021) Biochemical role of beneficial microorganisms: An overview on recent development in environmental and agro science. In: Adetunji, C.O., Panpatte, D.G., and Jhala, Y.K. (eds) *Microbial Rejuvenation of Polluted Environment. Microorganisms for Sustainability*, vol 27. Springer, Singapore. https://doi.org/10.1007/978-981-15-7459-7_2

Okamoto, M., Toyama, M. and Baba, M., 2020. The chemokine receptor antagonist cenicriviroc inhibits the replication of SARS-CoV-2 in vitro. *Antiviral Research*, 182, p.104902.

Parvathaneni, V., Kulkarni, N.S., Muth, A. and Gupta, V., 2019. Drug repurposing: A promising tool to accelerate the drug discovery process. *Drug Discovery Today*, 24(10), pp. 2076–2085.

Pushpakom, S., Iorio, F., Eyers, P.A., Escott, K.J., Hopper, S., Wells, A., Doig, A., Guilliams, T., Latimer, J., McNamee, C. and Norris, A., 2019. Drug repurposing: Progress, challenges and recommendations. *Nature reviews Drug discovery*, 18(1), pp. 41–58.

Rahimi, M.M., Jahantabi, E., Lotfi, B., Forouzesh, M., Valizadeh, R. and Farshid, S., 2021. Renal and liver injury following the treatment of COVID-19 by remdesivir. *Journal of Nephropathology,* 10(2).

Ren, Z., Luo, H., Yu, Z., Song, J., Liang, L., Wang, L., Wang, H., Cui, G., Liu, Y., Wang, J. and Li, Q., 2020. A randomized, open-label, controlled clinical trial of Azvudine tablets in the treatment of mild and common COVID-19: A pilot study. *Advanced Science*, 7(19), p.2001435.

Riva, L., Yuan, S., Yin, X., Martin-Sancho, L., Matsunaga, N., Pache, L., Burgstaller-Muehlbacher, S., De Jesus, P.D., Teriete, P., Hull, M.V. and Chang, M.W., 2020. Discovery of SARS-CoV-2 antiviral drugs through large-scale compound repurposing. *Nature*, 586(7827), pp.113–119.

Sadeghi, A., Ali Asgari, A., Norouzi, A., Kheiri, Z., Anushirvani, A., Montazeri, M., Hosamirudsai, H., Afhami, S., Akbarpour, E., Aliannejad, R. and Radmard, A.R., 2020. Sofosbuvir and daclatasvir compared with standard of care in the treatment of patients admitted to hospital with moderate or severe coronavirus infection (COVID-19): A randomized controlled trial. *Journal of Antimicrobial Chemotherapy*, 75(11), pp.3379–3385.

Sali, S., Yadegarinia, D., Abolghasemi, S., Tehrani, S., Gharaei, B., Khabiri, N., Sadeghi, S. and Ramezani, A. 2020. Comparison of the Efficacy of Sofosbuvir and Kaletra on the Outcome of COVID-19. Is Sofosbuvir A Potential Treatment For COVID-19? *Novelty in Biomedicine*, 9(1), pp. 17–23.

Shamsi, A., Mohammad, T., Anwar, S., AlAjmi, M.F., Hussain, A., Rehman, M., Islam, A. and Hassan, M., 2020. Glecaprevir and Maraviroc are high-affinity inhibitors of SARS-CoV-2 main protease: Possible implication in COVID-19 therapy. *Bioscience Reports,* 40(6).

Shinoda, M., Kamachi, K., Ota, S., Yoshimatsu, L., Boku, R., Yoshida, Y., Hirouchi, T., Shinada, K., Sato, T., Morikawa, M. and Iwata, K., 2021. Two cases of severe COVID-19 pneumonia effectively treated with extracorporeal membrane oxygenation in addition to favipiravir and corticosteroid. *Internal Medicine*, 60(1), pp. 123–130.

Simmons, B., Wentzel, H., Mobarak, S., Eslami, G., Sadeghi, A., Ali Asgari, A., Abbaspour Kasgari, H., Tirgar Fakheri, H., Merat, S. and Hill, A., 2021. Sofosbuvir/daclatasvir regimens for the treatment of COVID-19: An individual patient data meta-analysis. *Journal of Antimicrobial Chemotherapy*, 76(2), pp.286–291.

Tan, Q., Duan, L., Ma, Y., Wu, F., Huang, Q., Mao, K., Xiao, W., Xia, H., Zhang, S., Zhou, E. and Ma, P., 2020. Is oseltamivir suitable for fighting against COVID-19: In silico assessment, in vitro and retrospective study. *Bioorganic Chemistry*, 104, p.104257.

Tong, S., Su, Y., Yu, Y., Wu, C., Chen, J., Wang, S. and Jiang, J., 2020. Ribavirin therapy for severe COVID-19: A retrospective cohort study. *International Journal of Antimicrobial Agents*, 56(3), p.106114.

Thangadurai, D., Dabire, S. S., Sangeetha, J., Said Al-Tawaha, A. R. M., Adetunji, C. O., Islam, S., Shettar, A. K., David, M., Hospet, R., and Adetunji, J. B. (2021) Greener composites from plant fibers: Preparation, structure, and properties. In: Kharissova O.V., Martínez L.M.T., Kharisov B.I. (eds.), *Handbook of Nanomaterials and Nanocomposites for Energy and Environmental Applications*. Springer, Cham. https://doi.org/10.1007/978-3-030-11155-7_21-1

Udwadia Z. F., Singh P., Barkate H., Patil S., Rangwala S., Pendse A., Kadam J., Wu W., Caracta C. F., and Tandon M. (2021). Efficacy and safety of favipiravir, an oral RNA-dependent RNA polymerase inhibitor, in mild-to-moderate COVID-19: A randomized, comparative, open-label, multicenter, phase 3 clinical trial, *International Journal of Infectious Diseases*, 103, pp.62–71. https://doi.org/10.1016/j.ijid.2020.11.142

Wang, M., Cao, R., Zhang, L., Yang, X., Liu, J., Xu, M., Shi, Z., Hu, Z., Zhong, W. and Xiao, G., 2020. Remdesivir and chloroquine effectively inhibit the recently emerged novel coronavirus (2019-nCoV) in vitro. *Cell research*, 30(3), pp.269–271.

Wuerth, R., Thellung, S., Bajetto, A., Mazzanti, M., Florio, T. and Barbieri, F., 2016. Drug-repositioning opportunities for cancer therapy: novel molecular targets for known compounds. *Drug Discovery Today*, 21(1), pp. 190–199.

Yamazaki, S., Suzuki, T., Sayama, M., Nakada, T.A., Igari, H. and Ishii, I., 2021. Suspected cholestatic liver injury induced by favipiravir in a patient with COVID-19. *Journal of Infection and Chemotherapy*, 27(2), pp. 390–392.

Yu, B. and Chang, J., 2020. Azvudine (FNC): A promising clinical candidate for COVID-19 treatment. *Signal Transduction and Targeted Therapy*, 5(1), pp.1–2.

11 Corn Silk and Health Benefits

Charles Oluwaseun Adetunji
Edo State University Uzairue

Michael Olugbenga Samuel
Bowen University

Juliana Bunmi Adetunji
Osun State University

Olufemi Idowu Oluranti
Bowen University

CONTENTS

11.1 INTRODUCTION

Plants are sources of medicine for therapeutic management of illnesses, and they have received enormous attention because of their relatively adverse effects due to the fact that they are part of nature and sometimes some of them are consumed as food. The use of medicinal plants or herbs for healing in traditional medicine is as old as mankind itself (Bae et al., 2015). Medicinal plants application against diseases has come to stay, and it has gained relevance because many years of repression and sufferings caused by ailments or diseases prompted men to look for alternative treatment or therapeutic adventures in medicinal plants (Kelly, 2009; Biljana, 2012). Furthermore, the recent emergence of natural herbs with health benefits has been encouraged by the diminishing effectiveness of man-made medications, high cost, and the accumulating adverse reactions to their applications (Kelly, 2009; Biljana,

DOI: 10.1201/9781003178903-13

2012). A huge percentage of the global population depends on traditional medicine to boost and complement the progressively costly conventional drugs (WHO, 2000; Adetunji and Anani, 2021a,b,c; Adetunji et al., 2021; Dwivedi et al., 2021; Olaniyan and Adetunji, 2021, Mishra et al., 2020; Thangadurai et al., 2021; Anani et al., 2020; Islam et al., 2021). One of such plants with recognized medicinal applications in managing prevailing health challenges is corn silk.

Corn silk is an agricultural waste from maize plantation. It is a yellowish or greenish thread present in female flower of maize and it is in plenty supply across the globe. The main purpose of the corn silk is for trapping the pollens for pollination and it is collected with the corn cob. One kernel of corn is produced by the pollination of each corn silk. Corn silks are approximately 30 cm long with faint sweet flavor. Its composition includes protein, carbohydrates, vitamins, fixed and volatile oils, sodium, potassium, magnesium, alkaloids, crude fiber, polysaccharides, saponins, tannins and phytosterols, steroids, phenolic compounds, flavonoids (maysin, methoxymaysin, apimaysin, and luteolin derivatives), and anthocyanins (cyanidin, peonidin) which may be responsible for the interesting therapeutic potencies of the corn silk (Fossen et al., 2001; Ebrahimzadeh et al., 2008; Hu et al., 2011). Therefore, as a cheap medicinal plant, it is used as remedy for hypertension, gout, cancer, cystitis, and diabetes mellitus, across nations of the world such as Korea, Turkey, etc. (Maksimovic et al., 2004; Hasanudin et al., 2012; Saheed et al., 2015; Ha et al., 2018).

Numerous reports exist that described the extensive biological activities of corn silk including anticancer, anti-diabetic, antioxidant, antidepressant, anti-fatigue, and antibacterial effects (Ebrahimzadeh et al., 2009; Hu et al., 2010; Yang et al., 2014; Morshed and Islam, 2015; Žilić et al., 2016; Wang and Zhao, 2019). This health promoting physiological effects of corn silk will be discussed.

11.2 ANTIOXIDANT EFFECTS

Maksimovic and coworkers were the first to describe the antioxidative potentials of corn silk in an in vitro study by showing the association between polyphenol compounds and antioxidant molecules in corn silk extract (Maksimovic et al., 2005). Furthermore, corn silk extract was revealed to possess free radical scavenging activity and iron chelating effect (Liu et al., 2011; Wang and Zhao, 2019). The antioxidant effect of corn silk was reported in cancer and pancreatic β-cells exhibiting anti-glycation and prevention of toxicity associated hyperglycemia (Tian et al., 2013; Choi et al., 2014; Chang et al., 2016). This extract has been shown to improve glutathione peroxidase activity in high-salt fed rats (Oyabambi et al., 2020). Likewise, corn silk extract caused elevated renal reduced glutathione concentration after four weeks of treatment (Vranješ et al., 2016). Amelioration of oxidative damage caused by radiation (Bai et al., 2010) together with diabetes (Zhang et al., 2016) in rodents has been attributed to the activity of corn silk.

11.3 SUPPRESSIVE EFFECTS ON INFLAMMATION

Inflammation is a major part of most chronic diseases, and corn silk extract has been shown to suppress inflammatory exudate and leukocyte adhesion. The suppressive inflammatory response of corn silk extract is revealed through the reduction of some inflammatory cytokines which are key mediators of inflammatory processes (Stow et al., 2009; Wang et al., 2011).

Corn silk's anti-inflammatory potential has also been confirmed in rodent model of inflammation using lipopolysaccharide (LPS). LPS-mediated inflammatory tendencies typified by cytokine production and generalized systemic inflammation was reported to be suppressed by corn silk extract which provides a basis for the traditional use of corn silk in the management of inflammation-related diseases. The inflammation suppressive effect of corn silk extract was found to be through a bioactive molecule FK2 derived from the corn silk extract which exerts anti-inflammatory effect against LPS-induced inflammation by blocking the activity of NF-κB (Tin-Yun et al., 2017).

Silver nanoparticles synthesized from corn silk have been demonstrated to show anti-inflammatory effect though inhibition of the cyclooxygenase resulting in reduced edema size (Helmy et al., 2020). The obesity lowering activity of the corn silk might be through its anti-inflammatory pathway (Ahmed et al., 2016).

11.4 ANTI-HYPERTENSIVE EFFECT

Cardiac and blood vessel diseases remain the prominent cause of mortality globally. Also, hypertension represents huge predisposing factor for the pathogenesis of heart failure and cardiac damage. Extract of corn silk has been documented to possess hypotensive potential. However, this claim has not been confirmed by many experimental data in either humans or rodents. Report exists that corn silk extract when taken intravenously led to reduced diastolic blood pressure in dogs (Martín et al., 1991). In addition, extract of corn silk has potential to decrease intraocular pressure in systemic and non-systemic hypertension (George and Idu, 2015). Interestingly, corn silk extract exerts its cardioprotective effect via high potassium-induced diuretic and vasodilatation activities (Velazquez et al., 2005).

Furthermore, angiotensin converting enzyme (ACE) is a key drug target in treating vascular diseases. Reports exist that corn silk extract possess blood pressure lowering capacity using hypertensive rats model through suppression of ACE activity. Hence, corn silk extract can be said to possess ACE inhibitory activity (Li et al., 2019). More so, corn silk-derived bioactive phytopeptide (CSBp5) has been reported to cause dose-dependent systolic blood pressure reduction through its ACE inhibitory effect. The isolation of the bioactive constituent with ACE inhibitory properties from corn silk is a confirmation of its hypotensive effects and why it is used locally as a remedy for hypertension (Li et al., 2019).

Adedapo et al. (2016) performed the safety effectiveness of corn silk using histopathological alteration as well as electrocardiogram in albino rats. The concentration of corn silk extract that varies from 200 to 800 mg/kg was orally administered while control group was given distilled water. The level of ECG was determined in ketamine/xylazine-induced rats so as to evaluate alteration in the heart rate. The result obtained is an indication that there was a substantial alteration in the heart rate. Their study showed that corn silk could be applied as medicinal product for the treatment of patient suffering from heart diseases.

11.5 ANTI-DIABETIC EFFECT

Diabetes has been recognized as one of the most life-threatening global epidemic and effective novel healing stratagems and drugs for quick management of these diseases are straightaway needed. It has been observed that the Type 2 diabetes mellitus (T2DM) represents more than 90% of diabetes cases. It has been observed that there are many synthetic drugs that are utilized for effective management of T2DM clinically which includes oral anti-diabetics and insulin. Moreover, there are numerous of them that some adverse side effects and some limitations. Zhao and colleagues described the potentials of corn silk as a typical example of functional food and medicinal foods with hypoglycemic activity without any adverse effect (Zhao et al. 2012). The hypoglycemic influence of corn silk has been established in numerous ways (Guo et al., 2009; Zhang et al., 2015; Pan et al., 2017) which showed that corn silk prevents α-amylase and could prevent the process involved in the slowing down of starch digestion rate of food and portends the capability to confine the upsurge of post-meal blood sugar (Chen et al., 2013). It has been validated that they possess the potential to prevent the α-glucosidase activity that are involved in the regulation of glucose metabolism (Wang and Zhao, 2019). Moreover, it has been stated that corn silk exerts its hypoglycemic influence by targeting signal pathways to stimulate insulin activity and enhance glucose metabolism. Also, the corn silk polysaccharides possess potential to reduce body weight gain, lower glycemia as well as insulinemia, and cause improvement of glucose tolerance in T2DM mice (Pan et al., 2017). Interestingly, toxicological study established that corn silk is safe for human consumption without toxicity or adverse reaction (Wang et al., 2011). It has also been established that the corn silk could serve as herbal remedy and phytomedicine for the management of diabetes.

The application of herbal remedies which involves corn silk derived from *Zea mays* or *Stigma maydis* has been recognized as natural substitute to numerous standard Western allopathic medicine that have been applied for the management of several diseases such as obesity, prostate disorders, bedwetting, and a diuretic. Sani (2016) has also investigated the anti-diabetic capability of the methanolic extract of corn silk in alloxan-induced diabetic rats. The crude extracts of corn silk were administered to the rats that were having diabetes during the experiment. The effectiveness of the crude extract could be linked to the presence of glibenclamide. The result indicated that the administration of corn silk extract to diabetic mice could improve gluco-regulation in a dose-dependent routine. Moreover, it was observed that the methanol extract showed comparatively substantial improvement in the regulation of blood glucose in the diabetic mice after three days of administration. Phytochemical investigation of the extract revealed the following constituents like anthraquinones, alkaloids, cardiac glycoside, carbohydrates, steroids, flavonoids, and terpenoids.

Corn silk has been recognized as a renowned constituent that is regularly utilized in the traditional medicine. Zhang et al. (2016) performed an experiment to analyze the antioxidative, hypolipidemic, and hypoglyemic effectiveness of flavonoids derived from corn silk on streptozotocin (STZ)-induced diabetic mice. Result demonstrated the hypoglycemic effect, reduction in water consumption, and body weight gain, which establish the anti-diabetic effectiveness of the crude extract. The authors carried out malondialdehyde and serum total superoxide dismutase assays so as to establish the antioxidant effectiveness of the corn silk. Moreover, there was reduction in the dyslipidemia associated with diabetes. The hypolipidemic, anti-diabetic, and antioxidant effect indicated that corn silk is a typical example of natural compound that could be utilized for successful management of diabetic conditions.

The *Zea mays* L. derived from the dried husk which belongs to the family of Poaceae also recognized as corn husk has been applied for successful management of traditional illness most especially in some cultures which includes Ghana as a remedy for diabetes mellitus (DM, Type 2). Moreover, there is paucity of information concerning the phyto-constituents of corn husk as well as the scientific documentation on the anti-diabetic potential of corn husk. The aqueous and methanolic extract of corn husk was screened for phytochemicals, while that extract obtained from the corn husk was given orally to the human subjects (Brobbey et al., 2017). The level of the fasting blood sugar of the subject taking the recommended oral anti-diabetic substances was evaluated for 21 days, while the fasting blood sugar of the untreated group was determined for 6 weeks. The result obtained from the phytochemical examination showed the presence of glycosides, phenols, alkaloids, flavonoids, tannins, and saponins. It was further established that the corn husk possesses substantial amount of anti-diabetic features which established their application as anti-diabetic therapy (Brobbey et al., 2017).

Ahmed et al. (2016) evaluated the influence of methanol and aqueous corn silk extracts against obesity in animal model. The animals utilized for the experiment were nourished with a very high-cholesterol diet for a period of 12 weeks so as to stimulate obesity. The animals were later treated separately with aqueous or methanol corn silk extracts while Orlistat was also applied for a period of 6 weeks. The following observation were documented which include thoracic circumference, body mass index, and abdominal circumference. Moreover, some biochemical parameters like lipid profile, insulin resistance, and triglyceride-glucose index were investigated. The following parameters were further evaluated such as protein S100B levels, inflammatory cytokines visfatin, calprotectin, and haptoglobin. Their result indicated that substantial reduction in the level of Body mass index (BMI), thoracic circumference, and abdominal circumference was recorded in high-cholesterol diet (HCD)-fed groups that were given corn silk extract with respect to HCD-fed group through anthropometric measurement. The result obtained from the biochemical evaluation showed significant hypoglycemic and hypolipidemic influence of corn silk extracts. Administration of corn silk to HCD-fed rats caused momentous reduction of protein S100B, visfatin, calprotectin, endothelin-1, and haptoglobin. Corn silk was demonstrated to possess an obesity lowering effect that could be connected to numerous bioactive components of the plant having anti-inflammatory, hypolipidemic, and hypoglycemic activities (Ahmed et al., 2016). Obesity results in excess adiposity with increased production of cytokines known as adipokines which include pro-inflammatory interleukin 8 (IL-8), tumor necrosis factor-α, IL-6, monocyte chemoattractant protein-1, and IL-1 (Kalupahana et al., 2011). Therefore, obesity is an inflammation-mediated condition. Hence, anti-inflammatory effect of corn silk means that it has the capacity to reduce the inflammation and excess fat accumulation associated with obesity in the HCD-fed rats (Ahmed et al., 2016).

Vemuri et al. (2018) established the anti-diabetics and antioxidative effectiveness of some new dietary supplements derived from the following such as corn silk, black asphaltum, raw jackfruit, and black plum which was carried out using in vivo and in vitro models. The biological actions of phytoextracts were compared with synthetic known diabetic drugs that are available commercially such as insulin, glibenclamide, aminoguanidine. These drugs could be utilized for the management of streptozotocin-induced hyperglycemia, inhibition of inflammatory mediators (iNOS), advanced glycation end-products when performed in an in vivo assay. The activities were established using the following assays such as agarose gel electrophoresis, superoxide anion scavenging activity, and histopathology, body weights, hydroxyl radical scavenging experiments, quantification of glycated hemoglobin, total antioxidant capacity, and lipid profile. The authors discovered that the natural extract from the entire natural product utilized during this experiment portends a maximum antixoxidant effectiveness that could prevent the glycation of hemoglobin. The experimental diabetes induction with STZ in rats affirmed that crude extract exhibited a clear influence on lowering lipid accumulation, weight gain, and inflammation-related gene expression. The Histomorphological analysis from the pancreas of rats that were treated indicated further enhance and support the result obtained during in vivo assay. Their study showed that natural extracts could be a source of functional nutraceuticals, foods most especially for the treatment of diabetics and other related complications. They could also be adjuvant therapy or a complete substitute for anti-diabetes drugs.

Interestingly, extract derived from corn silk has been documented to include numerous level of maysin, which is a major source of flavonoid. There are quite a few investigations that have validated the influence of corn silk on cholesterol metabolism. Also, Cha et al. (2016) determined the effect of corn silk on the hypercholesterolemia in high-fat diet fed mice. The following parameters were evaluated which include insulin levels, blood glucose and hepatic glucose levels, triglycerides, serum free fatty acid, and total cholesterol. The outcome of the experiment indicated that oral administration of the extract derived from corn silk most especially with the enhancement in the level of insulin and serum glucose decreased fatty liver disease induced by high-fat diet. Moreover, it was detected that corn silk suggestively raised adipocytokines mRNA expression and decreased farnesoid X receptor (FXR) and acylcholesterol transferase (ACAT). Their study indicated that extract obtained from corn silk supplementation containing a high-fat diet portends the potential to enhance the level of glucose homeostasis and adipocytokine secretion. Also, it was established that corn silk could reduce the level of regulatory pool of hepatic cholesterol via variation of mRNA manifestation levels.

11.6 NEUROPROTECTIVE ACTIVITY

Herbs have been identified as one of the natural gifts to humanity for adequate sustainability and sustenance of health status which represent the main form of basic modern medicine. The whole universe is now searching for a typical example of foods that could serve as a typical example of diet having adequate nutrition and could serve as functional foods. A typical example of such food is corn silk which normally follows the ear of the corn and has a fiber-like growth in their structure. Therefore,

Vijitha and Saranya (2017) validated the application of corn silk for effective management of several diseases. Some examples of pharmacologically active constituents that are responsible for its biological activities are terpenoids and flavonoids.

Pharmacological studies have shown that traditional herbs have several biological activities such as anti-hyperlidemic, neuroprotective, antidepressant, anti-inflammatory, antioxidant, and anti-diabetic effects (Abu-Odeh and Talib, 2021). Numerous studies have validated the medicinal effectiveness of corn silk extract (Wang et al., 2011; Tin-Yun et al., 2017). Moreover, there is a need to validate the safe dose range and toxicity of the corn silk. In view of the aforementioned, Ha et al. (2018) performed the toxicity studies on the corn silk extract in rodents. In order to establish the acute toxic effect of the corn silk possessing a greater level of maysin was administered to mice in these experiments. Examples of some of the parameters evaluated involve alteration in weight, mortality, and clinical symptoms, which were documented after 14 days. In order to establish the sub-acute toxicity, the extract derived from the corn silk administered to the mice for 28 days, while food intake and organ weight were evaluated. Moreover, serum and urine were evaluated. The result obtained revealed that there is no abnormal symptom and mortality from all the groups most especially during these periods. The result obtained from the body weight indicated that there is no substantial change in comparison to the untreated group. Also, result observed after 4 weeks indicated that there is no toxic effect of corn silk containing maysin on organ weight, body weight, clinical chemistry, water intake, urine parameters, and food consumption. The histopathological investigation performed revealed absence of abnormality associated to the administered corn silk extract with the concentration of 500 mg/kg. The maximum non-noxious dose of maysin was discovered to exceed 500 mg/kg (Ha et al., 2018). Folk therapies have established that corn silk was utilized for diabetes treatment for several decades most especially when consumed as a tea for effective management of improvement of urinary tract diseases (Hasanudin et al., 2012).

Neurological diseases are connected with oxidative stress-mediated neural cell apoptosis, and oxidative damage plays significant role in the pathology of neurodegeneration, as a result of elevated cellular reactive oxygen species ROS generation (Contestabile, 2001; Moreira et al., 2005; Angelova and Abramov, 2018). Corn silk extract has been reported to possess butyrylcholinesterase and acetylcholinesterase inhibiting potentials (Kan et al., 2012). Choi and coworkers have demonstrated the neuroprotective capacity of corn silk-derived maysin in neurodegenerative disease. They found out that corn silk maysin mediate its neuroprotective action through increased antioxidant effects (Choi et al., 2014). Interestingly, the neuroprotective effect of corn silk maysin is mediated via decrease in ROS, apoptosis, neuronal cell death, and DNA oxidative damage. Hence, corn silk-derived maysin will be a good adjuvant for the amelioration of neurological disorders and associated cognitive decline (Choi et al., 2014).

11.7 ANTICANCER EFFECT

Cancer has been identified as one of the factors responsible for many deaths globally. Moreover, breast cancer has been recognized as one of the most common factors responsible for the high rate of death recorded among women. In spite of the recent trends towards the discovery of a more potent drug and diagnosis of cancer, it has been discovered that there is poor survival rate. Recently, it has been discovered that the application of plant-derived natural products could be an alternative pharmacologically active substance that could be applied for the management of cancer because of their nontoxic nature. In view of the aforementioned, Al-Oqail et al. (2019) evaluated the anti-neoplastic effectiveness of the extract derived from the corn silk most especially on normal human mesenchymal (hMSCTERT4) cells. The level of cytotoxicity was established 24 hours after treatment through 3-[4,5-dimethylthiazol-2-yl]-2,5 diphenyl tetrazolium bromide (MTT), morphological and neutral red uptake (NRU) assay.

Furthermore, manifestation of apoptotic marker genes include caspase-9, p53, caspase-3, Bax, lipid peroxidation (LPO), and glutathione (GSH) and generation of ROS and MMP alteration were carried out on MCF-7 cells when administered 250–1,000 μg/ml of extract derived from the corn silk for 24 h. The rate of ROS and LPO generation was discovered to be expressively higher, while the *mitochondrial membrane potential* (MMP) and the GSH level were observed to be reduced in corn silk extract (CSE)-treated MCF-7 cells. Moreover, the result obtained indicated that there was a substantial upregulation in the level of caspase-9, p53, Bax, and caspase-3, while there was downregulation in the mRNA expression of Bcl-2 genes in MCF-7 cells treated with CSE. The results obtained from the investigation showed that CSE could prompt cell death through ROS-regulated mitochondrial pathway (Al-Oqail et al., 2019).

Corn silk has been identified as a nutritionally and economically significant natural product which signifies a staple food most especially for bigger percentage of the globe. Guo et al. (2017) performed the antitumor effectiveness of extract obtained from corn silk most especially in human gastric cancer cells and human colon cells. The application of CSE indicated the manifestation of cell death-associated incidents which include damage of mitochondrial membrane prospective ($\Delta\Psi m$), prevention of cell propagation, and liberation of Ca^{2+}. The result obtained indicated that corn silk extract portends the capability to prevent the multiplying of cancerous cells and enhanced cell death. Their studies indicated that the extract obtained from corn silk may stimulate cell death through the mitochondria mediated pathway.

11.8 CONCLUSION

Corn silk extract is a promising anti-hypertensive, neuroprotective, and anti-diabetic agent, and this has been validated by numerous experimental studies (in vivo and in vitro). Interestingly, it is a cheap, nontoxic, natural product that has been widely considered as agricultural waste. Hence corn silk extract will offer a cost-effective alternative for treatment of chronic diseases as an adjuvant or a sole treatment. Further studies into the bioactivities of corn silk chemical constituents are warranted especially using in vivo models to ascertain the mechanisms involved in the corn silk extract bioactivity and its pharmacologic attributes. More so, clinical trials are necessary to increase confidence in the favorable therapeutic activity of corn silk for humans.

REFERENCES

Adetunji, C.O., Anani, O.A., and Panpatte, D. (2021). Mechanism of actions involved in sustainable ecorestoration of petroleum hydrocarbons polluted soil by the beneficial microorganism. In: Panpatte, D.G., and Jhala, Y.K. (eds.), *Microbial Rejuvenation of Polluted Environment. Microorganisms for Sustainability*, vol 26. Springer, Singapore. https://doi.org/10.1007/978-981-15-7455-9_8

Adetunji, C.O., and Anani, O.A. (2021a). Recent advances in the application of genetically engineered microorganisms for microbial rejuvenation of contaminated environment. In: Adetunji C.O., Panpatte D.G., and Jhala, Y.K. (eds.), *Microbial Rejuvenation of Polluted Environment. Microorganisms for Sustainability*, vol 27. Springer, Singapore. https://doi.org/10.1007/978-981-15-7459-7_14

Adetunji C.O., and Anani O.A. (2021b). Utilization of microbial biofilm for the biotransformation and bioremediation of heavily polluted environment. In: Panpatte, D.G., and Jhala, Y.K. (eds.), *Microbial Rejuvenation of Polluted Environment. Microorganisms for Sustainability*, vol 25. Springer, Singapore. https://doi.org/10.1007/978-981-15-7447-4_9

Anani, O.A., and Adetunji, C.O. (2021c). Bioremediation of polythene and plastics using beneficial microorganisms. In: Adetunji C.O., Panpatte, D.G., and Jhala, Y.K. (eds.), *Microbial Rejuvenation of Polluted Environment. Microorganisms for Sustainability*, vol 27. Springer, Singapore. https://doi.org/10.1007/978-981-15-7459-7_13

Anani, O.A., Mishra R.R., Mishra P., Enuneku A.A., Anani G.A., and Adetunji C.O. (2020). Effects of toxicant from pesticides on food security: Current developments. In: Mishra, P., Mishra, R.R., and Adetunji, C.O. (eds.), *Innovations in Food Technology*. Springer, Singapore. https://doi.org/10.1007/978-981-15-6121-4_22

Abu-Odeh, A.M. and Talib, W. H. (2012). Middle East medicinal plants in the treatment of diabetes: A review. *Molecules* 26: 742. https://doi.org/10.3390/molecules26030742.

Adedapo, A., Babarinsa, O., Oyagbemi, A., Adedapo, A., Omobowale, T. (2016). Cardiotoxicity study of the aqueous extract of corn silk in rats. *Macedonian Veterinary Review* 39(1): 43–49.

Ahmed, H.H, Hamza, A.H, Kotob, S.E. (2016). Corn silk offers multi-mechanistic approaches in mitigating obesity in rodents. *Asian Journal of Pharmaceutical and Clinical Research* 9(4): 292–298.

Al-Oqail, M.M., Al-Sheddi, E.S., Farshori, N.N., Al-Massarani, S.M., Al-Turki, E.A., Ahmad, J., Al-Khedhairy, A.A., Siddiqui, M.A. (2019). Corn silk (Zea mays L.) induced apoptosis in human breast cancer (MCF-7) cellsvia the ROS-mediated mitochondrial pathway. *Oxidative Medicine and Cellular Longevity* 2019:9789241. doi:10.1155/2019/9789241.

Angelova, P.R., Abramov, A.Y. (2018). Role of mitochondrial ROS in the brain: from physiology to neurodegeneration. *FEBS Letters* 592(5): 692–702.

Bae, J.W., Kim, D.H., Lee, W.W., Kim, H.Y., Son, C.G. (2015). Characterizing the human equivalent dose of herbal medicines in animal toxicity studies. *Journal of Ethnopharmacology*, 162: 1–6.

Bai, H., Hai, C., Xi, M., Liang, X., Liu, R. (2010). Protective effect of maize silks (maydis stigma) ethanol extract on radiation-induced oxidative stress in mice. *Plant Foods and Human Nutrition* 65: 271–276.

Biljana, B.P. (2012). Historical review of medicinal plants' usage. *Pharmacognosy Reviews* 6(11): 1–5.

Brobbey, A.A, Somuah-Asante, S., Asare-Nkansah, S., Boateng, F.O., Ayensu, I. (2017). Preliminary phytochemical screening and scientific validation of the anti-diabetic effect of the dried husk of Zea maysL. (Corn, Poaceae). *International Journal of Phytopharmacy* 7(1): 01–05.

Cha, J.H., Kim, S.R., Kang, H.J., Kim, M.H., Ha, A.W., Kim, W.K. (2016). Corn silk extract improves cholesterol metabolism in C57BL/6J mouse fed high-fat diets. *Nutrition Research and Practice* 10(5):501–506.

Chang, C.C., Yuan, W., Roan, H.Y., Chang, J.L., Huang, H.C., Lee, Y.C., Tsay, H.J., Liu, H.K (2016). The ethyl acetate fraction of corn silk exhibits dual antioxidant and anti-glycation activities and protects insulin-secreting cells from glucotoxicity. *BMC Complementary Alternative Medicine* 16: 432.

Chen, S., Chen, H., Tian, J., Wang, Y., Xing, L., Wang, J. (2013). Chemical modification, antioxidant and α-amylase inhibitory activities of corn silk polysaccharides. *Carbohydrate Polymers* 98(1):428–437. doi: 10.1016/j.carbpol.2013.06.011.

Choi, D.J., Kim, S.-L., Choi, J.W., Park, Y.I. (2014). Neuroprotective effects of corn silk maysin via inhibition of H_2O_2-induced apoptotic cell death in SK-N-MC cells. *Life Science* 109: 57–64.

Contestabile, A. (2001). Oxidative stress in neurodegeneration: mechanisms and therapeutic perspectives. *Current Topics in Medicinal Chemistry* 1(6):553–568. doi: 10.2174/1568026013394723.

Dwivedi, N., Dwivedi, S., and Adetunji, C. O. (2021). Efficacy of microorganisms in the removal of toxic materials from industrial effluents. In: Adetunji, C.O., Panpatte, D.G., and Jhala, Y.K. (eds.), *Microbial Rejuvenation of Polluted Environment. Microorganisms for Sustainability*, vol 27. Springer, Singapore. https://doi.org/10.1007/978-981-15-7459-7_15

Ebrahimzadeh, M.A., Mahmoudi, M., Ahangar, N., Ehteshami, S., Ansaroudi, F., Nabavi, S.F, Nabavis, M. (2009). Antidepressant activity of corn silk. *Pharmacologyonline* 3: 647–652.

Fossen, T., Slimestad, R., Andersen, Ø.M. (2001). Anthocyanins from maize (Zea mays) and reed canarygrass (Phalarisarundinacea). *Journal of Agricultural Food Chemistry* 49: 2318–2321.

George, G.O., Idu, F.K. (2015). Corn silk aqueous extracts and intraocular pressure of systemic and non-systemic hypertensive subjects. *Clinical and Experimental Optometry* 98: 138–149.

Guo, H., Guan, H., Yang, W., Liu, H., Hou, H., Chen, X., Liu, Z., Zang, C., Liu, Y., Liu, J. (2017). Pro-apoptotic and anti-proliferative effects of corn silk extract on human colon cancer cell lines. *Oncology Letters* 13: 973–978.

Ha, A. W., Kang H. J., Kim, S.L., Kim, M.H. and Kim, W.K. (2018). Acute and subacute toxicity evaluation of corn silk extract. *Preventive Nutrition and Food Science* 23(1): 70–76. doi.org/10.3746/pnf.2018.23.1.70.

Hasanudin, K., Hashim, P., Mustafa, S. (2012). Corn silk (Stigma maydis) in healthcare: a phytochemical and pharmacological review. *Molecules* 17(8): 9697–9715. doi: 10.3390/molecules17089697.

Helmy, A., El-Shazly, M., Seleem, A., Abdelmohsen, U., Salem, M., Samir, A., Rabeh, M., Elshamy, A., Singab, A.B. (2020). The synergistic effect of biosynthesized silver nanoparticles from a combined extract of parsley, corn silk, and gum arabic: in vivo antioxidant, anti-inflammatory and antimicrobial activities. *Materials Research Express* https://doi.org/10.1088/2053-1591/ab6e2d.

Ho, T.Y., Li, C.C., Lo, H.Y., Chen, F.Y., Hsiang, C.Y. (2017). Corn silk extract and its bioactive peptide ameliorated lipopolysaccharide-induced inflammation in mice via the nuclear factor-κB signaling pathway. *Journal of Agricultural Food Chemistry* 65(4): 759–768. doi: 10.1021/acs.jafc.6b03327.

Hu, Q. L., Zhang, L. J., Li, Y. N., Ding, Y. J., Li, F. L. (2010). Purification and anti-fatigue activity of flavonoids from corn silk. *International Journal of Physical Sciences* 5(4): 321–326.

Hu, Q., Deng, Z. (2011). Protective effects of flavonoids from corn silk on oxidative stress induced by exhaustive exercise in mice. *African Journal of Biotechnology* 10: 3163–3167.

Islam, S., Thangadurai, D., Adetunji, C. O., Nwankwo, W., Kadiri, O., Makinde, S., Michael, O. S., Anani, O. A., and Adetunji, J. B. (2021) Nanomaterials and nanocoatings for alternative antimicrobial therapy. In: Kharissova, O.V., Martínez, L.M.T., and Kharisov, B.I. (eds.), *Handbook of Nanomaterials and Nanocomposites for Energy and Environmental Applications*. Springer, Cham. https://doi.org/10.1007/978-3-030-11155-7_3-1

Kalupahana, N.S., Claycombe, K. J. and Moustaid-Moussa, N. (2011). (n-3) Fatty acids alleviate adipose tissue inflammation and insulin resistance: mechanistic insights. *Advances in Nutrition* 2:304–316. doi: 10.3945/an.111.000505.

Kan, A., Orhan, I., Coksari, G., Sener, B. (2012). In vitro neuroprotective properties of the Maydis stigma extracts from four corn varieties. *International Journal of Food Science and Nutrition* 63(1): 1–4.

Kelly, K. (2009). *History of Medicine*. New York: Facts on file. pp. 29–50

Li, C.C., Lee Y.C., Lo H.Y., Huang Y.W., Hsiang C.Y., Ho T.Y. (2019). Antihypertensive effects of corn silk extract and its novel bioactive constituent in spontaneously hypertensive rats: the involvement of angiotensin-converting enzyme inhibition. *Molecules* 24: 1886. doi:10.3390/molecules24101886.

Liu, J., Wang, C., Wang, Z., Zhang, C., Lu, S., Liu, J. (2011). The antioxidant and free-radical scavenging activities of extract and fractions from corn silk (Zea mays L.) and related flavone glycosides. *Food Chemistry* 126: 261–269.

Maksimovic, Z., Malencic, Đ., Kovacevic, N. (2005). Polyphenol contents and antioxidant activity of Maydis stigma extracts. *Bioresources Technology* 96: 873–877.

Martín, N., Pantoja, C., Chiang, L., Bardisa, L., Araya, C., Román, R. (1991). Hemodynamic effects of a boiling water dialysate of maize silk in normotensive anaesthetized dogs. *Journal of Ethnopharmacology* 31: 259–262.

Moreira, P.I., Smith, M.A., Zhu, X., Nunomura, A., Castellani, R.J., Perry, G. (2005). Oxidative stress and neurodegeneration. *Annals of the New York Academy of Sciences* 1043:545–552 doi: 10.1196/annals.1333.062.

Morshed, S. and Islam, S.S. (2015). Antimicrobial activity and phytochemical properties of corn (Zea mays L.) silk, SKUAST. *Journal of Research* 17(1): 8–14.

Mishra, P., Mishra, R. R., & Adetunji, C. O. (Eds.). (2020). *Innovations in Food Technology: Current Perspectives and Future Goals*. Springer Nature Singapore Pte Ltd. 2020. Springer, Singapore. Pp. 143-162. 1st Edition. DOI: 10.1007/978-981-15-6121-4

Olaniyan, O.T., and Adetunji, C.O. (2021) Biochemical role of beneficial microorganisms: An overview on recent development in environmental and agro science. In: Adetunji, C.O., Panpatte, D.G., and Jhala, Y.K. (eds.), *Microbial Rejuvenation of Polluted Environment. Microorganisms for Sustainability*, vol 27. Springer, Singapore. https://doi.org/10.1007/978-981-15-7459-7_2

Oyabambi, A.O., Areola, E.D., Olatunji, L.A., Soladoye, A.O. (2020). Uric acid is a key player in salt-induced endothelial dysfunction: the therapeutic role of *Stigma maydis* (corn silk) extract. *Applied Physiology, Nutrition and Metabolism* 45(1):67–71.

Pan, Y., Wang, C., Chen, Z., Li, W., Yuan, G., Chen, H. (2017). Physicochemical properties and antidiabetic effects of a polysaccharide from corn silk in high-fat diet and streptozotocin-induced diabetic mice. *Carbohydrate Polymers* 164:370–378 doi: 10.1016/j.carbpol.2017.01.092.

Saheed, S., Oladipipo, A.E., Abdulazeez, A.A., Olarewaju, S.A., Ismaila, N.O., Emmanuel, I.A., Fatimah, Q.D., Aisha, A.Y. (2015). Toxicological evaluations of *Stigma maydis* (corn silk) aqueous extract on hematological and lipid parameters in Wistar rats. *Toxicology Reports* 2:638–644.

Sani, U.M. (2016). Anti-diabetic potential of methanol extract of cooked corn silk (*stigma maydis)* on alloxan-induced diabetes in albino mice. *The Pharmaceutical and Chemical Journal* 3(4): 68–72.

Stow, J.L., Ching Low, P., Offenhauser, C. and Sangermani, D. (2009). Cytokine secretion in macrophages and other cells: pathways and mediators. *Immunobiology* 214: 601–614.

Tian, J., Chen, H., Chen, S., Xing, L., Wang, Y., Wang, J. (2013). Comparative studies on the constituents, antioxidant and anticancer activities of extracts from different varieties of corn silk. *Food and Function* 4: 1526–1534.

Thangadurai, D., Dabire, S. S., Sangeetha, J., Said Al-Tawaha, A. R. M., Adetunji, C. O., Islam, S., Shettar, A. K., David, M., Hospet, R., and Adetunji, J. B. (2021). Greener composites from plant fibers: Preparation, structure, and properties. In: Kharissova, O.V., Martínez, L.M.T., and Kharisov, B.I. (eds.), *Handbook of Nanomaterials and Nanocomposites for Energy and Environmental Applications.* Springer, Cham. https://doi.org/10.1007/978-3-030-11155-7_21-1

Velazquez, D.V., Xavier, H.S., Batista, J.E., de Castro-Chaves, C. (2005). *Zea mays* L. extracts modify glomerular function and potassium urinary excretion in conscious rats. *Phytomedicine* 12: 363–369.

Vemuri, S.K., Banala, R.R., Katragunta, K., Madhuri, V., Raju, K., Annapareddy, V.G.R., Goli, P.V.S. (2018). Antioxidant, anti-inflammatory and anti-diabetic efficiency of indian medicinal plants against streptozotocin induced diabetes in male Wistar rats. *Free Radicals and Antioxidants* 8(1): 141–148.

Vijitha, T.P., Saranya, D. (2017). Corn silk: a medicinal boon. *International Journal of ChemTech Research* 10(10): 129–137.

Vranješ, M., Popović, B.M., Štajner, D., Ivetić, V., Mandić, A., Vranješ, D. (2016). Effects of bearberry, parsley and corn silk extracts on diuresis, electrolytes composition, antioxidant capacity and histopathological features in mice kidneys. *Journal of Functional Foods* 21: 272–282.

Wang, C., Zhang, T., Liu, S., Zhang, C., Wang, E., Wang, Z., Zhang, Y., Liu, J. (2011). Subchronic toxicity study of corn silk with rats. *Journal of Ethnopharmacology* 137(1): 36–43.

Wang, G.Q., Xu, T., Bu, X.M., Liu, B.Y. (2011). Anti-inflammation effects of corn silk in a rat model of carrageenin-induced pleurisy. *Inflammation* 35: 822–827.

Wang, K.J., Zhao, J.L. (2019). Corn silk (*Zea mays* L.), a source of natural antioxidants with α-amylase, α-glucosidase, advanced glycation and diabetic nephropathy inhibitory activities. *Biomedicine and. Pharmacotherapy* 110: 510–517.

WHO (2000). Promoting the Role of Traditional Medicine in Health Systems: A Strategy for the African Region 2001–2010. Document Reference AFR/RC50/Doc. 9/R, World Health Organization, Harare, Zimbabwe, 2000.

Yang, J., Li, X., Xue, Y., Wang, N., Liu, W. (2014). Anti-hepatoma activity and mechanism of corn silk polysaccharides in H22 tumor-bearing mice. *International Journal of Biological Macromolecules* 64: 276–280.

Zhang, Y., Wu, L., Ma, Z., Cheng, J., Liu, J. (2015). Anti-diabetic, anti-oxidant and anti-hyperlipidemic activities of flavonoids from corn silk on STZ-induced diabetic mice. *Molecules* 21(1):E7. doi: 10.3390/molecules21010007.

Zhao, W., Yin, Y., Yu, Z, Liu, J, Chen, F. (2012). Comparison of anti-diabetic effects of polysaccharides from corn silk on normal and hyperglycemia rats. *International Journal of Biological Macromolecules* 50(4):1133–1137.

Žilić, S., Janković, M., Basić, Z., Vančetović, J., Maksimović, V. (2016). Antioxidant activity, phenolic profile, chlorophyll and mineral matter content of corn silk (Zea mays L): comparison with medicinal herbs. *Journal of Cereal Science* 69: 363–370.

12 Bioengineering of Pharmacologically-Active Metabolites for Effective Drug Nano-Formulations from the Callus and Metabolites of Medicinal Plants and Their Significant Application in Nanomedicine

Recent Advances

Charles Oluwaseun Adetunji and Akinola Samson Olayinka
Edo State University Uzairue

Denisa Ficai and Anton Ficai
University Politehnica of Bucharest

Michael Olugbenga Samuel
Bowen University

Wilson Nwankwo
Edo State University Uzairue

DOI: 10.1201/9781003178903-14

Muhammad Akram, Rumaisa Ansari, Tehreen Riaz, and Rida Zainab
Government College University Faisalabad

Chukwuebuka Egbuna
Chukwuemeka Odumegwu Ojukwu University
University of Port Harcourt

Oluwaseyi Paul Olaniyan
Osun State University

Hameed Shah
University of Swabi

Ruth Ebunoluwa Bodunrinde
Federal University of Technology

Juliana Bunmi Adetunji
Osun State University

Jonathan C. Ifemeje and Michael Chinedu Olisah
National Biotechnology Development Agency

Mohammad Ali Shariati
K.G. Razumovsky Moscow State University of Technologies and Management

Kingsley Eghonghon Ukhurebor
Edo State University Uzairue

Daniel Ingo Hefft
Reaseheath College

Wadzani Dauda Palnam
Federal University

CONTENTS

12.1 INTRODUCTION

Nanomedicine is the use of nanotechnology or nanoparticles (NPs) for analyzing, diagnosing/detecting and treatment of disorders and diseases such as angiogenesis inhibition (Meng et al., 2010), inflammation (Roursgaard et al., 2008), as well as cancer (Jiao et al., 2010). NPs or nanomaterials (NMs) are small substances in the size range of 1–100 nm with distinctive physiochemical properties. NPs have been found to have diverse applications in the fields of agriculture, medicine and biology (Adetunji and Anani, 2021a,b,c; Adetunji et al., 2021; Dwivedi et al., 2021; Olaniyan and Adetunji, 2021; Mishra et al., 2020; Thangadurai et al., 2021; Anani et al., 2020; Islam et al., 2021). The most famous and widely used NPs are magnesium, titanium, silicon, copper, silver, zinc and gold (Ruttkay-Nedecky et al., 2017).

Callus culture, amongst the different plant cell cultures, is an active contributor for the formation of bioactive metabolites (BMs) used in the production of flavors, fragrances, pesticides, food additives, dyes and pigments. The synthesis or production of such metabolites through microbial systems is difficult. Callus culture is used in the formation of metabolites like allicin, altamisine, saikosaponins, narigin, limonin, caffeine, isoquiniline alkaloids, plaunotol, sterols, phenolic compounds, triterpenes, camptothecine, acridone, furoquinoline, etc. These compounds are important molecules and have active roles against a number of pathogens such as fungal and bacterial strains. Callus culture have been described to possess several biological activities like anti-cancer, anti-diabetic, hepatoprotective and neuroprotective activity (Eibl et al., 2018).

Several challenges had been identified with most of the synthesized drugs used in the management of diseases including high level of resistance, toxicity, allergies and numerous other adverse effects. The callus of medicinal plants has been identified as a repository of several biologically active molecules of medicinal importance, the utilization of these pharmacologically active metabolites for the synthesis of nanodrugs is a move towards sustainability, eco-friendliness, potent and cost-effective drugs without inherent adverse effects. The implication of the foregoing development is the evolving of a new drug age, thus heralding the permanent exit of conventional synthetic drugs from the realm of pharmaco-therapeutics.

This review intends to provide current state of knowledge about the application of pharmacologically active metabolites for the fabrication of nanodrugs. The activities of these nanodrugs in medicine include anti-tumor, antibacterial, anti-microbial, antiviral, anti-cancer, anti-diabetic, hepatoprotective and neuroprotective activities. In addition, some recent *in vitro* and *in vitro* applications of these synthesized drugs as well as their clinical trials were provided.

12.2 MEDICINAL PLANTS WITH PHARMACOLOGICAL-ACTIVE METABOLITES FROM THEIR CALLUS AND THEIR GENERAL OVERVIEW

Herbs are a major form of modern medicine and most herbal therapeutic compounds and derivatives from plant-based extraction forms the foundation for the innovation of novel medications or drugs (Rodrigues et al., 2016). According to reports from the World Health Organization (WHO), close to 80 % of the developing nations focus on the use of herbal preparation or conventional medicine for basic medical needs (Robinson and Zhang, 2011). Beutler (2009) reported that researchers have focused extensively on the studies of the mechanism of bioactive composites, its structural configuration and the pharmacological capability of several species of plants in producing potent constituents with fewer side effects (Patra et al., 2018).

Most of the important plant bio-actives have been reported to be at the risk of annihilation in their natural environment (Ali et al., 2019). The extraction of active metabolites from plants is best described as natural biochemical factories of the world and constitutes the major source of medicine (Chinnadurai et al., 2016). The potential effects of active metabolites in medicinal plants reside in the complex molecular structures of the plant. However, the use of plant-based cultures has revealed more about the active metabolites and the need for conservation of medicinal plants.

Recently, elucidation and propagation of plant metabolites have been undertaken by means of specialized procedures such as metabolic engineering, cell control (immobilization), elicitation and in situ product exclusion (Adhikari and Thapa, 2014). Metabolic engineering is a highly purposeful alternation of metabolic pathways utilizing recombinant DNA technology (Verma et al., 2012). Plant-based cultures are cost-effective and promising for industrial production of plant metabolites and are useful for nano-medicinal applications. Thus, successful attempts in bioengineering of pharmacologically active metabolites for effective drug nano-formulations from the callus of medicinal plants can lead to efficacious biotechnological inventions of precise, valuable and as yet unknown plant chemicals for applications in the field of nanomedicine (Adhikari and Thapa, 2014).

Among the various cell culture techniques, the bioengineering of callus cultures is viable in producing the bioactive secondary metabolites (SMs) from plants and improvement of plants profile to withstand abiotic and biotic stresses (Yousaf et al., 2019). Over the years, extracts from plant-based cultures have been utilized as food constituents, phytochemicals, as well as colorants, flavors, nutraceuticals, vital oils, sweeteners and antioxidant. Research has focused globally to explore the utilization of biosynthetic gene clusters of natural products for effective drug nano-formulations (Weber and Kim, 2016). In vitro formation of active SMs by callus culture from numerous medicinal plants has been reported by Abdin and Kamaluddin (2006) to possess valuable chemical compounds for nanomedicine. The list of examples of various calluses that have been utilized for production of nanodrugs is presented in Table 12.1.

TABLE 12.1
Medicinal Plants with Pharmacological-Active Metabolites from Their Callus

S/N	Species Name	Metabolite(s)	Family	Uses	References
1	*Acorus calamus*	α-Asarone and β-asarone	*Acoraceae*	Antispasmotic, antihelminthic, antibacterial, insecticidal, antiulcerative	Pandit et al. (2011)
2	*Celestrus paniculatus*	Celastrine and panicukatin	*Celestraceae*	Antiarthritic, hypolipidemic, antitumor, Memory booster, depression, paralysis	Liu et al. (2005)
3	*Chomifora mukul*	Guggulsterone	*Burseraceae*	Antimicrobial, antiarthritic, anti-inflammatory, anticarcinogenic, astringent, expectorant, antispasmodic, anthelmintic	Siddiqui (2011)
4	*Bacopa monnieri*	Bacosides	*Scrophulariaceae*	Antiinflammatory, analgesic and antipyretic, anti-epileptic agent, mental function longevity, disease fatigue and depression, energize the CNS	Lai and Baraik (2019)
5	*Glycyrrhiza glabra*	Glycyrrhizin and glycyrrhetinic	*Fabaceae*	Ulcer, antispasmotic, asthma, cough	Ram et al. (2011)
6	*Aegle marmelos*	Monoterpenes and Sesquiterprenes	*Rutaceae*	Antidiarrheal, antihelminthics, antibacterial, astringent, antidysenteric, antipyretic, demulcent, antiinflammatory	Maity et al. (2009), Yadav and Singh (2011)
7	*Gymneme sylvestre*	Gymnemic acid	*Apocynaceae*	Antibiotic, antimicrobial, anticancer, anti-hyperlipidemic	Pragya et al. (2014)

(*Continued*)

TABLE 12.1 (*Continued*)
Medicinal Plants with Pharmacological-Active Metabolites from Their Callus

S/N	Species Name	Metabolite(s)	Family	Uses	References
8	*Catharanthus roseus*	Vanblastine and Vincrisine	*Apocynaceae*	Anticancer, antileukemic	Magnotta et al. (2006)
9	*Artemisia annua*	Artemisinin	*Asteraceae*	Antimalarial	Wang et al. (2015)
10	*Coptis japonica*	Berberine	*Ranunculales*	Intestinal ailment	Yashik et al. (2010)
11	*Camptotheca acuminata*	Camptothecin	*Nyssaceae*	Anti-umor	Linss et al. (2014)
12	*Capsicum frutescens*	Capsaicin	*Solanaceae*	Counterirritant	Joseph et al. (2012)
13	*Papaver somniferum*	Codeine, Morphine	*Papaveraceae*	Sedative	Ibrahim and Fulya (2013)
14	*Digitalis lanata or D. purpurea*	Digoxin or Digitoxin	*Plantaginaceae*	Heart stimulant	Bhushan et al. (2018)
15	*Plectranthus barbatus*	Forskolin	*Lamiaceae*	Bronchial asthma	Alasbahi and Meizig (2010)
16	*Podophyllum peltatum*	Podophyllotoxin	*Berberidaceae*	Antitumor	Lalaleo et al. (2018)
17	*Cinchona ledgeriana*	Quinine	*Rubiaceae*	Antimalarial	Pratiwi et al. (2018)
18	*Sanguinaria canadensis*	Sanguinarine	*Papaveraceae*	Antiplaque	Akhil (2019)
19	*Melaleuca alternifolia*	Tea tree oil	*Myrtaceae*	Antibacterial, antifungal	Brophy et al. (2013)
20	*Valeriana jatamansii*	Valmane and Valepotriates	*Valerianaceae*	Antitumor	Aabid et al. (2012)
21	*Withania somnifera*	Withanoides	*Solanaceae*	Antioxidant, anti-inflammatory, immunomodulatory	Pati et al. (2009)
22	*Stevia rebaudiana*	Stevioside	*Asteraceae*	Antibacterial, anti-diabetic	Janarthanam et al. (2010)
23	*Elettaria cardamomum*	Cardamon	*Zingiberaceae*	Antifungal, antibacterial, antiviral, antidiuretic, antiplatelet, antihypersensitive, anti-cancer	
24	*Allium chinense*	Allicin	*Amarylidaceae*	Anticancer, antidiabetic and immunomodulatory effects	Salehi et al. (2019)
25	*Commiphora gileadensis*	Balsam	*Burseraceae*	Antihypertensive	Luz et al. (2010)

12.3 ANTIMICROBIAL IN VITRO ACTIVITY OF NANOPARTICLES SYNTHESIZED

Panda et al. (2009) stated that the extracts from *D. bipinnata* could be employed as substitute medicinal composite for handling and treatment of several microbial infectious diseases. Considering its significance and the addition of innovative dimensions, silver NPs (AgNPs) could be produced from the extract of the leaf of Db as alternative for treating occurring and re-occurring infections due to resistance patterns of microorganisms to conventional antibiotics. They synthesized AgNPs by characterization using procedures such as Ultraviolet-Visible (UV-Vis) spectroscopy, X-ray diffraction (XRD), Fourier transform infrared spectroscopy (FTIR), scanning electron microscopy (SEM) and energy dispersive X-ray spectroscopy (EDAX) analysis. Additionally, AgNPs were applied during the in vitro investigation against some fungi and bacteria like *Escherichia coli for their* anti-microbial action against urinary tract poisonousness, *S. aureus* employed for blood stream as well as blood epidemics, *S. mutans* employed for tooth decays and *Candida albicans* employed for candidiasis. The researchers utilized green-synthesized AgNPs as an additional drug to antibiotics to combat intracellular pathogenic bacteria, which has been highlighted as drug-resilient bacteria (Johnson et al., 2014).

Panda et al. (2009) observed that the inhibition zone for *S. aureus* was remarkably distinct being a Gram-positive bacterium in comparison with *E. coli*, *S. mutans* and *C. albicans*. Consequently, they reported that the possible reason for this variation between microorganisms could be due to the cellular make up of bacteria, which is known to possess homogeneous peptidoglycan layer. This made it less resilient to the AgNPs when likened to bacteria with other membrane complexes (Panda et al., 2009; Pandey et al., 2013). The anti-microbial effects of the plants has been shown to be around 8–11 mm (Udaya and Hazeena, 2002; Mohan et al., 2012; Prasad and Swamy, 2013, Shownmyan et al., 2013).

Franci et al. (2015) reported that the extract of *Salvia spinosa* was used in the biosynthesis of AgNPs which was assayed against *B. vallismortis*, *E. coli* and *B. subtilis* at different concentrations. Hence, this specifies that Gram-positive bacteria (+GB) are more susceptible to the extract than with Gram-negative bacteria (-GB). The action of AgNPs ratifies multifaceted approach of AgNPs in the bacteria exposure to the NPs (Franci et al., 2015). The approach of bactericidal action of AgNPs could be as a result of the reaction of the AgNPs with the outer covering leading to the production of oxidative species. More so, the occurrence of AgNPs in bacteria cell membrane has been verified in some previous works (Oves et al., 2018). AgNPs possess the capability to interrupt the penetrability of the membrane through piercing the cell membrane and triggering the outflow of intracellular adenosine triphosphate (ATP) and apoptosis of the treated bacteria cell (Hajipour et al., 2012; Ajitha et al., 2014). Ag ions released from AgNPs could act as reservoir triggering antibacterial action of AgNPs (Ajitha et al., 2014). Hajipour et al. (2012) stated that Ag^+ in their investigation have an utmost propensity to perform with sulfur and phosphorus accessible in the biomolecules like deoxyribonucleic acid (DNA) and ribonucleic acid (RNA). This kind of attachment could result in the interruption of DNA and RNA functions (Umashankari et al., 2012).

As stated by Sapna et al. (2018), AgNPs produced from the extract of the leaf of *C. asiatica* exhibited antibacterial activity compared with all tested pathogenic microorganisms. *S. aureus*, *P. aeruginosa* and *E. coli* were found to have their zone of inhibitions at 15 mm, while the extract was found to be effective on *Klebsiella pneumoniae* at 12 mm. However, the result for the AgNPs produced from callus extract exhibited maximum activity in contrast to *S. aureus* at 16 mm and least in contrast to *Klebsiella pneumoniae* at 13 mm zone of inhibition. This could be as a result of NPs having larger surface range and minor size, making it easier to bind to the external of the cell membrane which plays a crucial role in the inhibition of the bacterial cell.

Gao et al. (2014) reported that AgNPs have been extensively utilized in various aspects of medical and environmental sciences. The authors applied the metabolites derived from the callus of the *Allophylus serratus* in order to establish their anti-microbial effectiveness. The synthesized AgNPs was tried for antibacterial action in contrast to both +GB "*Bacillus subtilis* and *Staphylococcus aureus*" and -GB "*Klebsiella pneumoniae* and *Pseudomonas aeruginosa*". The antibacterial action of manufactured AgNPs was assessed using the wall diffusion technique. The manufactured AgNPs confirmed for their antibacterial action showed zones of inhibition in contrast to the considered bacteria. The uppermost region of inhibition (14 mm) was recorded in contrast to *Klebsiella pneumoniae* and *P. aeruginosa*. The minimum inhibition zone (6 mm) was detected in contrast to *S. aureus*. The AgNPs exhibited antibacterial action as a result of their huge surface area which affords improved interaction with cell wall of bacteria (Ibrahim, 2015). Moreover, an inhibitory effect was observed in contrast to *E. coli* and *Staphylococcus* species was detected in the AgNPs produced by employing the extracts from *Mentha piperita* (Mubarak Ali et al., 2011). *Acalypha indica* and *Solanum torvum* extracts built on the produced AgNPs were found to exhibit high noxiousness to *E. coli* (Krishnaraj et al., 2010) as well as *Pseudomonas aeruginosa* and *Staphylococcus aureus* (Govindaraju et al., 2010), respectively.

The modes of action through which silver NPs demonstrate its antibacterial action might be linked to the presence of positive charge (+q) on the Ag^+ ion. This could ensue via the electrostatic attraction between +q of AgNPs and negative charges (−q) on the cell membrane of microbe (Fox, 1968; Hamouda et al., 2001; Dibrov et al., 2002; Raut et al., 2010). The antibacterial action of AgNPs in contrast to the −GB is greater than that against the +GB. This could be attributed to the high disparity in the composition of the cell wall between +GB and −GB. The AgNPs of leaf as well as the callus-derived leaf extracts of *Allophylus serratus* displayed uppermost antibacterial action in contrast to the *Klebsiella pneumoniae* followed by that of the *Staphylococcus aureus* and *Pseudomonas aeruginosa*.

Some research reports have also validated the anti-microbial action of numerous anti-microbial potentials of several NPs (Malabadi et al., 2007; Malabadi and Vijay, 2007; Malabadi et al., 2010, 2012a–c). The test pathogenic organisms that were utilized in the research study of antibacterial activity +GB "*Bacillus subtilis* and *Staphylococcus aureus*", −GB "*Escherichia coli* and *Klebsiella pneumoniae*" and fungal strain "*Candida albicans*" (Malabadi et al., 2007; Malabadi and Vijay, 2007, 2008; Malabadi et al., 2010, 2012a–c). The plates were incubated instantaneously at about 37°C. The optical density (OD) at 600 nm for every reduction were achieved by means of spectrophotometer, which was further employed to designate the quantity of bacterial cells in cultures to screen for antibacterials and determine the minimum inhibitory concentration (MIC) (Malabadi et al., 2012a–c).

The antimicrobial actions in relation to the zone of inhibition considerably differ based on the type of the tested pathogens and the type of the extracts. This variation observed in the various rates of anti-microbial physiognomies of the NPs synthesized from the *callus* of medicinal plants depends majorly on their shapes and various sizes (Jordan et al., 1991; Laflamme et al., 2001; Mahroug et al., 2006; Malabadi et al., 2009; Nabikhan et al., 2010; Satyavani et al., 2011; Malabadi et al., 2012a–c; Ghosh et al., 2012; Vankar and Shukla, 2012). Attachment of the AgNPs to the sulfur-containing proteins of the bacterial cell membrane causes the permeability of the membrane to be elevated, thereby enabling the penetration of the active components present in the NPs activity in the cell (Ghosh et al., 2012). It has been shown that AgNPs consist of Ag (0) atoms. AgNPs are greater in size than a Ag ion, which enhances their anti-microbial effectiveness (Shankar et al., 2003a,b; Shankar et al., 2004; Ankamwar et al., 2005; Chandran et al., 2006; Sharma et al., 2007; Thirumurgan et al., 2009; Saxena et al., 2010; Thirumurgan et al., 2010; Nabikhan et al., 2010; Elumalai et al., 2010; Geethalakshmi and Sarada, 2010; Khandelwal et al., 2010; Prabhu et al., 2010; Savithramma et al., 2011a,b; Mukunthan et al., 2011; Gnanadesigan et al., 2011; Kouvaris et al., 2012; Ghosh et al., 2012).

Also, Ag ion as an antimicrobial agent may be best linked to the fact that it possessed a high antimicrobial efficiency against pathogenic microorganisms without any level of toxicity when evaluated in the cells of animals in comparison with other metals (Ahmed et al., 2016). Furthermore, there has been a cumulative attention in the usage of Ag as NPs in the medical fields as a substitute for anti-microbial agent. Considering the issues of multidrug resistance to the conventional antibiotics, both +GB and −GB have recently been developed (Frieri et al., 2016); as efforts are being made to explore and produce more effective anti-microbial agents that could combat strains of multidrug-resilient bacteria (Kapil, 2005).

As reported by Singh et al. (2014), AgNPs formed by green synthesis by means of *A. rigidula* extract showed antimicrobial action against −GB (*E. coli* and *P. aeruginosa*), and the medical strain of *P. aeruginosa* and +GB (*B. subtilis*) verified in the study. Remarkably, the MIC values detected for *P. aeruginosa* are comparable to the attained values of AgNPs formed by *Phyllanthus amarus* in aqueous extracts. This is in contrast to many other clinical strains of *P. aeruginosa*, where MIC values ranged from 6.25 to 12.5 ppm. The MIC values attained for *B. subtilis* were significantly lower than the earlier obtained value of 12.5 ppm MIC by the AgNPs attained from rhizome aqueous extract of *Coptis chinensis* (Ahmad et al., 2017). Hitherto, it was also shown from the study that lesser MIC values for *E. coli*, where an MIC value of 25 ppm was obtained (Ahmad et al., 2017). A comparable minimum bactericidal concentration of the green produced AgNPs in contrast to the microbe tested was initiated by means of plant extracts as agents of reduction and stabilization. Consequently, it can be deduced that AgNPs formed via an *A. rigidula* extract exhibit active antibacterial action (both bactericidal and bacteriostatic) in contrast to +GB and −GB (Pattanayak et al., 2015; Dakal et al., 2016; Shankar et al., 2017; Nwankwo et al., 2020).

Sharifi-Rad et al. (2014) reported the anti-microbial effectiveness of allicin, AgNPs and their composites containing allicin and AgNPs. The authors tested the minimum bactericidal concentration (MBC), MIC and colony forming units (CFU) against *Methicillin-resilient Staphylococcus aureus* (MRSA) strains. The authors concluded that the combination of AgNPs and allicin has the best synergistic effects which can be used as an antimicrobial agent in dealing with skin infections that are instigated by MRSA strains.

Luo et al. (2009) reported the comparative study of pure allicin and polybutylcyanoacrylate (PBCA) NPs encumbered with allicin. Their results against the six fungal strains favor better results for the polybutylcyanoacrylate NPs loaded with allicin, which is attributed to the NP's virtuous lengthy release consequence and nano scale proportions.

Rabinkov et al. (1998) ascribed the biological potency of allicin to its instant reaction with thiol containing proteins. Hence, they consider allicin as a modulating agent that can regulate enzymatic activity of SH comprising of the enzymes by a thiol-disulphide exchange reaction. Thus, the SH-modifying properties of allicin may play a crucial role in its biological activity.

12.4 ANTIMICROBIAL IN VIVO ACTIVITY OF NPs OBTAINED FROM CALLUS OF MEDICINAL PLANTS

The in vivo cytotoxic influence of the AgNPs has been discovered to depend on some features like the NPs size and dose, the route or path of administration, the time or period of exposure and type of animal model. Reportedly, AgNPs accrue mostly in the targeted animal organs via several paths or routes of administration, thereby encouraging toxic manifestations like irritation, cell dysfunction, destruction of DNA and loss of lives of animal (Dziendzikowska et al., 2012; Gaiser et al., 2013; Lee et al., 2013; Wen et al., 2015; Zhang et al., 2015; Recordati et al., 2016; Fennell et al., 2017; Qui et al., 2017; Yang et al., 2017; Wen et al., 2017).

Liver is one of the vital organs considered during drug administration due to its role in biotransformation, detoxification and elimination roles including translocation of AgNPs in the blood circulation cycle (Wen et al., 2015; Recordati et al., 2016; Fennell et al., 2017; Yang et al., 2017; Wen et al., 2017). Accrued AgNPs in the liver could result in some adverse consequences like the development of reactive oxygen species (ROS), vicissitudes of pathological processes in the morphology of the hepatic cells and the enzyme activity. Dziendzikowska et al. (2012) intravenously vaccinated 20 and 200 nm AgNPs to male rats at a dosage of 5 mg/kg. The AgNPs was translocated from the veins/arteries to spleen, brain, liver, lungs and kidneys, with the foremost target organ being liver. The concentration of Ag ion present in this rat organ helps the preservation with AgNPs (20 nm) meaningfully greater than that of the preserved AgNPs (200 nm). Additionally, the concentrations of the Ag in these organs exhibit a time-dependent and size-dependent accrual means. Lee et al. (2013) induced AgNPs into Sprague-Dawley rats intraperitoneally, and stated that the AgNPs amassed primarily in the hepatic cells. The AgNPs resulted in substantial upsurge of caspase-3 level in preserved rat liver from the very 1st day up to the 30th day. Even though the autophagy was stimulated following injection on the 1st day, the inability of autophagy reservation in the subsequent days resulted in dysfunction of the hepatic cells and ensuing apoptosis.

Also, Recordati et al. (2016) reported that intravenously vaccinated CT- and PVP-coated AgNPs together with Ag acetate into CD-1 mice while Yang et al. (2017) utilized commercially available "CT-AgNPs and PVP-AgNPs" with sizes of 10, 40 and 100 nm. The cytotoxic effect of coated AgNPs has been established to be size-dependent, whereas the kind of coating (which could either be CT or PVP) could hardly have any impact on the bio-distribution of the AgNP present in the organ tissues (Yang et al. 2017). Ma et al. (2015) stated that the RT-qPCR study of the hepatocyte discovered considerable variations in the expression of gene outlines, for example up-regulation of some genes like *p*53, caspase-3, caspase-8, transferrin (Trf) and Bcl-2. It has been known that caspases are enzymes which could result in apoptosis by slicing the cellular proteins. Caspase 2, 8, 9 and 10, which are initiator caspases, recruit the apoptotic procedure, resulting in the instigation of the effector caspases such as caspase 3, 6 and 7.

Wen et al. (2017) in their study venously administered Sprague-Dawley rats with CT-AgNPs (6.3 nm) with a dose of 5 mg/kg body weight correspondingly. Reportedly, the Ag^+ ion concentration circulating in the organs from uppermost to lowermost was in the descending order: lung > spleen > liver > kidney > thymus > heart (Wen et al., 2017). Salem et al. (2015) established that AgNPs possess an antibacterial action which could cause a decrease of CFU in the set preserved with the AgNPs compared to the unprocessed group. The AgNPs were synthesized using *Caltropis procera* leaves and the fruit extract that was given to the toddler mice populated with *Vibrio cholerae* or enterotoxic *E. coli*, meaningfully condensed from 75 to 100 multiplied by the pathogenic bacteria population rates. In addition, nano-biocomposites of AgNPs produced using *Syzygium cumini* leaf into a medium of cellulose nanocrystals from *Dendrocalamus hamiltonii* and *Bambusa bambos* leaves exhibited antibacterial action and were active in locating an enhanced wound treatment during an in vivo study using the skin of male Swiss albino mice (Singla et al., 2017). Moreover, flavonoid-loaded AgNPs were synthesized using *Ricinus communis* leaf extract mixed with flavonoid, which exhibited a higher effectiveness as curative for *Staphylococcus aureus* contagion on silkworm *Bombyx mori* larvae. SMs, such as flavonoids, were utilized in the synthesis of AgNPs (Anjum et al., 2016; Hussain et al., 2017) similar to the observation from their study. Their existence on the surface can be a factor determining the anti-microbial action of these NPs.

One of the foremost differences between −GB and +GB is the width of the cell wall. −GB comprises of 2–3 nm thick, tiny peptidoglycan, between the external membrane and the cytoplasmic or internal membrane (Murray et al., 1965), whereas +GB consists of a denser layer of peptidoglycan of approximately 30 nm thick, with absence of external membrane (Shockman et al., 1983). Dibrov et al. (2002) stated that in *B. subtilis*, the growth inhibition could be accredited to the relations between the AgNPs in entrenched cell wall, instigating the membrane porousness due to the structural variations in the cell wall. While in −GB like *P. aeruginosa* and *Escherichia coli*, the external membrane is made up of primarily proteins and lipids serving as the primary structures bump into by the AgNPs. Consequently, the AgNPs can stick to these structures, changing their features and veracity-forming gaps inside the membrane, increasing its penetrability. Also, Pal et al. (2007) and Holt and Brad (2005) reported that the cell membrane comprises huge quantities of sulfur-comprising proteins with their inactivation could yield a part in the wear of the membrane formed by the AgNPs. Hence, these categories of proteins could be an alternative target of the AgNPs, deactivating enzymes and assisting in the weakening of the membrane. Again, Pal et al. (2007) and Holt and Brad (2005) also described in their studies that the antimicrobial action of colloidal AgNPs is affected by diverse factors such as the shape, the size and the capping agent (Kaviya et al., 2011). In this context, it has been reported that small sized AgNPs possess an improved bactericidal activity as a result of a superior binding surface when compared with the bigger AgNPs (Nwankwo et al., 2020). Furthermore, capping of the AgNPs fabricated comprises of OH functional groups (FGs) from polyphenol composites existing in the organic environment. Das et al. (2017) reported that the FGs could generate impairment and noxiousness in the bacteria membrane and cell wall by generating free radicals and ROS via oxidative stress. Nevertheless, AgNPs have been reported to be proficient in generating ROS for microbial growth inhibition (Kim et al., 2007). In reality, the sizes of AgNPs range from 20 to 80 nm mostly by the release of Ag^+ ions. Moreover, AgNPs of 22.46 nm are within the size of NPs described and possibly the Ag^+ ions release might be the cause for growth inhibition against the tested microorganism. Ivask et al. (2013) validated that Ag^+ ions can possibly interrelate with the cell wall and membrane or partake in the production of the ROS. Consequently, this machinery could result in the variations and destruction in the penetrability and composition of the bacteria cell. Furthermore, it has been revealed that the release of Ag^+ ions constitute the basis for the higher antibacterial effectiveness of AgNPs when tested against microorganisms (Ivask et al., 2013).

Koga et al. (1998) reported the plaunotol interactions with bacterial membranes and found that plaunotol (a diterpenoid) incorporated into a cell membrane led to an increase in cell membrane fluidity. Yeh and colleagues in 2009 isolated eight types of triterpenes from *Antrodia camphorata* and found that apoptosis induction was facilitated through the mitochondrial-dependent trail and also blocked the cell cycle development at the sub-G1 phase.

12.4.1 Wound Healing Properties of Some NPs

Makarov et al. (2014) reported that synthesis of green NPs from plant products or extracts are more cost-effective with minimal detrimental effect to the surroundings when compared to the physico-chemical procedures which have been applied before (Makarov et al., 2014). *Genipin* is being organized from geniposide via the enzyme β-glucosidase, which is generally taken out from *Penicillium nigricans*. Genipin forms a cross-link with chitosan, and PEG and AgNPs are milled into a nanocomposite for improved wound treatment and high antimicrobial efficacy (Liu et al., 2012). The AgNPs synthesized from *Coleus forskohlii* root extract was revealed to be effective in the treatment of wounds in male albino rats (Naraginti et al., 2016). AgNPs were formed in octadecylamine-modified montmorillonite clay mixed with *Homalomena aromatica* extracts and hyper-branched epoxy to generate AgNPs for wound treatment benefits (Barua et al., 2015). Though, this nanocomposite served as an effective wound treatment support in characteristic antimicrobial features.

Naraginti et al. (2016) reported that gold (Au), green-synthesized titanium dioxide (TiNPs), copper oxide NPs could serve as another source of wound treatment, which could result in swift wound treatment, prevention or reduction of infections as well as post-treatment side effects. AuNPs produced by *Coleus forskohlii* root extract strangely accelerated re-epithelialization and removal of the formed wound in rats, as well as amplified connective tissue development, and helped the propagation and relocation rate of epidermal cells (Naraginti et al. 2016). The synthesis of TiNPs from *Moringa oleifera* leaf has been reported to improve wound healing and condensed the excision site of the wound in albino rats (Sivaranjani et al., 2016). Sankar et al., (2015) also observed similar results in albino rats after treating the excision site of the wound with biosynthesized CO_2NPs employing *Ficus religiosa* leaf extract. These NSs in addition influenced insistent inhibitory potential of the extract in contrast to human pathogenic bacteria, and efficiently amplified the development of macrophages, fibroblast and collagen fibers (Sankar et al., 2015).

12.5 *IN VITRO* AND *IN VIVO* TRIAL AND BIOLOGICAL POTENCY OF NPs OBTAINED FROM CALLUS AND PLANT EXTRACT OF MEDICINAL PLANTS

Plant-derived SMs are composites with a diverse and sophisticated chemical structure that elucidate their importance in nanomedicine research. Plant-based cultures are the major complement to the traditional method of industrial production of bioactive plant metabolites (Dörnenburg, 2010). Although biochemical complexity of medicinal plants are quite challenging, a number of pharmacologically active metabolites are applied for effective drug nano-formulations from the callus; and extracts of medicinal plants have been designed exclusively for diagnostic purposes. Plant cells are potent and more convenient for extraction of pharmacologically active metabolites for production of nano-formulations than from the intact plants. Therefore, understanding the biosynthetic mechanisms of nano-formulations from the callus of medicinal plants and subsequently systemizing them into bioengineering framework may provide useful information during in vitro and in vivo trials (Du et al., 2011).

In 2019, Georgiev et al. stated that bioengineering of plant culture *in vitro* and in vivo is a key target for mass creation of therapeutically relevant SMs for drugs nano-formulations. The utilization of such plant-based *in vitro* callus cultures offers a wide potent usage in pharmaceutical sciences but also in agriculture and horticulture for the rapid propagation and short time production of SMs with limited space.

Medema and Fischbach (2015) asserted that metabolites of callus plant origin are a useful source of diverse drug molecules, among many others which include antibiotic, anti-carcinogenic, immunomodulatory, anti-malarial, anti-hypertensive, lipid-lowering agent, and anesthetic agents. Some drugs on the commercial and industrial treatment list include *Artemotil* resulting from *Artemisia annua* L. (*Asteraceae*), a local traditional or conventional Chinese herbal plant applied for handling of malaria, *Reminyl* (an acetylcholinesterase inhibitor) resulting from *Galanthus woronowii* Losinsk applied for the treatment and handling of Alzheimer's disease, *Paclitaxel* obtained from *Taxus brevifolia* plant used for cancer treatment and *silymarin* from *Silybum marianum* used for liver disease treatment (Atanasov et al., 2015).

The WHO statistics has shown that about 12 million people globally suffer from infections associated with diverse leishmanial species (Karunaweera and Ferreira, 2018). Leishmaniasis, a globally known life-threatening disease instigated by over 20 diverse leishmanial parasites is one of the utmost ignored tropical diseases (Hotez and Pecoul, 2010). Many attempts using different plant extracts such as *Khaya anthotheca* (hexane, petroleum), *Allium cepa* (aqueous extract), *Tylophora hirsute* (methanol extract), *Morinda citrifolia* (fruit extract), *Allium sativum* (methanol), *Lantana camara* (oil), *Maesopsi seminii* (dichloromethane) and *Porophyllum ruderale* (alkyl extract) were not potent to exert anti-leishmanial effect (Ullah et al., 2016). However, Yousaf et al.'s (2019) study on *Artemisia scoparia* using callus cultures elucidated the putative anti-leishmanial metabolites using MIT and cell apoptosis assay. Yousaf et al.'s (2019) protocol recommended for industrial production of effective phytomedicines in the treatment of leishmaniasis.

Another life-threatening disease is cancer which is an irregular uncontrollable development and propagation of cells, which causes over 6 million mortalities every year (Elisha et al., 2014; Rafik et al., 2014). Jafarain et al. (2014) had shown that callus extract of *Aervaja vanica* has auspicious anti-cancer efficacy than the leaf extracts of "*Moringa oleifera*" against Hela cells. A study by Kamalanathan and Natarajan (2018) showed the *in vitro* effect in contrast to the tested MCF7 breast cancer cells at lower concentration (about 11.89 and 22.45 μg/ml).

Mohammed and colleagues in 2015 also suggested the efficacy of callus cultures for the creation of plant active metabolites over the wild types. Their experiment using *Andrographis lineate*, a traditional or conventional medicine as a substitute for *Andrographis paniculate*, shows that the plant is a possible source of cancer chemotherapeutic agents. Researchers have been investigating on the angiogenesis inhibitors and promoters from plants for decades (Chinnadurai et al., 2016). Anti-angiogenesis and anti-cancer efficacy of the leaf and its callus extracts from *Baccharoides anthelmintica* (L.) Moench (*Asteraceae*) investigated by Chinnadurai et al. (2016) showed that both ethanol extract derived from the callus of *Baccharoides anthelmintica* effectively reserved the consequence of new blood vessels in dosage dependent means and tenor which is supported by Kalimuthu et al. (2014).

12.6 SOME PHYTOCHEMICALS OBTAINED FROM CALLUS AND METABOLITES OF DIFFERENT MEDICINAL PLANTS

Alkaloids are compounds produced by higher plants during their metabolism and contain nitrogen in their structure. Alkaloids play an irreplaceable role in the medicinal Chemistry attributed to their various beneficial characteristics including their anti-tumor, anti-microbial, and antiviral potentials, as well as agents for treating genetic disorders. Their vast and diverse actions are believed to be based on their highly specific but reversible and non-covalent binding to structures of nucleic acids leading to the modification of genetic material (Denny 1989; Hurley 2002; Maiti and Kumar 2009). SMs from *Centella asiatica* (L.) callus cultures have been described in an experimental study reported by Rao et al. (2015). Their study estimated the production of alkaloids, saponins, terpenoids, and flavonoids in the callus cultures of the plant. It was found that the alkaloid content increased in the callus cultures which had been originated from leaf extracts. Alkaloid content was found to be 4.2 mg/g dry weight in callus formed in medium complemented with 2 mg/l 2, 4-D + 0.5 mg/l BAP (Rao et al. 2015).

Several kinds of alkaloids are found in different plants, vinca alkaloids being some of the most common, and here including vincristine and vinblastine, isoquinolone alkaloids. The vinca alkaloids are used as important anti-cancer drugs in therapeutic formulations such as vincristine and vinblastine. The production of the vinca alkaloids is very low in plants, which makes them even more special and expensive in pharmaceutical drugs. Oxidation of the methyl group in vinblastine converts it into vincristine. Both vincristine and vinblastine are found in the callus cultures of the *Catharanthus roseus* (Iskandar and Iriawati, 2016) and they show significant anti-cancer activities. Microtubule stability plays a vital role in several cellular events like cell division, which cause microtubules a significant target for the anti-cancer drugs or chemotherapy (Jordan and Wilson, 1998). A wedge is introduced at the interface of two tubulin molecules by vincristine and vinblastine that averts the accumulation of heterodimers on the growing microtubules (Gigant et al. 2005). Mitotic spindle of dividing cells is not fashioned as a result of this, which caused the stimulation of c-Jun NH2-terminus kinase (JNK) (Gigant et al. 2005), phosphorylation of Bcl-2 (Ruvolo et al. 2001), arrest of G2/M and apoptosis (Wang et al., 2000). It has been described that proteins such as TCTP, FKBP59, L-plastin or HSP90 in CCRF-CEM leukemia cells are involved in initiating a cellular response to vincristine (Verrills et al. 2003). Moreover, it has also been reported that vincristine regulates the apoptosis of human tumor cells through the NF-kappaB/IkappaB pathway (Huang et al. 2004).

Reversible drug-enzyme-DNA cleavage complexes are formed by quinolones when they trap DNA gyrase which ultimately results in bacteriostasis. Chromosome fragmentation in pathways dependent or independent on protein synthesis could cause cell death (Cheng et al. 2013).

12.6.1 Essential Oils

Studies have suggested that some of the plant essential oils are involved in changing the penetrability of the cell membrane or its veracity. Examples of plants containing such essential oils include *Cinnamomum verum* J. Presl (Lambert et al., 2001), *Cinnamomum cassia* (L.) J. Presl (Oussalah et al., 2006), *Ocimum gratissimum* L. (Nguefack et al., 2004), *Syzygium aromaticum* (L.) Merrill & Perry (Rhayour et al., 2003), *Origanum vulgare* L. (Lambert et al., 2001; Rhayour et al., 2003; de Souza et al., 2010), *Rosmarinus officinalis* L. (Pérez-Fons et al., 2006), *Cymbopogon citratus* (DC.) Stapf (Nguefack et al., 2004), *Thymus eriocalix* (Ronniger) Jalas (Rasooli et al., 2006), *Corydothymus capitatus* (L.) Reichb (Pérez-Fons, Aranda et al., 2006), *Thymus vulgaris* L. (Horváth et al., 2009), *Satureja montana* L. (Oussalah et al., 2006), Thymus x-porlock (Rasooli, 2006), *Kaempferia pandurate* Roxb. (Jenie et al., 2008), *Origanum compactum* L. (Bouhdid et al., 2009), *Sinapis alba* L. (Turgis et al., 2009), *Cuminum cyminum* L. (Pajohi et al., 2011), *Coriandrum sativum* L. (Gill et al., 2002; Silva et al. 2011a, b), *Menthalongi folia* (L.) Huds. (Hafedh et al., 2010). A study performed on the callus cultures of *Ocimum basilicum* L. showed that under continuous illumination there was an upsurge in the percentage production of essential oils as well as several other important compounds (Purohit and Khanna, 1983).

12.6.2 Flavonoids

In legumes, common SMs are isoflavones having similarity with estradiol, a female sex hormone. Hence, they are also called as phytoestrogens. They display activities similar to that of estrogen while also possessing antioxidant properties. They work through inhibition of tyrosine kinases. These properties are said to be characteristic of their anti-cancer potential, and can also be used for women in their menopause or with the osteoporosis. Isoflavones found in red clover (*Trifolium pratensis*) and soybean (*Glycine max*) are marketed as nutraceuticals (Van Wyk and Wink, 2015).

12.6.3 Saponins

Saponins are the glycosides of steroids or triterpenes which represent the steroid parts of saponins which are characteristic of some categories of monocots, and are not common in dicots. At times, triterpenes, steroids and saponins structurally look similar to endogenously present anti-inflammatory hormones such as glucocorticoids. Corticomimetic effect could account for the anti-inflammatory effects of various medicinal plants. Glycyrrhizic acid which is found in *Glycyrrhiza glabra* shows a significant anti-inflammatory activity and is a triterpene saponin which is sweet in taste (Van Wyk and Wink 2015). Bidesmosidic compounds contain two sugar chains and some saponins may be stored as a compound in the vacuole. A β-glucosidase or an esterase cleaves them to monodesmosidic compounds which represent the active form. This is mediated by wounding-induced decompartmentalization that could be developed into a cholesterol complex in biomembranes and amphiphilic compounds. The water loving sugar portion remains external and interrelates with numerous carbohydrate conjugated proteins or glycolipids. Consequently, temporary or perpetual pores are produced in the membrane leading to the leakage which is easily manifested by the red blood cells that become leaky in the presence of saponins and caused the outward flow of hemoglobin. Monodesmosidic saponins could join in the surface glycolipids and glycoproteins with their side chain containing sugar (Teuscher and Lindequist 2010; Herrmann and Wink, 2011; Makky et al. 2014). This causes leakage and a serious tension in the biomembrane. Moreover, use of saponins as anti-infecting agents in traditional medicine has been well established (Woldemichael and Wink 2001; Van Wyk and Wink 2015).

12.6.4 Terpenes

Terpenes (mono-, sesqui-, di- and triterpenes) are stored in glandular trichomes or oil cells as well as resin ducts and they are substances which are highly hydrophobic in nature. Terpenes building blocks are C5-units and they can be further sectioned into polyterpenes, (C10) monoterpenes, (C15) sesquiterpenes, (C20) diterpenes, (C30) triterpenes and (C40) tetraterpenes. (C27) Steroids are derivatives of triterpenes (Dewick 2001; Teuscher and Lindequist 2010). Biomembranes are usually interacted by most of them. An uncontrolled outward flow of ions and metabolites can ensue as a result of amplified fluidity of the membranes. Tempering of receptors and membrane proteins or even cell leakage causing cell death can also be a consequence of this. This membrane activity of terpenes being very non-specific affects several living organisms which may be microbes such as fungi, bacteria and some vertebrates, insects and even including the membrane-enclosed viruses. The toxicity of several diterpenes has also been reported. Protein kinase C is stimulated by phorbol esters leading to the production of severe inflammation. Andromedotoxin or Grayanotoxin I is frequent in Ericaceae and is reportedly strong blocker of sodium medium and therefore potent neuron toxicants (Wink and Van Wyk, 2008).

12.7 INTRODUCING INFORMATICS AS A NECESSARY VEHICLE IN THE BIOFABRICATION, PHARMACEUTICAL NANO-FORMULATION AND NANOMEDICINE APPLICATIONS

The essence of medicinal drug production, notwithstanding the means of synthesis, is to aid in the prevention, control and management as well as possible elimination of infections, diseases and other health complications that have become encumbered to humans generally. With the ancient traditions of doing things, no innovative results could be rationally expected. The fabrication of pharmacologically active metabolites and their use in pharmaceutical nano-formulation for the synthesis and use in nanomedicine could be enhanced with the help of some of the recent advances in technology. Consequently, the role of information technology has been documented in every sector and domain across all socioeconomic and politico-cultural structures of the society. Moreover, it has become clearer that moderate or no success can be recorded in recent times without the use of technology.

Consequently, informatics has evolved in the last decade as an interdisciplinary scientific field involving procedures, progressions and means through which adequate support could be rendered to both healthcare delivery and all other arrangements that could assist in the provision of resolutions to health complications generally. Hence, biomedical informatics (IF) also called bioinformatics (BIF), cheminformatics (CIF), nanoinformatics (NIF) and health informatics have all become aspects of dissertation presently as a result of their significance in all segments of the medical and allied sectors. It should be noted that the various areas of IF aforementioned are intertwined and application choice is habitually a matter of perception. In this chapter, due to the simple fact that all areas of informatics mentioned above are relevant, the generic term informatics is adopted and the reader should construe the term in its general context. The relationship between these various spheres of informatics is presented in Figure 12.1. From Figure 12.1, it is easily deducible that BIF and CIF are relatively corresponding whereas NIF is a hybrid of the two, an exceedingly differentiated constituent of nanotechnology (NT). Health IF appears to be the endpoint of all other alternates. Hence, its purpose discovers the outcomes from the other scopes and employs such outcomes in managing and improving medical delivery mechanisms.

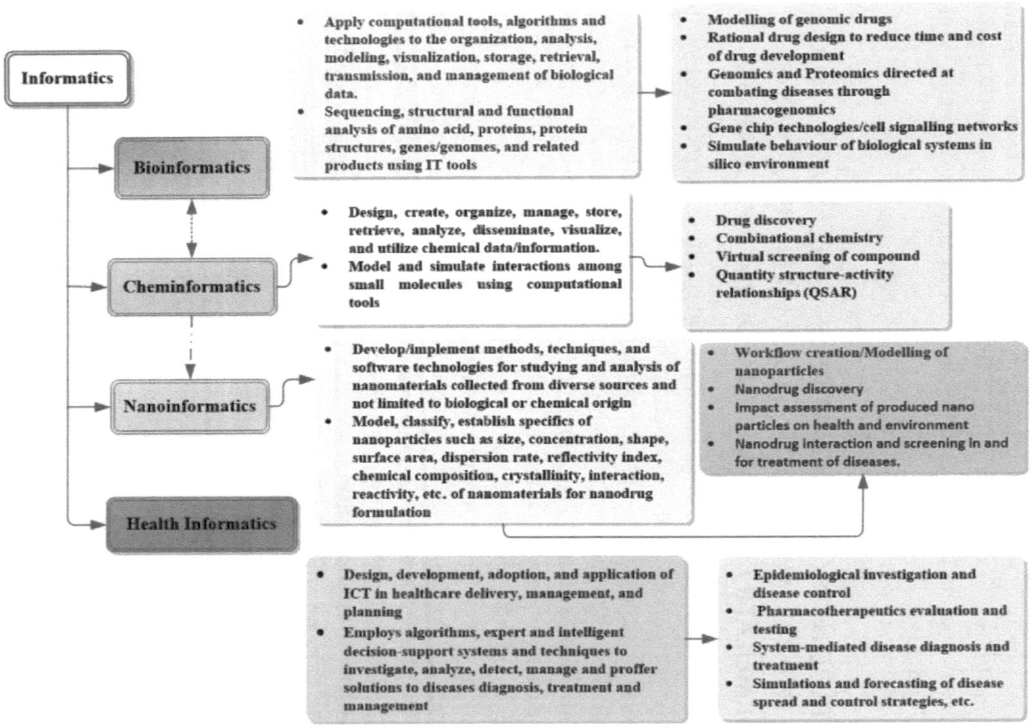

FIGURE 12.1 Informatics and applications to the realm of nanodrug discovery, formulation, production and treatment of diseases.

12.7.1 Role of IF in Biofabrication and Pharmaceutical Nano-formulation for Nanomedicine Applications

There are different techniques that could be applied in characterization of the most active chemical compositions available in molecules while the application of models as well as other chemical statistical techniques could be applied in subsequent storage of chemical in various databases for imminent referencing and investigation (Nwankwo et al., 2020).

In silico modeling of chemical compositions and related multidimensional imaginings permit the virtual investigation and examination of the connections amongst composites at diverse levels. Interestingly, in drug preparation, the anticipated results are enhanced composite that produces the required threshold effectiveness. Frequently, the anticipated composite is a combination of several composites which have been revealed to display some preservative consequences. The preceding explains the multidimensional chemical modeling and classification. Consequently, preceding to removal and alteration of the diverse metabolites from the callus of selected herbal plants, a superior stimulated model of recognized metabolites can be employed for understudying of the innovative composites and as well make estimates on their biomedical physiognomies (Adetunji et al., 2021a–c).

CIF offers opportunity for virtual transmission; comprising the involvement of powerful computational means for the examination of bulky bank of chemical composites *in silico* for the discovery of the several composites that display stimulating physiognomies such as anti-fungal, antiviral, antibacterial and anti-carcinogenic. This suggests that huge chemical composites are not just limited to plant metabolites but can be investigated in the absence of conducting actual physical laboratory tests, which implies that it is likely to determine innovative drugs from normal screening of huge and varied data banks of natural and artificial/synthetic composites (Adetunji et al., 2021a–c).

Another benefit of CIF to pharmaceutical nano-formulation is the use of scientific means for the establishment of the relations amongst the quantitative activity, structure and importance of chemical and biochemical composites like metabolites. The use of such procedures could result in the precise estimation of biochemical values of metabolites employing only their compositions (Leach and Gillet, 2007).

Bioinformatics has in most spheres been identified as a hybrid scientific and technology domain comprising biology, computer science and information technology. Bioinformatics plays a key function in precision drug design; hence, lessening the time and cost involved in drug formulations. Certainly, the synthesis of drugs as BIF offers all-inclusive investigative means and massive databases for modeling of genes, proteins and interrelated biological composites together with the assessment of dealings, behavior with functionalities and compatibilities (Nwankwo et al., 2020). This implies that it offers a low-cost in silico setting for the simulation of the activities of drug and drug-related composites in biological arrangements.

The relations amongst the various segments of IF and results of the exploitation are illustrated in Figure 12.1. The scope of IF in the figure is also appropriate from the design of nanodrugs, the detection to formulation, the production of nanomedicine and the treatment of infections/diseases. It should be noted that the relations that occurs and how these relations can be discovered for the promotion of further inquiry, modification and production of nanodrugs from the natural resources (raw materials) such as medicinal plants as earlier stated.

12.8 CONCLUSIONS AND FUTURE WORK

This chapter provided an exhaustive report on the application of pharmacologically active metabolites derived from the callus (undifferentiated mass of tissues formed over wounded parts of plants) of medicinal plants for the synthesis of nanodrugs. It was also established in this study that these metabolites could be used in the management of several diseases especially those that have been documented as prevalent cases resulting to high rate of mortality and morbidity in most countries particularly the developing countries. Nowadays, the use of these metabolites has gained much relevance due to their comparatively excellent efficacy and limited toxicity. Metabolites are phytochemical constituents of medicinal plants, and due to their pharmacological value, they have been used in traditional medicine, most especially the ones derived from the callus of medicinal plants. Natural medicine is considered a better therapeutic alternative due to its ubiquity and reduced side effects. The application of active metabolites of plants to nanomedicine had also been linked to their high potency as well as well-defined mechanisms of action. Some emphasis was also laid on some phytochemicals obtained from the callus of medicinal plants such as alkaloids, saponins and terpenes because of their relevance in pharmaco-therapeutics and disease control especially in their use as anti-microbial, anti-tumor, antibacterial, antiviral, anti-inflammatory or anti-osteoporotic activity. They also play imperative roles in the management of genetic diseases. Special emphasis was laid on the application of essential oils in the synthesis of nanodrugs which is linked to their efficacy. In view of the aforementioned, there is a need to apply some biotechnological techniques like metabolomics, structural as well as advanced synthetic biology that employs recombinant DNA technology, towards improving the quality of biological effects of pharmacologically active metabolites. Conclusively, collaborative interdisciplinary programs and researches into suitable technology-driven methods that would support amplified target medicinal plant production, extraction and mass production of these pharmacologically active metabolites for effective biosynthesis of pharmacologically active nanodrugs should be encouraged, premeditated and trailed as this would be a lead to innovation and advances in therapeutics, disorders monitoring and management, as well as healthcare delivery.

REFERENCES

Aabid, M. R., Irshad, A. N., Aijaz, H. G., Harpal, S., Bhupender, D. and Aijaz, A. W. (2012). Bioactive Compounds and Medicinal properties of ValerianajatamansiJones – A review. *Life Science Journal* 9(2):847–850.

Abdin, M. Z., and Kamaluddin, A. (2006). *Traditional Systems of Medicine: Improving the Quality of Medicinal Herbs through Physico-Chemical and Molecular Approaches.* India Publishing House Pvt. Ltd, New Delhi.

Adhikari, P., and Thapa, G. (2014). A note on feasibility and optimality of transportation problem. *Journal of the Institute of Engineering* 10(1): 59–68. https://doi.org/10.3126/jie.v10i1.10879.

Adetunji, C.O., Anani, O.A., and Panpatte, D. (2021). Mechanism of actions involved in sustainable ecorestoration of petroleum hydrocarbons polluted soil by the beneficial microorganism. In: Panpatte, D.G., and Jhala, Y.K. (eds.), *Microbial Rejuvenation of Polluted Environment. Microorganisms for Sustainability*, vol 26. Springer, Singapore. https://doi.org/10.1007/978-981-15-7455-9_8

Adetunji, C.O., and Anani, O.A. (2021a). Recent advances in the application of genetically engineered microorganisms for microbial rejuvenation of contaminated environment. In: Adetunji, C.O., Panpatte, D.G., and Jhala, Y.K. (eds.), *Microbial Rejuvenation of Polluted Environment. Microorganisms for Sustainability*, vol 27. Springer, Singapore. https://doi.org/10.1007/978-981-15-7459-7_14

Adetunji, C.O., and Anani, O.A. (2021b). Utilization of microbial biofilm for the biotransformation and bioremediation of heavily polluted environment. In: Panpatte, D.G., and Jhala, Y.K. (eds.), *Microbial Rejuvenation of Polluted Environment. Microorganisms for Sustainability*, vol 25. Springer, Singapore. https://doi.org/10.1007/978-981-15-7447-4_9

Adetunji, C.O., and Anani, O.A. (2021c). Bioremediation of polythene and plastics using beneficial microorganisms. In: Adetunji C.O., Panpatte, D.G., and Jhala, Y.K. (eds.), *Microbial Rejuvenation of Polluted Environment. Microorganisms for Sustainability*, vol 27. Springer, Singapore. https://doi.org/10.1007/978-981-15-7459-7_13

Adetunji, C. O., Olaniyan, O., Adetunji, J. B. and Roli, I. (2020a). Phytochemical-based nanoparticles as foes and friends. In: Mtewa, A. G., Egbina, C. and Rao, G. M. N. (eds) - *Poisonous Plants and Phytochemicals in Drug Discovery*. https://doi.org/10.1002/9781119650034.ch14.

Adetunji, C. O., Mitembo, W. P., Egbuna, C., Rao, G. M. N. (2020b). In silico modelling as a tool to predict and characterize plant toxicity. In: Mtewa, A. G., Egbina, C. and Rao, G. M. N. (eds) - *Poisonous Plants and Phytochemicals in Drug Discovery*. https://doi.org/10.1002/9781119650034.ch14.

Adetunji, C. O., Anani, O. A. and Michael, O. S. (2021c). *Application of Nanoceutical Technology for Fast and Efficient Control of Illness*. Phytochemistry, the Military and Health; Phytotoxins and Natural Defenses, pp. 497–508.

Ahmed, S., Saifullah, A. M., Swami, B. L., and Ikram, S.(2016). Green synthesis of silver nanoparticles using Azadirachta indica aqueous leaf extract. *Journal of Radiation Research and Applied Sciences* 9(1): 1–7.

Ahmad, A., Wei, Y., and Syed, F. (2017). The effects of bacteria-nanoparticles interface on the antibacterial activity of green synthesized silver nanoparticles. *Microbial Pathogenesis* 102: 133–142.

Ajitha, B., Reddy, Y.A.K., Reddy, P. S. (2014). Biogenic nano-scale silver particles by Tephrosiapurpurea leaf extract and their inborn antimicrobial activity. *Spectrochimica Acta, Part A: Molecular and Biomolecular Spectroscopy* 121: 164–172.

Akhil, K. (2019). Beneficial plant product, their uses their evolutionary perspectives: A review. *International Journal of Chemical Studies* 6(5): 970–975.

Alasbahi, R. H. and Meizig, M.F. (2010). Plectranthus barbatus: A review of Phytochemistry, ethnobotanical uses and pharmacology. *Plant Medica* 76(7): 653–661.

Ali, A., Mohammad, S., Khan, M. A., Raja, N. I., Arif, M., Kamil, A. and Mashwani, Z. (2019). Silver nanoparticles elicited in vitro callus cultures for the accumulation of biomass and secondary metabolites in Caralluma tuberculate. *Artificial cells, Nanomedicine, and Biotechnology* 47(1): 715–724.

Anani, O.A., Mishra R.R., Mishra P., Enuneku A.A., Anani G.A., and Adetunji C.O. (2020). Effects of toxicant from pesticides on food security: Current developments. In: Mishra P., Mishra R.R., Adetunji C.O. (eds.), *Innovations in Food Technology*. Springer, Singapore. https://doi.org/10.1007/978-981-15-6121-4_22

Anjum, S. and Abbasi, B. H. (2016). Thidiazuron-enhanced biosynthesis and antimicrobial efficacy of silver nanoparticles via improving phytochemical reducing potential in callus culture of Linumusitatissimum L. *International Journal of Nanomedicine* 11:715.

Ankamwar, B., Chaudhary, M., and Sastry, M. (2005). Gold nanotriangles biologically synthesized using tamrind leaf extract and potential application in vapor sensing. *Synthetic Reaction Inorganic Metal Oroganic Nano Metal Chemistry* 35(1): 19–26.

Atanasov, A. G., Waltenberger, B., Pferschy-Wenzig, E. M., Linder, T., Wawrosch, C., Uhrin, P., Temml, V., Wang, L., and Schwaiger, S. (2015). Discovery and resupply of pharmacologically active plant-derived natural products: A review. *Biotechnology Advances* 33: 1582–1614.

Barua, S., Chattopadhyay, P., Aidew, L., Buragohain, A. K., and Karak, N. (2015). Infection-resistant hyperbranched epoxy nanocomposite as a scaffold for skin tissue regeneration. *Polymer International* 46: 303–311.

Beutler, J. A. (2009). Natural products as a foundation for drug discovery. *Current Protocols in Pharmacology* 46(1): 9–11.

Bhushan, P. B., John, C.K., Bhatt, V.P., and Nikam T.D. (2018). In vitro propagation of Digitalis lanataEhrh. through direct shoot regeneration – A source cardiotonic glycoside. *Industrial Crops and Products* 121: 313–319.

Bouhdid, S., Abrini, J., Zhiri, A., Espuny, M. J., and Manresa, A. (2009). Investigation of functional and morphological changes in Pseudomonas aeruginosa and Staphylococcus aureus cells induced by Origanum compactum essential oil. *Journal of Applied Microbiology* 106(5): 1558–1568.

Brophy, J. J., Craven, L. A. and Doran J. C. (2013). *Melalecuas: Their Botany, Essential Oils and Uses.* Australian Centre for international Agriculture Research, Canberra, p. 73. ISBN 9781922137517.

Chandran, S. P., Chaudhury, M., Pasricha, R., Ahmad, A. and Sastry, M. (2006). Synthesis of gold nanotriangles and silver nanoparticles using Aloevera plant extract. *Biotechnology Progress* 22(2): 577–583.

Cheng, G., Hao, H., Dai, M., Liu, Z. and Yuan, Z. (2013). Antibacterial action of quinolones: From target to network. *European Journal of Medicinal Chemistry* 66: 555–562.

Chinnadurai, R., Copland, I. B., Garcia, M. A., Petersen, C. T., Lewis, C. N., Waller, E. K., Kirk, A. D., and Galipeau, J. (2016). Cryopreserved mesenchymal stromal cells are susceptible to T-cell mediated apoptosis which is partly rescued by IFNγ licensing. *Stem cells (Dayton, Ohio)* 34(9): 2429–2442. https://doi.org/10.1002/stem.2415.

Dakal, T. C., Kumar, A., Majumdar, R. S., and Yadav, V. (2016). Mechanistic basis of antimicrobial actions of silver nanoparticles. *Frontiers in Microbiology* 7:1831.

Das, B., Dash, S. K., and Mandal, D. (2017). Green synthesized silver nanoparticles destroy multidrug resistant bacteria via reactive oxygen species mediated membrane damage. *Arab Journal of Chemistry* 10(6):862–876.

Denny, W. (1989). DNA-intercalating ligands as anti-cancer drugs: Prospects for future design. *Anti-Cancer Drug Design* 4(4): 241–263.

de Souza, E. L., de Barros, J. C., de Oliveira, C. E. and da Conceição, M. L. (2010). Influence of Origanum vulgare L. essential oil on enterotoxin production, membrane permeability and surface characteristics of Staphylococcus aureus. *International Journal of Food Microbiology* 137(2–3): 308–311.

Dewick, P. (2001). The acetate pathway: Fatty acids and polyketides. *Medicinal Natural Products: A Biosynthetic Approach* 3.

Dibrov, P., Dzioba, J., Gosink, K. K. and Häse, C. C. (2002) Chemiosmotic mechanism of antimicrobial activity of Ag+ in Vibrio cholera. *Antimicrobial Agents and Chemotherapy*, 46(8): 2668–2670.

Dörnenburg, H. (2010). Cyclotide synthesis and supply: From plant to bioprocess. *Biopolymers and Peptide Science* 94(5):602–610.

Du, W., Sun, Y., Ji, R., Zhu, J., Wu, J., Guo, H. (2011). TiO_2 and ZnO nanoparticles negatively affect wheat growth and soil enzyme activities in agricultural soil. *Journal of Environmental Monitoring* 13: 822–828.

Dwivedi, N., Dwivedi, S., and Adetunji, C. O. (2021). Efficacy of microorganisms in the removal of toxic materials from industrial effluents. In: Adetunji, C.O., Panpatte, D.G., and Jhala, Y.K. (eds.), *Microbial Rejuvenation of Polluted Environment. Microorganisms for Sustainability*, vol 27. Springer, Singapore. https://doi.org/10.1007/978-981-15-7459-7_15

Dziendzikowska, K., Gromadzka-Ostrowska, J., Lankoff, A., Oczkowski, M., Krawczynska, A., Chwastowska, J., Sadowska-Bratek, M., Chajduk, E., Wojewodzka, M., and Kruszewski, M. (2012). Time-dependent biodistribution and excretion of silver nanoparticles in male Wistar rats. *Journal of Applied Toxicology* 32: 920–928.

Elisha, S., Michal, L., Sarah, S., Helena, P., Elaine, S., and Haya, L. G. (2014). Evaluating medicinal plants for anticancer activity. *Scientific World Journal* 1–12.

Elumalai, E. K., Prasad, T.N.V.K.V., Hemachandran, J., Viviyam, T. S., Thirumalai, T. and David, E. (2010). Extracellular synthesis of silver nanoparticles using leaves of Euphorbia hirta and their antibacterial activities. *Journal of Pharmaceutical Science Research* 2: 549–554.

Fennell, T. R., Mortensen, N.P., Black, S.R., Snyder, R.W., Levine, K.E., Poitras, E., Harrington, J.M., Wingard, C.J., Holland, N.A., and Patmasiri, W. (2017). Disposition of intravenously or orally administered silver nanoparticles in pregnant rats and the effect on the biochemical profile in urine. *Journal of Applied Toxicology* 37: 530–544.

Fox, C. L. (1968). Silver sulfadiazine: A new topical therapy for Pseudomonas in burns. Therapy of Pseudomonas infection in burns. *Archives of Surgery* 96(2): 184–188.

Franci, G., Falanga, A., Galdiero, S., Palomba, L., Rai, M., Morelli, G. and Galdiero, M. (2015). Silver nanoparticles as potential antibacterial agents. *Molecules* 20: 8856–8874.

Frenkel, N., Makky, A., Sudji, I. R., Wink, M. and Tanaka, M. (2014). Mechanistic investigation of interactions between steroidal saponin digitonin and cell membrane models. *The Journal of Physical Chemistry B* 118(50): 14632–14639.

Frieri. M., Kumar, K. and Boutin, A. (2017). Antibiotic resistance. *Journal of Infection and Public Health* 10(4):369–378.

Furuya, T., Ikuta, A. and Syono, K. (1972). Alkaloids from callus cultures of Papaver somniferum. *Phytochemistry* 11:3041–3044.

Gaiser, B.K., Hirn, S., Kermanizadeh, A., Kanase, N., Fytianos, K., Wenk, A., Harbel, N., Brunelli, A., Kreyling, W. G. and Stove, V. (2013). Effects of silver nanoparticles on the liver and hepatocytes in vitro. *Toxicology Science* 131: 537–547.

Gao, X., Yourick, J. J. and Topping, V. D. (2014). Toxicogenomic study in rat thymus of F1 generation offspring following maternal exposure to silver ion. *Toxicology Reports* 2: 341–350.

Geethalakshmi, R. and Sarada, D. V. L. (2010). Synthesis of plant-mediated silver nanoparticles using Trianthemadecandra extract and evaluation of their anti-microbial activities. *International Journal of Engineering Science and Technology* 2(5): 970–975.

Georgiev, M. I., Weber, J. and Macuik, A. (2009). Bioprocessing of plant cell cultures for mass production of targeted compounds. *Applied Microbial Biotechnology* 83(5): 809–823.

Ghosh, S., Patil, S., Ahire, M., Kitture, R., Kale, S., Pardesi, K., Cameotra, S. S., Bellare, J., Dhavale, D. D., Jabgunde, A. and Chopade, B. A. (2012). Synthesis of silver nanoparticles using Dioscoreabulbiferatuber extract and evaluation of its synergistic potential in combination with antimicrobial agents. *International Journal of Nanomedicine* 7: 483–496.

Gigant, B., Wang, C., Ravelli, R. B., Roussi, F., Steinmetz, M. O., Curmi, P. A., Sobel, A. and Knossow, M. (2005). Structural basis for the regulation of tubulin by vinblastine. *Nature* 435(7041): 519.

Gill. A. O., Delaquis, P., Russo, P. and Holley, R. A. (2002). Evaluation of antilisterial action of cilantro oil on vacuum packed ham. *International Journal of Food Microbiology* 73(1): 83–92.

Gnanadesigan, M., Anand, M., Ravikumar, S., Maruthupandy, M., Vijayakumar, V., Selvam, S., Dhineshkumar, M. and Kumaraguru, A. K. (2011). Biosynthesis of silver nanoparticles by using mangrove plant extract and their potential mosquito property. *Asian Pacific Journal of Tropical Medicine* 799–803.

Govindaraju, K., Tamilselvan, S., Kiruthiga, V. and Singaravelu, G. (2010). Biogenic silver nanoparticles by Solanum torvum and their promising antimicrobial activity. *Journal of Biopesticides* 3(1): 394–399.

Hafed, H., Fethi, B. A., Mejdi, S., Emira, N. and Amina, B. (2010). Effect of Mentha longifolia L. ssp longifolia essential oil on the morphology of four pathogenic bacteria visualized by atomic force microscopy. *African Journal of Microbiology Research* 4(11): 1122–1127.

Hajipour, M. J., Fromm, K. M., Ashkarran, A. A., de Aberasturi, D. J., de Larramendi, I. R., Rojo, T., Serpooshan, V., Parak, W. J. and Mahmoudi M. (2012). Antibacterial properties of nanoparticles. *Trends Biotechnology* 30: 499–511.

Hamouda, T., Myc, A., Donovan, B., Shih, A. Y., Reuter, J. D. and Baker, J. R. Jr. (2001). A novel surfactant nanoemulsion with a unique non-irritant topical antimicrobial activity against bacteria, enveloped viruses and fungi. *Microbiological Research* 156(1): 1–7.

Herrmann, F. and Wink, M. (2011). Synergistic interactions of saponins and monoterpenes in HeLa cells, Cos7 cells and in erythrocytes. *Phytomedicine* 18(13): 1191–1196.

Holt, K. B. and Bard, A. J. (2005). Interaction of silver (I) ions with the respiratory chain of Escherichia coli: An electrochemical and scanning electrochemical microscopy study of the antimicrobial mechanism of micromolar $Ag+$. *Biochemistry* 44(39):13214–13223.

Horváth, G., Kovács, K., Kocsis, B. AndKustos, I. (2009). Effect of thyme (Thymus vulgaris L.) essential oil and its main constituents on the outer membrane protein composition of Erwinia strains studied with microfluid chip technology. *Chromatographia* 70(11–12): 1645–1650.

Hotez, P. J. and Pecoul, B. (2010). Manifesto for advancing the control and elimination of neglected tropical diseases. *PLoS Neglected Tropical Diseases* 4:718.

Huang, Y., Fang, Y., Wu, J., Dziadyk, J. M., Zhu, X., Sui, M. and Fan, W. (2004). Regulation of Vinca alkaloid-induced apoptosis by NF-κB/IκB pathway in human tumor cells. *Molecular Cancer Therapeutics* 3(3): 271–277.

Hurley, L. H. (2002). DNA and its associated processes as targets for cancer therapy. *Nature Reviews Cancer* 2(3): 188.

Hussain, M., Raja, N. I., Iqbal, M. and Aslam, S. (2017). Applications of plant flavonoids in the green synthesis of colloidal silver nanoparticles and impacts on human health. *Iran Journal of Science and Technology Transaction A: Science* 3: 1381–1392.

Ibrahim, H. M. (2015). Green synthesis and characterization of silver nanoparticles using banana peel extract and their antimicrobial activity against representative microorganisms. *Journal of Radiation Research and Applied Sciences* 8(3): 265–275.

Islam, S., Thangadurai, D., Adetunji, C. O., Nwankwo, W., Kadiri, O., Makinde, S., Michael, O. S., Anani, O. A., and Adetunji, J. B. (2021) Nanomaterials and nanocoatings for alternative antimicrobial therapy. In: Kharissova, O.V., Martínez, L.M.T., and Kharisov, B.I. (eds.), *Handbook of Nanomaterials and Nanocomposites for Energy and Environmental Applications*. Springer, Cham. https://doi.org/10.1007/978-3-030-11155-7_3-1

Iskandar, N. N. and Iriawati, I. (2016). Vinblastine and Vincristine production on Madagascar Periwinkle (Catharanthus roseus (L.) G. Don) callus culture treated with polethylene glycol. *Makara Journal of Science* 7–16.

Ivask, A., ElBadawy, A. and Kaweeteerawat, C. (2013). Toxicity mechanisms in Escherichia coli vary for silver nanoparticles and differ from ionic silver. *ACS Nano* 8(1):374–386.

Jafarain, A., Asghari, G. and Ghassami, E. (2014). Evaluation of cytotoxicity of Moringa oleifera Lam. callus and leaf extracts on Hela cells. *Advance Biomedical Research* 3:194.

Janarthanam, B., Gopalakrishan, M. and Sekar, T. (2010). Secondary Metabolites production in callus culturea of Stevia rebaudianabertoni. *Bangladesh Journal of Scientific and Industrial Research* 45(3):243–248.

Jenie, B. S. L., Priosoeryanto, B. P., Syarief, R. and Rekso, G. T. (2008). Mode of action Temu kunci (Kaempferia pandurata) essential oil on E. coli K1. 1 cell determined by leakage of material cell and salt tolerance assays. *HAYATI Journal of Biosciences* 15(2): 56–60.

Jiao, F., Liu, Y., Qu, Y., Li, W., Zhou, G., Ge, C., Sun, B. And Chen, C. (2010). Studies on anti- tumor and antimetastatic activities of fullerenol in a mouse breast cancer model. *Carbon* 48(8): 2231–2243.

Johnson, A. S., Obot, I. B. and Ukpong, U. S. (2014). Green synthesis of silver nanoparticles using Artemisia annua and Sidaacuta leaves extract and their antimicrobial, antioxidant and corrosion inhibition potentials. *Journal of Materials and Environmental Science* 5:899–906.

Jordan, M.A., Thrower, D. and Wilsen, L. (1991). Mechanism of inhibition of cell proliferation by Vinca alkaloids. *Cancer Research* 51: 2212–2222.

Jordan, M. A. and Wilson, L (1998). Microtubules and actin filaments: Dynamic targets for cancer chemotherapy. *Current Opinion in Cell Biology* 10(1): 123–130.

Joseph, E., Pizzorno, Jr. and Murray, M. T. (2012). *Textbook of Natural Medicine (4th ed.).* Churchill Liivngstone, Edinburgh, p. 634.

Kalimuthu, K., Prabakaran, R. and Saraswathy M. (2014). In vitro anticancer activity of ethanolic extract of in vitro and in vivo tuber of CeropegiapusillaWight and Arn. *International Journal of Pure & Applied Bioscience* 2(4):82–87.

Kamalanathan, D. and Natarajan, D. (2018). Anticancer potential of leaf and leaf-derived callus extracts of Aervajavanicaagainst MCF-7 breast cancer cell line. *Journal of Cancer Research and Therapeutics* 14:321–327.

Kapil, A. (2005). The challenge of antibiotic resistance: need to contemplate. *The Indian Journal of Medical Research* 121(2), 83–91.

Karunaweera, N. D. and Ferreira, M. U. (2018). Leishmaniasis: current challenges and prospects for elimination with special focus on the South Asian region. *Parasitology* 145:425–429.

Kaviya, S., Santhanalakshmi, J. and Viswanathan, B. (2011) Green synthesis of silver nanoparticles using Polyalthialongifolia leaf extract along with D-sorbitol: Study of antibacterial activity. *Journal of Nanotechnology* 5.

Khandelwal, N., Singh, A., Jain, D., Upadhyay, M. K and Verma, H. N. (2010). Green synthesis of silver nanoparticles using Argimonemexicana leaf extract and evaluation of their antimicrobial activities. Digest. *Journal of Nanomaterials and Biostructure* 5: 483–489.

Kim. J. S., Kuk, E. and Yu, K. N. (2007). Antimicrobial effects of silver nanoparticles. *Nanomedicine* 3(1):95–101.

Koga, T., Watanabe, H., Kawada, H., Takahashi, A., Utsui, Y., Domon, H., Ishii, C., Narita, T. and Yasuda, H. (1998). Interactions of plaunotol with bacterial membranes. *Journal of Antimicrobial Chemotherapy* 42: 133–140.

Kouvaris, P., Delimitis, A., Zaspalis V., Papadopoulos D., Tsipas S. A., Michailidis N. (2012). Green synthesis and characterization of silver nanoparticles produced using Arbutus unedo leaf extract. *Materials Letters* 76:18–20.

Krishnaraj, C., Jagan, E. G., Rajasekar, S., Selvakumar, S., Kalaichelvan, P. T. and Mohan, N. (2010). Synthesis of silver nanoparticles using Acalyphaindica leaf extracts and its antibacterial activity against water borne pathogens. *Colloids and Surfaces B: Biointerfaces* 76(1): 50–56.

Kumar, V. and Staden, J. V. (2016). A review of Swertiachirayita (Gentianaceae as a traditional medicinal plant. *Frontiers in Pharmacology* 306:308.

Laflamme, P., St-Pierre, B. and De Luca, V. (2001). Molecular and biochemical analysis of a Madagascar periwinkle root specific minovincinine-19-hydroxy-o-acetyltransferase. *Plant Physiology* 125: 189–198.

Lai, D. Y. and Woo, Y. T. (1987). Naturally occurring carcinogens: an overview. *Journal of Environmental Science and Health Part C* 5(2): 121–173.

Lai, S. and Baraik, B. (2019). Phytochemicals and pharmacological profile of Bacopa monnieri – an ethnomedicinal plant. *International Journal of Pharmacological Science and Research* 10(3):1001–1013.

Lambert, R., Skandamis, P. N., Coote, P. J. and Nychas G.J. (2001). A study of the minimum inhibitory concentration and mode of action of oregano essential oil, thymol and carvacrol. *Journal of Applied Microbiology* 91(3): 453–462.

Leach, A.R. and Gillet, V.J. (2007). *An Introduction to Chemoinformatics.* Springer, p. 256. DOI: 10.1007/978-1-4020-6291-9.

Lee, T.Y., Liu, M.S., Huan, L. J., Lue, S.I., Lin, L.C., Kwan, A. L. and Yang, R.C. (2013). Bioenergetic failure correlates with autophagy and apoptosis in rat liver following silver nanoparticle intraperitoneally administration. Part. *Fibre Toxicology* 10: 40.

Lalaleo, L., Testillano, T., Risueno, M.-C., Cusido, R. M., Alazon, J., Alcazar, R., and Bonfil, M. (2018). Effect of intro morphogenesis on the production of podophylotoxin derivatives in callus cultures of Linum album. *Journal of Plant Physiology* 228: 47–58.

Lin, C. S., Chen, P. C., Wang, C. K., Wang, C. W., Chang, Y. J. and Tai, C. J. (2014). Antitumor effects and Biological mechanism of action of the aqueous extract of the Camptotheca acuminate fruit in human endometrial Carcinoma cell. *Evidence-Based Complementary and Alternative Medicine* 1–10. http://dx.doi.org/10.1155/2014/564810.

Liu, M. J., Li, J. X., Lee, K. M., Qin, L. and Chan, K. M. (2005). Oxidative stress after muscle damage from immobilization and remobilization occurs locally. *Clinical Orthopaedics and Related Research* 246–250.

Liu, Y., Chen, W., and Kim, H.-I. (2012). Antibacterial activity of pH-sensitive genipin cross-linked chitosan/poly(ethylene glycol)/silver nanocomposites. *Polymers for Advanced Technologies* 23: 8–14.

Luo, D. Q., Guo, J. H., Wang, F. J., Jin, Z. X., Cheng, X. L., Zhu, J. C., Peng, C. Q., Zhang, C. (2009). Anti-fungal efficacy of polybutylcyanoacrylate nanoparticles of allicin and comparison with pure allicin. *Journal of Biomaterials Science*, Polymer Edition. 20(1): 21–31.

Luz, D., Hoffman, M., Gilboa-Garber, N. and Amar, Z. (2010). Medicinal properties of Commiphoragieadensis. *Africa Journal of Pharmacy and Pharmacology* 4(8): 516–520.

Ma, W., Jing, L., Valladares, A., Mehta, S. L., Wang, Z., Li, P. A., Bang, J. J. (2015). Stress, caspase-3 activation and cell death: Amelioration by sodium selenite. *International Journal of Biological Science* 11: 86–87.

Magnotta, M., Murata, J., Chen, J., and De Luca, V. (2006). Identification of low vindoline accumulating cultivar of Catharanthus roseus (L) G Don. by alkaloid and enzymatic profiling. *Phytochemistry* 67: 1758–1764.

Mahroug, S., Courdavault, V., Thiersault, M., St-Pierre, B., Burlat, V. (2006) Epidermis is a pivotal site of at least four secondary metabolic pathways in Catharanthus roseus aerial organs. *Planta* 223: 1191–1200.

Maiti, M. and Kumar, G. S. (2009). Biophysical aspects and biological implications of the interaction of benzophenanthridine alkaloids with DNA. *Biophysical Reviews* 1(3): 119–129.

Maiti, M., Nandi, R., Chaudhuri, K. (1982). Sanguinarine: A monofunctional intercalating alkaloid. *FEBS Letters* 142(2): 280–284.

Maity, P., Hansda, D., Bandyopadhyay, U., and Mishra, D.K. (2009). Biological activities of crude extracts and chemical constituents of bael, Aegemarmeos(L.) Correa. *Indian Journal of Experimental Biology* 47: 849–861.

Makarov, V. V., Love, A. J., Sinitsyna, O. V., Makarova, S. S., Yaminsky, I. V., Taliansky, M. E., et al. (2014). "Green" nanotechnologies: Synthesis of metal nanoparticles using plants. *Acta naturae* 6: 35–44.

Malabadi, R. B., Lokare-Naik, S., Meti, N. T., Mulgund, G. S., Nataraja, K., Vijaya, K. S. (2012a). Synthesis of silver nanoparticles from in vitro derived plants and callus cultures of Clitoriaternatea; Evaluation of antimicrobial activity. *Research in Biotechnology* 3(5): 26–38.

Malabadi, R. B., Meti, N. T., Mulgund, G. S., Nataraja, K., Vijaya, K. S. (2012c). Synthesis of silver nanoparticles from in vitro derived plants and callus cultures of Costusspeciosus (Koen.); Assessment of antibacterial activity. *Research in Plant Biology* 2(4): 32–42.

Malabadi, R. B., Mulgund, G. S., Meti, N. T., Nataraja, K., Vijaya, K. S. (2012b). Antibacterial activity of silver nanoparticles synthesized by using whole plant extracts of Clitoriaternatea. *Research in Pharmacy* 2(4): 10–21.

Malabadi, R. B., Mulgund, G. S., Nataraja, K. (2007). Ethanobotanical survey of medicinal plants of Belgaum district, Karnataka, India. *Journal of Medicinal and Aromatic Plant Sciences* 2: 70–77.

Malabadi, R. B., Mulgund, G. S., Nataraja, K. (2009). Triacontanol induced somatic embryogenesis and plantlet regeneration in Catharanthus roseus. *Journal of Medicinal and Aromatic Plant Sciences* 31: 147–151.

Malabadi, R. B., Vijay K. S. (2007). Assessment of antifungal activity of some medicinal plants. *International Journal of Pharmacology* 6:499–504.

Malabadi, R. B., Vijay, K. S. (2008). Evaluation of antifungal property of medicinal plants. *Journal of Phytological Research* 1: 139–142.

Malabadi, R. B., Mulgund, G. S., Nataraja, K. (2010). Evaluation of antifungal activity of selected medicinal plants. *Journal of Medicinal and Aromatic Plant Sciences* 1: 42–45.

Medema, M.H. and Fischbach, M.A. (2015). Computational approaches to natural product discovery. *Nature of Chemistry Biology* 11(9): 639–648. doi: 10.1038/nchembio.1884. PMID: 26284671; PMCID: PMC5024737.

Meng, H., et al. (2010). Potent angiogenesis inhibition by the particulate form of fullerene derivatives. *ACS Nano* 4(5): 2773–2783.

Mishra, P., Mishra, R. R., and Adetunji, C. O. (Eds.). (2020). *Innovations in Food Technology: Current Perspectives and Future Goals* Springer Nature Singapore Pte Ltd. 2020. Springer, Singapore. https://doi.org/10.1007/978-981-15-6121-4.

Mohammed, A., Chiruvella, K. K., Rao, Y. K., Geethangili, M., Raghavan, S. C., Ghanta, R. G. (2015). In vitro production of echioidinin, 7-O-methywogonin from callus cultures of andrographis lineata and their cytotoxicity on cancer cells. *PLoS ONE* 10(10): e0141154. doi:10.1371/journal.pone.0141154.

Mohan, K.., Sinha, M., Shridhara, R. P. (2012). Green synthesis of silver nanoparticles using Terminalia chebula extract at room temperature and their antimicrobial studies. *Spectrochim Acta A* 91:228–233.

Mori, R., Kato, A., Komenoi, K., Hurasaki, H., Ijima, T., Kawagoshi, M., Kiran, Y. B., Takeda, S., Sakai, N., Konakahara, T. (2014). Synthesis and in vitro antitumor activity of novel 2-alkyl-5-methoxycarbonyl-11-methyl-6H-pyrido [4, 3-b] carbazol-2-ium and 2-alkylellipticin-2-ium chloride derivatives. *European Journal of Medicinal Chemistry* 82: 16–35.

MubarakAli, D., Thajuddin, N., Jeganathan, V. and Gunasekaran, M. (2011). Plant extract mediated synthesis of silver and gold nanoparticles and its antibacterial activity against clinically isolated pathogens. *Colloids and Surfaces B: Biointerfaces* 85(2) 360–364.

Mukunthan, K. S., Elumalai, E. K., Patel, T. N., Murthy, V. R (2011). Catharanthus roseus: A natural source for the synthesis of silver nanoparticles. *Asian Pacific Journal of Tropical Biomedicine* 270–274.

Murray, R., Steed, P., Elson, H. (1965). The location of the mucopeptide in sections of the cell wall of Escherichia coli and other Gram-negative bacteria. *Canadian Journal of Microbiology* 11(3): 547–560.

Nabikhan, A., Kandaswamy, K., Ra, A., Alikunhi, N. B. (2010). Synthesis of antimicrobial silver nanoparticles by callus and leaf extracts from saltmarsh plant, Sesuviumportulacastrum L. *Colloids and Surfaces B: Biointerfaces* 79: 488–493.

Naraginti, S., Kumari, P.L., Das, R.K., Sivakumar, A., Patil, S.H., Andhalkar, V.V. (2016). Amelioration of excision wounds by topical application of green synthesized, formulated silver and gold nanoparticles in albino Wistar rats. *Materials Science and Engineering: C Materials for Biological Applications* 62: 293–300.

Nguefack, J., V. Leth, Zollo P. H. A., Mathur, S. B. (2004). Evaluation of five essential oils from aromatic plants of Cameroon for controlling food spoilage and mycotoxin producing fungi. *International Journal of Food Microbiology* 94(3): 329–334.

Nwankwo, W., Olayinka, A.S., and Ukhurebor, K.E. (2020). Nanoinformatics: Why Design of Projects on Nanomedicine Development and Clinical Applications may Fail? *Proceeding of the 2020 International Conference in Mathematics, Computer Engineering and Computer Science (ICMCECS)*, Lagos, Nigeria, IEEE Xplore: 1–7.

Oussalah, M., Caillet, S., Lacroix, M.(2006). Mechanism of action of Spanish oregano, Chinese cinnamon, and savory essential oils against cell membranes and walls of Escherichia coli O157: H7 and Listeria monocytogenes. *Journal of Food Protection* 69(5): 1046–1055.

Olaniyan, O.T., and Adetunji, C.O. (2021) Biochemical role of beneficial microorganisms: An overview on recent development in environmental and agro science. In: Adetunji, C.O., Panpatte, D.G., and Jhala, Y.K. (eds.), *Microbial Rejuvenation of Polluted Environment. Microorganisms for Sustainability*, vol 27. Springer, Singapore. https://doi.org/10.1007/978-981-15-7459-7_2

Oves, M., Aslam, M., Rauf, M. A., Qayyum, S., Qari, H.A., Khan, M. S., Alam, M. Z., Tabrez, S., Pugazhendhi, A., Ismail, I. M. (2018). Antimicrobial and anticancer activities of silver nanoparticles synthesized from the root hair extract of Phoenix dactylifera. *Material and Science Engineering* 89: 429–443.

Pajohi, M., Tajik, H., Farshid, A. A., Hadian, M. (2011). Synergistic antibacterial activity of the essential oil of Cuminum cyminum L. seed and nisin in a food model. *Journal of Applied Microbiology* 110(4): 943–951.

Pal, S., Tak, Y. K., Song, J. M. (2007). Does the antibacterial activity of silver nanoparticles depend on the shape of the nanoparticle? A study of the Gram-negative bacterium Escherichia coli. *Applied Environmental Microbiology* 73(6): 1712–1720.

Panda, S., Choudhury, N. S. K., Patro, V. J., Pradhan, D. K., Jan, G. K. (2009). Analgesic, antipyretic and anti-inflammatory effect of the whole plant extract of DesmostachyabipinnataStapf (Poaceae) in albino rats. *Drug Invent Today* 1:150–153.

Pandey, A., Sharma, S. K., Singh, L., Singh, T. (2013). An overview on Desmostachyabipinnata. *Journal of Drug Discovery Therapy* 7: 67–68.

Pandit, S., Mukherjee, P. K., Ponnusankar, S., Venkatesh, M., and Srikanth, N. (2011). Metabolism mediated interaction of α-asarone and Acorus calamus with CYP3A4 and CYP2D6. *Fitoterapia* 82(3): 369–374. https://doi.org/10.1016/j.fitote.2010.11.009.

Pati, P. K., Sharma, M., Salar, R. K., Sharma, A., Gupta, A. P., Singh, B. (2009). Studies on leaf spot disease of with Aniasomnifera and its impact on secondary metabolites. *Indian Journal of Microbiology* 48(4): 432–437.

Patra, J. K., Das, G., Fraceto, L. F., Campos, E., Rodriguez-Torres, M., Acosta-Torres, L. S., Diaz-Torres, L. A., Grillo, R., Swamy, M. K., Sharma, S., Habtemariam, S., and Shin, H. S. (2018). Nano based drug delivery systems: recent developments and future prospects. *Journal of Nanobiotechnology* 16(1): 71. https://doi.org/10.1186/s12951-018-0392-8.

Pattanayak, S., Mollick, M. M. R., Maity, D. (2015). Butea monosperma bark extract mediated green synthesis of silver nanoparticles: Characterization and biomedical applications. *Journal of Saudi Chemistry Society* 21(6): 673–684.

Pérez-Fons, L., F. J. Aranda, J. Guillén, J. Villalaín, V. Micol (2006). Rosemary (Rosmarinus officinalis) diterpenes affect lipid polymorphism and fluidity in phospholipid membranes. *Archives of Biochemistry and Biophysics* 453(2): 224–236.

Prabhu, N., Divya, T. R., Yamuna, G. (2010). Synthesis of silver phyto nanoparticles and their antibacterial efficacy. Digest. *Journal of Nanomaterials and Biostructure* 5: 185–189.

Pragya, T., Mishra, B. N. and Neelam, S. S. (2014). Phytochemical and Pharmacological Properties of Gymnemasylvestre: An Important Medicinal Plant. *BioMedical Research International* 18. http://dx.doi.org/10.1155/2014/830285.

Prasad, R., Swamy, V. S. (2013). Antimicrobial activities of silver nanoparticles synthesized by the bark extract of Syzygium cumin. *Journal of Nanoparticles* 6.

Pratiwi. D. R., Sari. P. T., and Ratnadewi (2018). Cinchona cells performance in in vitro culture: quinine alkaloid production with the application of different elicitors. *IOP Conference Series: Earth and Environmental Science* 185(1): 012029.

Purohit, P. and Khanna, P. (1983). Production of essential oil from callus cultures of Ocimum basilicum L. *Plant Cell Culture in Crop Improvement*, Springer: 377–380.

Rabinkov, A., Miron, T., Konstantinovski, L., Wilchek, M., Mirelman, D., Weiner, L. (1998). The mode of action of allicin: trapping of radicals and interaction with thiol containing proteins. *Biochimica et Biophysica Acta* 1379: 233–244.

Rafik, S., Mahesh, P., Ashwini, D., and Sayyed, I. (2014). Evaluation of anticancer, antioxidant, and possible anti-inflammatory properties of selected medicinal plants used in Indian traditional medication. *Journal of Traditional Complementary Medicine* 4: 253–257.

Rajasekharreddy, P., Rani, P. U., Mattapally, S., Banerjee, S. K. (2017). Ultra-small silver nanoparticles induced ROS activated toll-pathway against Staphylococcus aureus disease in silkworm model. *Material Science and Engineering C* 77: 990–1002.

Ram, A., Mabalirajan, U., Das, M., Bhattacharya, I., Dinda, A. K., Gangal, S. V., and Ghosh, B. (2006). Glycyrrhizin alleviates experimental allergic asthma in mice. *International Immunopharmacology* 6(9): 1468–1477. doi:10.1016/j.intimp.2006.04.020.

Rao, S., K. Usha, et al. (2015). Production of secondary metabolites from callus cultures of Centella asiatica (L.) Urban. *Annals of Phytomedicine-An International Journal* 4(1): 74–78.

Raut, R. W., Kolekar, N. S., Lakkakula, J. R., Mendhulkar, V. D. and Kashid, S. B. (2010). Extracellular synthesis of silver nanoparticles using dried leaves of Pongamiapinnata (L) pierre. *Nano-Micro Letters* 2(2): 106–113.

Recordati, C., De Maglie, M., Bianchessi, S., Argentiere, S., Cella, C., Mattiello, S., Cubadda, F., Aureli, F., D'Amato, M., Raggi, A. (2016). Tissue distribution and acute toxicity of silver after single intravenous administration in mice: Nano-specific and size-dependent effects. *Part. Fibre Toxicology* 13:12.

Rhayour, K., Bouchikhi, T., Tantaoui-Elaraki, A., Sendide, K. Remmal, A. (2003). The mechanism of bactericidal action of oregano and clove essential oils and of their phenolic major components on Escherichia coli and Bacillus subtilis. *Journal of Essential Oil Research* 15(5): 356–362.

Robinson, M., and Zhang, X. (2011). *The World Medicines Situation. Traditional Medicines: Global Situation, Issues and Challenges.* World Health Organization, Geneva, pp. 1–12.

Rodrigues, T., Reker, D., Schneider, P., and Schneider, G. (2016). Counting on natural products for drug design. *Nature Chemistry* 8:531.

Rasooli, I., Rezaei, M.B. and Allameh, A. (2006) Growth inhibition and morphological alterations of Aspergillus niger by essential oils from Thymus eriocalyx and Thymus x-porlock. *Food Control* 17: 359–364.

Roursgaard, M., Poulsen, S. S., Kepley, C. L., Hammer, M., Nelson, G. D. (2008). Polyhydroxylated C60 fullerene (fullerenol) attenuates neutrophilic lung inflammation in mice. *Basic & Clinical Pharmacology & Toxicology* 103(4): 386–388.

Ruttkay-Nedecky, B., Krystofova, O., Nejdl, L. and Vojtech, A. (2017). Nanoparticles based on essential metals and their Phytotoxicity. *Journal of Nanobiotechnology* 15(1): 33.

Ruvolo, P., Deng, X., May, W. S. (2001). Phosphorylation of Bcl2 and regulation of apoptosis. *Leukemia* 15(4): 515.

Sharifi-Rad, J., Alfatemi, S. M. H., Sharifi-Rad, M., Iriti, M. (2014). Antimicrobial synergic effect of allicin and silver nanoparticles on skin infection caused by methicillin-resistant staphylococcus aureus spp. *Annals of Medical Health Sciences Research* 4(6): 863–868.

Salehi, B., Zucca, P., Orhan, I. E., Azzini, E., Adetunji, C. O, Mohammed, S. A, Banerjee, S. K., Sharopov, F., Rigano, D., Sharifi-Rad, J., Armstrong, L., Martorell, M., Sureda, A., Martins, N., Selamoğlu, Z., and Ahmad, Z. (2019). Allicin and health: A comprehensive review. *Trends in Food Science & Technology* 86: 502–516.

Salem, W., Leitner, D. R., Zingl, F. G. (2015). Antibacterial activity of silver and zinc nanoparticles against Vibrio cholerae and enterotoxic Escherichia coli. *International Journal of Medical Microbiology* 305(1): 85–95.

Sapna, T., Shweta, T., Mamta, S., and Radheshyam, R. (2018). Synthesis of silver nanoparticles from Centellaasiatica (L.) plant and in vitro derived callus culture: Assessment of antibacterial activity. *Academia Journal of Biotechnology* 6(3): 066–073.

Sankar, R., Baskaran, A., Shivashangari, K.S., Ravikumar, V. (2015). Inhibition of pathogenic bacterial growth on excision wound by green synthesized copper oxide nanoparticles leads to accelerated wound healing activity in Wistar Albino rats. *Journal of Materials Science: Materials in Medicine* 26(7): 1–7.

Satyavani, K., Gurudeeban, S., Ramanathan, T., Balasubramanian, T. (2011). Biomedical potential of silver nanoparticles synthesized from calli cells of Citrullus colocynthis (L.) Schrad. *Journal of Nanotechnology* 9(43): 1–8.

Savithramma, N., Linga Rao, M., Rukmini, K., Suvarnalathadevi P. (2011a). Antimircrobial activity of silver nanoparticles synthesized by using medicinal plants. *International Journal of Chemistry and Technology Research* 3(3): 1394–1402.

Savithramma, N., Lingarao, M., Basha, S. K. M. (2011b). Antifungal efficacy of silver nanoparticles synthesized from the medicinal plants. *Der Pharma Chemica* 3: 364–372.

Saxena A., Tripathi R. M., Singh R. P. (2010). Biological synthesis of silver nanoparticles by using Onion (Allium cepa) extract and their antibacterial activity. *Digest Journal of Nanomaterial Biostructure* 5: 427–432.

Shankar, T., Karthiga, P., Swarnalatha, K., Rajkumar, K. (2017). Green synthesis of silver nanoparticles using Capsicum frutescence and its intensified activity against E. coli. *Resource- Efficient Technologies* 3(3): 303–308.

Shankar, S. S., Ahmad, A., Pasricha, R., Sastry, M. (2003a). Bioreduction of chloroaurate ions by geranium leaves and its endophytic fungus yields gold nanoparticles of different shapes. *Journal of Material Chemistry* 13: 1822–1826.

Shankar, S. S., Ahmed, A., Sastry, M. (2003b). Geranium leaf assisted biosynthesis of silver nanoparticles. *Biotechnology Progress* 19: 1627–1631.

Shankar, S. S., Rai, A., Ahmad, A., Sastry, M. (2004). Biosynthesis of silver and gold nanoparticles from extracts of different parts of the geranium plant. *Applied Nano Science* 1: 69–77.

Sharma, N. C., Sahi, S. V., Nath, S., Parsons, J. G., Gardea-Torresdey, J. L., Pal, T. (2007). Synthesis of plant-mediated gold nanoparticles and catalytic role of biomatrix-embedded nanomaterials. *Environmental Science Technology* 41(14): 5137–5142.

Shockman, G. D. and Barren, J. (1983). Structure, function, and assembly of cell walls of Gram-positive bacteria. *Annual Review of Microbiology* 37(1): 501–527.

Shownmyan, J. J., Harini, K., Pradeepa, M., Thyagarajan, M., Manikandan, R., Venkatachalan, P. (2013). Rapid green synthesis of silver nanoparticles using seed extract of Foeniculum vulgare and screening of its antibacterial activity. *Plant Cell Biotechnology and Molecular Biology* 13: 31–38.

Siddiqui, M. Z. (2011). Guggul: an excellent herbal panacea. *Asian Journal of Pharmacology and Health Sciences* 1: 35–39.

Sivaranjani, V., Philominathan, P. (2016). Synthesize of titanium dioxide nanoparticles using Moringa oleifera leaves and evaluation of wound healing activity. *Wound Medicine* 12: 1–5.

Silva, F., Ferreira, S., Duarte, A., Mendonça, D. I., Domingues, F. C. (2011a). Antifungal activity of Coriandrum sativum essential oil, its mode of action against Candida species and potential synergism with amphotericin B. *Phytomedicine* 19(1): 42–47.

Silva, F., Ferreira S., Queiroz J. A. (2011b). Coriander (Coriandrum sativum L.) essential oil: Its antibacterial activity and mode of action evaluated by flow cytometry. *Journal of Medical Microbiology* 60(10): 1479–1486.

Singh, K., Panghal, M., Kadyan, S., Chaudhary, U., Yadav, J. P. (2014). Green silver nanoparticles of Phyllanthus amarus: As an antibacterial agent against multi drug resistant clinical isolates of Pseudomonas aeruginosa. *Journal of Nanobiotechnology* 12(1):40.

Singla, R., Soni, S., Kulurkar, P. M. (2017). In situ functionalized nanobiocomposites dressings of bamboo cellulose nanocrystals and silver nanoparticles for accelerated wound healing. *Carbohydrare Polymer* 155: 152–162.

Stich, H. F. (1991). The beneficial and hazardous effects of simple phenolic compounds. *Mutation Research/Genetic Toxicology* 259(3–4): 307–324.

Tanveer, T., Ali, S., Khan, D. Z. (2011). Appraisal of secondary metabolites in in vitro cultures of citrulluscolocynthis (L) Schard. *Science International* 23(4): 289–294.

Teuscher, E. and Lindequist, U. (2010). Biogene Gifte: Biologie-Chemie-Pharmakologie-Toxikologie, Wissenschaftliche Verlagsgesellschaft Stuttgart.

Thirumurgan, A., Tomy, N. A., Jai, G. R., Gobikrishnan, S. (2010). Biological reduction of silver nanoparticles using plant leaf extracts and its effect on increased antimicrobial activity against clinically isolated organism. *Der Pharma Chemica* 2: 279–284.

Thirumurgan, G., Shaheedha, S. M., Dhanaraju, M. D. (2009). In vitro evaluation of antibacterial activity of silver nanoparticles synthesized by using Phytothorainfestan. *International Journal of Chemical Technology Research* 1: 714–716.

Thangadurai, D., Dabire, S. S., Sangeetha, J., Said Al-Tawaha, A. R. M., Adetunji, C. O., Islam, S., Shettar, A. K., David, M., Hospet, R., and Adetunji, J. B. (2021) Greener composites from plant fibers: Preparation, structure, and properties. In: Kharissova, O.V., Martínez, L.M.T., and Kharisov, B.I. (eds.), *Handbook of Nanomaterials and Nanocomposites for Energy and Environmental Applications*. Springer, Cham. https://doi.org/10.1007/978-3-030-11155-7_21-1

Turgis, M., Han, J., Caillet, S., Lacroix, M. (2009). Antimicrobial activity of mustard essential oil against Escherichia coli O157: H7 and Salmonella typhi. *Food Control* 20(12): 1073–1079.

Udaya, K. R., Hazeena, B. V. (2002). Antimicrobial studies of some selected medicinal plants. *Ancient Science Life* 4: 230–239.

Ullah, N., Nadhman, A., Siddiq, S., Mehwish, S., Islam, A., Jafri, L. Hamayun, M. (2016). Plants as antileishmanial agents: Current scenario. *Phytotherapeutic Research* 30: 1905–1925.

Umashankari, J., Inbakandan, D., Ajithkumar, T. T., Balasubramanian, T. (2012). Mangrove plant, Rhizophora mucronata (Lamk, 1804) mediated one pot green synthesis of silver nanoparticles and its antibacterial activity against aquatic pathogens. *Aquatic Biosystems* 8: 1–8.

Vankar, P. S., Shukla, D. (2012). Biosynthesis of silver nanoparticles using lemon leaves extract and its applications for antimicrobial finish on fabric. *Applied Nanoscience* 2: 163–168.

Verma, P., Mathur, A. K., Srivastava, A., and Mathur, A. (2012) Emerging trends in research on the spatial and temporal organization of terpenoid indole alkaloid pathway in Catharanthus roseus: Literature update. *Protoplasma* 249(2): 255–268.

Verrills, N. M., Walsh, B. J., Cobon, G. S., Hains, P. G., Kavallaris, M. (2003). Proteome analysis of vinca alkaloid response and resistance in acute lymphoblastic leukemia reveals novel cytoskeletal alterations. *Journal of Biological Chemistry* 278(46): 45082–45093.

Wang, T, Wang, H, Soong, Y. (2000) Paclitaxel-induced cell death: where the cell cycle and apoptosis come together. *Cancer* 88(11): 2619–2628.

Wang, J., Zhang, C. J., Chia, W. N., Loh, C. C., Li, Z., Lee, Y. M., He, Y., Yuan, L. X., Lim, T. K., Liu, M., Liew, C. K., Lee, Y. Q., Zhang, J., Lu, N., Lim, C. T., Hua, Z. C., Liu, B., Shen, H. M., Tan, K. S., and Lin, Q. (2015). Heam-activated promiscuous targeting of artemisinin in Plasmodium falciparum. *Nature Communications* 6: 10111, Doi 10.1038/ncomms10111.

Weber, T. and Kim, H. U. (2016). The secondary metabolite bioinformatics portal: Computational tools to facilitate synthetic biology of secondary metabolite production. *Synthetic and Systems Biotechnology* 1(2): 69–79.

Wen, H., Dan, M., Yang, Y., Lyu, J., Shao, A., Cheng, X., Chen, L., Xu, L. (2017). Acutetoxicity and genotoxicity of silver nanoparticle in rats. *PLoS One* 12: e0185554.

Wen, H., Yang, X., Hu, L., Sun, C., Zhou, Q., Jiang, G. (2015). Brain-targeted distribution and high retention of silver by chronic intranasal instillation of silver nanoparticles and ions in Sprague–Dawley rats. *Journal of Applied Toxicology* 36: 445–453.

Woldemichael, G. M. and Wink, M. (2001). Identification and biological activities of triterpenoid saponins from Chenopodium quinoa. *Journal of Agricultural and Food Chemistry* 49(5): 2327–2332.

Yang, L., Kuang, H., Zhang, W., Aguilar, Z. P., Wei, H., Xu, H. (2017). Comparisons of the biodistribution and toxicological examinations after repeated intravenous administration of silver and gold nanoparticles in mice. *Science Report* 7: 3303.

Yashiki, T, Kiso, K, Zhou, A. Y., Iwasaki, Y., and Kenji, (2010). The effects of Coptis japonica root ectract and its key component, berberine, on Human subcutaneous Adipocytes. *International Journal of Cosmetic Science* 32: 292–392.10.111/j.1468–2494.2010.00583_5.

Yeh, C. T., Rao, Y. K., Yao, C. J., Yeh, C. F., Li, C. H., Chuang, S. E., Luong, J. H.T., Lai, G. M., and Tzeng, Y. M. (2009). Cytotoxic triterpenes from Antrodia camphorata and their mode of action In HT-29 human colon cancer cells. *Cancer Letters* 285: 73–79.

Yousaf, R., Khan, M. A., Ullah, N., Khan, I., Hayat, O., Shehzad, M. A., Khan, I., Taj, F., Din, N. U., Khan, A., Naeem, I. and Ali, H. (2019). Biosynthesis of anti-leishmanial natural products in callus cultures of Artemisia scoparia. *Artificial Cells, Nanomedicine, and Biotechnology* 47 (1): 1122–1131.

Zhang, X. F., Park, J. H., Choi, Y. J., Kang, M. H., Gurunathan, S., Kim, J.H. (2015) Silver nanoparticles cause complications in pregnant mice. *International Journal of Nanomedicine* 10:7057–7071 doi: 10.2147/IJN.S95694. PMID: 26622177; PMCID: PMC4654536.

Part 3

Omics and Bioinformatics

Handling Editor:
Charles Oluwaseun Adetunji

13 Bioinformatics
A Tool for Biotechnological Advancement[1]

O. M. Oyawoye
Adeleke University

H. A. Olorunsola
Covenant University

A. O. Oluduro
Obafemi Awolowo University

O. O. Bamigboye, E. K. Oladipo, T. M. Olotu, I. J. Adeosun, and M. O. Kaka
Adeleke University

O. O. Obembe
Covenant University

CONTENTS

[1] O. M. Oyawoye and H. A. Olorunsola have contributed equally to the paper and are therefore co-first authors.

DOI: 10.1201/9781003178903-16

13.1 INTRODUCTION

The term "bioinformatics" is an abbreviation for "Biological Informatics." It is a combination of biological sciences and computer science, and many scientists now refer to it as computational biology. It became more recognized following the establishment of the human genome project (Kumar et al., 2017). Bioinformatics is a field that brings together computer science, information technology, and biology into one. Genomics, proteomics, genetics, transcriptomics, and evolution are only a few of the biological sciences covered. The goal of the field is to allow for the discovery of new biological ideas as well as the creation of a global context from which to comprehend unifying biological principles.

Bioinformatics is a multidisciplinary science and has been developed as an essential scientific discipline causing a paradigm change in various fields such as comparative genomics and molecular biology. A variety of bioinformatics tools, software, and databases are available to help researchers better understand biological complexities and interpret and store biological data. In recent years, biological data has evolved at an exponential rate, resulting in two problems. The first concern is effective data storage, while the second is how valuable knowledge can be extracted from the data. The second problem can be solved through machine learning, which can produce information from heterogeneous data (Shastry et al., 2020).

Bioinformatics, according to NCBI (National Centre for Biotechnology Information), is a division of science that incorporates computer science, genetics, and information technology into one field. As a result, it covers two topics: regulation of computing resources and biological data (genes, robots, proteins, cells, artificial intelligence, genomes, medical data, and ecological systems) (Wani et al., 2018). Bioinformatics tools have proven to be effective in data analysis because of changes brought about by the pre-genomic era. As a result, many breakthroughs in biotechnology have occurred, especially in the areas of personalized medicine, proteomics and transcriptomics, vaccines, diagnostics, and genetic engineering.

13.2 BIOINFORMATICS IN GENOMICS ERA

13.2.1 Interspecies Comparison

Comparing genome organization across cell types and species may reveal important details about the processes that underpin these genomic structures, such as their biochemical pathways (Potgieter et al., 2020). To identify the genes of interest, syntenic regions of related species considered to house a gene of known role were used at the genome structural stage. The mechanism between various species has proven to be such a powerful tool that databases have been created to capitalize on. C-InterSecture is a bioinformatics tool that was recently developed and used to identify regions that alter their spatial organization during vertebrate evolution.

To establish evolutionary relationships, all these methods may be used between various species as well as phylogenetic trees to be established. With rising emergence intervals, there is going to be a remarkable increase in sequence variations between homologous and increasingly different genomic structures. These distinctions can also make it possible to detect horizontally transferred genes between organisms.

13.2.2 Intra-species Comparison

Normal population size differences are predicted to affect the rate of aggregation of advantageous and marginally deleterious gene variants, affecting the relative contributions of selection and migration to gene variations (Xu, 2020). Although selective mating and genetic analysis can be used to compare phenotype and genotype in many organisms, this comparison is not feasible in many parasites. Only *Plasmodium falciparum*, which requires passage by mosquitoes and chimpanzees, has yielded useful genetic crosses among the widely studied human parasites. The sequence information presented by parasite genome projects will fill this void, enhancing nucleotide variations over phenotypic variance. On a small scale, this has been achieved successfully with genes implicated in drug resistance and antigenic heterogeneity. We will be able to differentiate genes that are difficult to identify by comparing sequence variants to phenotypic differences, due to our understanding of whole-genome sequences and the technical capability to search several genes one at a time depending on the feature.

13.2.3 Patterns of Gene Expression

Exon identification from genomic sequences in selected organisms has also had a host of functional uses, including the possibility of explaining gene expression patterns. If an organism's DNA sequence is documented, all genes will eventually be detectable. This will also give researchers a better understanding of the changes parasites go through as they transition from one life or host to the next, as well as the biochemical pathways required for the organism's survival in the host environment.

New techniques for chemotherapeutic intervention are likely to emerge now that the host and parasite sequences have been completely determined, as well as their expression patterns. Because comparing homologous pathways in the host and parasites can reveal some specific parasite molecules, novel therapeutic targets may emerge.

13.2.4 From Sequence Databases to Parasite Biology

In 1965, the first bioinformatic database was developed, dataset stored, and made available to the scientific community all previously mentioned DNA and protein sequences. The current focus on genomic structure and sequence data will soon turn to the relationship between genomics and proteomics, as well as the biological significance of this knowledge. Protein databases are important for modeling molecular relationships. To the sequence results in yeast, the genome complete sequences make scientists tag genes to add function and cellular localization. This method can be used in organisms where transformation and homogeneous integration are simple. Bioinformatics has aided in the understanding of past drug resistance pathways, such as chloroquine, as well as the merging of resistance to artemisinin-based combination therapies (ACT). Scientists can use sequence data to search for genes, gain clues to gene functions, analyze genetic variation, and investigate evolutionary relationships by using the right computer software.

Viruses have complicated methods for maximizing the coding ability of genomes and evaluating them due to their limited genome size (Gautam et al., 2019). In the meantime, genomics and bioinformatics have made important contributions to our understanding of infectious disease pathogenesis, mechanisms, and antimicrobial resistance transmission, as well as host immune responses (Bah et al., 2018; Ray et al., 2021).

13.2.5 Bioinformatics as a Tool in Viral Genomics

Bioinformatic approaches could be able to help inform epidemiology and approaches to transboundary viruses by synthesizing diverse information more effectively and systematically (Bauer et al., 2020). Bacteriophage and archaeal virus sequences, eukaryotic virus sequences, or variations of bacteriophages, archaeal viruses, and massive eukaryotic virus sequences can now be recognized, and this is because of the development of several bioinformatics tools and protocols (Roux et al., 2018). These approaches use a few parameters to determine if a genome is viral, such as whether it is significantly similar to identified viruses (in terms of genes encoding or nucleotide ingestion pattern) or whether it is unrelated to any recognized virus or cellular genome but contains single or multiple characteristic viral genome (Roux et al., 2018); this is now possible, thanks to bioinformatics. Bioinformatic analysis of viruses includes gene recognition, gene functional annotation, and phylogenetic association analysis related to the analysis of any novel sequences. As the viral genome is of small size, large numbers of isolates can be sequenced, which necessitate the use of complex analytical techniques (Angelin and Georrge, 2019).

Merging bioinformatics, epidemiological, and observational results can aid in the selection of animal models for effective preclinical responses (Callaway, 2020). Changes in viral genomics have resulted in a rapid increase in viral genome sequences. Scientists can now learn more about pathogenic viruses, thanks to advances in bioinformatics. The understanding of core genes present in all *Poxviridae* family members has been enabled by a series of genomes obtained through novel advances in viral genome sequencing (Ali et al., 2017). Poorly activated pathogen lysates may have adverse effects, so there was a need to introduce a new approach to develop universally applicable and safe vaccines, hence, reverse vaccinology.

13.2.6 Reverse Vaccinology

Reverse vaccinology is the identification of antigens using genome knowledge as a starting point. This technique, which was first used to produce vaccines against *Neisseria meningitidis* group B, has eventually progressed and is now recognized as a popular tool of vaccine discovery, as it can be used to develop vaccines against a wide range of pathogens. Antigens, which are only found in pathogenic isolates but not found in beneficial strains, can be identified using comparative *in silico* analysis of different genome sequences to distinguish conserved antigens within a heterogeneous pathogen population (Tordello et al., 2017).

It used a directed, computer data-based method to diagnose target vaccine antigens that were highly sensitive but not reliable, also most notably, it was not postulate-driven. The conventional laboratory-based, hypothesis-driven analysis of microorganisms to distinguish components that may elicit immune responses, on the other hand, is no longer used. As Dalsass et al. (2019) argued and detailed, to identify vaccine candidates for further experimentation, reverse vaccinology depends on mathematical tools and techniques, with refinement being crucial for their optimal use (Moxon et al., 2019).

The advantages of reverse vaccinology are numerous, which has brought several advancements to the field of biotechnology and vaccinology. Different numbers of bioinformatics tools have proven to be applicable in providing functional details on the genes that exist in microbial species' genetic data. These methods make it possible to identify the role of a wide range of genes and the factors that influence them (Ohyanagi et al., 2016).

The vaccine design and production process have been significantly aided by the increasing genomic and proteomic results. Reverse vaccinology is a cost-efficient and effective method of screening the whole infectious organism's genome (Naz et al., 2019).

A few of the bioinformatics tools used in reverse vaccinology and other *in silico* analysis similar to immunology include NetMHC, CTLPred, nHLAPred, Propred 1, Propred, and RankPrepb used in the prediction of major histocompatibility complex (MHC) binding proteins (Andreatta et al., 2015).

13.2.7 OMICS of Bacterial Pathogens and Host Responses

Epidemiology using genomic data campaigns has also served an important public position in confirming the SARS-CoV-2 population spread within the United States, China, and other parts of the world (Butler et al., 2021). The Human Genome Project's success has paved the way for the significant research, making structural projects became a possibility, thus allowing the field of systems biology to grow at a rapid pace. Some features of Gram-negative bacterial recognition, host intracellular transfer, and resistant adaptation have been discovered, thanks to systems biology approaches to host-pathogen interactions (Khan et al., 2019).

The discovery of drug resistance-conferring mutations and monitoring also greatly benefited from genomic research. Newly discovered genetic mutations in *Mycobacterium tuberculosis* genomes have also been linked to the emergence of anti-mycobacterial drug resistance. Anti-mycobacterial tolerance has been predicted using genomic analysis, which has shown variations in lineage mutation rates.

Genomic research has also been used to assess the evolutionary origins and disease monitoring, such as the distribution of mycobacterial strains like the Beijing strain, indicating its isolation from the Far East, and genomics has also been used to study TB outbreaks, genotyping of epidemic lineages, and their development during the outbreak (Sobkowiak et al., 2018). Indeed, different analysis methods for predicting *M. tuberculosis* spoligotypes from raw sequence reads have been created. Antibiotic tolerance and propagation dynamics can be detected using this instrument in conjunction with other bioinformatics research methods. Some genome-based approaches have recently been extended to clinical isolates from Malawi to distinguish mixed infections and infections caused by a single strain (Sobkowiak et al., 2018).

Omics methods use high-throughput analytical tools to describe the structure, activity, and dynamics of proteins (proteomics), low-molecular-weight metabolites (metabolomics), genes (genomics), and expressed genes (transcriptomics) in a biological sample. Most omics are powerful instruments in immune studies because of their sensitivity and precision (Nguyen et al., 2020).

13.2.8 Bioinformatics in Parasites/Protozoan Genomics

Rapid advances in genomics and gene editing techniques would almost inevitably result in the constant implementation of new integrated approaches for interrogating and altering genomes, helping us to better understand parasite activity and, ultimately, rationally formulate better methods for the control of animal infectious diseases (Suarez et al., 2017). Sundar et al. (2018) looked at how different proteomic technologies will help with disease intervention, as we need to better understand universal protein profiling and identify parasite development, longevity, pathogenicity, and drug resistance processes. Recent advances in proteomics have revealed information about the limitations of traditional chemotherapeutic methods, as well as *Leishmania* genetics, host-parasite interactions, and the emergence of new curative methods. A considerable number of protozoan parasite genomes have been sequenced, annotated, and made publicly available in the last 15 years. This abundance of data has sparked and fueled new studies into the vaccine and drug development, high-throughput experimentation, and the genetic engineering of important protozoan parasites. The World Health Organization's strategic committee on parasite genome research has integrated genome projects for seven human parasites (Toxoplasma, Filarial, Schistosomes, *Leishmania*, African and American trypanosomes, and *P. falciparum*), except for toxoplasma and *Plasmodium*.

13.2.9 Bioinformatics in Fungal Genomics

Since the late 1990s, phylogenetic experiments based on DNA sequences at single or multiple loci have been progressively used to classify fungal species (Schoch et al., 2012). Among a growing number of instances, DNA sequence or genotype data is being used as the primary or only means of dividing strains into different species (Hagen et al., 2015). As a result of this process, many new and enigmatic fungal species have been discovered among previously recognized fungal groups. According to a recent study of 51 studies involving fungi from 16 genera belonging to different taxonomic groups, each of the previously described species contained an average of three cryptic species (Matute and Sepulveda, 2019; Xu, 2020).

In the past, some strategies for the identification of fungi have depended mostly on conventional procedures. These procedures have their limitations due to multi-linking species phenotypic features closely related to other species. Next-generation genome sequencing supplies multiple opportunities to compare genomes of fungi, which are either pathogen or mycorrhizae, intending to gain an understanding of the genes involved in beneficial plant-fungus interactions. Gene structure is described by transcribed sequences and proteins. PASA (program to arrange spliced alignments) recognizes alternative splicing patterns, novel genes/exons, and transcription start/endpoints to boost gene calls using data from proteomics and transcriptomics. This method has been used to build and modify fungal genetic prototypes because fungi are essential antimicrobial drug discovery sources, identifying novel secondary metabolites which has been the subject of several recent metabolomics experiments. More fungal genomes are being available, thanks to modern sequencing methods and collaborative effort.

Soverini et al. (2019) developed a bioinformatics tool called the Human Mycobiome Scan used to profile the mycobiome's taxonomic diversity directly from raw NGS (next-generation sequencing) data. The tool makes use of hierarchical fungi databases to assign reads to fungal species >10,000 times faster than most related approaches. The fungal genomics community at the Large Institute has sequenced over 100 fungi, including plant and human pathogens and fungi that act as basic models for cellular and molecular biology (Broad Institute, 2017).

When the first fungal genome sequences were made available, it became clear that fungal biology was much more complex than previously thought. Since then, more genomes have been available, allowing for more comprehensive comparative studies, particularly when combined with postgenomic tools like proteomics and transcriptomics (de Vries et al., 2018).

13.3 RECENT ADVANCES IN BIOINFORMATICS APPLICATIONS

13.3.1 Drug Discovery

More accurate drugs that work on the cause, not just the effects, of the disease, will be produced through a greater comprehension of disease pathways alongside the use of statistical methods to define and validate experimental drug targets (Aryal, 2021). Relevant applications such as gene sequencing, molecular statistics, and calculating gene expression levels have improved the dose-response, toxicity profile, and overall efficacy of drugs used to treat a range of genetic diseases (Cuffari, 2021).

Researchers will not only offer instant relief to patients but also reduce rising healthcare costs by speeding up the drug discovery process. Patients with autoimmune disorders like inflammatory bowel disease (IBD) have also benefited from such interventions, with a gene-level screening of IBD subtypes and their interactions with other diseases being used to identify new drug candidates for repositioning (Cuffari, 2021).

Drug design has been successfully made more reliable using bioinformatics tools and reliable drug design processes, which have enormously reduced the amount of time and resources it takes to produce effective pharmaceuticals (Reuben-kalu, 2015;

Edu, 2020). To predict drug-likeness, computational approaches are used to classify and exclude candidate molecules that are not going to make it through the selection and growth stages (Dahiya, 2017).

13.3.2 Personalized Medicine

Personalized medicine refers to medical treatment that is specific to each patient's genetic profile. This will remove much of the negative side effects associated with current therapies, as well as mitigate or eliminate allergic reactions, lower hospital costs, and decrease patient pain through more successful treatments. Each patient's genome must first be converted into digital data, which is then analyzed, preserved, and extracted as required to practice personalized medicine. As a result, a three-pronged approach to personalized medicine, genomics, and bioinformatics is needed (BioCode, 2020).

Drugs are often given in combination to improve the therapy success rate. The interaction between a patient's molecular data and their disease phenotype is intricate and impossible to establish manually (Ray, 2018). Hence, personalized medicine science aims to find unique options focused on each person's susceptibility profile. These fields are believed to enable new approaches to personalized therapy, diagnosis, and drug production (Valeska et al., 2019).

13.3.3 Waste Removal

The primary goal of waste removal using bioinformatics approach is to classify and analyze the DNA sequencing of microbes to use them for sewage cleaning, hazardous waste removal, oil spill clean-up, and other applications (Edu, 2020). Bacteria and other species assist in the clean-up of debris in bioinformatics. *Deinococcus radiodurans*, for example, can isolate degraded DNA and small fragments from chromosomes and repair them (Dahiya, 2017). By isolating damaged segments clustered region, this bacterium may fix affected DNA and small chromosome fragments. This is because it has many versions of the genome. To clear up the area, *D. radiodurans* has had genes from other bacteria introduced into it. It was used to break down organic compounds, solvents, and heavy metals in hazardous waste facilities (Dahiya, 2017).

Microbial depletion is one of the most cost-effective ways to clean up a contaminated environment. The research fields of applied microbiology and biotechnology are bioremediation/biodegradation and bioinformatics. Several databases and prediction tools/approaches are currently available to aid in the growth and implementation of bioremediation. The use of an *in silico* methodology saves time and by eliminating the need for routine experimental research, and it aids in the prediction of degradation pathways (Kumari et al., 2021).

13.3.4 Changes in the Climate

Carbon dioxide emissions are believed to be contributing to global climate change, mostly due to the expanding use of fossil fuels for oil (Waldvogel et al., 2020). Bioinformatics has allowed researchers to study the genomes of bacteria that use carbon dioxide as their primary carbon source (Aryal, 2018). Bioinformatics research is being undertaken regarding sequencing the microbial genome of certain microbes that may decrease carbon dioxide and other greenhouse gas levels and could potentially play a larger role in stabilizing global climate change (Sinha, 2015). The radiation-tolerant microbe *D. radiodurans*, for example, is used to clean up radioactive hazardous waste areas (Abberton, 2015).

13.3.5 Crop and Livestock Improvement

Bioinformatics allows efficient use of proteomics, metabolomics, genomics in crop production to produce high-yielding, more drought-resistant, and insect-resistant crops, improving livestock quality and disease resistance (Edu, 2020). *Arabidopsis thaliana* (watercress) and *Oryza sativa* (rice) have complete genomes that are currently accessible (Aryal, 2018). At 430 million nucleotides, *O. sativa* has the lowest genome of all the cereals. It can, however, be used as a reference genome for monocotyledons, one of the two main groups of flowering plants (Hao, 2016).

Scientists recently succeeded in translating genes into rice, resulting in increased levels of vitamin A, iron, and other micronutrients. This study could help to reduce the prevalence of vitamin A and iron deficiency-related blindness and anemia. Scientists have successfully introduced a yeast gene into the tomato plant, resulting in fruit that lasts longer on the vine and has a longer shelf life (Aryal, 2018).

13.4 CONCLUSION

Bioinformatics is a useful tool for the enhancement of biotechnology, thereby inventing useful products and creating solutions to problems in aspects including utilization of the genome attributes of viruses to acquire a better understanding of the biology of pathogenic viruses, reverse vaccinology, drug discovery, personalized medicine, waste clean-up, climate change, and agricultural improvement.

REFERENCES

Angelin, G. and Georrge, J.J. (2019). Viroinformatics: Databases and Tools. *Proceedings of 11th National Science Symposium on Recent Trends in Science and Technology* (pp. 117–126. ISBN: 9788192952147. Rajkot, Gujarat, India: Christ Publications.

Bauer, D. C., Tay, A. P., Wilson, L. O. W., Reti, D., Hosking, C., McAuley, A. J., Pharo, E., Todd, S., Stevens, V., Neave, M. J., Tachedjian, M., Drew, T. W., and Vasan, S. S. (2020). Supporting pandemic response using genomics and bioinformatics: A case study on the emergent SARS-CoV-2 outbreak. *Transboundary and Emerging Diseases* 67(4): 1453–1462. https://doi.org/10.1111/tbed.13588.

BioCode (2020). Role of bioinformatics in personalized medicine. Biocodekb.Com. https://www.biocodekb.com/2020/11/07/role-of-bioinformatics-in-personalized-medicine/.

Broad Institute (2017). Fungal Genomics. Broad Institute. https://www.broadinstitute.org/fungal-genome-initiative.

Butler, D., Mozsary, C., Meydan, C. et al. (2021). Shotgun transcriptome, spatial omics, and isothermal profiling of SARS-CoV-2 infection reveal unique host responses, viral diversification, and drug interactions. *Nature Communication* 12, 1660. https://doi.org/10.1038/s41467-021-21361-7.

Cuffari, B. M. (2021). Role of bioinformatics in drug discovery. *News Medical.* https://www.azolifesciences.com/article/Role-of-Bioinformatics-in-Drug-Discovery.aspx.

Dahiya, P. B. (2017). Bioinformatics impacts on medicine, microbial genome, and agriculture. *Journal of Pharmacognosy and Phytochemistry* 6(4): 1938–1942.

Dalsass, M., Brozzi, A., Medini, D. and Rappuoli, R. (2019). Comparison of open-source reverse vaccinology programs for bacterial vaccine antigen discovery. *Frontiers in Immunology* 10:113. doi: 10.3389/fimmu.2019.00113.

de Vries, R. P., Grigoriev, I. V., and Tsang, A. (2018). Introduction: Overview of fungal genomics. *Methods in Molecular Biology* 1–7. https://doi.org/10.1007/978-1-4939-7804-5_1.

Del Tordello, E., Rappuoli, R., and Delany, I. (2017). Reverse vaccinology. *Human Vaccines* 65–86. https://doi.org/10.1016/b978-0-12-802302-0.00002-9.

Edu, T. L. (2020). Application of bioinformatics in various sectors. Leverage Edu. https://leverageedu.com/blog/application-of-bioinformatics/.

Khan, M. M., Ernst, O., Manes, N. P., Oyler, B. L., Fraser, I. D. C., Goodlett, D. R., and Nita-Lazar, A. (2019). Multi-Omics Strategies Uncover Host-Pathogen Interactions. *ACS Infectious Diseases* 5(4): 493–505. https://doi.org/10.1021/acsinfecdis.9b00080.

Kumari, P., and Kumar, Y. (2021). Bioinformatics and computational tools in bioremediation and biodegradation of environmental pollutants. *Bioremediation for Environmental Sustainability* 421–444. https://doi.org/10.1016/b978-0-12-820318-7.00019-8.

Moxon, R., Reche, P.A. and Rappuoli, R. (2019). Editorial: Reverse vaccinology. *Frontiers in Immunology* 10:2776. doi: 10.3389/fimmu.2019.02776.

Naz, K., Naz, A., Ashraf, S.T., Rizwan, M., Ahmad, J., Baumbach, J., Ali, A.(2019). PanRV: Pangenome-reverse vaccinology approach for identifications of potential vaccine candidates in microbial pangenome. *BMC Bioinformatics* 20: 123. https://doi.org/10.1186/s12859-019-2713-9.

Nguyen, T. V., and Alfaro, A. C. (2020). Applications of omics to investigate responses of bivalve haemocytes to pathogen infections and environmental stress. *Aquaculture* 518: 734488. https://doi.org/10.1016/j.aquaculture.2019.734488.

Potgieter, L., Feurtey, A., Dutheil, J.Y. and Stukenbrock, E.H. (2020). On variant discovery in genomes of fungal plant pathogens. *Frontiers in Microbiology* 11:626. doi: 10.3389/fmicb.2020.00626.

Ray, T. (2018). Applications of bioinformatics in medicine and biotechnology. *Tanmoy Ray.* https://tanmoyray.com/blog/biotech-life-sciences/applications-of-bioinformatics-in-medicine-and-biotechnology/.

Reuben-kalu,J.(2015).Applicationsofbioinformaticsindrugdiscovery.10.13140/2.1.1099.8569.

Roux, S., Adriaenssens, E. M., Dutilh, B. E., Koonin, E. V., Kropinski, A. M., Krupovic, M., Kuhn, J. H., Lavigne, R., Brister, J. R., Varsani, A., Amid, C., Aziz, R. K., Bordenstein, S. R., Bork, P., Breitbart, M., Cochrane, G. R., Daly, R. A., Desnues, C., Duhaime, M. B., . . . Eloe-Fadrosh, E. A. (2018). Minimum Information about an Uncultivated Virus Genome (MIUViG). *Nature Biotechnology* 37(1): 29–37. https://doi.org/10.1038/nbt.4306.

Sinha, S. (2015). Role of bioinformatics in climate change studies. *Journal of Science* 1: 1–9.

Soverini, M., Turroni, S., Biagi, E., Brigidi, P., Candela, M., Rampelli, S. (2019). Human Mycobiome Scan: A new bioinformatics tool for the characterization of the fungal fraction in metagenomic samples. *BMC Genomics* 20: 496 https://doi.org/10.1186/s12864-019-5883-y.

Suarez, C. E., Alzan, H. F., and Cooke, B. M. (2017). Genomics and genetic manipulation of protozoan parasites affecting farm animals. *Parasitic Protozoa of Farm Animals and Pets* 413–438. https://doi.org/10.1007/978-3-319-70132-5_16.

Sundar, S. and Singh, B. (2018). Understanding Leishmania parasites through proteomics and implications for the clinic. *Expert Review of Proteomics* 15(5): 371–390. https://doi.org/10.1080/14789450.2018.1468754.

Themes, U. (2019). Host response to infections: The "-omics" revolution. Obgyn Key. https://obgynkey.com/host-response-to-infections-the-omics-revolution/.

Valeska, M. D., Adisurja, G. P., Bernard, S., Wijaya, R. M., Hafidzhah, M. A., Parikesit, A. A. (2019). The role of bioinformatics in personalized medicine. *Your Future Medical Treatment* 46: 785–788.

Waldvogel, A., Feldmeyer, B., Rolshausen, G., Exposito-Alonso, M., Rellstab, C., Kofler, R., Mock, T., Schmid, K., Schmitt, I., Bataillon, T., Savolainen, O., Bergland, A., Flatt, T., Guillaume, F., and Pfenninger, M. (2020). Evolutionary genomics can improve prediction of species' responses to climate change. *Evolution Letters* 4(1): 4–18. https://doi.org/10.1002/evl3.154.

Wikipedia Contributors. (2021). Reverse vaccinology. Wikipedia. https://en.wikipedia.org/wiki/Reverse_vaccinology.

Xu, J. (2020). Fungal species concepts in the genomics era. *Genome* 63(9): 459–468. https://doi.org/10.1139/gen-2020-0022.

14 Roles of Nucleotide Sequence Analysis in Human Genetics and Genomics

S. S. D. Mohammed
Nile University of Nigeria

I. Abraham
National Biotechnology Development Agency (NABDA)

D. Enoma
Covenant University

L. E. Okoror
Federal University

CONTENTS

DOI: 10.1201/9781003178903-17

14.1 INTRODUCTION

The consistent exponential increase in the rate of amino acid- and nucleotide sequence data sets arising from the development of high-throughput sequencing methods is currently revolutionizing genome study and analysis giving birth to accurate insight to the study of diseases and drug formulation (Durbin et al., 1998). There has also been an increase in the comparison of sequences to find similarity, identification of inherent sequence features such as the open reading frames (orf), post-translational modification sites (pms), active sites, and gene structures. There is also the need to identify the distribution of introns, exons, and regulatory elements (Sanbi, 2018). The DNA sequence determines the sequence of RNA product and sequences of amino acid. In DNA, there are only four nucleotides, namely: Adenine (A), Thymine (T), Guanine (G), and Cytosine (C), and between the two strands, the rule of pairing states that A will be paired with T and G with C. The synthesis and sequencing of DNA strands enable an accurate rapid generation of genetic parts. The elimination of unwanted secondary structures of RNA and restriction sites in addition to the optimization of the expression of gene are prerequisites to the designing of a rational novel DNA sequence (Liang et al., 2011; Rodrigues and Kluskens, 2011). The understanding of human genetics is currently facilitating the fight against racial abuse. Various findings have reviewed that all human races are different results of adaptation to unique geographical and environmental conditions. The difference is not based on the possession of "good" or "bad" genes but on how the general genes common to all humans are distributed (Great Soviet Encyclopedia 2010). Human genetics studies play a regulatory role in the discipline of genetics. Mutation and recombination can bring about various phenotypes in humans. The mutation rate is estimated at 1 per 100,000 to 1 per 10,000,000 gametes per generation. Population genetics, a term used to describe the frequency of a particular gene in a group of people, also reveals the process by which mutations are distributed among populations. This yields charts and distributions that can show the distribution of genes which are involved in the normal development and in diseased states (Science Daily, 2017). The source of this interest is based upon the need to understand who humans are and why we act the way we do. Practically, an understanding of human heredity is relevant to the projection, detection, and treatment of diseases which have a genetic basis. This endeavor to understand the genetic basis of human health gives rise to the field of medical genetics. Generally, medicine has also added more focus and direction to the purpose of medical genetics, hence the phrases medical genetics and human genetics are used interchangeably. Medical genetics has become a major component of preventive medicine such as genetic counseling and prenatal diagnosis. Genetic-based diseases may be divided into three classes: chromosomal, Mendelian, and multifactorial diseases.

Genomics studies the entire genomic sequence of an individual and the interactions between genes with each other and with their environment (Brief Guide to Genomics, 2019). A genome is a human complete DNA sequence including all of its genes (Science Daily, 2017). In a normal human, the nucleus of the cells comprises 23 pairs of chromosomes (46 in total). Of each chromosome pair, one is of maternal origin and the other is paternal in origin. One member of each pair is passed down from the egg and sperm cells by Great Soviet Encyclopedia (2010). There are 24 different types of chromosomes in humans: 22 autosomal and the gender determining X and Y chromosomes (Science Daily, 2017). Somatic cells have one copy each from each parent including an X chromosome from the mother, while the father contributes an X or Y chromosome. The estimated number of protein coding genes is at 20,000–25,000. This estimate has been constantly revised down from the first predictions of about 100,000. This is as a result of the improvement of sequence quality and gene identification methods and this estimate might continue to drop (Science Daily, 2017).

14.2 NUCLEOTIDE SEQUENCE ANALYSIS: AN OVERVIEW

Nucleotide sequence is the order of individual nucleotide in a nucleic acid molecule. This sequence of nucleotides within a gene sequence informs the eventual amino acid sequence of the protein or RNA product. Furthermore, the precise sequence of DNA molecules is made known through methodologies that involve DNA sequencing. These sequence analyses also use sequence alignment, database searches, phylogenetics, and other techniques to achieve their functionality and provide quality and reliable results in genetics and genomics studies. For example, in comparison of normal and mutated DNA sequences, an investigator may be able to obtain useful information for the diagnosis of different genetic diseases including cancer (Chmielecki and Meyerson, 2014).

14.2.1 Sequence Alignments

Sequence alignment is an in silico technique that carries out a comparative alignment search between the query nucleotide (DNA or RNA) and or amino acid sequence to the subject sequence of the database. The resulting similarity index between the query and the subject sequence can be a good source for drawing relational inference in function, structure, or transmutative origin (Mount, 2004).

14.2.1.1 Global and Local Alignment

The main objective of carrying out a global alignment search is to compare each residue in all the sequences. This approach is the most important if the length of the sequences in question are equal and looks similar. The Needleman–Wunsch algorithm is a popular global alignment technique that is based on the dynamic programming technique. On the other hand, local alignment is more applied to dissimilar sequences which possibly contain similar regions or sequence motifs in the larger sequence context. The algorithm of Smith–Waterman technique is a dynamic programming with additional options to start and end within any position of the sequence (Polyanovsky et al., 2011).

14.2.1.2 Multiple Sequence Alignments

Multiple sequence alignment extends local alignment in that it allows incorporation of more than one concurrently. They are generally applied for the identification of conserved reads (sequence segments) across a set of sequences which are possibly related evolutionarily (Wang and Jiang, 1994 and Elias, 2006).

14.2.1.3 Structural Alignments

Structural alignments are generally exclusive to amino acid sequences although they also include RNA sequences. They use the data collected from the secondary and tertiary structural protein or RNA molecule to align the sequences. Two or more sequences can be aligned simultaneously using the structural alignment method, though they are dependent on available structural information, thereby being limited to sequences whose structures have been modeled through X-ray crystallography or NMR spectroscopy (Chothia and Lesk, 1986).

14.2.1.4 Some Software for Sequence Alignments

The Molecular Evolutionary Genetics Analysis **(MEGA)**: Multiple options with distance calculation, does multiple alignments of query sequences and an output of a phylogenetic tree that will give a reliable base identification. Clustal Omega **(Clustalo)**: multiple alignment format + profile alignment, is a new multiple sequence alignment program that uses seeded guide trees and HMM profile-profile techniques to generate alignments between three or more sequences. **Clustalx**: nt + protein alignment in windows interface (alignment exploration): It provides an integrated environment for performing multiple sequence and profile alignments and analyzing the results. **Mauve**: use for genome alignment, muscle: large data set alignments. **MAFFT**: deals with nucleotide sequences with high variation and **Tcoffee,** is a multiple sequence alignment server. It can align Protein, DNA and RNA sequences (EMBL-EBI, 2010).

14.2.2 Computational Nucleotide Sequence Analysis

14.2.2.1 Algorithm Insights

The basic algorithm that runs as the foundation of the program is: Inputting of sequence of nucleotides from the user or from any external source, the nucleotides that are accepted from the user follows two criteria – (A) They should be in the upper case (A, U, G, C) and (B) They should be in triplets; as per the nucleotide, the program runs through the codes, gives as output the corresponding amino acid in the order of inputted sequence. In essence, the triplet formation of nucleotides yields the corresponding amino acids. This process continues until a stop codon is reached and the translation process is completed.

14.2.3 Restriction Analysis of DNA Molecules

Restriction analysis is the application of specific restriction enzymes to analyze the DNA molecules (Kessler and Vicentiu, 1990). Fragments of the DNA molecules that have been generated as a result of the cleaving by the restriction enzyme are generally separated by using the technique of gel electrophoresis (Churchill, 2012). The pattern of restriction fragments produced can be used for various purposes: to detect fragments of a specific gene using a nucleic acid probe, to detect the presence of mutations, to compare DNA from different individual, and to construct physical maps, called restriction maps, of DNA molecules that indicate the position of restriction sites in the molecules. Restriction enzymes can scan an entire length of DNA in search for particular sequence of bases which they can recognize (Kessler and Vicentiu, 1990; Churchill, 2012). The site for the recognition of the sequence is normally between 4 and 6 base pairs in size. Immediately after the identification has been done, the restriction enzyme attaches to precise location of the sequence and excise it. The repeated occurrence of this process throughout the entire length of the double-stranded DNA molecule eventually leads to the production of DNA fragments (Churchill, 2012).

14.2.4 Regulatory DNA Sequence

DNA sequences that are responsible for regulating gene expression like promoters or operators are called regulatory DNA sequences. Despite the genome's large size, it is only about 2% of our genome that encodes for proteins. Moreover, the remaining 98% comprises non-coding regions that control where and when the protein coding genes are expressed (Science Daily, 2017). The non-coding regions of the genomes have been identified in genetic and genomic studies as potential drivers of human diseases such as cancers. An improved understating of these regulatory regions and also the basic mechanisms which control how the genes are turned on or off are necessary to delineate how these diseases start or progress and also to discover new potential treatments (Jian et al., 2017). Unfortunately, the tools which are necessary to identify these non-coding regions and also to uncover their mechanisms of action are limited. The tools need the identification of protein factors which regulate these regions. This depends on the reagents (e.g., antibodies) which are available and they need advanced genetic manipulations. These systems are named Clusters of Regularly Interspaced Short Palindromic Repeats (CRISPR) and Affinity Purification in situ of Regulatory Elements (CAPTURE). The CAPTURE provides a method to concurrently isolate genome sequence-associated proteins and their RNA/DNA interactions. This method was created by repurposing the CRISPR genomic editing system (Science Daily, 2017; Jian et al., 2017). It is a specific region of DNA with two unmistakable attributes: with nucleotide repeats and spacers present. The distribution of DNA building blocks and DNA sequence repeats, which are normally interspaced with bits of DNA called spacers, is normally affixed all through a CRISPR region. The principle of operation for the CAPTURE is the application of guide RNAs that serve as a directional molecule for deactivated Cas9 to the targeted DNA elements of interest (Jian et al., 2017). Using the CAPTURE system, novel telomere-associated proteins associated with human were identified as a proof of principle (Science Daily, 2017). Telomeres are short repetitive sequences at the end of chromosomes that protect chromosomes from joining or fusing with the neighboring chromosomes. Furthermore, investigators discovered novel mechanisms that control malfunctioning beta-globin gene expression in human blood cells. This protein is part of a larger protein complex known as hemoglobin which is responsible for oxygen and carbon dioxide exchange between our lungs and body tissues (Science Daily,

2017; Jian et al., 2017). This aberrant beta-globin gene expression is associated with hemoglobin disorders which are inherited like sickle cell disease that affects 5% of the world's population. Unbiased analysis of the genome by CAPTURE gives scientists a very useful tool in understanding the regulatory principles of gene expression. This tool will increase our understanding of the human genome and genetic variations in various disease conditions (Science Daily, 2017).

14.2.5 Human Molecular Phylogenetics

Molecular phylogenetics can be defined as the process by which nucleotide and amino acid sequences are used to generate and elucidate the evolutionary relationship between two or more organisms (Human Evolution, 2020). The continuous increase in genomic sequence data set of diverse organisms has revolutionized the subject of phylogenetic with its impacts being appreciated globally. For example, the application of molecular phylogenetics has enabled research scientists to reveal the close relationship between *Homo sapiens* and *Gorilla gorilla* (2020). The paradigm shifts in human molecular phylogenetics that haze the line between population genetics and phylogenetics have occurred in recent years (Ting and Sterner, 2012; Jeffrey, 1993). The protein and DNA sequence give the most detailed and complete data for molecular phylogenetics. However, methods for sequencing proteins did not become regularly used until the late 1960s. Fast DNA sequencing also was not developed until about ten years after (Brown, 2002). The evolutionary relationships between humans and other primates have been clarified by the application of DNA phylogenetics. The detailed molecular data collected from the earliest immunological studies in 1960s (Goodman, 1962; Sarich and Wilson, 1967) has revealed that there is a single clade between gorillas, chimpanzees, and humans.

14.2.5.1 Conditions for the Human Molecular Phylogenetic Studies

The ability of a chosen gene of interest to display variability in the experimental organism is a key requirement in any application of molecular phylogenetics. It is impossible to generate any phylogenetic information without variability of the gene of interest (Ting and Sterner, 2012). Genetic variability presents a great challenge in intraspecific research due to the genetic similarity of organisms being compared arising from the fact that they belong to the same species. Consequently, it means that the most variable DNA sequences must be used during phylogenetic analysis if maximum results must be obtained. The multiallelic genes (including members of the HLA family, which occur in various dissimilar sequence forms), microsatellites (which evolve by replication slippage), and mitochondrial DNA (such as the haplotypes) are the three main possibilities in humans as reported by Ting and Sterner (2012).

14.2.6 Gene Location and Identification

For decades now, geneticists apply maps in describing the exact location of a specific gene on its chromosome. One map type uses cytogenetic location in order to describe a gene's specific position (Genetics Home Reference, 2020). The location is dependent on a unique pattern of bands that are generated after the staining of the chromosomes have been done with suitable chemicals. Another map type utilizes molecular location (the exact location of a gene on a chromosome). This molecular location depends on the DNA sequence that constitutes the chromosome.

14.2.6.1 Gene Mapping

Identification of gene location helps progressing study and manipulation of particular genes. Genes close together are more likely to be carried along together (Foxman, 2012). Genetic markers like single-nucleotide polymorphism (SNP), may be associated with a particular disease or function due to its location near the targeted gene. The human genome has genes in fixed physical locations but bacterial genomes often do not. Knowledge of a gene's location is instrumental for the designing of primers. Genetic sequences and maps have increased the chances of designing a specific primers. For an example, if an investigator intends to find variations within an entire gene, multiple primers may be created that start at the inside and outside of the gene region. Genomes can be very large, and vary in content, so there are two different types of genome maps: genetic maps that give landmark locations or markers and physical maps that show both the markers and the genetic sequence between the landmarks (Foxman, 2012). Relative location can be determined by examining genetic sequence available in public databases for the presence of restriction sites or other markers. Knowledge of a gene's specific location helps studies of determination of the exact gene sequence, gene function, and variation. Biological information that is related to markers in the same physical region may provide more details regarding the function of a gene (Foxman, 2012).

14.2.6.2 Molecular Location

The Human Genome Project was an international research effort which was completed in the year 2003. The objective was to determine the nucleotide sequence for every human chromosome (Genetics Home Reference, 2020). The sequence information allows for provision to researchers of a more specific address of the gene of interest than the cytogenetic location. The molecular address of a gene pinpoints the gene's location in terms of base pairs. This address describes the exact position on a chromosome and shows the size of the gene. This knowledge of the molecular location also gives researchers the ability to find out precisely the distance a gene is from other genes on the same chromosome. Moreover, different research groups have presented different values for the molecular location of a gene (Genetics Home Reference, 2020). Investigators interpret the human genome sequence with the use of various methods and this might cause variations in the gene's molecular address. Genetics Home Reference presents data for National Center for Biotechnology Information for the molecular location of genes using the Basic Local Alignment Search Tool (BLAST) program or DNA Data Base of Japan (DDBJ), European Molecular Biology Laboratory-European Bioinformatics Institute (EMBL-EBI), and SWISS-PROT and Tr-EMBL for all living species.

14.3 REPETITIVE AND SATELLITE DNA

Repetitive and satellite DNA are two forms of DNA that are found in most eukaryotic genomes. The major distinction between the two types of DNA is that the repetitive DNA comprises small sequences of DNA which are repeated 100–1,000 times.

Moreover, satellite DNA is highly repetitive and constitutes a major part of the human genome. The two forms of repeats which are found in the genome are tandem and interspersed repeats according to Bhagavan and Chung-Eun (2015). Satellite DNA comprises many tandem repeats of a short basic repeating unit. They are repeated sequences that are found close to centrosomes, telomeres, and Y chromosomes. They possess a repeat length of about 171 base pairs. A significant number of these sequences are confined in centromeres and telomeres; however, they are additionally scattered all through the genome. These were achievable through the nucleotide sequence analysis of the entire genome (John, 2012).

14.4 SEQUENCE ANALYSIS OF HUMAN DISEASED DNA

Human disease is not left out of this myriad of applications as diseases often have a molecular genetic basis. Genomics is a relatively new scientific discipline, having DNA sequencing as its core technology. This in turn has enabled enquiry into diseases at a genome-wide scale (Koboldt et al., 2013). Genotyping by Sequencing Techniques and standardized ontological annotation both plays functional roles in methodically analyzing the effects of human genomic variation and phenotypes of model organisms (Belizário and Hilliker, 2013). These can play key roles in finding candidate genes and new clues for disease's etiology and treatment (Drmanac et al., 2010). Sanger sequencing, genome-wide association analysis, and Next-Generation Sequencing (NGS) analysis, respectively, is the order in which sequencing has evolved at the whole genome and exome sequencing levels (Hitomi and Tokunaga, 2017; Petersen et al., 2017). Sequence analysis, whole genome phylogenetic analysis, and genetic phylogenetic analysis were employed in the current covid-19 disease outbreak (Suzuki et al., 2020). They aimed to understand the current evolutionary dynamics of bovine coronaviruses (BCoVs) through whole genome analysis of multiple Japanese BCoV isolates collected in recent years. This is significant as coronavirus can cross the species barrier and infect humans with a severe respiratory syndrome (Qiang et al., 2020). Genomic analysis also enabled scientists identify SARS-COv-2 to have greatest similarity with the bat coronavirus (Li et al., 2020; Zhou et al., 2020). Sequence analysis is also used in the study of cancerous cells' DNA. This is to give an insight into the molecular features of the disease (Li et al., 2020). As the cancerous phenotype is complex, it is only fitting that we turn to advances in sequences to address this problem. GWAS is widely used in understanding complex non-communicable diseases. They are useful for detecting variants at genomic loci which are associated with complex traits in the population (Visscher et al., 2012). These traits include complex diseases such as heart disease, diabetes, auto-immune diseases, and psychiatric disorders. Specifically, NGS has been very useful in the fight against cancer as it has made patient selection for targeted drugs possible, some of which have shown remarkable efficacy in cancers that have the cognate molecular signatures (Adashek et al., 2020). In general, analysis of sequences (coding and non-coding) is playing a major role in current clinical practice. They will continue to grow as the cost of sequencing continuously reduces and eventually will become a key component of healthcare systems.

14.4.1 Chromosomal Abnormalities

The string-like structures called chromosomes that are primarily made up of DNA and proteins are situated in all the nucleus of the cell. The uniqueness of the functional roles played by these string-like structures in serving as carriers of genetic material make them very delicate molecules that is capable of determining the health state of any individual and, as a result, can lead to disease at any slight aberration in chromosome number. Slight or major alteration during cell division is often the major reason for chromosomal abnormalities. A common instance is the genetic disorder called Down syndrome (also called Trisomy 21) that occurs when all the cells in an individual has three copies of chromosome 21 instead of two (Stöppler, 2019).

14.4.1.1 Symptoms and Signs of Down Syndrome

The sequence of chromosome 21 was a turning point for the understanding of Down syndrome. Persons that have Down syndrome might have a number of defects from birth. Approximately half of the children that are affected are given birth to with a defect of the heart. Other disorders such as digestive abnormalities (e.g., intestinal blockage) are less prominent (Antonarakis et al., 2004 and Stöppler, 2019). A child born with Down syndrome has a unique appearance, although all aspects of the child's appearance need not be present as the phenotype. The combinations of genes might make the child look different from other patients. Popular symptoms of Down syndrome include: having a flat face and small head. Also having upward slanting eyes, short neck, and ears are flat and positioned lower than "normal." There is tongue protrusion and it seems to be too big for the mouth, wide hands, muscles lacking tone, flexible joints, short fingers, and a single flexion crease in the palm. There are a lot of other abnormalities of the chromosome and they include: Klinefelter syndrome (47, XXY), Turner syndrome (45, X0), and Cri du chat syndrome. Some of these abnormalities might occur due to chromosomal translocation where in parts of two chromosomes are exchanged (Stöppler, 2019).

14.4.2 Mitochondrial Genetic Inheritance Disorders

These are genetic disorders that are caused by mutations in the non-nuclear DNA of mitochondria. Mitochondria are rod-like or small round organelles which are important for cellular respiration. They are located in the cytoplasm of animal and plant cells. Every mitochondrion might have 5–10 circular pieces of DNA. Furthermore, because human egg cells, and not sperm cells, retain their mitochondria during fertilization, the mitochondrial DNA is always inherited from the female parent. Some diseases of the mitochondria include Leber's hereditary optic atrophy (LHON) (a disease of the eye), myoclonic epilepsy with ragged red fibers (MERRF) and mitochondrial encephalopathy, lactic acidosis, and stroke-like episodes (MELAS), a rare form of dementia (Stöppler, 2019). The several applications and results obtained are evident of the role of nucleotide sequence analysis in genetics and genomics.

14.5 CONCLUSION

The importance of nucleotide sequence analysis in the current era of human genetics and genomics cannot be overemphasized. Nucleotide sequence alignment has enabled us to make significant progress in life sciences and medicine. From the earliest innovation of sequence alignment to the present time, the improvements have continued. Restriction fragment, regulatory sequence, and phylogenetic analysis can be performed on these sequences which are majorly generated through NGS platforms. Identification and location of genes on repetitive or satellite DNA analysis are also useful analysis to be done on DNA sequences. Finally, human diseases with a strong genetic component can only be approached through the analysis of nucleotide sequences. This includes monogenic and multifactorial diseases. The role of nucleotide sequences will only continue to rise as high-throughput sequences are being generated with technology and will provide tremendous insights into human genetics and genomics.

REFERENCES

Adashek, J. J., Kato, S., Lippman, S. M., and Kurzrock, R. (2020). The paradox of cancer genes in non-malignant conditions: implications for precision medicine. *Genome Medicine* 12. https://doi.org/10.1186/S13073-020-0714-Y.

Antonarakis, S.E., Lyle, R., Dermitzakis, E.T., Reymond, A. and Deutsch, S. (2004). Chromosome 21 and downsyndrome: from genomics to pathophysiology. *Natural Review of Genetics* 10:725–738.

Belizário, J. E., and Hilliker, A. J. (2013). The humankind genome: from genetic diversity to the origin of human diseases. *Genome* 56 (12): 705–715. https://doi.org/10.1139/gen-2013-0125.

Bhagavan, N.V. and Chung-Eun, H., (2015). Structure and Properties of DNA. In *Essentials of Medical Biochemistry*. 2nd edition. https://www.sciencedirect.com/topics/neuroscience/interspersed-repeat. Retrieved on 22/04/2020.

Brief Guide to Genomics (2019). https://www.genome.gov/about-genomics/fact-sheets/A-Brief-Guide-to-Genomics. Retrieved on 18/03/2020.

Brown, T.A. (2002). Chapter 16, Molecular Phylogenetics. In *Genomes*. 2nd edition. Oxford: Wiley. Available from: https://www.ncbi.nlm.nih.gov/books/NBK21122/.

Chmielecki, J., Meyerson, M. (2014). DNA sequencing of cancer: what have we learned?. *Annual Review of Medicine* 65 (1): 63–79. doi:10.1146/annurev-med-060712-200152.

Chothia, C. Lesk, A.M. (1986). The relation between the divergence of sequence and structure in proteins. *EMBO* Journal 5 (4): 823–826. doi:10.1002/j.1460-2075.1986.tb04288.

Darwin, C. (1871). *The Descent of Man, and Selection in Relation to Sex*. Princeton University Press, Princeton, NJ.

Drmanac, R., Sparks, A. B., Callow, M. J., Halpern, A. L., Burns, N. L., Kermani, B. G., Carnevali, P., Nazarenko, I., Nilsen, G. B., Yeung, G., Dahl, F., Fernandez, A., Staker, B., Pant, K. P., Baccash, J., Borcherding, A. P., Brownley, A., Cedeno, R., Chen, L., … Reid, C. A. (2010). Human genome sequencing using unchained base reads on self-assembling DNA nanoarrays. *Science* 327(5961): 78–81. https://doi.org/10.1126/science.1181498.

Durbin, R. M., Eddy, S. R., Krogh., A. and Mitchison, G. (1998). *Biological Sequence Analysis: Probabilistic Models of Proteins and Nucleic Acids*. 1st edition. Cambridge, New York: Cambridge University Press, doi:10.2277/0521629713, ISBN 0-521-62971-3, OCLC 593254083.

EBI (2020). Human genetic variation (I): an introduction | EMBL-EBI Train online. https://www.ebi.ac.uk/training/online/course/human-genetic-variation-i-introduction-2019.

Elias, I. (2006). Settling the intractability of multiple alignment. *Journal of Computational Biology* 13 (7): 1323–1339. CiteSeerX 10.1.1.6.256. doi:10.1089/cmb.2006.13.1323.

Foxman, B. (2012). Omics Analyses in Molecular Epidemiologic Studies. In *Molecular Tools and Infectious Disease Epidemiology*. https://www.sciencedirect.com/topics/medicine-and-dentistry/gene-location. Retrieved on 08/05/2020.

Genetics Home Reference (2020). How geneticists indicate the location of a gene. https://ghr.nlm.nih.gov/primer/howgeneswork/genelocation. Retrieved on 30/04/2020.

Goodman, M. (1962). Immunochemistry of the primates and primate evolution. *Annals of the New York Academy of Science* 102: 219–234.

Great Soviet Encyclopedia (2010). Human Genetics Medical. https://encyclopedia2.thefreedictionary.com/Human+Genetics. The 3rd Edition (1970–1979). Retrieved on 21/04/2020.

Hitomi, Y. and Tokunaga, K. (2017). Significance of functional disease-causal/susceptible variants identified by whole-genome analyses for the understanding of human diseases. *Proceedings of the Japan Academy Series B: Physical and Biological Sciences* 93 (9): 657–676. Japan Academy. https://doi.org/10.2183/pjab.93.042.

Human Evolution (2020). Molecular phylogenetics. https://www.ucl.ac.uk/human-evolution/molecular-phylogenetics. Retrieved on 04/04/2020.

Jeffrey, C. (1993). Long human molecular phylogenetics. *Annual Review of Anthropology* 22: 251–272. Accessed: 05-05-2020 21:09 UTC.

Jian, X., et al. (2017). In situ capture of chromatin interactions by biotinylated Cas9. *Cell* doi: 10.1016/j.cell.2017.08.003.

John, W.P. (2012). Organization, Synthesis, and Repair of DNA. In *Elsevier's Integrated Review Biochemistry*. 2nd edition. https://www.sciencedirect.com/topics/medicine-and-dentistry/repetitive-dna. Retrieved on 08/04/2020.

Kessler, C. and Vicentiu, M.(1990). Specificity of restriction endonucleases and DNA modification methyltransferases: a review (Edition 3). *Gene* 92(1):1–240.

Koboldt, D. C., Steinberg, K. M., Larson, D. E., Wilson, R. K., and Mardis, E. R. (2013). The next-generation sequencing revolution and its impact on genomics. *Cell* 155(1): 27–38. https://doi.org/10.1016/j.cell.2013.09.006.

Li, J. Y., You, Z., Wang, Q., Zhou, Z. J., Qiu, Y., Luo, R., and Ge, X. Y. (2020). The epidemic of 2019-novel-coronavirus (2019-nCoV) pneumonia and insights for emerging infectious diseases in the future. *Microbes and Infection* 22(2): 80–85. https://doi.org/10.1016/j.micinf.2020.02.002.

Liang, C, Fioroni, M, Rodríguez-Ropero, F, Xue, Y, Schwaneberg, U and Ma, Y.(2011). Directed evolution of a thermophilic endoglucanase (Cel5A) into highly active Cel5A variants with an expanded temperature profile. *Journal of Biotechnology* 154:46–53.

Mount, D.M. (2004). *Bioinformatics: Sequence and Genome Analysis*. 2nd edition. Cold Spring Harbor Laboratory Press: Cold Spring Harbor, NY. ISBN 978-0-87969-608-5.

Petersen, B. S., Fredrich, B., Hoeppner, M. P., Ellinghaus, D., and Franke, A. (2017). Opportunities and challenges of whole-genome and -exome sequencing. *BMC Genetics* 18 (1). BioMed Central Ltd. https://doi.org/10.1186/s12863-017-0479-5.

Polyanovsky, V. O., Roytberg, M. A. and Tumanyan, V. G. (2011). Comparative analysis of the quality of a global algorithm and a local algorithm for alignment of two sequences. *Algorithms for Molecular Biology* 6 (1): 25.

Qiang, X. L., Xu, P., Fang, G., Liu, W. Bin, and Kou, Z. (2020). Using the spike protein feature to predict infection risk and monitor the evolutionary dynamic of coronavirus. *Infectious Diseases of Poverty* 9(1). https://doi.org/10.1186/s40249-020-00649-8.

Sanbi, D.A. (2018). Bioinformatics: Sequence analysis. http://biocourse.wp.sanbi.ac.za/wp-content/uploads/sites/21/2018/03/Sequence-analysis_2018-1.pdf. Retrieved:26/04/2020.

Sarich, V.M. and Wilson, A.C. (1967). Immunological time scale for hominid evolution. *Science* 158:1200–1203.

Science Daily (2017). Scientists characterize regulatory DNA sequences responsible for human diseases. https://www.sciencedaily.com/releases/2017/08/170824141210.htm. Retrieved on 18/04/2020.

Stöppler, M. C. (2019). Genetic Diseases. https://www.medicinenet.com/genetic_disease/article.htm. Retrieved on 26/04/2020.

Suzuki, T., Otake, Y., Uchimoto, S., Hasebe, A., and Goto, Y. (2020). Genomic characterization and phylogenetic classification of bovine coronaviruses through whole genome sequence analysis. *Viruses* 12(2). https://doi.org/10.3390/v12020183.

Ting, N. and Sterner, K.N. (2012). Primate molecular phylogenetics in a genomic era. *Molecular Phylogenetics and Evolution* http://dx.doi.org/10.1016/j.ympev.2012.08.021.

Visscher, P. M., Brown, M. A., McCarthy, M. I., & Yang, J. (2012). Five years of GWAS discovery. *American Journal of Human Genetics* 90 (1): 7–24. https://doi.org/10.1016/j.ajhg.2011.11.029.

Wang, L. and Jiang, T. (1994). On the complexity of multiple sequence alignment. *Journal of Computational Biology* 1 (4): 337–48. CiteSeerX 10.1.1.408.894. doi:10.1089/cmb.1994.1.337. PMID 8790475.

Zhou, P., Yang, X.-L., Wang, X.-G., Hu, B., Zhang, L., Zhang, W., Si, H.-R., Zhu, Y., Li, B., Huang, C.-L., Chen, H.-D., Chen, J., Luo, Y., Guo, H., Jiang, R.-D., Liu, M.-Q., Chen, Y., Shen, X.-R., Wang, X., … Shi, Z.-L. (2020). A pneumonia outbreak associated with a new coronavirus of probable bat origin. *Nature* 579 (7798): 270–273. https://doi.org/10.1038/s41586-020-2012-7.Types.

15 Application of Computational and Bioinformatics Techniques in Drug Repurposing for Effective Development of Potential Drug Candidate for the Management of COVID-19

Charles Oluwaseun Adetunji and Olaniyan Tope Olugbemi
Edo State University Uzairue

Muhammad Akram and Umme Laila
Government College University Faisalabad

Michael Olugbenga Samuel and Ayomide Michael Oshinjo
Bowen University

Juliana Bunmi Adetunji
Osun State University

Gloria E. Okotie
University of Ibadan

Nwadiuto (Diuto) Esiobu
Florida Atlantic University

DOI: 10.1201/9781003178903-18

Omotayo Opemipo Oyedara
Universidad Autónoma de Nuevo León
Osun State University

Folasade Muibat Adeyemi
Osun State University

CONTENTS

15.1 INTRODUCTION

Over the hundred years, outbreaks of both bacterial diseases such as cholera and viral diseases like Spanish flu and severe acute respiratory syndrome virus-1 (SARS-CoV-1) infection have undermined the confidence of the common man in the ability of science and technology to determine and predict the future waves of disease outbreaks and pandemics (Adetunji and Anani, 2021a,b,c; Adetunji et al., 2021; Dwivedi et al., 2021; Olaniyan and Adetunji, 2021; Mishra et al., 2020; Thangadurai et al., 2021; Anani et al., 2020; Islam et al., 2021). Several investigations and studies have shown the patterns of epidemics and pandemics which could help us for effective management of future pandemic diseases (Altay et al., 2020). Viral diseases have been discovered to be one of the most dangerous diseases that could affect humans, and coronavirus disease-19 (COVID-19) is currently one of the most dreadful illnesses across the world. In December 2019, an outbreak of an unknown pneumonia-like illness was reported to have occurred in Wuhan, China. Soon after, coronavirus was identified as a causative agent of the illness. The World Health Organization (WHO) named the disease Coronavirus Disease-2019 (COVID-19) and declared it a global pandemic on March 11, 2020. The etiological agent of the novel COVID-19 was named severe acute respiratory syndrome coronavirus-2 (SARS-CoV-2) following its resemblance with the existing virus SARS-CoV. The SARS-CoV-2 affects the human respiratory tracts leading to respiratory tract infections that could range from mild to severe. The COVID-19 pandemic disease is a major public health issue globally. It has affected millions of people and caused billions of people to stay at home, under a lockdown situation. This novel illness has claimed over 2 million lives and posed a huge burden on the government and health and public sectors. COVID-19 has affected people all over the world, and there is no approved drug for its treatment.

Therefore, there are intense efforts to identify and develop potent medicines for the prevention and treatment of COVID-19. Considering the present situation and urgent need to discover COVID-19 antidote, drug repositioning or repurposing offers an immediate approach for anti-COVID-19 drugs discovery. Drug repurposing is an effective and efficient method that could help toward the discovery of drugs to combat disease outbreaks. Moreover, repurposing of already existing drugs has been identified as a promising solution to the COVID-19 pandemic, because it is time saving and cost effective in producing novel therapeutic drugs (Cheng, 2019; Cheng et al., 2017).

Several considerable efforts have been devoted to the development of novel inhibitors or employ drug repurposing approach to identify anti-COVID-19 drugs, which may act as good inhibitors against SARS-CoV-2 drug targets such as proteases (Khan et al., 2020; Sarma et al., 2020). Bioinformatics and computational techniques could be utilized in drug repurposing for the discovery of potent drugs against COVID-19. Bioinformatics encompasses in silico and computational techniques to understand chemical information and pharmacological properties of drug-like molecules for the purpose of drug development. Moreover, repurposed drug to drug-target interaction and drug to drug interaction can be studied through in silico and virtual molecular docking and simulation. The process involved in the computational and bioinformatics techniques in drug repurposing for effective development of potential drug candidate for the management of COVID-19 is shown in Figure 15.1.

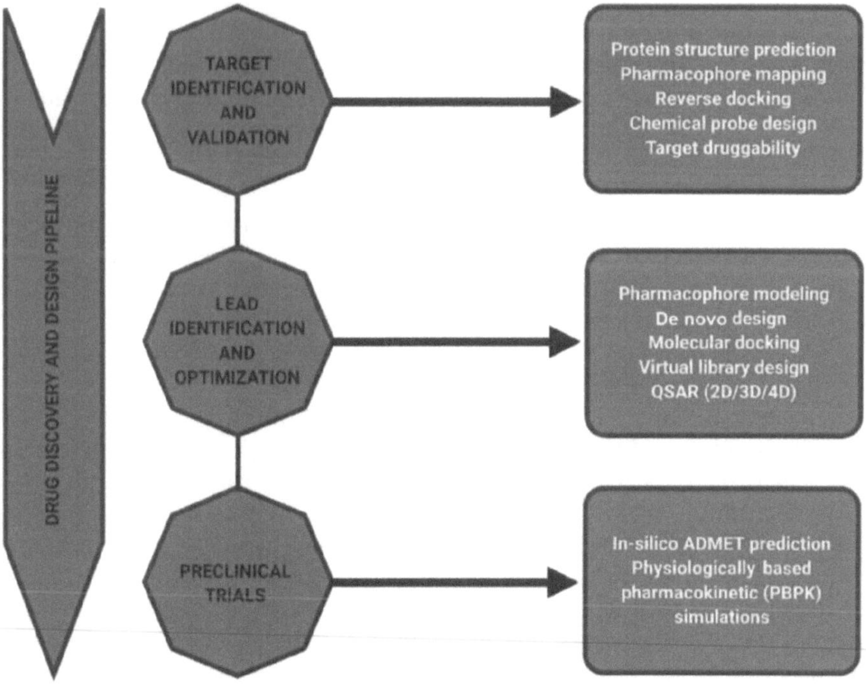

FIGURE 15.1 The computational and bioinformatics techniques in drug repurposing.

Therefore, this chapter intends to provide detailed information on the application of computational and bioinformatics techniques in drug repurposing for the effective development of potential drug candidates for COVID-19 management and treatment.

15.1.1 Important Tools and Techniques for Drug Repositioning

15.1.1.1 Bioinformatics

Bioinformatics is a biological science that deals with statistics, mathematics, information engineering, and computer science. It has been found very helpful in understanding, integrating, and analyzing data using different software and databases. It is used widely in research and in many other fields. It is applied for the purpose of developing diagnostic kits for SARS-CoV-2, study viral mutation, and predict drug targets, which are relevant for COVID-19 treatment. Biological databases and software tools are two important aspects of bioinformatics. Genomic and Proteomic information on SARS-CoV-2 or COVID-19 archived in several databases such as PubChem, GenBank, UniProt, PubMed, among others, can be accessed and processed using bioinformatics tools to generate meaningful information that can be applied in the treatment and management of COVID-19.

15.1.1.2 Machine Learning Method

The integration of machine learning into bioinformatics is helpful for the therapeutic effectiveness of clinical compounds against pathogens. Machine learning is an efficient method to check the effectiveness of a drug. This technique provides a time-saving and promising new way to conduct more studies and analyze the information (Cheng, 2019). Machine learning method suggests anti-virals based on the protein segments present in a proteome. This method gives the best prediction to know the actual antiviral associating with the protein segments.

15.1.1.3 Computational Techniques

The computational approach to drug repurposing entails signature-based drug repurposing, artificial intelligence method, computer-aided drug design, and cloud computing, big data analytics, internet of things, supercomputer calculation, homology modeling, molecular dynamic simulation, and molecular docking.

15.1.1.3.1 Signature-Based Drug Repurposing Approach

This method is often referred to as the "signature reversion" method. It utilizes the profiles of the gene expression of various drugs, which are analyzed for comparisons between drugs or drug-disease comparisons or the utilization of the clinical phenotypes by operating through the omics data, which refers to the comprehensive analysis of the molecular and genetics profile including glycomics, lipomics, transcriptomic, proteomic, and metabolomics. By studying these gene expressions, it could lead to the construction of a map that could help in the study of interrelations and connections between drugs and diseases. The recent advances in technology and molecular biology have led to novel and powerful methods which have been applied in computational biology for drug repurposing. The recent LINCS (Library of Integrated Network-Based Cellular Signatures) datasets, connectivity maps (the bank of transcription data cultured from human cells and various other organisms), and the NCBI Gene Expression database are extensively studied and explored in the drug repositioning (Cheng, 2019; Stephenson et al., 2019).

15.1.1.3.2 Artificial Intelligence Technique

This technique involves the application of bioinformatics techniques with a promising result of streamlining the discovery of novel drugs (Stephenson et al., 2019). The ability of humans to build artificial intelligence in computers and their applications in machine learning is a great achievement of this century. With the help of artificial neural network, deep learning techniques is also helpful in creating computer models for the purpose of comparing X-ray crystallography structures of novel drugs that can be effective against SARS-CoV-2. This technique could play a crucial role in tracking variations in the genetics of coronavirus. With artificial intelligence, an important tool or platform named COVID-19 genotyping tool (CGT) has been designed. By using this platform, hospitals can compare SARS-CoV-2 sequences with global strain sequences and also analyze the sequences within minutes. Moreover, viral strains and their evolving character can be assessed by using this artificial intelligence. The software building team stated that the CGT will be helpful in tracking SARS-CoV-2 mutations immediately and provide information on future COVID-19 re-occurrence. (Maan et al., 2020).

15.1.1.3.3 Computer-Aided Drug Design

Novel drug discovery and development is time taking, and it involves the investment of significant resources, money, efforts, and time. However, drug discovery can be aided by computer in a process known as computer-aided drug design or in silico drug designing (Bisht and Singh, 2018). The key protein molecule of SARS-CoV-2 in the case of COVID-19 can be modeled to obtain a three-dimensional structure that can be a drug molecule target either by drug repurposing or novel drug development. Moreover, computer are utilized to conduct large numbers of such experiments.

15.1.1.3.4 Cloud Computing

Cloud computing technique is an important technique that plays an important role in the treatment of COVID-19 disease. It accelerates the development of a new vaccine against coronavirus disease. This technique is also used with artificial intelligence for the advanced manufacturing process to develop vaccine and drugs for animal testing and human trials.

15.1.1.3.5 Big Data Analysis

Big data is a systematic analysis of complex and large (big) data. It is applicable as a result of limitations of traditional methods of analysis in calculating proper results effectively for large amount of data. It is used widely for analyzing data in the banking, education, agricultural, insurance, intelligence, and business sectors. It has been used widely in the health sector to record large amount of patients' data. During the COVID-19 crisis, large amount of data ("Big data") was generated in hospitals, and the analysis of such data using different computational techniques, including artificial intelligence, yielded valuable information. Big data analysis is also beneficial in predicting the possible changes occurring in our life style, economy, employment, and other aspects of life.

15.1.1.3.6 Internet of Things

It is a unified network system that connects devices to internet to collect and exchange data without human intervention in many cases. It is used widely in safety issues. For instance, in traffic control, security applications, agricultural monitoring, and weather forecast. Generally, it is used in monitoring electronic and mechanical devices. Internet of things has helped out, in many ways, especially in the development of the healthcare sectors during COVID-19 crisis. For contact tracing and quarantine purpose, the foremost applications deployed were known as geofencing. These include WI-Fi-GPS, Bluetooth, cellular networking, and many other wireless technologies which were used to track suspected and infected COVID-19 patients. By using internet of things, various new devices are controlled. For example, remotes are used and deployed for cleaning and disinfecting surfaces (Yang et al., 2020). The internet of things-based devices have been used to prevent the spread of diseases through fomites such as doorknobs and smart switches. Furthermore, drones with artificial intelligence were used during quarantine for supplying medicines, disinfection, sample transport, and temperature monitoring. The internet of things smart devices played big role during the COVID-19 crisis and quarantine period.

15.1.1.3.7 Supercomputer Calculation

A supercomputer is used for highly computational task in molecular modeling. It plays important role in analyzing, storing, and remodeling data of repurposed drugs, which will not be possible to manage in a short time without the supercomputer.

15.1.1.3.8 Homology Modeling

The lack of information about the crystal structure of proteins hinders drug development by pharmaceutical companies or scientists. Hence, homology modeling is a widely used application to build reliable sequence and structure of amino acids and proteins (Vyas et al., 2012). This technique is also known as comparative modeling of proteins. Homology modeling is considered as a reliable and successful technique for comparing sequences of amino acid and identifying the three-dimensional structure of proteins. Different tools that can be used for homology modeling include MODELLER, Phyre2, and web-based programs like SWISS-MODEL (Xu et al., 2000).

15.1.1.3.9 Molecular Dynamic Simulation

Molecular dynamic simulation is a computational approach that provides information on the structural stability, atom behavior, and conformational changes occurring in a protein-ligand complex at the atomic level. During molecular dynamic simulation, atoms and molecules interact for a specific period of time, followed by the determination of trajectories, forces, and potential energies between the particles. Molecular dynamics simulation has been applied in studying and confirming the stability behavior of SARS-CoV-2 protease-inhibitors complexes. Several compound inhibitors against SARS-CoV-2 proteases such as 3CL hydrolase (Mpro), SARS-CoV-2 papain-like protease (PLpro), SARS-CoV-2 Guanine-N7 methyltransferase, and SARS-CoV-2 binding sites in humans such as Angiotensin Converting Enzyme 2 (ACE2) have been studied using molecular dynamic simulation (Choudhary et al., 2020; Selvaraj et al., 2020; Surti et al., 2020; Khelfaoui et al., 2020). Software such as GROMACS, CHARMM, DESMOND, and AMBER can be employed for molecular dynamic stimulation in COVID-19 studies (Abraham et al., 2015).

15.1.1.3.10 Molecular Docking

Molecular docking is an approach in drug repurposing which is used to study and analyze different interactions of various amino acids in the proteins and ligands. This technique is thus applied in order to anticipate the confirmation and accurate orientation of a molecule interacting and binding to another molecule resulting in a stable adduct. There are several programs to carry out molecular docking and these include Glide, Autodock Vina, and Gold, among others (Pagadala et al., 2017). In Glide Standard Precision protocol, for instance, the binding affinity of the ligand and predicting activity of a drug suitable against an infection can be carried out flexibly. At the end of docking analysis, the final free energy assessment is concluded by calculating what is known as the dock score, and the lower the dock scores, the better and efficient is the drug to fight against the infection. The docked ligands can also be visualized using applications such as Maestro interface, PyMOL, and Discovery studio (Kannadasan et al., 2016).

15.1.2 Studies Related to the Application of Computational and Bioinformatics Techniques in Drug Repurposing for COVID-19

Drug repositioning is a method of identifying new therapeutic drugs from previously existing ones as well as determining remedies for untreated pathogens (Jarada et al., 2020). This method is time saving and cost effective for developing therapies in basic medical sciences compared to the conventional drug discovery de novo methods. Drug repositioning requires large public health, biological, and biomedical data together with highly skilled computational physiologists to hasten the process (Jarada et al., 2020). Furthermore, network-based approaches can be utilized to predict drug target for repurposing and repositioning; thus, integrated data set could be used for evaluation metrics for research and treatment outcomes (Lotfi et al., 2017).

There are several data mining programs for drug discovery and development in medical sciences, but there is a need to implement rapid analytical tools and applications for decision-making, particularly against COVID-19. Hodos et al. (2016) showed that computational physiology and pharmacology utilize molecular techniques such as in silico techniques to predict and understand the biological effects of many drugs so as to improve their clinical utilization. The authors also showed that one of the important uses of computational biology is for drug repurposing to discover new important applications and efficacy for the drug molecules, particularly those of rare diseases such as COVID-19. They suggested that the use of matrix factorization data modality method can integrate several data platforms to enhance drug performance and predictive action. Dyall et al. (2014) revealed that several molecular techniques are available such as in silico methods which can be utilized to accumulate information from previously existing drugs for the purpose of repositioning. They also demonstrated that molecular docking of protein data bank, artificial intelligence algorithms, enzyme inhibition assays, genomics as well as computational physiology for novel receptor drug interaction can be used for drug discovery. The authors demonstrated that single base drug repurposing is insufficient; hence, cocktail approach of combined molecules will offer more usefulness.

Xu et al. (2020) revealed the relevance of repositioning existing drugs for possible treatment options for COVID-19. They reported that through physiological profiling, many molecules can be identified from repurposing platforms for possible drug targets. They further analyzed the genomic sequence of the SARS-CoV-2 to generate 34 genes related to COVID-19 utilizing AutoSeed pipeline database and built an algorithm and molecular network to discover 24 diseases linked to the human COVID-19 pathway with about 78 possible drugs for repositioning. In another study, Cava et al. (2020) showed the role of ACE2 in the pathophysiological mechanisms of COVID-19. They utilized in silico mechanisms to provide new reports of ACE2 in the lung as a possible drug target for COVID-19 treatment. The study involved using gene expression obtained from various datasets to analyze the results of ACE2-linked genes using enrichment pathway analysis. They discovered that steroidal biosynthetic, aryldialkylphosphatase activity, trialkylsulfonium hydrolase, adenosylhomocysteinase, Coenzyme A ligase, and acetate-Coenzyme A processes are associated with ACE2 activity with a network of about 193 genes, 222 interactions, and about 36 potential drug target.

Narkhede et al. (2020) demonstrated that scientists across the globe are urgently adapting and utilizing different applications of machine learning and computational biology to investigate the pathophysiology of SARS-CoV-2 so as to understand the mechanisms and provide a target drug candidate for the treatment. The authors measured the interaction between SARS-CoV-2 protease and some antimalarial as well as anticancer agents demonstrating their docking mechanism and the binding energy of many target drugs such as ritonavir, favipiravir, oseltamivir, hydroxychloroquine, ribavirin, chloroquine, and remdesivir, suggesting a high affinity rate of these drugs against SARS-CoV-2 and spike proteins through ACE2 complex interaction. They revealed that these drugs demonstrated a very high potential value for use against SARS-CoV-2.

In another study, Loganathan et al. (2020) adopted a transcriptome-based host-drug repurposing approach from public data utilizing gene expression profiling to identify upregulated viral factors. They identified several drugs that can suppress these pro-viral mediators, particularly PTGS2 gene product for SARS-CoV-2 infection. Thus, computational physiology could be adopted for repurposing against COVID-19.

Pawełczyk and Zaprutko (2020) revealed that due to the potential threat caused by COVID-19 pandemic and lack of standard drug for the specific treatment of COVID-19, computational chemistry strategies can utilize integrated modification approach of target molecules possessing antiviral properties, converted into a potential compound for the treatment of COVID-19, as therapeutic assisted drugs. By utilizing computational physiology and bioinformatics as well as machine learning, they were able to predict bioactive molecules against SARS-CoV-2 infection. Also, Zhou et al. (2020) showed the unprecedented mortality and morbidity caused by SARS-CoV-2 across the globe and concluded that drug repositioning and repurposing offers a unique advantage for drug target and discovery for the SARS-CoV-2 since no specific treatment is currently recognized. The authors investigated an integrated antiviral drug molecule and SARS-CoV-2 in host cells. The whole genomes showed increased nucleotide sequence characterized with SARS-CoV by network proximity analyses interactome for 16 anti-HCoV (human coronavirus) molecules such as sirolimus, melatonin, and mercaptopurine. The authors later described three useful drug combinations based on HCoV-stimulated transcriptomics as well as analyses of drug-gene signatures obtained from cell lines to identify repurposable candidate drugs for SARS-CoV-2.

Wang (2020) revealed that in global market search has been on to identify novel molecular drug target for the COVID-19 disease. Promising drug-target molecules utilizing computational physiology have been designed to combat COVID-19 by molecular docking screening. Receptor ligand interaction were analyzed and identified for corona virus to facilitate drug design targeting COVID-19 protease. He and Garmire (2020) discovered that till date lack of adequate treatment protocol for COVID-19 has caused significant setback in the management of the pandemic; thus, the authors carried out drug repurposing utilizing bioinformatics analysis to discover drug candidate for COVID-19 virus. The findings revealed two important drug candidate, cyclin-dependent kinase inhibitor and chemically modified tetracycline, through bioinformatics to evaluate different pathways involved in the pathophysiology of the disease mechanism.

Nam et al. (2020) revealed that drug repositioning involves the process of reusing already approved old drugs as a means of resolving medical condition so as to save time and reduce cost. They revealed that in the nearest future, many clinical trials will be conducted to identify therapeutic molecules that are very effective to combat COVID-19 pandemic. Thus, the authors applied the knowledge of COVID-19 to develop a formidable network of drug-based repurposing framework for disease-gene-drug network against COVID-19. They revealed 30 different drug candidates as repurposed drugs applicable for COVID-19 treatment and has been placed under clinical trials utilizing data-based computational physiology methods to reduce errors.

The typical examples of drugs which have been studied for repurposing in the treatment of COVID-19 are shown in Table 15.1.

TABLE 15.1
Specific Examples of Drugs Which Have Been Docked against COVID-19

S/N	Drugs	Modes of Action
1	ABT450 is a drug approved for the Hepatitis C infection	Inhibition of the serine protease
2	Acyclovir is a drug approved for the Cytomegalovirus infections	Inhibition viral DNA polymerase
3	Amprenavir is a drug approved for the HIV infection	Protease inhibition
4	Arbidol (Umifenovir) is a drug approved for the Influenza virus	Inhibits membrane fusion
5	Adefovir approved for the HIV infection	Reverse transcriptase inhibitor
6	Azidothimidine is approved for HIV infection	Inhibits reverse transcriptase
7	Asunaprevir is approved for the Hepatitis C infection	Protease inhibitor
8	Baricitinib is a drug for the Rheumatoid arthritis	Inhibition of Janus kinase
9	Beclabuvir drug against the Hepatitis C infection	Inhibits nonstructural protein 5B (NS5B)
10	Boceprevir is a drug approved for the Hepatitis C infection	NS3/4A protease inhibition
11	Camostat is an approved drug for Pancreatitis	Inhibits serine protease
12	Darunavir is a drug approved for the HIV infection	Inhibition of HIV protease enzyme
13	Didanosine is a drug approved for the HIV infection	Involves the nucleoside reverse transcriptase inhibitor
14	CGP42112A is used for the treatment of vasodilation and blood pressure reduction	Involved in the angiotensin AT2 receptor agonist
15	Dasabuvir is approved for the Hepatitis C	Involves the inhibition of the action of NS5B polymerase
16	Danoprevir is a drug approved for the Hepatitis C infection	Involves the NS3/4A protease inhibitor
17	Daclatasvir is drug used for Hepatitis C infection	Involves the NS5A inhibitor
18	Elvitegravir is a drug approved for HIV	Integrase inhibition
19	Dihydroxypropyl adenine treats the Herpes Virus	Inhibition of viral replication
20	Efavirenz is approved for HIV	Inhibition of viral replication
21	Faldaprevir is a drug approved for the Hepatitis C infection	Protease inhibition
22	Famciclovir is approved for the Hepatitis B infection	Inhibition viral DNA polymerase
23	Galidesivir is used in the treatment of Ebola	RNA polymerase inhibition
24	Entecavir is approved for treating Hepatitis B infection	Inhibition of reverse transcription
25	Elbasvir is a drug approved for the Hepatitis C infection	Involves the NS5A inhibition
26	Ganciclovir is used for treating the Cytomegalovirus	Inhibition of viral DNA polymerases
27	Grazoprevir is a drug approved for the Hepatitis C infection	Involved in the blockage of the NS3

(*Continued*)

TABLE 15.1 (*Continued*)
Specific Examples of Drugs Which Have Been Docked against COVID-19

S/N	Drugs	Modes of Action
28	Idoxuridine is approved for treating the Herpes Simplex virus	Involves the interruption of the viral DNA replication
29	Ritonavir is a drug approved for the HIV infection	Involves the HIV protease inhibition
30	Indinavir is a drug approved for HIV infection	Protease inhibition
31	Marboran/methisazone is approved for treating small pox virus	Inhibition of mRNA and protein synthesis
32	Lopinavir is used for treating HIV	The protease inhibition
33	Mericitabine is used for treating Hepatitis C	An inhibitor of RdRp
34	Radalbuvir is used for treating Hepatitis C	HIV-l integrase inhibition
35	Nevirapine is approved for treatment of HIV	Non-nucleoside reverse transcriptase inhibition.
36	Foscarnet is a drug approved for treating the hepatitis B infection, HIV	Inhibition of the pyrophosphates to DNA polymerases.
37	Saquinavir is used for treating HIV	Protease inhibition
38	Simeprevir is approved for treating Hepatitis C	Hepatitis C virus protease inhibition
39	Ombitasvir is used to treat Hepatitis C	Involved in the NS5A inhibition
40	Stavudine is a drug approved for treating HIV	Inhibition of HIV reverse transcriptase enzyme
41	Zanamivir is used for treating influenza viruses	Neuraminidase inhibition
42	Telbivudine is approved for treating Hepatitis B	Involved in reverse transcriptase inhibition
43	Ravidasvir is used for treating Hepatitis C	Involves the NS5A inhibition
44	Zalcitabine is approved for treating HIV	Involves the nucleoside reverse transcriptase inhibition
45	Telaprevir is a drug approved for treating Hepatitis C	Involved in NS3/4a protease inhibition
46	Velpatasvir is a drug of choice for Hepatitis C	Involves the NS5A protein inhibition
47	Uprifosbuvir is used for treating Hepatitis C	Involves the NS5B polymerase inhibition

15.2 CONCLUSION AND FUTURE COLLABORATION

This chapter has provided information on the application of computational and bioinformatics techniques for effective development of potential drug candidates for the management of COVID-19. To fight off the life-threatening, contagious COVID-19 infection, several studies and researches are being carried out utilizing antiviral drugs to come up with an efficient treatment to combat the disease and to ease the global burden. Due to the intense need, repurposing of drugs is considered to lead the way for the development of anti-COVID-19 drugs by using bioinformatics and computational and laboratory techniques. Computational techniques and bioinformatics are certainly helping us to understand this pandemic situation scientifically and aid the screening of several promising repurposed drugs out of thousands of drugs available in market.

REFERENCES

Abraham, M.J., Murtola, T., Schulz, R., Páll, S., Smith, J.C., Hess, B. and Lindahl, E. (2015). GROMACS: High performance molecular simulations through multi-level parallelism from laptops to supercomputers. *SoftwareX* 1:19–25.

Adetunji, C.O., Anani, O.A., and Panpatte, D. (2021). Mechanism of actions involved in sustainable ecorestoration of petroleum hydrocarbons polluted soil by the beneficial microorganism. In: Panpatte, D.G., and Jhala, Y.K. (eds.), *Microbial Rejuvenation of Polluted Environment. Microorganisms for Sustainability*, vol 26. Springer, Singapore. https://doi.org/10.1007/978-981-15-7455-9_8

Adetunji, C.O., and Anani, O.A. (2021a). Recent advances in the application of genetically engineered microorganisms for microbial rejuvenation of contaminated environment. In: Adetunji C.O., Panpatte D.G., Jhala Y.K. (eds.), *Microbial Rejuvenation of Polluted Environment. Microorganisms for Sustainability*, vol 27. Springer, Singapore. https://doi.org/10.1007/978-981-15-7459-7_14

Adetunji, C.O., and Anani, O.A. (2021b). Utilization of microbial biofilm for the biotransformation and bioremediation of heavily polluted environment. In: Panpatte, D.G., and Jhala, Y.K. (eds.), *Microbial Rejuvenation of Polluted Environment. Microorganisms for Sustainability*, vol 25. Springer, Singapore. https://doi.org/10.1007/978-981-15-7447-4_9

Adetunji, C.O., and Anani, O.A. (2021c). Bioremediation of polythene and plastics using beneficial microorganisms. In: Adetunji, C.O., Panpatte, D.G., and Jhala, Y.K. (eds.), *Microbial Rejuvenation of Polluted Environment. Microorganisms for Sustainability*, vol 27. Springer, Singapore. https://doi.org/10.1007/978-981-15-7459-7_13

Altay, O., Mohammadi, E., Lam, S., Turkez, H., Boren, J., Nielsen, J., Uhlen, M. and Mardinoglu, A., (2020). Current status of COVID-19 therapies and drug repositioning applications. *Iscience* 101303.

Anani, O.A., Mishra R.R., Mishra P., Enuneku A.A., Anani G.A., and Adetunji C.O. (2020). Effects of toxicant from pesticides on food security: Current developments. In: Mishra P., Mishra R.R., Adetunji C.O. (eds.), *Innovations in Food Technology*. Springer, Singapore. https://doi.org/10.1007/978-981-15-6121-4_22

Bisht, N. and Singh, B.K., (2018). Role of computer aided drug design in drug development and drug discovery. *International Journal of Pharmaceutical Sciences and Research* 9(4): 1405–1415.

Cava, C., Bertoli, G. and Castiglioni, I., (2020). In silico discovery of candidate drugs against Covid-19. *Viruses* 12(4): 404. https://doi.org/10.3390/v12040404

Cheng, F., (2019). *In silico* oncology drug repositioning and polypharmacology. *Cancer Bioinformatics* 243–261.

Cheng, F., Hong, H., Yang, S. and Wei, Y., (2017). Individualized network-based drug repositioning infrastructure for precision oncology in the panomics era. *Briefings in Bioinformatics* 18(4): 682–697.

Choudhary, M.I., Shaikh, M., tul-Wahab, A. and ur-Rahman, A. (2020). *In silico* identification of potential inhibitors of key SARS-CoV-2 3CL hydrolase (Mpro) via molecular docking, MMGBSA predictive binding energy calculations, and molecular dynamics simulation. *Plos One* 15(7): e0235030.

Dwivedi, N., Dwivedi, S., and Adetunji, C. O. (2021). Efficacy of microorganisms in the removal of toxic materials from industrial effluents. In: Adetunji, C.O., Panpatte, D.G., and Jhala, Y.K. (eds.), *Microbial Rejuvenation of Polluted Environment. Microorganisms for Sustainability*, vol 27. Springer, Singapore. https://doi.org/10.1007/978-981-15-7459-7_15

Dyall, J., Coleman, C.M., Hart, B.J., Venkataraman, T., Holbrook, M.R., Kindrachuk, J., Johnson, R.F., Olinger, G.G., Jahrling, P.B., Laidlaw, M. and Johansen, L.M. (2014). Repurposing of clinically developed drugs for treatment of Middle East respiratory syndrome coronavirus infection. *Antimicrobial Agents and Chemotherapy* 58(8): 4885–4893.

He, B. and Garmire, L. (2020). Prediction of repurposed drugs for treating lung injury in COVID-19. *F1000Research* 9.

Hodos, R.A., Kidd, B.A., Khader, S., Readhead, B.P. and Dudley, J.T., 2016. Computational approaches to drug repurposing and pharmacology. *Wiley Interdisciplinary Reviews. Systems Biology and Medicine* 8(3): 186.

Islam, S., Thangadurai, D., Adetunji, C. O., Nwankwo, W., Kadiri, O., Makinde, S., Michael, O. S., Anani, O. A., and Adetunji, J. B. (2021). Nanomaterials and nanocoatings for alternative antimicrobial therapy. In: Kharissova, O.V., Martínez, L.M.T., and Kharisov B.I. (eds.), *Handbook of Nanomaterials and Nanocomposites for Energy and Environmental Applications*. Springer, Cham. https://doi.org/10.1007/978-3-030-11155-7_3-1

Jarada, T.N., Rokne, J.G. and Alhajj, R., (2020). A review of computational drug repositioning: strategies, approaches, opportunities, challenges, and directions. *Journal of Cheminformatics* 12(1): 1–23.

Kannadasan, R., Saleembasha, M.S. and Emerson, I.A. (2016). A framework for learning drug designing through molecular modelling software techniques and biological databases for protein-ligand interactions. *International Journal of Engineering Research in Africa* 27: 111–118. Trans Tech Publications Ltd.

Khan, S.A., Zia, K., Ashraf, S., Uddin, R. and Ul-Haq, Z. (2020). Identification of chymotrypsin-like protease inhibitors of SARS-CoV-2 via integrated computational approach. *Journal of Biomolecular Structure and Dynamics* 1–10.

Khelfaoui, H., Harkati, D. and Saleh, B.A. (2020). Molecular docking, molecular dynamics simulations and reactivity, studies on approved drugs library targeting ACE2 and SARS-CoV-2 binding with ACE2. *Journal of Biomolecular Structure and Dynamics* 1–17.

Li, X., Yu, J., Zhang, Z., Ren, J., Peluffo, A.E., Zhang, W., Zhao, Y., Yan, K., Cohen, D. and Wang, W., (2020). Network bioinformatics analysis provides insight into drug repurposing for COVID-2019. Preprints 2020, 2020030286 (doi: 10.20944/preprints202003.0286.v1).

Loganathan, T., Ramachandran, S., Shankaran, P., Nagarajan, D. and Mohan S.S. (2020). Host transcriptome-guided drug repurposing for COVID-19 treatment: a meta-analysis based approach. *PeerJ* 8: e9357. https://doi.org/10.7717/peerj.9357

Lotfi Shahreza, M., Ghadiri, N., Mousavi, S.R., Varshosaz, J. and Green, J.R., 2018. A review of network-based approaches to drug repositioning. *Briefings in Bioinformatics* 19(5): 878–892.

Maan, H., Mbareche, H., Raphenya, A. R., Banerjee, A., Nasir, J. A., Kozak, R. A., and Wang B. (2020) COVID-19 Genotyping Tool (CGT) (Version v0.2.2).

Mishra, P., Mishra, R. R., & Adetunji, C. O. (Eds.). (2020). *Innovations in Food Technology: Current Perspectives and Future Goals*. Springer Nature Singapore Pte Ltd. 2020. Springer, Singapore. Pp. 143-162. 1st Edition. DOI: 10.1007/978-981-15-6121-4

Nam, Y., Yun, J.S., Lee, S.M., Park, J.W., Chen, Z., Lee, B., Verma, A., Ning, X., Shen, L. and Kim, D., 2020. Network reinforcement driven drug repurposing for COVID-19 by exploiting disease-gene-drug associations. *arXiv*. https://doi.org/10.48550/arXiv.2008.05377.

Narkhede, R.R., Cheke, R.S., Ambhore, J.P. and Shinde, S.D., 2020. The molecular docking study of potential drug candidates showing anti-COVID-19 activity by exploring of therapeutic targets of SARS-CoV-2. *Eurasian Journal of Medicine and Oncology* 4(3): 185–195.

Olaniyan, O.T., and Adetunji, C.O. (2021). Biochemical role of beneficial microorganisms: An overview on recent development in environmental and agro science. In: Adetunji, C.O., Panpatte, D.G., and Jhala, Y.K. (eds.), *Microbial Rejuvenation of Polluted Environment. Microorganisms for Sustainability*, vol 27. Springer, Singapore. https://doi.org/10.1007/978-981-15-7459-7_2

Pagadala, N.S., Syed, K. and Tuszynski, J., (2017). Software for molecular docking: A review. *Biophysical Reviews* 9(2): 91–102.

Pawełczyk, A. and Zaprutko, L., 2020. Anti-COVID drugs: repurposing existing drugs or search for new complex entities, strategies and perspectives. *Future Medicinal Chemistry* 12(19): 1743–1757.

Sarma, P., Shekhar, N., Prajapat, M., Avti, P., Kaur, H., Kumar, S., Singh, S., Kumar, H., Prakash, A., Dhibar, D.P. and Medhi, B. (2020). In-silico homology assisted identification of inhibitor of RNA binding against 2019-nCoV N-protein (N terminal domain). *Journal of Biomolecular Structure and Dynamics* 1–9.

Selvaraj, C., Dinesh, D.C., Panwar, U., Abhirami, R., Boura, E. and Singh, S.K. (2020). Structure-based virtual screening and molecular dynamics simulation of SARS-CoV-2 Guanine-N7 methyltransferase (nsp14) for identifying antiviral inhibitors against COVID-19. *Journal of Biomolecular Structure and Dynamics* 1–12.

Stephenson, N., Shane, E., Chase, J., Rowland, J., Ries, D., Justice, N., Zhang, J., Chan, L. and Cao, R., (2019). Survey of machine learning techniques in drug discovery. *Current Drug Metabolism 20*(3): 185–193.

Surti, M., Patel, M., Adnan, M., Moin, A., Ashraf, S.A., Siddiqui, A.J., Snoussi, M., Deshpande, S. and Reddy, M.N. (2020). Ilimaquinone (marine sponge metabolite) as a novel inhibitor of SARS-CoV-2 key target proteins in comparison with suggested COVID-19 drugs: Designing, docking and molecular dynamics simulation study. *RSC Advances* 10(62): 37707–37720.

Thangadurai, D., Dabire, S. S., Sangeetha, J., Said Al-Tawaha, A. R. M., Adetunji, C. O., Islam, S., Shettar, A. K., David, M., Hospet, R., and Adetunji, J. B. (2021) Greener composites from plant fibers: Preparation, structure, and properties. In: Kharissova O.V., Martínez L.M.T., Kharisov B.I. (eds.), *Handbook of Nanomaterials and Nanocomposites for Energy and Environmental Applications*. Springer, Cham. https://doi.org/10.1007/978-3-030-11155-7_21-1

Vyas, V.K., Ukawala, R.D., Ghate, M. and Chintha, C., (2012). Homology modeling a fast tool for drug discovery: current perspectives. *Indian Journal of Pharmaceutical Sciences* 74(1): 1.

Wang, J. (2020). Fast identification of possible drug treatment of coronavirus disease-19 (COVID-19) through computational drug repurposing study. *Journal of Chemical Information and Modeling* 60(6): 3277–3286.

Xu, D., Xu, Y. and Uberbacher, C.E., (2000). Computational tools for protein modeling. *Current Protein and Peptide Science* 1(1): 1–21.

Yang, G.Z., Nelson, B.J., Murphy, R.R., Choset, H., Christensen, H., Collins, S.H., Dario, P., Goldberg, K., Ikuta, K., Jacobstein, N. and Kragic, D., (2020). Combating COVID-19-The role of robotics in managing public health and infectious diseases. *Science Robotics* 5: eabb5589.

Zhou, Y., Hou, Y., Shen, J., Huang, Y., Martin, W. and Cheng, F., 2020. Network-based drug repurposing for novel coronavirus 2019-nCoV/SARS-CoV-2. *Cell Discovery* 6(1): 1–18.

16 Computational Intelligence Techniques for Combating COVID-19

Charles Oluwaseun Adetunji, Wilson Nwankwo, Akinola Samson Olayinka, and Olaniyan Tope Olugbemi
Edo State University Uzairue

Muhammad Akram and Umme Laila
Government College University Faisalabad

Michael Olugbenga Samuel and Ayomide Michael Oshinjo
Bowen University

Juliana Bunmi Adetunji
Osun State University

Gloria E. Okotie
University of Ibadan

Nwadiuto (Diuto) Esiobu
Florida Atlantic University

CONTENTS

DOI: 10.1201/9781003178903-19

16.1 INTRODUCTION

The outbreak of COVID-19 pandemic has been described as one of the most chaotic public health concerns bedeviling humanity in recent history with a very high level of mortality globally. Since the outbreak of the pandemic, several researchers developed various approaches to unravel the mystery behind this disease and possible control mechanisms have been explored (Ahuja et al., 2020; Barstugan et al., 2020; Olaniyan et al., 2020; Shen et al., 2020). This disease has affected mostly aged people, people with underlining ailments as well as people with very low level of immunity. Several thousands of people have been documented as disease sufferers with over 2,719,163 deaths while the number of confirmed cases of infected persons with this illness stands at 123,419,065. However, in March 2021, over 400 thousand doses of various vaccines have been administered in order to combat the pandemic globally amid the rising infections (WHO, 2021). In a bid to contain the impact of coronavirus, the governments of different countries have introduced several interventions and policies that could lead to total eradication of coronavirus disease or at least reduce its impact to the barest minimum. During this chaotic period, the role of innovative techniques and tools in the implementation of meaningful health policies have been appraised (Adetunji and Anani, 2021a,b,c; Adetunji et al., 2021; Dwivedi et al., 2021; Olaniyan and Adetunji, 2021; Mishra et al., 2020; Thangadurai et al., 2021; Anani et al., 2020; Islam et al., 2021).

Robotic technologies have been intelligently deployed with remote control capabilities in hospitals to interact with patients in dispensing medication and food. This helps to reduce the contact period between patients and healthcare personnel (Shripad et al., 2013; Calvaresi et al., 2019; Gofuku, 2020). Disinfection of contaminated area most especially in the public areas and streets could be carried out using drones. Prior to the evolution of drones, this would require human intervention, and for a large contaminated area, it would be an almost impossible task amid some consequences which may include infection of persons doing the disinfection. With drones, the huge efforts and expenditure may be grossly reduced and safety assured.

Several researchers globally have devoted considerable energy and applied many resources toward development of prospective drugs, which could effectively cure this disease, while numerous processes have been introduced to drive the discovery of vaccines that could prevent or reduce the spread of the virus. Intelligent computing tools are very crucial in these processes as they could support critical investigations even at the early stage of infection in patients. These tools could be integrated into several investigative techniques like computed tomography and resonance imaging. These technologies include: natural language processing, deep learning, visualization tools and speech recognition (Nguyen, 2020; Shi et al., 2021). The applications of these technologies are extensive. For instance, in drug discovery, Figure 16.1 shows the different stages where these tools could be deployed. This chapter is aimed at providing a detailed review of these technologies and their respective applications which could be utilized to support the management of the COVID-19 public health crisis.

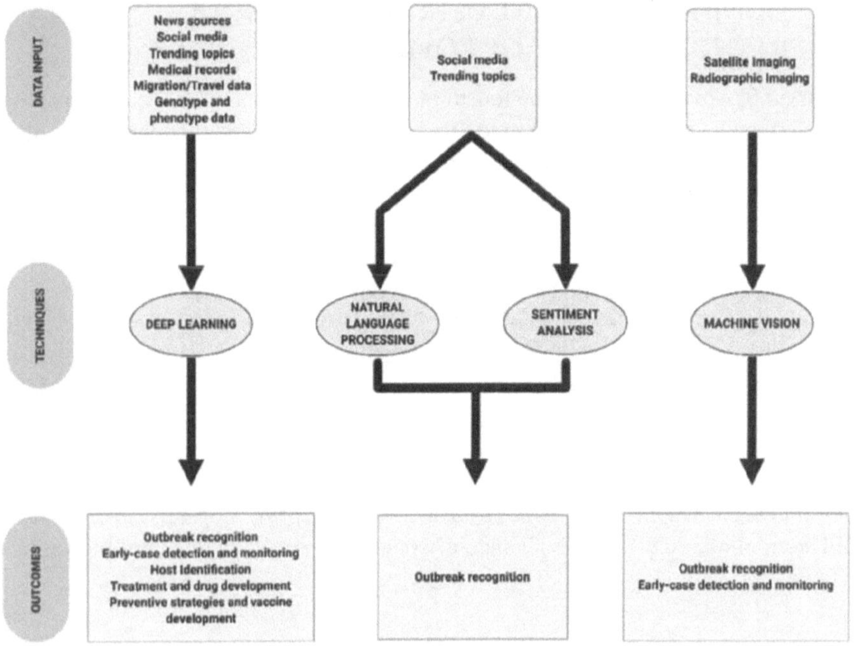

FIGURE 16.1 The computational and bioinformatics techniques in drug repurposing.

16.2 VARIOUS APPLICATION AND TECHNIQUES FOR CONTAINING COVID-19

In the bid to detect and investigate the coronavirus disease, scientists have used various Artificial Intelligence (AI) techniques. Such techniques include but not limited to simulation and modeling of the virus and its mode of action. Data science, Internet of Things (IoT), AI, machine learning, deep learning and data mining are some of the computational intelligence techniques that have been applied to demystify the coronavirus pandemic. Researchers from different scientific backgrounds such as Computational Biology, Computational Physics, Computer Science, Biotechnology and Mathematics have worked in multidisciplinary domains using these techniques to unravel the virus and proffer solutions to the challenges posed by the pandemic. Data is key to effective application of computational intelligence techniques and various data have been used to model and train machine learning algorithms (Javor et al., 2020; Yang et al., 2020). Images that are of medical relevance, diagnostic data, data on mode of transmission, etc. are some of the data that have been applied in the computational intelligence approach to COVID-19 research.

Many doctors and radiologists use computed tomography for the examination of the chest during their investigation of the progression of the disease in coronavirus patients. They utilize these techniques for differentiation of the disease from other disease conditions like pneumonia (Shen et al., 2020; Shi et al., 2021).

16.3 SIGNIFICANCE OF INDUSTRY 4.0 TECHNIQUE IN THE TREATMENT OF COVID-19

Industry 4.0, which represents the fourth industrial revolution, provides some strategies, methodologies, tooling and guidelines on how to manage this pandemic infection (Bai et al., 2020; Cheng et al., 2017; Zhou et al., 2016). With the prevailing pandemic, there has been much pressure on companies to develop solutions that could revolutionize the way people live especially as it affects the reduction of the impacts of coronavirus (Ahmed et al., 2020). These technologies also provide insight on how best people could perform their daily activities during this period without necessarily being exposed to risk (Haleem and Javaid, 2019; Javaid et al., 2020; Ren et al., 2020; Xu and Zhang, 2020). By using some of the technologies that have evolved over the pandemic period which conform to the industry 4-0 specifications, it may be easier to detect the symptoms of corona virus (Lawrence, 2020; Naudé, 2020). Some industry 4.0 compliant solutions and techniques would permit effective management of the coronavirus, thereby decreasing the workload of healthcare personnel (Baldwin and Tomiura, 2020; Gupta et al., 2020; Moeslund et al., 2006; Moeslund and Granum, 2001; Ren et al., 2020). The relevance of some of these techniques are highlighted in Table 16.1.

16.4 COVID-19 AND ARTIFICIAL INTELLIGENCE

The role of AI in tackling the menace associated with coronavirus disease cannot be overemphasized. AI applications are very diverse and are relevant at the various levels of medical care. Basic applications include diagnosis and screening of the viral diseases as well as the treatment of infected patients. AI solutions such as AI-enhanced health informatics could provide data and information such as clinical notes, imaging, clinical research repository, gadgets, case histories, diagnostic assessment histories, treatment histories, etc. (Gillies et al., 2016; Lambin et al., 2017; Li et al., 2016).

In medical radiology and radiography, many radiologists have documented several advances in the application of AI (Shin et al., 2010). The effectiveness of AI in providing solutions for combating COVID-19 has been largely documented (Ahuja et al., 2020; Hussain et al., 2020; Karakulah et al., 2014; McCall, 2020). Table 16.2 shows some of the applications of AI in the management of COVID-19 disease. Table 16.3 shows the application of modern technologies to the management of COVID-19.

16.5 SPECIFIC EXAMPLES OF AUTHORS THAT HAVE WORKED ON INTELLIGENT CLINICAL TRIALS FOR COVID-19

Zhu et al. (2020) revealed that due to the pandemic caused by coronavirus across the globe, business, pharmaceuticals and health sectors are undergoing unprecedented challenges for prevention and treatment options. It is revealed that COVID-19 has caused significant setback in clinical trials for COVID-19 and other related conditions.

TABLE 16.1
Relevance of Industry 4.0 Techniques in the Treatment of COVID-19

Technology	Description of Technology	Work
Printing 3D	Used in the production of ventilators	This technology is applicable to facemask design
Scanning 3D	Used for imaging of the body	It is a non-contact technology that is used to investigate the chest of coronavirus patient
Robots	This technology can work without any external influence. They could be used for hazardous tasks	In COVID-19 period, this technology helps to know the people following the rule or not. This technology is also helpful for doctors in performing their duties
Cloud computing	This technology includes networking, database storage, servers and intelligence	Cloud computing could be used to support large volume of data collection, IoT, data analysis, visualization, etc. (Lawrence, 2020)
Holography	This technology represents the 3D views of objects. It reports the complex and phase altitude of wave	It decreases the exposure of coronavirus. People can attend live streaming at home
Virtual reality	This technology supports training of medical personnel, 3D games and education	During this chaotic corona period, virtual reality provides opportunity for video calling. It also promotes social distancing (Irvine 2020)
Big Data	It provides technologies to accommodate huge data on patients	It is beneficial for forecasting and analysis as well as visualization
Internet of Things	It is a good automated solution. This includes data collection, transfer usually over the internet, data storage and analytic for intelligent decision-making	IoT could ensue a new revolution in healthcare management. Patients could be attached with IoT-enabled devices and left unattended. The IoT devices could measure physiologic and chemical parameters and relay them to a remote server or consultant where analysis could be done to support the management of the patient
Artificial intelligence	AI provides a conglomerate of technologies for supporting the management of coronavirus disease	Machine learning, deep learning, analytics, expert systems could be designed to support the management of COVID-19 patients (Haleem and Javaid, 2019; Javaid et al., 2020)

Kumar (2020) conducted a study on the impact of COVID-19 on pharmaceutical industry and clinical trials with the aim of providing latest report on different drugs and disruptive effects of COVID-19 on clinical trials. Results showed central nervous system and cancer research clinical trials are the worst hit therapeutic areas. The author emphasized the need to start new clinical trials amid the existing ones so as to maintain the healthcare sector to develop innovative and sustainable approach; thus, deliberate action must be taken to guarantee scientific research output for clinical trials and treatment of patients.

TABLE 16.2
AI in the Management of COVID-19 Disease

Application	Data Type	Method
Real-time surveillance and tracking of disease spread. Geographical location alerts and early warning (Bogoch et al., 2016)	Vary	Common methods include deep learning and other machine learning algorithms (Bird et al., 2009, 2016; Howard and Ruder, 2018). Extraction, summarization, automation and classification
Intervention strategies and response such as lockdown and physical distancing	Data from travels, reports from health care, outlets news and data from around the world which are usually posted on social apps using various languages	
Incubation, genomic analysis, evaluation, origin, transmission mode, medical care and non-pharmaceutical intervention	Data taken from some articles on coronavirus disease (Initiative, 2020; L. L. Wang et al., 2020)	
Vaccination and therapeutics	Data from vaccination, treatment and therapeutics	Data Mining, Text mining and machine learning techniques
Preventions of fake information related to corona virus; campaigns and awareness programs and communication	Information and guideline related to medical experts	
Trial and screening, individual investigation and treatment, evaluation of risk, prediction for survival and planning for medical and healthcare resources	This includes examination of blood test from lab, record of electronic-health, rate of respiration and heart and associated clinical symptoms	Various techniques of learning like machine and deep learning which include regression logistic, Bayes naïve, regression net-elastic, optimization method and techniques which are deep learning (Hochreiter and Schmidhuber, 1997; Rasmussen and Williams, 2006, 2018; Zou and Hastie, 2005)
Investigation of incidence and prevalence rate and trend of spreading	Demographic, population and series of cases data	
Medical imaging during early management of infection	Computed tomography and x-ray	Convolutional neural networks, facial recognition and computer vision (Zeiler and Fergus, 2014)
Tertiary and secondary structure of protein, track the changes that exist genetically and physiologically. Viral genomic analysis	For this, data was obtained from protein and via genome of virus	In this, several methods like probabilistic and heuristic method, clustering algorithms and programming dynamic and alignment sequence used

(*Continued*)

TABLE 16.2 (*Continued*)
AI in the Management of COVID-19 Disease

Application	Data Type	Method
Biochemical drug and vaccine tests	Interaction of target drug, interaction of P-P (protein protein), protein and their crystal structure, ligand co-crystalized	Multitask, reinforcement and deep learning, machine learning which usually include auto-encoders adversarial, combinatorics, theory of graph, algorithms heuristic (Bung et al., 2020; Gao et al., 2020; Nguyen et al., 2020; Santana and Silva-Jr, 2021; Shi et al., 2021; Tang et al., 2020)
Disinfection: use of robots and drones for effective cleaning, medication delivery and treatment	Demonstrate data and stimulate the environment by using autonomous training agent	Control and optimization, vision of computer, Deep Reinforcement Learning (DRL) and demonstration learning (Nguyen et al., 2020; Tavakoli et al., 2020)
Predict the recovery of economy	By using Global Positioning System (GPS) data and satellite images	Deep learning (Hou et al., 2019; Nguyen et al., 2020)

TABLE 16.3
Application of Modern Technology against COVID-19

Application	Description
Social awareness control through internet	To measure the COVID-19 patient temperature using thermometer, smart phone is very authentic (Maddah and Beigzadeh, 2020; Nemati et al., 2019)
Drug discovery	For the synthesis of important drug component, AI is very effective (Makhzani et al., 2015; Zhavoronkov et al., 2020)
Prediction of protein structure	This technique could be applied for assessment of the protein structure. Learning residual framework could also be performed (He et al., 2016; Senior et al., 2019; Sheng Wang et al., 2017; Yu and Koltun, 2016)
Medicine, Telemedicine, Nanoinformatics and computational perspective	AI benevolent could be applied in the predicting the effectiveness of baricitinib in the reduction of viral infection (Richardson et al., 2020). Telemedicine is effective in rural and remote outreach (Melton et al., 2020; Olayinka et al., 2018). Nanoinformatics is essential in synthesis and prognostic approach associated with disease prevention, control and possibly eradication (Nwankwo et al., 2020; Olayinka and Chiemeke, 2019)
Patient's condition of health and outcome	One of important tests which quickly identifies the infected patient is xgboost (Qi et al., 2020; Yang et al., 2020)
Tracking of diseases	To know about the infected person, Susceptible, infectious, and recovered compartments (SIR) model which is time dependent is very effective (Wang et al., 2020)
Radiological image for diagnosis	To achieve correct information regarding to corona patient, AI is very effective (Gunraj et al., 2020;. Kumar et al., 2020; Narin et al., 2020; Shuai Wang et al., 2020)

Chen et al. (2020) revealed that the emergence of candidate drugs for clinical trials for the purpose of combating COVID-19 has rapidly gained significant global responses from different scientists to test their safety and efficacy. Report shows that many significant candidate drugs have been registered for national and international registry for clinical trials, and findings revealed that hydroxychloroquine, lopinavir–ritonavir have undergone clinical trials against supportive care and standard care, respectively. In addition, favipiravir matched with arbidol and several others including non-randomized such as hydroxychloroquine against hydroxychloroquine together with azithromycin. The authors revealed that further investigation with enrollment of trials, drugs such as IL-6 inhibitors, e.g., sarilumab or tocilizumab, remdesivir, stem-cell transfusion, convalescent plasma therapy, traditional Chinese medicines as well as vaccine candidates are subjected to rigorous research evaluation. It is believed that a successful randomized control trail will give an evidence-based support for critical intervention while most candidates will just offer relative effects. Owning to the increased rate of investigation on clinical trials, duplication of efforts should be avoided by collating data on the International Clinical Trials Registry Platform from every national clinical trial system. Based on this, the authors reported that AI was developed to enhance routine navigation of trial registries seen to save about 50%–80% of time. Data mapping for clinical trials are done based on patients, geographical location, interventional properties and trials to encourage evidence-based decision-making by policy makers, scientists and medical researchers through the synthesis of trials to combat COVID-19.

Stebbing et al. (2020) revealed that AI can be carefully utilized to monitor COVID-19 pandemic through anti-cytokine action of baricitinib, a Janus kinase (JAK)1/JAK2 inhibitor of host cell viral transmission. The authors evaluated the physiological role of baricitinib against COVID-19 diseases and cytokines mediated infections. AI was utilized to predict the biochemical interaction of baricitinib on numb-linked kinase members with significant reduction in viral infectivity, rapid radiologic recovery and significant decline in IL-6 levels, inflammatory markers and severe acute respiratory syndrome *coronavirus 2* (SARS-CoV-2) viral load. Thus, this evidence supported further research on the clinical trial of baricitinib for its antiviral action in COVID-19 patients.

Bragazzi et al. (2020) revealed that significant advancement has been demonstrated in the field of computational physiology, AI, information and communication technologies, big data and molecular biology which can be adapted to investigate COVID-19 pandemic for data gathering, health monitoring and surveillance, forecasting and policy-making processes by government, stakeholders and researchers so as to manage the pandemic. Ho (2020) also make a report that the outburst of COVID-19 pandemic has resulted into substantial crisis in healthcare sector and global economy; thus, efforts are put in place to identify candidate therapy, vaccine or antibody to combat the pathogen with a phase 1 trial of mRNA-1273 for COVID-19 treatment which has been estimated to last for about 18 months are underway. The author revealed that search for AI approach for combination therapy repurposing design must be utilized.

Cao et al. (2020) revealed that combination therapy is among the most investigated and explored treatments for clinical trials and identified different challenges associated with drug repurposing and traditional combined approach. The authors carried out an investigation utilizing the combination of lopinavir and ritonavir against standard of care therapy like ventilation, antibiotics, supplemental oxygen and other approaches. The findings revealed no significant effects, thus combinational approach is suggested. The clinical trial was carried using the combination of azithromycin and hydroxychloroquine respectively. Thus, it is shown that combinatorial therapy repurposing is the fastest way of identifying candidate drugs for COVID-19 disease involving co-delivery and synergistic mode of actions. Combination therapy regimen is an area that can be seriously harnessed using AI to optimize response to COVID-19 intervention and treatment particularly against future pandemics.

Abdulla et al. (2020) revealed that AI can be utilized for optimizing and enhancing combined therapy development. Application of AI in infectious diseases treatment is undoubtedly becoming a conventional way or pathways of validating candidate drug for clinical trials such as drug design and development, plus matching of patients for trials. The authors revealed that AI-based drug design during pandemic is an important strategy for quick intervention and optimizing drug within few days. Several methods have been utilized in the past for drug combination against infectious diseases and cancer, which are based on drug interaction for enhancing treatment efficacy through synergistic multidrug regimes. Studies have revealed that the most potent solution for optimizing drug interaction for complicated diseases without complex data platform is the AI-based solution strategy with any big data training to implement. AI is a neural network for drug design and treatment efficacy and dosage.

Ho et al. (2020) demonstrated progress made in the global approach in the adoption of technology to combat COVID-19 pandemic. Thus, various combination therapy is being proposed for clinical trials and possible solution even though many will fail to yield desired result due to insufficient dosing and combinatory permutations.

Torequl et al. (2020) revealed that the genome of corona virus is very large with unusual proliferative mechanism for replications characterized by protruding spikes on the surface for penetration. The authors summarize strategies for treatment, pathogenicity and replication coupled with prevention protocol. They revealed that due to lack of approved or centralized option for treatment and prevention strategy, an alternative method is urgently needed to combat the pandemic globally. Thus, management of the virus, clinical trials for proposed treatment and vaccine and drug development utilizing AI are the major focal point of the authors.

Rosa and Santos (2020) in their work showed that *Betacoronavirus* genus of the coronavirus is RNA virus being transmitted from birds, mammals and humans, and more than 80 clinical trials, repositioning or drug repurposing have been done to combat COVID-19. Some of the clinical trials list in the clinicaltrials.gov database highlighted by the authors is interferons, human immunoglobulin, favipiravir, arbidol, chloroquine, oseltamivir, remdesivir, bevacizumab, lopinavir, methylprednisolone, ritonavir, traditional Chinese medicines and hydroxychloroquine. The authors revealed that although there are several challenges in drug repurposing, it still represents the best alternative option for drug discovery and development which might be linked to its low cost

Wong et al. (2020) revealed that due to the lack of well-developed treatment protocol for COVID-19 pandemic, the utilization of traditional public health approaches remains the available option to control the pandemic. The authors demonstrated that routine checkup such as temperature monitoring and symptoms surveillance is necessary which can be done with the advancement in technology such as biosensor. Many physiological parameters and variables can be effectively monitored for early diagnosis of COVID-19 utilizing AI and biosensor system. Thus, the authors carried out a randomized clinical control trial involving 200–1,000 asymptomatic participants with COVID-19 at the isolation center in Hong Kong. Remote sensing devices and standard strategy where randomly given to patients and grouped into experimental and control to monitor fever and other physiological parameters such as respiratory rate, blood oxygen saturation, blood pressure, skin temperature, pulse rate and daily activity by continuously wearing wearable biosensors device. These physiological parameters are subsequently transferred to smart phone application for analysis utilizing cloud-based physiological multivariate analytics system referred to as Biovitals for any physiological changes and clinical decision-making. This represents a smart way of quick diagnosis of COVID-19.

McCall (2020) revealed that new emerging advanced technology can be deployed to combat COVID-19 pandemic such as AI. The author showed that AI has now become a well-recognized application for improving healthcare services such as prediction and surveillance. Benevolent AI, Blue Dot, AI algorithms are some of the few examples of AI technology developed by scientists to curtail the spread of COVID-19, monitor successful public health response, understanding of transmissibility, prediction of vulnerable populations, climate change and transmission as well as epidemiological modeling. Through machine learning technology, corona virus spread and location can be predicted for public health emergency. The author revealed that China which is the epicenter is engaging the use of AI in the management of COVID-19 by utilizing Infervision's AI application to monitor and rapid diagnosis. Rapid computed tomography (CT) scanners and AI algorithm are deployed to increase efficiency and accuracy toward combating the virus by providing images for change in density, volume, shape and lesions.

Vaishya et al. (2020) revealed that new technologies must be deployed to healthcare sector such as IoT, AI, Machine Learning and Big Data to curtail the coronavirus disease. Through rigorous research, the authors were able to identify many studies that have utilized various technologies to fight against COVID-19. It has been revealed that these technologies can predict, monitor, analyze, screen, track and diagnose pathogenic organisms effectively for decision-making process; thus, it was suggested that proper adoption of these technologies must be deployed or utilized to help curtail the spread of the virus.

Saxena (2020) revealed that the coronavirus pandemic has caused serious havoc on health, economy and security of many populations across the globe; thus, rapid identification of effective candidate drugs to combat the virus seems necessary. He opined that smart strategies should be deployed for drug discovery against COVID-19. Heiser et al. (2020) revealed that in the process of identifying target drug for COVID-19 virus, evaluation of compounds from library against SARS-CoV-2 was profiled utilizing deep learning. The authors showed that from their findings remdesivir showed significant effects against SARS-CoV-2 in human cell model. In addition, other drugs such as β-blockers, Vitamin D analogs, and mTOR/PI3K inhibitors not given consideration displayed interesting results.

Gordon et al. (2020) demonstrated that rapid identification of candidate drug for COVID-19 is important through the deployment of deep learning approach to drug discovery platform for clinical trials on human cells. Profiling and image generation across multiple therapeutic areas are important such as genetic diseases, inflammatory diseases, and infectious diseases particularly SARS-CoV-2. Technological approaches provide rapid high-quality throughput phenotypic screening of many cell types and detect traditional inhibitors against the virus. Their work also revealed that innovations and advancement in computational physiology, AI, data science and machine learning provide a central and broad platform for the fight against COVID-19 pandemic affecting socioeconomic, biomedical and epidemiological issues. The author revealed that many ethical issues are also involved in the adoption of technologies in health care such as data sharing, public trust and security issues. Thus, many steps are provided by the authors to encourage innovations and research for equitable, accountable, AI/machine learning design and discovery centered which are governed by integrity.

Alabool et al. (2020) demonstrated the various advantages offered by AI deployments against COVID-19 pandemic. Even though they differ in terms of purpose, approach and methodology, the authors suggested various ways to avoid shortcomings in the utilization of AI techniques for COVID-19. Similarly, Vaishya et al. (2020) presented a comprehensive study on the various applications and utilization of AI technologies against SARS-CoV-2 suggesting how to avoid errors in the deployment of AI against SARS-CoV-2. Altay et al. (2020) in their study showed that the SARS-CoV-2 has impacted negatively on the health and economic sector globally resulting to the evolution of emergency approaches to curtail the spread of the disease. The authors suggested that drug repositioning is the latest novel approach that is attracting massive attention by scientists particularly in the pharmaceutical industry globally to facilitate and enhance drug discovery.

16.6 CONCLUSIONS

In this chapter, we set out to review some landmark tools, methods and techniques that have been proposed, experimented or clinically tested and/or deployed. The emphasis was on relevant computational intelligence techniques and tools, which are for either improving the infrastructure for public health management of the crisis or supporting the caregivers in their bid to provide meaningful services to patients and other concerned persons who may not be infected but need vital public health awareness to remain protected amid the rising incidences of COVID-19 infections. From various published sources, we have established that computational intelligence has emerged the preferred innovative approach(es) to support the sustained fight against the menace of the disease. Unarguably, computational intelligence tooling such as machine learning models, analytics and other AI solutions have been emphasized by many researchers as promising game changer that would provide excellent solutions to drive intelligent clinical trials, public health awareness campaigns, surveillance, pharmacovigilance, drug discovery, diagnosis, treatment, recovery and post-treatment management, among others. We conclude that while the traditional method of clinical management of the coronavirus infection is appreciated, prospective and reliable outcomes could be obtained with the entrenchment of computational intelligence as part of the healthcare architecture of every district globally.

REFERENCES

Adetunji, C.O., Anani, O.A., and Panpatte, D. (2021). Mechanism of actions involved in sustainable ecorestoration of petroleum hydrocarbons polluted soil by the beneficial microorganism. In: Panpatte, D.G., and Jhala, Y.K. (eds.), *Microbial Rejuvenation of Polluted Environment. Microorganisms for Sustainability*, vol 26. Springer, Singapore. https://doi.org/10.1007/978-981-15-7455-9_8

Adetunji, C.O., and Anani, O.A. (2021a) Recent advances in the application of genetically engineered microorganisms for microbial rejuvenation of contaminated environment. In: Adetunji C.O., Panpatte, D.G., and Jhala, Y.K. (eds.), *Microbial Rejuvenation of Polluted Environment. Microorganisms for Sustainability*, vol 27. Springer, Singapore. https://doi.org/10.1007/978-981-15-7459-7_14

Adetunji C.O., and Anani O.A. (2021b) Utilization of microbial biofilm for the biotransformation and bioremediation of heavily polluted environment. In: Panpatte, D.G., and Jhala, Y.K. (eds.), *Microbial Rejuvenation of Polluted Environment. Microorganisms for Sustainability*, vol 25. Springer, Singapore. https://doi.org/10.1007/978-981-15-7447-4_9

Adetunji, C.O., and Anani, O.A. (2021c) Bioremediation of polythene and plastics using beneficial microorganisms. In: Adetunji, C.O., Panpatte, D.G., and Jhala, Y.K. (eds.), *Microbial Rejuvenation of Polluted Environment. Microorganisms for Sustainability*, vol 27. Springer, Singapore. https://doi.org/10.1007/978-981-15-7459-7_13

Abdulla, A., Wang, B., Qian, F., Kee, T., Blasiak, A., Ong, Y. H., Hooi, L., Parekh, F., Soriano, R., Olinger, G. G., Keppo, J., Hardesty, C. L., Chow, E. K., Ho, D., & Ding, X. (2020). Project IDentif.AI: Harnessing artificial intelligence to rapidly optimize combination therapy development for infectious disease intervention. *Advanced Therapeutics*, *3*(7), 2000034. https://doi.org/10.1002/adtp.202000034.

Ahmed, S. F., Quadeer, A. A., & McKay, M. R. (2020). Preliminary identification of potential vaccine targets for the COVID-19 Coronavirus (SARS-CoV-2) Based on SARS-CoV Immunological Studies. *Viruses*. https://doi.org/10.3390/v12030254.

Ahuja, A. S., Reddy, V. P., & Marques, O. (2020). Artificial intelligence and COVID-19: A multidisciplinary approach. *Integrative Medicine Research*. https://doi.org/10.1016/j.imr.2020.100434.

Alabool, H., Alarabiat, D., Abualigah, L., Habib, M., Khasawneh, A., Alshinwan, M., & Shehab, M. (2020). *Artificial intelligence techniques for Containment COVID-19 Pandemic: A Systematic Review*. https://doi.org/10.21203/rs.3.rs-30432/v1.

Altay, O., Mohammadi, E., Lam, S., Turkez, H., Boren, J., Nielsen, J., Uhlen, M., & Mardinoglu, A. (2020). Current status of COVID-19 therapies and drug repositioning applications. *iScience*. https://doi.org/10.1016/j.isci.2020.101303.

Anani, O.A., Mishra R.R., Mishra P., Enuneku A.A., Anani G.A., and Adetunji C.O. (2020) Effects of toxicant from pesticides on food security: Current developments. In: Mishra P., Mishra R.R., Adetunji C.O. (eds.), *Innovations in Food Technology.* Springer, Singapore. https://doi.org/10.1007/978-981-15-6121-4_22

Bai, H. X., Hsieh, B., Xiong, Z., Halsey, K., Choi, J. W., Tran, T. M. L., Pan, I., Shi, L. B., Wang, D. C., Mei, J., Jiang, X. L., Zeng, Q. H., Egglin, T. K., Hu, P. F., Agarwal, S., Xie, F. F., Li, S., Healey, T., Atalay, M. K., & Liao, W. H. (2020). Performance of radiologists in differentiating COVID-19 from non-COVID-19 viral pneumonia at chest CT. *Radiology.* https://doi.org/10.1148/radiol.2020200823.

Baldwin, R., & Tomiura, E. (2020). Thinking ahead about the trade impact of COVID-19. *Economics in the Time of COVID-19.* 59.

Barstugan, M., Ozkaya, U., & Ozturk, S. (2020). Coronavirus (COVID-19) classification using CT images by machine learning methods. *arXiv.*

Bird, S., Bird, S., & Loper, E. (2016). NLTK : The natural language toolkit NLTK : The Natural Language Toolkit. *Proceedings of the ACL-02 Workshop on Effective Tools and Methodologies for Teaching Natural Language Processing and Computational Linguistics-Volume 1.*

Bird, S., Loper, E., & Klein, E. (2009). Natural Language ToolKit (NLTK) Book. *O'Reilly Media Inc.*

Bogoch, I. I., Brady, O. J., Kraemer, M. U. G., German, M., Creatore, M. I., Kulkarni, M. A., Brownstein, J. S., Mekaru, S. R., Hay, S. I., Groot, E., Watts, A., & Khan, K. (2016). Anticipating the international spread of Zika virus from Brazil. In *The Lancet.* https://doi.org/10.1016/S0140-6736(16)00080-5.

Bragazzi, N. L., Dai, H., Damiani, G., Behzadifar, M., Martini, M., & Wu, J. (2020). How big data and artificial intelligence can help better manage the covid-19 pandemic. *International Journal of Environmental Research and Public Health.* https://doi.org/10.3390/ijerph17093176.

Bung, N., Krishnan, S. R., Bulusu, G., & Roy, A. (2020). De novo design of new chemical entities (NCEs) for SARS-CoV-2 using artificial intelligence. *ChemRxiv.* https://doi.org/10.26434/chemrxiv.11998347.v2.

Calvaresi, D., Marinoni, M., Dragoni, A. F., Hilfiker, R., & Schumacher, M. (2019). Real-time multi-agent systems for telerehabilitation scenarios. *Artificial Intelligence in Medicine.* https://doi.org/10.1016/j.artmed.2019.02.001.

Cao, B., Wang, Y., Wen, D., Liu, W., Wang, J., Fan, G., Ruan, L., Song, B., Cai, Y., Wei, M., Li, X., Xia, J., Chen, N., Xiang, J., Yu, T., Bai, T., Xie, X., Zhang, L., Li, C., … Wang, C. (2020). A trial of lopinavir–ritonavir in adults hospitalized with severe covid-19. *New England Journal of Medicine.* 382(19), 1787–1799. https://doi.org/10.1056/NEJMoa2001282.

Chauhan, N. S., Bathula, S., Vishwakarma, A., Bhardwaj, R., Johari, K. K., Gahtori, B., & Dhar, A. (2019). Enhanced thermoelectric performance in p-type ZrCoSb based half-Heusler alloys employing nanostructuring and compositional modulation. *Journal of Materiomics.* https://doi.org/10.1016/j.jmat.2018.11.003.

Chen, C., Zhang, Y., Huang, J., Yin, P., Cheng, Z., Wu, J., Chen, S., Zhang, Y., Chen, B., Lu, M., Luo, Y., Ju, L., Zhang, J., & Wang, X. (2020). Favipiravir versus Arbidol for COVID-19: A randomized clinical trial. *medRxiv.* https://doi.org/10.1101/2020.03.17.20037432.

Cheng, G. J., Liu, L. T., Qiang, X. J., & Liu, Y. (2017). Industry 4.0 development and application of intelligent manufacturing. *Proceedings -2016 International Conference on Information System and Artificial Intelligence, ISAI 2016.* https://doi.org/10.1109/ISAI.2016.0092.

Dwivedi, N., Dwivedi, S., and Adetunji, C. O. (2021) Efficacy of microorganisms in the removal of toxic materials from industrial effluents. In: Adetunji C.O., Panpatte, D.G., and Jhala, Y.K. (eds.), *Microbial Rejuvenation of Polluted Environment. Microorganisms for Sustainability*, vol 27. Springer, Singapore. https://doi.org/10.1007/978-981-15-7459-7_15

Gao, K., Nguyen, D. D., Wang, M., & Wei, G. W. (2020). MathDL: Mathematical deep learning for D3R Grand Challenge 4. *Journal of Computer-Aided Molecular Design*. https://doi.org/10.1007/s10822-019-00237-5.

Gillies, R. J., Kinahan, P. E., & Hricak, H. (2016). Radiomics: Images are more than pictures, they are data. *Radiology*. 278(2), 563–577. https://doi.org/10.1148/radiol.2015151169.

Gofuku, A. (2020). Applications of information and robot technologies to improve medical therapies. *ACM International Conference Proceeding Series*. https://doi.org/10.1145/3427423.3427466.

Gordon, D. E., Jang, G. M., Bouhaddou, M., Xu, J., Obernier, K., White, K. M., O'Meara, M. J., Rezelj, V. V., Guo, J. Z., Swaney, D. L., Tummino, T. A., Hüttenhain, R., Kaake, R. M., Richards, A. L., Tutuncuoglu, B., Foussard, H., Batra, J., Haas, K., Modak, M., … Krogan, N. J. (2020). A SARS-CoV-2 protein interaction map reveals targets for drug repurposing. *Nature*. 583(7816), 459–468. https://doi.org/10.1038/s41586-020-2286-9.

Gunraj, H., Wang, L., & Wong, A. (2020). COVIDNet-CT: A tailored deep convolutional neural network design for detection of COVID-19 cases from chest CT images. *Frontiers in Medicine*. https://doi.org/10.3389/fmed.2020.608525.

Gupta, R., Ghosh, A., Singh, A. K., & Misra, A. (2020). Clinical considerations for patients with diabetes in times of COVID-19 epidemic. *Diabetes and Metabolic Syndrome: Clinical Research and Reviews*. 14(3), 211–212. Elsevier Ltd. https://doi.org/10.1016/j.dsx.2020.03.002.

Haleem, A., & Javaid, M. (2019). Additive manufacturing applications in industry 4.0: A review. *Journal of Industrial Integration and Management*. https://doi.org/10.1142/s2424862219300011.

He, K., Zhang, X., Ren, S., & Sun, J. (2016). Deep residual learning for image recognition. *Proceedings of the IEEE Computer Society Conference on Computer Vision and Pattern Recognition*. https://doi.org/10.1109/CVPR.2016.90.

Heiser, K., McLean, P. F., Davis, C. T., Fogelson, B., Gordon, H. B., Jacobson, P., Hurst, B., Miller, B., Alfa, R. W., Earnshaw, B. A., Victors, M. L., Chong, Y. T., Haque, I. S., Low, A. S., & Gibson, C. C. (2020). Identification of potential treatments for COVID-19 through artificial intelligence-enabled phenomic analysis of human cells infected with SARS-CoV-2. *bioRxiv*. https://doi.org/10.1101/2020.04.21.054387.

Ho, D., Quake, S. R., McCabe, E. R. B., Chng, W. J., Chow, E. K., Ding, X., Gelb, B. D., Ginsburg, G. S., Hassenstab, J., Ho, C. M., Mobley, W. C., Nolan, G. P., Rosen, S. T., Tan, P., Yen, Y., & Zarrinpar, A. (2020). Enabling technologies for personalized and precision medicine. *Trends in Biotechnology*. 38(5), 497–518. Elsevier Ltd. https://doi.org/10.1016/j.tibtech.2019.12.021.

Hochreiter, S., & Schmidhuber, J. (1997). Long short-term memory. *Neural Computation*. https://doi.org/10.1162/neco.1997.9.8.1735.

Hou, X., Wang, B., Hu, W., Yin, L., & Wu, H. (2019). SolarNet: A deep learning framework to map solar power plants in China from satellite imagery. *arXiv*.

Howard, J., & Ruder, S. (2018). Universal language model fine-tuning for text classification. *ACL 2018–56th Annual Meeting of the Association for Computational Linguistics, Proceedings of the Conference (Long Papers)*. https://doi.org/10.18653/v1/p18-1031.

Hussain, A. A., Bouachir, O., Al-Turjman, F., & Aloqaily, M. (2020). AI techniques for COVID-19. *IEEE Access*. https://doi.org/10.1109/ACCESS.2020.3007939.

Initiative, A. I. for A. in partnership with the C. Z. (2020). COVID-19 open research dataset challenge (CORD-19). *Semantic Scholar*.

Islam, S., Thangadurai, D., Adetunji, C. O., Nwankwo, W., Kadiri, O., Makinde, S., Michael, O. S., Anani, O. A., and Adetunji, J. B. (2021) Nanomaterials and nanocoatings for alternative antimicrobial therapy. In: Kharissova O.V., Martínez L.M.T., Kharisov B.I. (eds.), *Handbook of Nanomaterials and Nanocomposites for Energy and Environmental Applications*. Springer, Cham. https://doi.org/10.1007/978-3-030-11155-7_3-1

Javaid, M., Haleem, A., Vaishya, R., Bahl, S., Suman, R., & Vaish, A. (2020). Industry 4.0 technologies and their applications in fighting COVID-19 pandemic. *Diabetes and Metabolic Syndrome: Clinical Research and Reviews*. https://doi.org/10.1016/j.dsx.2020.04.032.

Javor, D., Kaplan, H., Kaplan, A., Puchner, S. B., Krestan, C., & Baltzer, P. (2020). Deep learning analysis provides accurate COVID-19 diagnosis on chest computed tomography. *European Journal of Radiology*. https://doi.org/10.1016/j.ejrad.2020.109402.

Karakulah, G., Dicle, O., Kosaner, O., Suner, A., Birant, C. C., Berber, T., & Canbek, S. (2014). Computer based extraction of phenoptypic features of human congenital anomalies from the digital literature with natural language processing techniques. *Studies in Health Technology and Informatics*. https://doi.org/10.3233/978-1-61499-432-9-570.

Kumar, A., Gupta, P. K., & Srivastava, A. (2020). A review of modern technologies for tackling COVID-19 pandemic. *Diabetes and Metabolic Syndrome: Clinical Research and Reviews*. https://doi.org/10.1016/j.dsx.2020.05.008.

Kumar, V. (2020). *Title: COVID-19 Crisis: An Update on the Disruption of Clinical trials*. https://doi.org/10.20944/preprints202005.0411.v1.

Lambin, P., Leijenaar, R. T. H., Deist, T. M., Peerlings, J., De Jong, E. E. C., Van Timmeren, J., Sanduleanu, S., Larue, R. T. H. M., Even, A. J. G., Jochems, A., Van Wijk, Y., Woodruff, H., Van Soest, J., Lustberg, T., Roelofs, E., Van Elmpt, W., Dekker, A., Mottaghy, F. M., Wildberger, J. E., & Walsh, S. (2017). Radiomics: The bridge between medical imaging and personalized medicine. *Nature Reviews Clinical Oncology*. 14(12), 749–762. Nature Publishing Group. https://doi.org/10.1038/nrclinonc.2017.141.

Lawrence, C. (2020). *Is Cloud Computing the Superhero of Covid-19?* Codemotion Magazine. https://www.codemotion.com/magazine/dev-hub/cloud-manager/cloud-computing-covid-19/.

Li, C. Y., Liang, G. Y., Yao, W. Z., Sui, J., Shen, X., Zhang, Y. Q., Peng, H., Hong, W. W., Ye, Y. C., Zhang, Z. Y., Zhang, W. H., Yin, L. H., & Pu, Y. P. (2016). Integrated analysis of long non-coding RNA competing interactions reveals the potential role in progression of human gastric cancer. *International Journal of Oncology*. 48(5), 1965–1976. https://doi.org/10.3892/ijo.2016.3407.

Maddah, E., & Beigzadeh, B. (2020). Use of a smartphone thermometer to monitor thermal conductivity changes in diabetic foot ulcers: A pilot study. *Journal of Wound Care*. https://doi.org/10.12968/jowc.2020.29.1.61.

Makhzani, A., Shlens, J., Jaitly, N., Goodfellow, I., & Frey, B. (2015). *Adversarial Autoencoders*. http://arxiv.org/abs/1511.05644.

McCall, B. (2020). COVID-19 and artificial intelligence: protecting health-care workers and curbing the spread. *The Lancet Digital Health*. https://doi.org/10.1016/s2589-7500(20)30054–6.

Melton, S., Simmons, S. C., Smith, B. A., & Hamilton, D. R. (2020). Telemedicine. *Principles of Clinical Medicine for Space Flight*. https://doi.org/10.1007/978-1-4939-9889-0_8.

Moeslund, T. B., & Granum, E. (2001). A survey of computer vision-based human motion capture. *Computer Vision and Image Understanding.* 81(3), 231–268. https://doi.org/10.1006/cviu.2000.0897.

Moeslund, T. B., Hilton, A., & Krüger, V. (2006). A survey of advances in vision-based human motion capture and analysis. *Computer Vision and Image Understanding.* 104(2–3 SPEC. ISS.), 90–126. Academic Press. https://doi.org/10.1016/j.cviu.2006.08.002.

Mishra, P., Mishra, R. R., & Adetunji, C. O. (Eds.). (2020). Innovations in Food Technology: Current Perspectives and Future Goals. Springer Nature Singapore Pte Ltd. 2020. Springer, Singapore. Pp. 143–162. 1st Edition. DOI: 10.1007/978-981-15-6121-4.

Narin, A., Kaya, C., & Pamuk, Z. (2020). Automatic detection of coronavirus disease (COVID-19) using X-ray images and deep convolutional neural networks. *arXiv.*

Naudé, W. (2020). Artificial intelligence against COVID-19: An early review. *IZA Discussion Paper.*

Nemati, E., Rahman, M. M., Nathan, V., Vatanparvar, K., & Kuang, J. (2019). Poster Abstract: A Comprehensive Approach for Cough Type Detection. *Proceedings -4th IEEE/ACM Conference on Connected Health: Applications, Systems and Engineering Technologies, CHASE 2019,* 15–16. https://doi.org/10.1109/CHASE48038.2019.00013.

Nguyen, T. T. (2020). Artificial intelligence in the battle against Coronavirus (COVID-19): A survey and future research directions. *arXiv.*

Nguyen, T. T., Nguyen, N. D., & Nahavandi, S. (2020). Deep reinforcement learning for multiagent systems: A review of challenges, solutions, and applications. *IEEE Transactions on Cybernetics.* https://doi.org/10.1109/TCYB.2020.2977374.

Nwankwo, W., Olayinka, A. S., & Ukhurebor, K. E. (2020). Nanoinformatics: Why Design of Projects on Nanomedicine Development and Clinical Applications may fail? *2020 International Conference in Mathematics, Computer Engineering and Computer Science, ICMCECS 2020.* https://doi.org/10.1109/ICMCECS47690.2020.246992.

Olaniyan, O. T., Dare, A., Okotie, G. E., Adetunji, C. O., Ibitoye, B. O., Bamidele, O. J., & Eweoya, O. O. (2020). Testis and blood-testis barrier in Covid-19 infestation: role of angiotensin-converting enzyme 2 in male infertility. *Journal of Basic and Clinical Physiology and Pharmacology.* 31(6). https://doi.org/10.1515/JBCPP-2020-0156.

Olaniyan, O.T., and Adetunji, C.O. (2021) Biochemical role of beneficial microorganisms: An overview on recent development in environmental and agro science. In: Adetunji, C.O., Panpatte, D.G., and Jhala, Y.K. (eds) *Microbial Rejuvenation of Polluted Environment. Microorganisms for Sustainability,* vol 27. Springer, Singapore. https://doi.org/10.1007/978-981-15-7459-7_2

Olayinka, T.., Olayinka, A. S., & Oladimeji, O. R. (2018). Framework for web-based and desktop-based telemedicine system. *Journal of Computer Engineering.*

Olayinka, T. C., & Chiemeke, S. C. (2019). Predicting paediatric malaria occurrence using classification algorithm in data mining. *Journal of Advances in Mathematics and Computer Science.* https://doi.org/10.9734/jamcs/2019/v31i430118.

OT, O., A, D., GE, O., CO, A., BO, I., OJ, B., & OO, E. (2020). Testis and blood-testis barrier in Covid-19 infestation: Role of angiotensin-converting enzyme 2 in male infertility. *Journal of Basic and Clinical Physiology and Pharmacology.* 31(6). https://doi.org/10.1515/JBCPP-2020-0156.

Qi, X., Jiang, Z., Yu, Q., Shao, C., Zhang, H., Yue, H., Ma, B., Wang, Y., Liu, C., Meng, X., Huang, S., Wang, J., Xu, D., Lei, J., Xie, G., Huang, H., Yang, J., Ji, J., Pan, H., … Ju, S. (2020). Machine learning-based CT radiomics model for predicting hospital stay in patients with pneumonia associated with SARS-CoV-2 infection: A multicenter study. *medRxiv.* https://doi.org/10.1101/2020.02.29.20029603.

Rasmussen, C. E., & Williams, C. K. I. (2006). *Gaussian Processes for Machine Learning.* The MIT Press, Cambridge, MA.

Rasmussen, C. E., & Williams, C. K. I. (2018). Gaussian processes for machine learning. *Gaussian Processes for Machine Learning.* https://doi.org/10.7551/mitpress/3206.001.0001.

Ren, J. ling, Zhang, A. H., & Wang, X. J. (2020). Traditional Chinese medicine for COVID-19 treatment. *Pharmacological Research.* https://doi.org/10.1016/j.phrs.2020.104743.

Richardson, P., Griffin, I., Tucker, C., Smith, D., Oechsle, O., Phelan, A., & Stebbing, J. (2020). Baricitinib as potential treatment for 2019-nCoV acute respiratory disease. *The Lancet.* https://doi.org/10.1016/S0140-6736(20)30304-4.

Santana, M. V. S., & Silva-Jr, F. P. (2021). De novo design and bioactivity prediction of SARS-CoV-2 main protease inhibitors using recurrent neural network-based transfer learning. *BMC Chemistry.* https://doi.org/10.1186/s13065-021-00737-2.

Saxena, A. (2020). Drug targets for COVID-19 therapeutics: Ongoing global efforts. *Journal of Biosciences.* https://doi.org/10.1007/s12038-020-00067-w.

Senior, A. W., Evans, R., Jumper, J., Kirkpatrick, J., Sifre, L., Green, T., Qin, C., Žídek, A., Nelson, A. W. R., Bridgland, A., Penedones, H., Petersen, S., Simonyan, K., Crossan, S., Kohli, P., Jones, D. T., Silver, D., Kavukcuoglu, K., & Hassabis, D. (2019). Protein structure prediction using multiple deep neural networks in the 13th Critical Assessment of Protein Structure Prediction (CASP13). *Proteins: Structure, Function and Bioinformatics.* 87(12), 1141–1148. https://doi.org/10.1002/prot.25834.

Shen, C., Yu, N., Cai, S., Zhou, J., Sheng, J., Liu, K., Zhou, H., Guo, Y., & Niu, G. (2020). Quantitative computed tomography analysis for stratifying the severity of Coronavirus Disease 2019. *Journal of Pharmaceutical Analysis.* https://doi.org/10.1016/j.jpha.2020.03.004.

Shi, F., Wang, J., Shi, J., Wu, Z., Wang, Q., Tang, Z., He, K., Shi, Y., & Shen, D. (2021). Review of artificial intelligence techniques in imaging data acquisition, segmentation, and diagnosis for COVID-19. *IEEE Reviews in Biomedical Engineering.* https://doi.org/10.1109/RBME.2020.2987975.

Shin, H., Kim, K. H., Song, C., Lee, I., Lee, K., Kang, J., & Kang, Y. K. (2010). Electrodiagnosis support system for localizing neural injury in an upper limb. *Journal of the American Medical Informatics Association.* 17(3), 345–347. https://doi.org/10.1136/jamia.2009.001594.

Shripad, S. C., Sagar, P. K., & Chaitanya, B. B. (2013). Robotics in medicine applications. *Journal of Engineering Research and Applications (IJERA)* www.ijera.com.

Stebbing, J., Krishnan, V., Bono, S., Ottaviani, S., Casalini, G., Richardson, P. J., Monteil, V., Lauschke, V. M., Mirazimi, A., Youhanna, S., Tan, Y., Baldanti, F., Sarasini, A., Terres, J. A. R., Nickoloff, B. J., Higgs, R. E., Rocha, G., Byers, N. L., Schlichting, D. E., … Corbellino, M. (2020). Mechanism of baricitinib supports artificial intelligence-predicted testing in COVID-19 patients. *EMBO Molecular Medicine.* https://doi.org/10.15252/emmm.202012697.

Tang, B., He, F., Liu, D., Fang, M., Wu, Z., & Xu, D. (2020). AI-aided design of novel targeted covalent inhibitors against SARS-CoV-2. *BioRxiv : The Preprint Server for Biology.* https://doi.org/10.1101/2020.03.03.972133.

Tavakoli, M., Carriere, J., & Torabi, A. (2020). Robotics, smart wearable technologies, and autonomous intelligent systems for healthcare during the COVID-19 pandemic: An analysis of the state of the art and future vision. *Advanced Intelligent Systems.* https://doi.org/10.1002/aisy.202000071.

Thangadurai, D., Dabire, S. S., Sangeetha, J., Said Al-Tawaha, A. R. M., Adetunji, C. O., Islam, S., Shettar, A. K., David, M., Hospet, R., and Adetunji, J. B. (2021) Greener composites from plant fibers: Preparation, structure, and properties. In: Kharissova O.V., Martínez L.M.T., and Kharisov B.I. (eds.), *Handbook of Nanomaterials and Nanocomposites for Energy and Environmental Applications.* Springer, Cham. https://doi.org/10.1007/978-3-030-11155-7_21-1

Torequl Islam, M., Nasiruddin, M., Khan, I. N., Mishra, S. K., Kudrat-E-Zahan, M., Alam Riaz, T., Ali, E. S., Rahman, M. S., Mubarak, M. S., Martorell, M., Cho, W. C., Calina, D., Docea, A. O., & Sharifi-Rad, J. (2020). A perspective on emerging therapeutic interventions for COVID-19. *Frontiers in Public Health*. https://doi.org/10.3389/fpubh.2020.00281.

Vaishya, R., Javaid, M., Khan, I. H., & Haleem, A. (2020). Artificial Intelligence (AI) applications for COVID-19 pandemic. *Diabetes and Metabolic Syndrome: Clinical Research and Reviews*. https://doi.org/10.1016/j.dsx.2020.04.012

Viveiros Rosa, S. G., & Santos, W. C. (2020). Clinical trials on drug repositioning for COVID-19 treatment. *Revista Panamericana de Salud Publica/Pan American Journal of Public Health*. https://doi.org/10.26633/RPSP.2020.40.

Wang, L. L., Lo, K., Chandrasekhar, Y., Reas, R., Yang, J., Eide, D., Funk, K., Kinney, R., Liu, Z., Merrill, W., Mooney, P., Murdick, D., Rishi, D., Sheehan, J., Shen, Z., Stilson, B., Wade, A. D., Wang, K., Wilhelm, C., … Kohlmeier, S. (2020). CORD-19: The COVID-19 open research dataset. *arXiv*.

Wang, Sheng, Sun, S., Li, Z., Zhang, R., & Xu, J. (2017). Accurate De Novo prediction of protein contact map by ultra-deep learning model. *PLoS Computational Biology*. 13(1). https://doi.org/10.1371/journal.pcbi.1005324.

Wang, Shuai, Kang, B., Ma, J., Zeng, X., Xiao, M., Guo, J., Cai, M., Yang, J., Li, Y., Meng, X., & Xu, B. (2020). A deep learning algorithm using CT images to screen for Corona Virus Disease (COVID-19). *medRxiv*. https://doi.org/10.1101/2020.02.14.20023028.

Wang, Y., Hu, M., Li, Q., Zhang, X. P., Zhai, G., & Yao, N. (2020). Abnormal respiratory patterns classifier may contribute to large-scale screening of people infected with COVID-19 in an accurate and unobtrusive manner. *arXiv*.

Wong, C. K., Ho, D. T. Y., Tam, A. R., Zhou, M., Lau, Y. M., Tang, M. O. Y., Tong, R. C. F., Rajput, K. S., Chen, G., Chan, S. C., Siu, C. W., & Hung, I. F. N. (2020). Artificial intelligence mobile health platform for early detection of COVID-19 in quarantine subjects using a wearable biosensor: Protocol for a randomised controlled trial. *BMJ Open*. https://doi.org/10.1136/bmjopen-2020-038555.

World Health Organization. (2021). *WHO Coronavirus Disease Dashboard*. WHO.Int.

Xu, J., & Zhang, Y. (2020). Traditional Chinese Medicine treatment of COVID-19. *Complementary Therapies in Clinical Practice*. https://doi.org/10.1016/j.ctcp.2020.101165.

Yang, S., Jiang, L., Cao, Z., Wang, L., Cao, J., Feng, R., Zhang, Z., Xue, X., Shi, Y., & Shan, F. (2020). Deep learning for detecting corona virus disease 2019 (COVID-19) on high-resolution computed tomography: A pilot study. *Annals of Translational Medicine*. https://doi.org/10.21037/atm.2020.03.132.

Yu, F., & Koltun, V. (2016). Multi-scale context aggregation by dilated convolutions. *4th International Conference on Learning Representations, ICLR 2016-Conference Track Proceedings*.

Zeiler, M. D., & Fergus, R. (2014). Visualizing and Understanding Convolutional Networks arXiv:1311.2901v3 [cs.CV] 28 Nov 2013. *Computer Vision–ECCV 2014*. https://doi.org/10.1007/978-3-319-10590-1_53.

Zhavoronkov, A., Aladinskiy, V. A., Zhebrak, A., Zagribelnyy, B., Terentiev, V., Bezrukov, D. S., Polykovskiy, D., Shayakhmetov, R., Filimonov, A., Orekhov, P., Yan, Y., Popova, O., Vanhaelen, Q., Aliper, A., & Ivanenkov, Y. A. (2020). Potential 2019-nCoV 3C-like protease inhibitors designed using generative deep learning approaches. *ChemRxiv*. https://doi.org/10.26434/chemrxiv.11829102.v2.

Zhou, K., Liu, T., & Zhou, L. (2016). Industry 4.0: Towards future industrial opportunities and challenges. *2015 12th International Conference on Fuzzy Systems and Knowledge Discovery, FSKD 2015*. https://doi.org/10.1109/FSKD.2015.7382284.

Zhu, N., Zhang, D., Wang, W., Li, X., Yang, B., Song, J., Zhao, X., Huang, B., Shi, W., Lu, R., Niu, P., Zhan, F., Ma, X., Wang, D., Xu, W., Wu, G., Gao, G. F., & Tan, W. (2020). A novel coronavirus from patients with pneumonia in China, 2019. *New England Journal of Medicine*. https://doi.org/10.1056/nejmoa2001017.

Zou, H., & Hastie, T. (2005). Regularization and variable selection via the elastic net. *Journal of the Royal Statistical Society. Series B: Statistical Methodology*. https://doi.org/10.1111/j.1467-9868.2005.00503.x.

17 Machine Learning and Behaviour Modification for COVID-19

Charles Oluwaseun Adetunji, Wilson Nwankwo, Akinola Samson Olayinka, and Olaniyan Tope Olugbemi
Edo State University Uzairue

Muhammad Akram and Umme Laila
Government College University Faisalabad

Michael Samuel Olugbenga and Ayomide Michael Oshinjo
Bowen University

Juliana Bunmi Adetunji
Osun State University

Gloria E. Okotie
University of Ibadan

Nwadiuto (Diuto) Esiobu
Florida Atlantic University

CONTENTS

DOI: 10.1201/9781003178903-20

17.1 INTRODUCTION

The prevailing coronavirus disease (christened COVID-19) is associated with the severe acute respiratory syndrome coronavirus 2 (SARS-CoV2) also known as the 2019 coronavirus and the human coronavirus, respectively. Within a matter of months, the disease had spread across most countries of the world (World Health Organization 2020). Most hospitals in Europe and the Americas were overcrowded, most especially at the pandemic's peak, when the outpatient, emergency and intensive care units overextended their capacities and resources (Evans and Armour, 2020).

This novel disease has exerted so much socioeconomic pressure and burdens on many patients, families and healthcare providers necessitating rapid health management decisions amid inadequate information. Typical COVID-19 case varies comparatively and may be asymptomatic. In some cases, accompanying symptoms and signs may include moderate to severe weakness, multi-organ dysfunction, acute respiratory failure and death (Umezurike et al., 2020). In recent times, it has been uncovered that different strains of the virus are emerging uncontrollably, a situation which is downplaying the earlier prowess recorded in vaccine production. The issue of absence of reliable and quality health solution (e.g. Infrastructure, trained and well-motivated healthcare workers, diagnostic kits, medicaments, etc.) is creating unparalleled workload for public health physicians and other healthcare professionals (Adetunji and Anani, 2021a,b,c; Adetunji et al., 2021; Dwivedi et al., 2021; Olaniyan and Adetunji, 2021; Mishra et al., 2020; Thangadurai et al., 2021; Anani et al., 2020; Islam et al., 2021). The result of such unusual workloads has been linked to poor care, which would necessarily aggravate the numbers of COVID-19 infections that further spiral into more pressing demands for critical care, provisions and techniques. These unprecedented challenges have proved overwhelming even in developed jurisdictions such as the United States where thousands of deaths have been recorded within few days. The rate and mode of infection and disease patterns appear not to adhere to any of the popular viral diseases, which have been managed in the past. These paradoxical presentations of the coronavirus disease in various districts necessitate the adoption of techniques, which do not strictly adhere to the traditional stochastic patterns of disease management. Like in many other stochastic and dynamic systems with paradoxical behavioural patterns, Machine Learning (ML) and Artificial Intelligence (AI) have become indispensable in providing useful solutions (Nwankwo and Ukhurebor, 2020; Olayinka et al., 2020); hence, this chapter reviews the applicable roles and contribution of ML to the management of critical care occasioned by the Covid-19 pandemic.

17.2 USEFULNESS OF MACHINE LEARNING IN THE MANAGEMENT OF PANDEMICS

ML finds useful applications in diverse subdomain of healthcare. First, ML may be utilized to provide proper insights into subgroups of patients which ultimately enhance the decision-making process in medicine, and thereby promoting patient-centred care devoid of parlous operational consequences (Albahri et al., 2020; Debnath et al., 2020). Consequently, ML and AI are rapidly evolving as necessary components of the healthcare toolset. For instance, these pervasive technologies constitute major entrants and influencers in constructing modern bioinformatics tools that could be applied for proper investigation. This toolset may involve various programmes that collectively provide self-adjusting (self-propagated intelligence) components that translate biomedical data to unconventional prognostic tools deployed for predictive purposes, amidst other numerous health consequences, and boost clinical resolutions (Churpek and Edelson, 2016; Kipnis et al., 2016; Brekke et al., 2019). Moreover, the dynamics of COVID-19 pandemic necessitates urgent awakening to harness diverse measures towards improving decision-making.

Towards ensuring swift reactions to COVID-19 demands, scientists have been reviewing the existing ML algorithms, evolving new hybrid algorithms, conducting comparative experiments on the precision and accuracy of prediction models, incremental launches, prospective validation, retrospective testing and training (Muhammad et al., 2021; Fernandes et al., 2021; Zoabi et al., 2021). It has been discovered that some clinical results show that ML algorithms could forecast the spread of the infection as well as the prevalence (Muhammed et al., 2021). In addition, it has been documented that some clinical data that entails numerous patient cohort studies and varying results could provide valuable input for modelling intelligent solutions for diagnostic assistance in containing the menace of the disease (Santosh, 2020).

Several scientists have applied ML for the diagnosis and modelling for detection of coronavirus disease particularly in patients with symptoms (Wang et al., 2019; Tomasev et al., 2019; Zampieri et al., 2019; Mousavizadeh and Ghasemi, 2020). Accordingly, the adoption of ML as well as application of training algorithms most importantly to mechanisms directed at exceptional patient population is vital to creating operative prediction models. It has been observed that access to correct information on the outbreaks and the utilization of such to build prediction models have emerged a necessity to enhance insights on the rate of spread and the impairment that the infectious disease might cause on the affected person. Under such circumstances, the legislative authorities and other government agencies may depend on experts who provide and/or understand the workability of some outstanding prediction models that could draw inferences on future developments using epidemiologic datasets particularly when such would influence policy formulation, enunciation and implementation (Remuzzi and Remuzzi, 2020).

Uninterestingly, the scourge of COVID-19 has painfully affected several people around the globe in different dimensions. Many have lost their breadwinners to the dreaded disease. On the other side, many families have been deprived of their sources of sustenance and economic wellbeing owing to excessive and unplanned spending on procuring reliable healthcare for their loved ones caught up in the trap of the disease. The third case is a peculiar situation that has kept the lower and middle classes in our society in a ‘reset’ mode, that is, the protracted lockdowns on businesses and other socioeconomic activities across the world, which inadvertently occasioned severe hardship on the masses across all strata. According to Ivanov (2020), the COVID-19 pandemic has showed a complex and nonlinear nature with its far-reaching complexities. It has been observed that the viral strain is not the same across various regions and districts. It appears there is a kind of mutation that occurs along the line as the disease progresses within a district. The possibility of rapid mutation is ultimately a complicated global emergency in itself, as extensive inquiries would need to be conducted to ascertain the capability of standard models to effectively convey accurate outcomes (Koolhof et al., 2020). Additionally, aside hidden variables that influence the spread and transformations in containment stratagems, the involvement of population-wide comportment in numerous geopolitical regions may melodramatically amplified model vagueness particularly the non-adaptive ones (Darwish et al., 2020). Moreover, it has been observed that typical epidemiological models may lack the capability to convey better outcomes on novel trials.

Therefore, to nip these challenges in the bud, numerous new models have been developed which present prospective results to modelling. For instance, emphasis on the adoption of controlled social behavioural conduct and/or restrictions such as quarantine, social distancing and curfew situations may be used to generate data that could yield more reliable results in future predictions (Rypdal and Sugihara, 2019; Scarpino and Petri, 2019; Zhan et al., 2019). This is consistent with Koike (2009) who posits that an efficient instrumentation could cope with epidemiological dynamics provided the modelling policy is designed with the postulation that reflects the following: infection is through contacts, the susceptibility of various populations to infection (class S), infected (class I) and the removed population (class R) representing the group that have recovered, established immunity and those that have been subjected to isolation or mortality. The assumption is that the infection is transmitted from class I to class S and the amount of likely transmissions is dependent on contacts (Dallas et al., 2019; de Groot and Ogris, 2019; Kelly et al., 2019). A typical strategy that applies ML model as a diagnostic tool for COVID-19 is shown in Figure 17.1.

17.3 PROPOSED CLASSIFICATION-BASED IMAGE METHOD

The deployment of ML for the detection of patients suffering from COVID-19 has proven to be effective in some circumstances. The imaging technique appears very promising in this regard. This new imaging technique is based on polynomial techniques. In a polynomial technique, the role of image moment is very effective as it does involve the use of localized features and global extract (Salah et al., 2020). Image moments represent a feature extraction method in which precise and potentially discriminative features capable of enveloping particular features of some essentially defined pattern are used to differentiate similar as well as completely different entities whether biological or physical. The use of digital projection of image on polynomials orthogonal with the order of friction represents orthogonal moments in the fractional order which would show useful enveloped information from the digital image (Singh and Singh, 2018). In addition, another variant of this technique through which the images could be processed is the orthogonal image moments, which could entail separation and processing of digital images from both high and low orders. This orthogonal moment shows transformation of geometric entity with high significance characteristics for recognition of images. Some of its unique features include cost effectiveness and easy mode of processing data during computation (Qin et al., 2012; Hu et al., 2014; Hosny and Darwish, 2019).

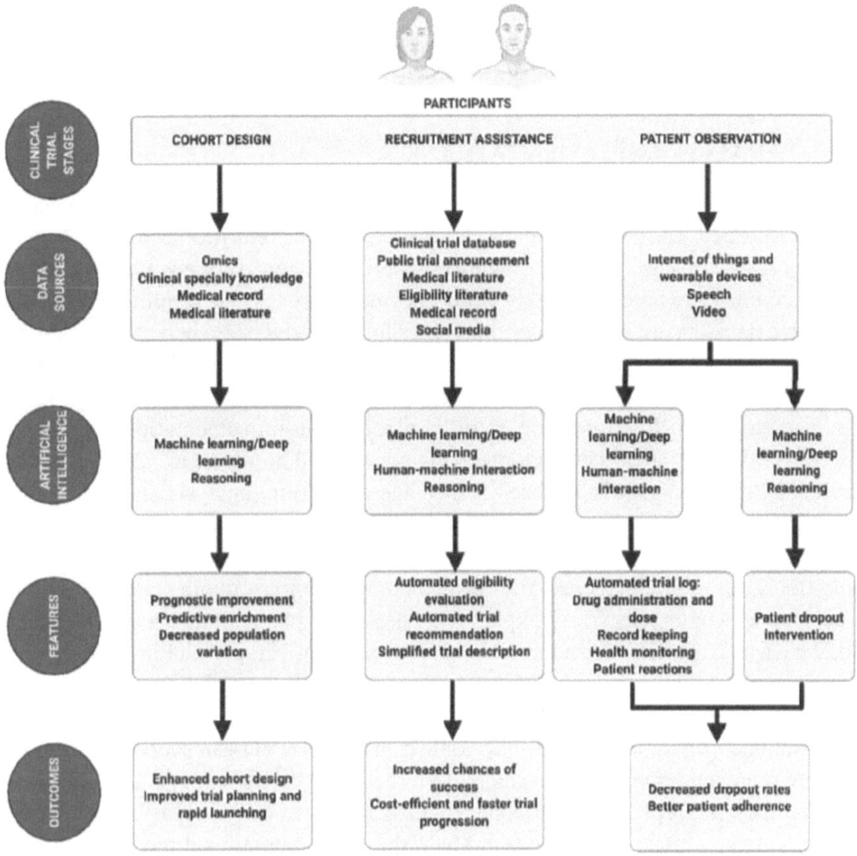

FIGURE 17.1 ML strategy for managing diseases.

17.4 FEATURE EXTRACTION

Generally, feature extraction is a dimensionality reduction strategy wherein we divide and reduce the raw data into manageable groups to ease the processing. Note that a large dataset may inherently contain many variables, which would ultimately necessitate much computing power and resources during processing. Ordinarily, feature extraction would help to isolate the desirable properties from the raw dataset. In some cases, one or more variable may be combined into a single feature. This would greatly reduce the size of the original data as well as the features. Feature extraction significantly introduce the exponential moment. Several scientists have applied various multichannel image processing strategies and some have used colour images as inputs and using the RGB model for extraction and processing (Elaziz et al., 2020; Tuli et al., 2020; Zame et al., 2020). Wang et al. (2015) have demonstrated that that orthogonal circular moment are involved in invariance scaling where the mapping of colour image is present in a circle unit (Debnath et al., 2020). In most cases, the fitting of the unit into domain circle through the interpolation of input image occurs and the scaling variants are computed using frames (Flusser, Suk et al., 2016), and the computation could be achieved through the kernel approaches (Wang et al., 2015).

17.5 PARALLEL IMPLEMENTATIONS

As datasets and algorithms get more complex, they often do require intensive computing resources. Large image datasets collected from real-life scenarios such as the COVID-19 disease spread, diagnostics and treatment prognostications across several countries would absolutely pose enormous processing requirements. Note that image datasets are often larger than other forms of datasets such as textual dataset; hence, the storage, processor capacity and memory resources on any computer may undermine the success of such ML model construction using image processing. To circumvent this limitation, parallel algorithmic implementations are necessary. Several researchers have adopted this approach (Cuomo et al., 2018; Suk and Flusser, 2009). The goal is to accelerate image extraction, analysis and processing. For example, the image feature extraction and image processing of a clinical dataset comprising diagnostic scans of a monitored progressive respiratory disease process using the Zernike moments (usually the orthogonal complex moments or transforms of an image) would effectively exploit the parallel implementation method (Hosny and Darwish, 2018; Khotanzad and Hong, 1990). Moreover, parallel implementation has also been shown to be friendly with other forms of moments where image feature extraction and processing are involved (Cuomo et al., 2018). Parallelism is also important in computation of polar harmonic transforms (PHT) an evolving technique in image representation, analysis and pattern recognition (Liu et al., 2020; Singh and Kaur, 2015). PHT represent orthogonal rotation invariant transformations with more stable features, which may be susceptible to digital computational methods as these features could be easily captured using integer matrices (Miao et al., 2012). This has a greater application in processing and implementation dataset derived from chest x-ray (Hosny and Darwish, 2019).

17.6 PREDICTION MODEL AND PERFORMANCE COMPARISON

ML plays an effective role in the management, analysis and visualization of epidemiological datasets. In recent times, at the centre of these datasets is the COVID-19 disease spread, diagnostics, treatment and prognostic datasets. Several detrimental effects connected with the coronavirus pandemic have been noted globally. From daily experience and statistics, we believe that enormous data on the various aspects of COVID-19 exist though they have not been condensed and structured into a databank. For countries in Europe and the Americas where the incidence and prevalence rates are overwhelmingly high while it has been established thatenormous datasets could drive the building of intelligent and adaptive ML models with relevant algorithms. Comprehensive datasets may also be obtained from the World Health Organization (WHO) and National Centres for Disease Control in different countries. With such daily updates as provided by these organizations, functional and reliable prediction models have been developed. However, it has been reported above, the complex manifestation of the coronavirus genome appears to limit the number of prediction models that could be developed. The implication is that effort should be either channelled to adopting self-adaptive algorithms for the development of relevant models or the creation of hybrid or mixed models that could tilt towards any axis to accommodate the intricacies occasioned by the rapid evolution of variants of the virus. Nevertheless, the interest is not only on the virus but also on the potency and efficacy of the drugs and vaccines used, quality of care and a host of other socioeconomic factors that flow naturally from the care of a given infected patient. In any case, it does not end with building prediction models. These models should be subjected to intensive testing using real-life data from clinics, hospitals, etc. Comparative analysis is very important across board. For instance, it is important to establish the performance of the various models whether a model uses a single, multiple or modified/hybridized model. The outcome is what would be used for evaluating the effectiveness of the ML model (Tuli et al., 2020).

17.7 MACHINE LEARNING METHOD

The rate at which coronavirus is spreading with its attendant economical and several health impairments globally is quite alarming. Several scientists have initiated diverse methods, and there is a continued search for more innovative techniques that could mitigate the high risk of coronavirus pandemic as well as stem up several discoveries that could control the spread. Governments of different countries have closed academic programmes and almost all countries globally now adopt a consistent stiff advocacy for social distancing and other precautionary measures that could lower the transmission of this viral infection. The application of AI and ML is contributing significantly to the management of COVID-19. Some of these techniques entail the role of automatic diagnosis, analysis of route of infection, trend, social analysis and assistance in drug discovery, analysis of spread of infection by using both AI and ML. For the analysis of the spread of infection, ML technologies have been rated as being very effective. It has been observed that ML solutions could support the following:

a. Assist public healthcare workers in high incidence areas to promote residents' social behaviours especially compliance with protocols towards the pandemic;
b. Help decision makers to determine places where socioeconomic activities could resume following any incidence of coronavirus disease outbreak;
c. Where it is a requirement to investigate and profile infectious diseases with emphasis on those factors, which would help in determining the correlation amongst outcomes of COVID-19 care and future prognosis.

The scope of ML is quite robust owing to the availability of a plethora of statistical and numerical algorithms each of which could be applied to one or more problem domains. For instance, multi-variate linear regression is very useful in predicting the mortality rates as well as incidence rates over a given period given disease spread data from a given district (Suzuki and Suzuki, 2020; Pinter et al., 2020; Onovo et al., 2020; Baldé, 2020; Kumar et al., 2020).

17.8 MACHINE LEARNING AND BEHAVIOUR MODIFICATION FOR COVID-19

Tanne et al. (2020) revealed that due to COVID-19 pandemic, healthcare supplies and equipment have drastically reduced globally with unprecedented adverse effects on the economy of many countries. The authors noted that several findings generated from diagnostic imaging such as computer tomography (CT) scan and chest x-ray revealed that large amounts of data can be generated for COVID-19 for adaptation into ML or deep learning algorithms. Bachtiger et al. (2020) reported that huge data on COVID-19 related symptoms are coming from the Republic of China as well as Europe for utilization in AI applications and ML. The authors suggested that training should be given to users of the data to increase their knowledge base especially on ML and AI.

Studies have shown that radiotherapy technology for inner eye scanning and ML device for 3-D heart modelling have received massive attention by many clinical scientists for coronary angiography. Again, the attention of many scientists has been drawn to the risk associated with COVID-19 related ML applications and models, and their potential biases such as accuracy and speed that may not truly reflect human performance or clinical practice. An ML tool may be designed to detect a dysfunction like pulmonary emboli/nodules. The same tool may be altered to detect and screen COVID-19. In ML, it is possible that such an ML prototype would give a better performance than the conventional real time PCR. While evolving an algorithmic tool, the researchers should ensure that the model is designed to accommodate the full spectrum of disease including origin, manifestations, early diagnosis, treatment, prognosis, etc. in both symptomatic and asymptomatic groups. Additional clinical strategy may be incorporated into the design of ML algorithms such as ventilation requirements, mortality rate, recovery rate and intensive care routine, etc. Studies have shown that short phase disease behaviour is a major predictor of long phase outcome. In European countries, massive pools of CT images are pulled in a multi-centre European project into an ML algorithm for research. Public health crisis can effectively be resolved through a broad transformative clinical impact utilizing ML and AI.

Fragkou et al. (2020) revealed that randomized clinical controlled trials are still lacking regardless of the enormous rise in the COVID-19 cases; thus, the authors highlighted all the current clinical trials for the management as well as treatment of coronavirus. Reports have shown that ML algorithms can be utilized for enhancing clinical trials and management of large data during pandemics particularly COVID-19 related complications. Alimadadi et al. (2020) revealed that many collaborative research and calls have been rolled out by the US government to handle the health issues related to COVID-19 through Big data, ML and AI programmes to assist with combating COVID-19. The data would be collected through various sources including Internet of Things (IoT), electronic medical databases, databanks from the centre for disease control and verified published sources. Following the collection, data analysis and model building would be done using various ML algorithms while implementation may be done using appropriate general-purpose system development platforms like python, C# and Java. The constructed ML models would offer intelligent and graphical understanding as to the pathophysiology of the virus, diagnostic approaches, disease progression, prognostications, effective therapeutic methods and recovery projections. Research has shown that advanced ML tools can be utilized for various physiochemical processes like drug discovery, testing, survival prediction of severe COVID-19 patients, genomic analysis of COVID-19, Clustered regularly interspaced short palindromic (CRISPR)-based COVID-19 detection bioassay. Patient's bio-data, physiological characteristics (genetic polymorphism and clinical data) from asymptomatic patients, could be collected, analysed and visualized with ML models. Important forecasts/predictions such as patient's predisposition and vulnerability, therapeutic prowess, drug resistance, therapeutic combinations, treatment regimen, etc. are possible with advanced ML-based informatics software. Similarly, ML-enhanced medical informatics software could be used for detection of infection, automated monitoring patients under clinical management, as well as support the protection of healthcare workers. These systems provide a total quality toolset for boosting healthcare delivery efficiency.

Senior et al. (2020) revealed that there is no proven therapeutics for coronavirus infection management. Currently, many preparations from different countries are undergoing various verification and validation tests. In October 2020, the National Agency for Food Drug Administration and Control (NAFDAC), the principal regulator of drugs in Nigeria, issued a guideline that enlisted procedures to be followed to licencing and accessing preparations on COVID-19 vaccine. The guidelines indicated the essence of pharmacovigilance, surveillance and emergency approvals. With the rising spate of fake drugs and vaccines, we submit that majority of the challenges identified by agencies like NAFDAC, which hampers their success in their operations including surveillance, drug trials, verification, etc., could be resolved using ML solutions. Thus, we suggest that there is need for these agencies to collaborate with research institutions and experts in ML and AI. Globally, these pervasive technologies are currently employed to extract and analyse important parameters from COVID-19 data. The intention is to develop therapeutic products that would contain the various health issues posed by the pandemic. To support such innovative model building and deployments, various levels of clinical data aggregations at the levels of community, districts, zones, states, national and regional levels would be necessary.

Zame et al. (2020) posit that sufficient knowledge on COVID-19 and immunity will help to reduce risks involved in vaccine development. The authors showed that numerous challenges are presented with the adoption of ML for clinical trials in COVID-19 related drug development. Some of the highlighted challenges include repurposing of therapeutic target of COVID-19 and clinical trials for novel drugs for COVID-19 treatment and clinical trials for non-COVID-19 related issues. Yadav et al. (2020) submitted that scientists and some governments have put various measures or models in place during pandemic. They noted that some of the infrastructure included AI and ML solutions. Some of the approaches they highlighted include multi-layered perceptron, neural networks, adaptive network-based fuzzy inference systems, etc. and other notable analytics.

Khan et al. (2020) noted that due to the massive devastating effects of coronavirus, there is urgent need to search for alternative and faster ways to develop drugs and vaccines. The authors suggested that repurposing potential compounds would be the alternative approach to immediately reduce death and introduce novel drugs against COVID-19 disease and the virus through molecular docking for drug discovery. Chellapandi and Saranya (2020) revealed that to date no drug or vaccine have received official approval to combat the pandemic. Thus, more research should be conducted to evaluate the mode of action and the pathogenesis of the disease utilizing bioinformatics and ML for drug repurposing and repositioning strategies against COVID-19. The authors revealed that protease activity, spike physiology and RNA polymerase of the virus can be analysed with ML and computational physiology solutions. Several system physiology, pharmacology and molecular docking methods are currently being investigated for possible drug target against COVID-19 based on the pathophysiology.

Binti Hamzah et al. (2020) revealed that in order to limit coronavirus transmission at community level, different platform models and applications can be deployed such as the Corona Tracker to forecast and predict cases, recoveries, deaths and share public opinions and sentiment on COVID-19 through Susceptible-Exposed-Infectious-Recovered (SEIR) predictive modelling. Individual's behaviour or attitude towards the virus could also be predicted. Albahri et al. (2020) revealed that there has been an increased interest and attention in the development of AI applications towards addressing the pandemic particularly COVID-19. The authors revealed that these platforms and applications are currently not sufficient to tackle the virus as data mining and information sharing network for ML and other models for diagnosing the COVID-19 disease are still evolving.

Naudé (2020) opined that the use of ML and AI against coronavirus infection need urgent revisiting as no significant impact have been felt due to lack of sufficient data that would drive the various modelling processes. Iwendi et al. (2020) demonstrated that integration of ML and AI devices is highly demanding to support forecasts on pandemic occurrences particularly COVID-19 due to the rapidly increasing number of cases and mortality. The authors used an algorithm to forecast the severity of coronavirus cases. They showed that a correlation exists between patient's data and mortality cases. Arulprakash and Aruldoss (2020) revealed that the demand for technological intervention for combating COVID-19 spread in India and other parts of the world is on the increase. The authors observed that ML and AI are presently utilized to support rapid responses to the situation by engaging data analysts and programmers in churning out solutions that could support detection, treatment and diagnostic procedures in radiological imaging thereby improving process efficiency and effectiveness. Kaushik and Raj (2020) demonstrated that due to the havoc wrecked by coronavirus globally, there is urgent need to develop alternative approaches and techniques for evaluating candidate drugs and vaccines which would allow asymptomatic patients to be treated over the pandemic period. They discovered that AI tools and applications could predict drug receptor interaction and disease severity though feasibility and safety must be tested and verified before utilization.

Farhat et al. (2020) revealed that deep learning algorithms have been utilized for imaging analysis particularly medical imaging; therefore machine and deep learning tools can be deployed for diagnosis, detection, classification and segmentation of different tissue damage due to COVID-19. They summarized the current trends in the application of advanced technologies in the understanding of the pathogenesis of COVID-19. Whitelaw et al. (2020) submitted that the early detection, rapid testing, quick surveillance, quarantine and contract tracing are strategies deployed by countries who have recorded significantly reduced number of cases of COVID-19. This is possible through data management and application of digital technologies, which are incorporated into health and government policy programmes. The authors thus provided a summary of the framework in the adoption of digital technologies for rapid response against pandemic particularly COVID-19.

Ou et al. (2020) showed that COVID-19 has exerted some remarkable negative impact on all sectors of the economy, necessitating rapid planning, projections and prediction utilizing ML solutions as an alternative approach to combat the pandemic particularly in the energy sector. Khanday et al. (2020) revealed that technological advancement has had positive footprints on every sector with AI and ML tools showing promising results particularly in the decision-making processes. The authors enunciated the role of ensemble ML algorithms in supporting and improving the traditional approaches. They believe that ML solutions provide high level of accuracy and effectiveness. Ong et al. (2020) submitted that to combat COVID-19 pandemic, drug design and vaccine development must be carried out with the adoption and application of ML algorithms and tools. The authors demonstrated the application of a special tool called Vaxign-ML reverse vaccinology in forecasting the coronavirus drugs candidate and molecular pathways with a potential role against COVID-19. Punn et al. (2020) showed the importance of deep and ML models as tools for combating the transmission and deleterious effects of coronavirus infection. ML solutions could also offer intelligent analytics that enable professionals understand the behaviour of the virus.

Randhawa et al. (2020) posit that COVID-19 pandemic requires special classification and taxonomy categorization. The implication is that ML and digital signal processing could be effectively used for genomic analysis and validation. According to Rustam et al. (2020) decision-making on COVID-19 outcomes could be reasonably improved through ML modelling, analytics and visualization provided relevant datasets are available. They submitted that these predictive tools have demonstrated usefulness in offering insights on results following the implementation of extant policy management and health strategies. However, recent studies show that such technologies are still at their infancy; hence, more studies should be done to evolve more sophisticated solutions that would support various campaigns to contain the menace of the COVID-19 pandemic. On that note, they enlisted effective planning and forecasting as areas of concern for which AI and ML solutions should be tailored. They had reasoned that effective planning and timely prediction would provide real insights on the state of affairs for policy makers, healthcare institutions and the response of the larger society to the pandemic.

17.9 CONCLUSION

In this chapter, we provided detailed discourse on the domain of ML and how this evolving cutting-edge albeit pervasive technology could support the paradigm shift in the management of chaotic or near chaotic health challenges such as that posed by the prevailing COVID-19 disease. We have surveyed various studies on several researches and used case studies including proposals, sociological, experimental and pragmatic submissions. From these studies, we have established a position that ML and its variants are critical tools that could be integrated into appropriate healthcare delivery arsenal to drive the global fight against the menace of the coronavirus disease. Conclusively, ML when integrated into the healthcare delivery architecture would affect the behaviour of government machineries especially the hospitals, healthcare professionals across their various ranks, as well as all stakeholders in the society. This is because not only would the new architecture involve a new stream of professionals (data scientists, programmers and medical sociologists), but would also require a significant adjustment in the process of diagnosis treatment and post-treatment follow-ups.

REFERENCES

Adetunji, C.O., Anani, O.A., and Panpatte, D. (2021) Mechanism of actions involved in sustainable ecorestoration of petroleum hydrocarbons polluted soil by the beneficial microorganism. In: Panpatte, D.G., and Jhala, Y.K. (eds.), *Microbial Rejuvenation of Polluted Environment. Microorganisms for Sustainability*, vol 26. Springer, Singapore. https://doi.org/10.1007/978-981-15-7455-9_8

Adetunji, C.O., and Anani, O.A. (2021a) Recent advances in the application of genetically engineered microorganisms for microbial rejuvenation of contaminated environment. In: Adetunji, C.O., Panpatte, D.G., and Jhala, Y.K. (eds.), *Microbial Rejuvenation of Polluted Environment. Microorganisms for Sustainability*, vol 27. Springer, Singapore. https://doi.org/10.1007/978-981-15-7459-7_14

Adetunji, C.O., and Anani, O.A. (2021b) Utilization of microbial biofilm for the biotransformation and bioremediation of heavily polluted environment. In: Panpatte, D.G., and Jhala, Y.K. (eds.), *Microbial Rejuvenation of Polluted Environment. Microorganisms for Sustainability*, vol 25. Springer, Singapore. https://doi.org/10.1007/978-981-15-7447-4_9

Adetunji C.O., and Anani O.A. (2021c) Bioremediation of polythene and plastics using beneficial microorganisms. In: Adetunji C.O., Panpatte D.G., and Jhala, Y.K. (eds.), *Microbial Rejuvenation of Polluted Environment. Microorganisms for Sustainability*, vol 27. Springer, Singapore. https://doi.org/10.1007/978-981-15-7459-7_13

Anani, O.A., Mishra, R.R., Mishra, P., Enuneku A.A., Anani G.A., and Adetunji, C.O. (2020) Effects of toxicant from pesticides on food security: Current developments. In: Mishra P., Mishra R.R., and Adetunji C.O. (eds.), *Innovations in Food Technology*. Springer, Singapore. https://doi.org/10.1007/978-981-15-6121-4_22

Akyol, K. and Karaci, A. (2019) Predicting the critical temperature of super conductors with XGBoost. *UMYMK ICAIAME* 2019: 72–76.

Alimadadi, A., Aryal, S., Manandhar, I., Munroe, P. B., Joe, B. and Cheng, X. (2020) Artificial intelligence and machine learning to fight COVID-19. *Physiological Genomics* 52: 200–202, doi:10.1152/physiolgenomics.00029.

Albahri, A. S., Hamid, R. A., Alwan, J. K., Al-qays, Z. T., Zaidan, A. A., Zaidan, B. B., Albahri, A. O. S., AlAmoodi, A. H., Khlaf, J. M., Almahdi, E. M., Thabet, E., Hadi, S. M., Mohammed, K. I., Alsalem, M. A., Al-Obaidi, J. R. and Madhloom, H. T. (2020) Role of biological data mining and machine learning techniques in detecting and diagnosing the novel coronavirus (COVID-19): A systematic review. *Journal of Medical Systems* 44:122, https://doi.org/10.1007/s10916-020-01582-x.

Arulprakash, E. and Aruldoss, M. (2020) A study on fight against COVID 19 from latest technological intervention. *SN Computer Science* 1:277, https://doi.org/10..1007/s42979-020-00301-0.

Baldé, M. A. M. T. (2020). Fitting SIR model to COVID-19 pandemic data and comparative forecasting with machine learning. *medRxiv*. https://doi.org/10.1101/2020.04.26.20081042.

Bachtiger, P., Peters, N. and Walsh, S. L. F. (2020). https://doi.org/10.1016/ S2589-7500(20)30162-X. Vol 2. 391–392. www.thelancet.com/digital-health.

Binti Hamzah, F. A., Lau, C. H., Nazri, H., Ligot, D. V., Lee, G., Tan, C. L., Shaib, M. K. B. M., Binti Zaidon, U. H., Binti Abdullah, A., Chung, M. H., Ong, C. H., Chew, P. Y. and Salunga, R. E. (2020). CoronaTracker: World-wide COVID-19 outbreak data analysis and prediction. *Bulletin World Health Organ.* http://dx.doi.org/10.2471/ BLT.20.251561.

Brekke, I. J., Puntervoll, L. H., Pedersen, P. B., Kellett, J. and Brabrand, M. (2019). The value of vital sign trends in predicting and monitoring clinical deterioration: a systematic review. *PLoS One* 14(1):e0210875, doi: 10.1371/journal.pone.0210875.

Chellapandi, P. and Saranya, S. (2020). Genomics insights of SARS-CoV-2 (COVID-19) into target-based drug discovery. *Medicinal Chemistry Research*, https://doi.org/10.1007/ s00044-020-02610-8.

Churpek, M. M. and Edelson, D. P. (2016). Moving beyond single-parameter early warning scores for rapid response system activation. *Critical Care Medicine* 44:2283–2285, doi: 10.1097/CCM.0000000000002105.

Dallas, T.A., Carlson, C. J. and Poisot, T. (2019). Testing predictability of disease outbreaks with a simple model of pathogen biogeography. *Royal Society Open Science* 6, doi:10.1098/rsos.190883.

Darwish, A., Rahhal, Y. and Jafar, A. A. (2020) Comparative study on predicting influenza outbreaks using different feature spaces: Application of influenza-like illness data from Early Warning Alert and Response System in Syria. *BMC Research Notes* 13: 33, doi:10.1186/s13104-020-4889-5.

de Groot, M. and Ogris, N. (2019) Short-term forecasting of bark beetle outbreaks on two economically important conifer tree species. *Forestry Ecology and Management* 450, doi:10.1016/j.foreco.2019.117495.

Dwivedi, N., Dwivedi, S., and Adetunji, C. O. (2021) Efficacy of microorganisms in the removal of toxic materials from industrial effluents. In: Adetunji C.O., Panpatte D.G., Jhala Y.K. (eds.), *Microbial Rejuvenation of Polluted Environment. Microorganisms for Sustainability*, vol 27. Springer, Singapore. https://doi.org/10.1007/978-981-15-7459-7_15

Debnath, S., Barnaby, D. P., Coppa, K., Makhnevich, A., Kim, E. J., Chatterjee, S., Tóth, V., Levy, T. J., Paradis, M. D., Cohen, S. L., Hirsch, J. S., Zanos, T. P. and Northwell COVID-19 Research Consortium (2020). Machine learning to assist clinical decision-making during the COVID-19 pandemic. *Bioelectronic Medicine* 6: 14. https://doi.org/10.1186/s42234-020-00050-8.

Debnath, S., D. P. Barnaby, K. Coppa, A. Makhnevich, E. J. Kim, S. Chatterjee, V. Tóth, T. J. Levy, M. d Paradis and S. L. Cohen (2020). Machine learning to assist clinical decision-making during the COVID-19 pandemic. *Bioelectronic Medicine* 6(1): 1–8.

Deng, A.-W., Wei C.-H, and Gwo, C.-Y, (2015). Algorithms for computing zernike moments and image reconstruction in parallel process. *2nd International Conference on Information Science and Control Engineering, IEEE.*

Elaziz, M. A., Hosny, K. M., Salah, A., Darwish, M. M., Lu, S. and Sahlol, A. T. (2020). New machine learning method for image-based diagnosis of COVID-19. *Plos One* 15(6): e0235187.

Evans, M. and Armour, S. (2020) Hospital capacity crosses tipping point in U.S. coronavirus hot spots. *The Wall Street Journal*. New York: Dow Jones & Company, Inc.

Fernandes, F.T., de Oliveira, T.A., Teixeira, C.E., Teixeira, C. E., Batista, A. F. De M., Costa, G. D., Filho, A. D. P. C. (2021) A multipurpose machine learning approach to predict COVID-19 negative prognosis in São Paulo, Brazilian Scientific Reports 11, 3343. https://doi.org/10.1038/s41598-021-82885-y.

Flusser, J., Suk, T. and Zitová, B. (2016). *2D and 3D Image Analysis by Moments*, John Wiley & Sons.

Fragkou, P. C., Belhadi, D., Peiffer-Smadja, N., Moschopoulos, C. D., Lescure, F. X., Janocha, H., Karofylakis, E., Yazdanpanah, Y., Mentre, F., Skevaki, C., Laouenan, C. and Tsiodras, S. (2020) Review of trials currently testing treatment and prevention of COVID-19 on behalf of the ESCMID Study Group for Respiratory Viruses. *Clinical Microbiology and Infection* 26: 988–998.

Farhat, H., Sakr, G. E. and Kilany, R. (2020) Deep learning applications in pulmonary medical imaging: recent updates and insights on COVID-19. *Machine Vision and Applications* 31:53 https://doi.org/10.1007/s00138-020-01101-5.

Hosny, K. M. and Darwish, M. M. (2018). New set of quaternion moments for color images representation and recognition. *Journal of Mathematical Imaging and Vision* 60(5): 717–736.

Hosny, K. M. and Darwish, M. M. (2019). A Kernel-Based method for Fast and accurate computation of PHT in polar coordinates. *Journal of Real-Time Image Processing* 16(4): 1235–1247.

Hosny, K. M. and Darwish, M. M. (2019). New set of multi-channel orthogonal moments for color image representation and recognition. *Pattern Recognition* 88: 153–173.

Hu, H.- t., Zhang, Y.- d., Shao, C. and Ju, Q. (2014). Orthogonal moments based on exponent functions: Exponent-Fourier moments. *Pattern Recognition* 47(8): 2596–2606.

Ivanov, D. (2020) Predicting the impacts of epidemic outbreaks on global supply chains: A simulation-based analysis on the coronavirus outbreak (COVID-19/SARS-CoV-2) case. *Transportation Research. Part E: Logistics and Transportation Review* 136, doi:10.1016/j.tre.2020.101922.

Islam, S., Thangadurai, D., Adetunji, C. O., Nwankwo, W., Kadiri, O., Makinde, S., Michael, O. S., Anani, O. A., and Adetunji, J. B. (2021) Nanomaterials and nanocoatings for alternative antimicrobial therapy. In: Kharissova O.V., Martínez L.M.T., Kharisov B.I. (eds.), *Handbook of Nanomaterials and Nanocomposites for Energy and Environmental Applications*. Springer, Cham. https://doi.org/10.1007/978-3-030-11155-7_3-1

Iwendi, C., Bashir, A. K., Peshkar, A., Sujatha, R., Chatterjee, J. M., Pasupuleti, S., Mishra, R., Pillai, S. and Jo, O. (2020). COVID-19 Patient Health Prediction Using Boosted Random Forest Algorithm. *Frontier Public Health* 8:357, doi: 10.3389/fpubh.2020.00357.

Kaushik, A. C. and Raj, U. (2020) AI-driven drug discovery: A boon against COVID-19? *AI Open* 1: 1–4.

Kelly, J. D., Park, J., Harrigan, R. J., Hoff, N. A., Lee, S. D. Wannier, R., Selo, B., Mossoko, M., Njoloko, B., Okitolonda-Wemakoy, E., Mbala-Kingebeni, P., Rutherford, G. W., Smith, T. B., Ahuka-Mundeke, S., Muyembe-Tamfum, J. J., Rimon, A. W. and Schoenberg, F. P. (2019). Real-time predictions of the 2018–2019 Ebola virus disease outbreak in the Democratic Republic of the Congo using Hawkes point process models. *Epidemics* 28, doi:10.1016/j.epidem.2019.100354.

Khanday, A. M. U. D., Rabani, S. T., Khan, Q. R., Rouf, N. and Din, M. M. U. (2020). Machine learning based approaches for detecting COVID-19 using clinical text data. *International Journal of Information Technology*, https://doi.org/10.1007/s41870-020-00495-9.

Khan, A., Mahmud, S., Rubayet Ul Alam, A. S. M., Ekhtiar Rahman, Md., Ahmed, F. and Rahmatullah. A. (2020). Comparative molecular investigation of the potential inhibitors against SARSCoV-2 main protease: A molecular docking study Md. *Journal of Biomolecular Structure and Dynamics*, https://doi.org/10.1080/07391102.2020.1796813.

Kipnis, P., Turk, B. J., Wulf, D. A., LaGuardia, J. C., Liu, V., Churpek, M. M., Romero-Brufau, S. and Escobar, G. J. (2016). Development and validation of an electronic medical record-based alert score for detection of inpatient deterioration outside the ICU. *Journal of Biomedical Informatics* 64:10–19, doi: 10.1016/j.jbi.2016.09.013.

Koike, F., Morimoto, N. (2018). Supervised forecasting of the range expansion of novel non-indigenous organisms: Alien pest organisms and the 2009 H1N1 flu pandemic. *Global Ecology and Biogeography* 27: 991–1000, doi:10.1111/geb.12754.

Koolhof, I. S., Gibney, K. B., Bettiol, S., Charleston, M., Wiethoelter, A., Arnold, A. L., Campbell, P.T., Neville, P. J., Aung, P., Shiga, T., Caver, S., and Firestone, S. M. (2020). The forecasting of dynamical Ross River virus outbreaks: Victoria, Australia. *Epidemics* 30, doi:10.1016/j.epidem.2019.100377.

Kumar, A., F. M. Khan, R. Gupta and H. Puppala (2020). Preparedness and Mitigation by projecting the risk against COVID-19 transmission using Machine Learning Techniques. *medRxiv*.

Ledford, H. (2020). Coronavirus reinfections: three questions scientists are asking. *Nature* 585 (7824): 168–169, doi:10.1038/d41586-020-02506-y.

Mishra, P., Mishra, R. R., & Adetunji, C. O. (Eds.). (2020). *Innovations in Food Technology: Current Perspectives and Future Goals.* Springer Nature Singapore Pte Ltd. 2020. Springer, Singapore. Pp. 143-162. 1st Edition. DOI: 10.1007/978-981-15-6121-4

Mousavizadeh, L. and Ghasemi, S. (2020). Genotype and phenotype of COVID-19: Their roles in pathogenesis. *Journal of Microbiology, Immunology and Infection*, https://doi.org/10.1016/j.jmii.2020.03.022.

Muhammad, L. J., Algehyne, E. A., Usman, S. S., Ahmad, A., Chakraborty, C., and Mohammed, I. A. (2021). Supervised Machine Learning Models for Prediction of COVID-19 Infection using Epidemiology Dataset. *SN Computer Science* 2: 11, https://doi.org/10.1007/s42979-020-00394-7.

Naudé, W. (2020). Artificial intelligence vs COVID-19: Limitations, constraints and pitfalls. *AI & Society* 1–5. Advance online publication. https://doi.org/10.1007/s00146-020-00978-0.

Nwankwo, W. and Ukhurebor, K.E. (2020). Web forum and social media: A model for automatic removal of fake media using multilayered neural networks. *International Journal of Scientific and Technology Research* 9(1).

Olayinka, A. S., Nwankwo, W., and Olayinka, T. C. (2020). Model based machine learning approach to predict thermoelectric figure of merit. *Archive of Science and Technology* 1(1).

Ong, E., Wong, M. U., Huffman, A and He, Y. (2020). COVID-19 coronavirus vaccine design using reverse vaccinology and machine learning. *Frontiers in Immunology* 11:1581, doi: 10.3389/fimmu.2020.01581.

Onovo, A., Atobatele, A., Kalaiwo, A., Obanubi, C., James, E., Gado, P., Odezugo, G., Ogundehin, D., Magaji, D. and Russell, M. (2020). Using supervised machine learning and empirical Bayesian Kriging to reveal Correlates and Patterns of COVID-19 Disease outbreak in sub-Saharan Africa: Exploratory Data Analysis. Available at SSRN 3580721.

Ou, S., He, X., Ji, W., Chen, W., Sui, L., Gan, Y., Lu, Z., Lin, Z., Deng, S., Przesmitzki, S. and Bouchard, J. (2020). Machine learning model to project the impact of COVID-19 on US motor gasoline demand. *Nature Energy* 5: 666–673.

Olaniyan, O.T., and Adetunji, C.O. (2021) Biochemical role of beneficial microorganisms: An overview on recent development in environmental and agro science. In: Adetunji, C.O., Panpatte, D.G., and Jhala, Y.K. (eds.), *Microbial Rejuvenation of Polluted Environment. Microorganisms for Sustainability*, vol 27. Springer, Singapore. https://doi.org/10.1007/978-981-15-7459-7_2

Pinter, G., Felde, I., Mosavi, A., Ghamisi, P. and Gloaguen, R. (2020). COVID-19 Pandemic prediction for hungary; a hybrid machine learning approach. *Mathematics* 8(6): 890.

Punn, N. S., Sonbhadra, S. K. and Agarwal, S. (2020). COVID-19 Epidemic analysis using machine learning and deep learning algorithms. doi: https://doi.org/10.1101/2020.04.08.20057679.

Qin, H., Qin, L., Xue, L. and Yu, C. (2012). A parallel recurrence method for the fast computation of Zernike moments. *Applied Mathematics and Computation* 219(4): 1549–1561.

Randhawa, G. S, Soltysiak, M. P. M., El Roz, H., de Souza, C. P. E., Hill, K. A, and Kari, L (2020). Machine learning using intrinsic genomic signatures for rapid classification of novel pathogens: COVID-19 case study. *PLoS ONE* 15(4): e0232391, https://doi.org/10.1371/journal.pone.0232391.

Remuzzi, A. and Remuzzi, G. (2020). COVID-19 and Italy: What next? *Lancet.*

Rustam, F., Reshi, A. A., Mehmood, K., Ullah, S., On, B.-W., Aslam, W. and Choi, G. S. (2020). COVID-19 Future forecasting using supervised machine learning models. *IEEE Access* 8: 101489–101499. Digital Object Identifier 10.1109/ACCESS.2020.2997311.

Rypdal, M. and Sugihara, G. (2019). Inter-outbreak stability reflects the size of the susceptible pool and forecasts magnitudes of seasonal epidemics. *Nature Communications* 10, doi:10.1038/s41467-019-10099-y.

Salah, A., K. Li, K. M. Hosny, M. M. Darwish and Q. Tian (2020). Accelerated CPU–GPUs implementations for quaternion polar harmonic transform of color images. *Future Generation Computer Systems* 107: 368–382.

Santosh, K.C. (2020). AI-driven tools for coronavirus outbreak: need of active learning and cross-population train/test models on multitudinal/multimodal data. *Journal of Medical Systems* 44(5):93, doi: 10.1007/s10916-020-01562-1.

Scarpino, S.V., Petri, G. (2019). On the predictability of infectious disease outbreaks. *Nature Communication* 10, doi:10.1038/s41467-019-08616-0.

Senior, A.W., Evans, R., Jumper, J., Kirkpatrick, J., Sifre, L., Green, T., Qin, C., Žídek, A., Nelson, A.W.R., Bridgland, A., Penedones, H., Petersen, S., Simonyan, K., Crossan, S., Kohli, P., Jones, D.T., Silver, D., Kavukcuoglu, K., Hassabis, D. (2020). Improved protein structure prediction using potentials from deep learning. *Nature* 577: 706–710, doi:10.1038/s41586-019-1923-7.

Singh, C. and Singh, J. (2018). Multi-channel versus quaternion orthogonal rotation invariant moments for color image representation. *Digital Signal Processing* 78: 376–392.

Suk, T. and Flusser, J. (2009). Affine moment invariants of color images. *International Conference on Computer Analysis of Images and Patterns*, Springer.

Suzuki, Y. and Suzuki, A. (2020). Machine learning model estimating number of COVID-19 infection cases over coming 24 days in every province of South Korea (XGBoost and MultiOutputRegressor). *medRxiv.*

Tanne, J. H., Hayasaki, E., Zastrow, M., Pulla, P., Smith, P. and Garcia Rada, A. (2020). Covid-19: how doctors and healthcare systems are tackling coronavirus worldwide. *BMJ* 368: m1090.

Tomasev, N., Glorot, X., Rae, J. W., Zielinski, M., Askham, H., Saraiva, A., et al. (2019). A clinically applicable approach to continuous prediction of future acute kidney injury. *Nature* 572(7767):116–119, doi: 10.1038/s41586-019-1390-1.

Tuli, S., Tuli, S., Tuli, R. and Gill, S. S. (2020). Predicting the growth and trend of COVID-19 pandemic using machine learning and cloud computing. *Internet of Things* 100222.

Thangadurai, D., Dabire, S. S., Sangeetha, J., Said Al-Tawaha, A. R. M., Adetunji, C. O., Islam, S., Shettar, A. K., David, M., Hospet, R., and Adetunji, J. B. (2021) Greener composites from plant fibers: Preparation, structure, and properties. In: Kharissova, O.V., Martínez, L.M.T., and Kharisov, B.I. (eds.), *Handbook of Nanomaterials and Nanocomposites for Energy and Environmental Applications.* Springer, Cham. https://doi.org/10.1007/978-3-030-11155-7_21-1

Umezuruike, C., Nwankwo, W., Tibenderana, P., John Patrick Assimwe, Muhirwa Ronald (2020). Corona virus disease (COVID 19): Analysis and design of an alert and real-time tracking system. *International Journal of Emerging Trends in Engineering Research* 8(5).

Wang, S., Kang, B., Ma, J., Zeng, X., Xiao, M., Guo, J., Cai, M., Yang, J., Li, Y., Meng. X. And Xu, B. O. (2020). A deep learning algorithm using CT images to screen for Corona virus disease (COVID-19). *medRxiv* 2020.02.14.20023028.

Wang, X.-y., Li, W.-y., Yang, H.-y., Wang, P. and Li, Y.-w. (2015). Quaternion polar complex exponential transform for invariant color image description. *Applied Mathematics and Computation* 256: 951–967.

Whitelaw, S., Mamas, M. A., Topol, E. and van Spall, H. G. C. (2020). Applications of digital technology in COVID-19 pandemic planning and response. thelancet.com/digital-health 2: 435–440.

World Health Organization (WHO). (2020). Director-General's opening remarks at the media briefing on COVID-19–20 March 2020. Geneva: World Health Organization.

Yadav, D., Chhabra, D., Gupta, R. K., Phogat, A. and Ahlawat, A. (2020). Modeling and analysis of significant process parameters of FDM 3D printer using ANFIS. *Materials Today: Proceedings* 21: 1592–1604.

Zame, W. R., Bica, I., Shen, C., Curth, A., Lee, H.-S., Bailey, S., Weatherall, J., Wright, D., Bretz, F. and van der Schaar, M. (2020). Machine learning for clinical trials in the era of COVID-19. *Statistics in Biopharmaceutical Research* 1–20, doi: 10.1080/19466315.2020.1797867.

Zampieri, F. G., Costa, E. L., Iwashyna, T. J., Carvalho, C. R. R., Damiani, L. P., Taniguchi, L. U., Amato, M. B. P. and Cavalcanti, A. B. (2019). Heterogeneous effects of alveolar recruitment in acute respiratory distress syndrome: a machine learning reanalysis of the alveolar recruitment for acute respiratory distress syndrome trial. *British Journal of Anaesthesia* 123(1):88–95, doi: 10.1016/j.bja.2019.02.026.

Zhan, Z., Dong, W., Lu, Y., Yang, P., Wang, Q. and Jia, P. (2019). Real-time forecasting of hand-foot-and-mouth disease outbreaks using the integrating compartment model and assimilation filtering. *Scientific Reports* 9, doi:10.1038/s41598-019-38930-y.

Zoabi, Y., Deri-Rozov, S. and Shomron, N. (2021). Machine learning-based prediction of COVID-19 diagnosis based on symptoms. *npj Digital Medicine* 4(3). https://doi.org/10.1038/s41746-020-00372-6.

Part 4

Working Group 12

Medicinal and Underutilized Plants

Handling Editor: Prof. Abubakar Gidado

18 Underutilized Plant Species

A Huge Potential for Economic Development in Nigeria

Friday Nwalo Nweke
Alex Ekwueme Federal University

Godwill Azeh Engwa
Godfrey Okoye University

Oluchi Cynthia Abiamere
National Biotechnology Development Agency (NABDA)

CONTENTS

DOI: 10.1201/9781003178903-22

18.1 INTRODUCTION

Nigeria is a country blessed with abundant natural resources, conducive environment, and green pastures. Among all these wonders of nature are loads of plants and animal species. It is no longer news that a lot of these plant species in Nigeria have been underutilized. Presently, the rate of reliance on a few plant species globally is alarming. Different species, especially those known as 'underutilized,' can meaningfully contribute toward better nutrition and health.

The economy of Nigeria is a mixed one, comprising different sectors, middle income earners, and so on. When we focus on the utilization of some group of plant species, there could be some adverse effects on the economy, development, ecosystem, and the environment at large. Many communities today suffer problems of food and nutrition security including sustainability of environment as a result of much dependence on a particular plant classes. Harnessing unexploited and underutilized plant species provides huge potential that contributes toward the attainment of Sustainable Development Goals, principally in fighting hidden hunger and providing pharmaceuticals and income giving opportunities. They are related to ethnic backgrounds, and play an essential part in supporting different cultures and lifestyle.

18.1.1 Underutilized Plant Species

According to Anon. (1986), underutilized plants are those species which are generally underexploited but which have capacity to add to food production, healthcare, money generation, and conservation of the environment. They have inherent potential to contribute to socioeconomic and industrial development and poverty alleviation. For many centuries, most of these underused crops have served as food, fiber, and medicines, but their importance can be reduced over time. They are also called neglected crops. Plant species that are not classified as major staple crops are usually described as 'Neglected and Underutilized Species' (NUS) and are occasionally called 'orphan' crops.

The Northern region of Nigeria has 39 endemic species, while the West and Central parts of the country have 38. It has been estimated that the highest degree of endemism occurs in low land forests of the South-Eastern part which has 128 endemic species (NEST, 1992). The underutilized industrial plants in Nigeria involve so much species with old-style values as food, medicine, and for various native uses. A number of these plants are compiled in definite parts of the country (Aju, 2010).

18.1.2 Economic Development in Nigeria

Development is all about making people to have their freedom and avoid obstacles in life. Economic development is a broader concept than economic growth. A nation's economic growth is usually based on increase in citizen's quality of life. Economic development means increasing actual income per head – GDP per capita, with increase in the reading ability level and standard of education, enhanced quality of housing, life expectancy, and environmental standards. Development relieves people from low-living standards to proper occupation with appropriate accommodation.

Nigeria is a developing country with 35% of the population living under the poverty level. The economy relies much on agriculture. Nigeria as a nation is still contending with economic challenges like unemployment, shortage of food, burden of external debt, and bad infrastructures irrespective of her huge nonrenewable and renewable resources. The problems of refining one sector or province at the expense of another as well as presenting policy alterations coupled with inadequacies in resource sharing to benefit a particular assemblage are Nigeria's developmental challenges. These led to increased poverty for some people in the past. However, if social service-oriented and growth policies was adopted, it would have enabled all Nigeria's occupants to improve their welfare.

Regardless of Nigeria's momentary economic hindrances, the country needs ample investment in social capital, human, and physical infrastructure. However, to change the economic situation of the country, we should intensify effort on hi-tech drive in the activities of the economy, and have a long-term plan on global economic development.

18.1.3 Underutilized Plant Species' Contribution to the Sustainable Development Goals

The short- and medium-term guidelines for research and development were provided by the UN Millennium Development Goals (MDGs5), while the International Plant Genetic Resources Institute (IPGRI), in conjunction with the M. S. Swaminathan Research Foundation (MSSRF) and the Global Facilitation Unit for Underused Species, revealed the contributions of underutilized plants toward achieving the MDGs, and the vital role of biodiversity in solving global poverty and hunger problems. It was adopted that stopping hunger and malnutrition is needful to meeting other MDGs (FAO, 2005).

The reduction in hunger that exists nationwide is a great deed. Nevertheless, we have some dietary concerns that needed to be addressed. The issue of hidden hunger or malnutrition resulting from deficiency of micronutrients and essential dietary components has been highlighted (Frison, 2004). Over 120 million teenagers are affected by vitamin A deficiency and other related diseases. Iron deficiency has affected majority of the world's population leading to anemia. The dietary supplements that could combat these deficiencies cannot be afforded by the poor people. Meanwhile, sicknesses like obesity and diabetes are rampant in developing countries due to poor diets.

The MDG's challenge is to stop hunger by 2015 and also to reduce hidden hunger through nutritional diversification. Several underutilized crops can play a central role in providing the solution in both developing countries as well as high-income countries. Similarly, most of the rural population in African and Asian countries uses traditional medicinal plants to treat diseases. Developing underutilized plants offer openings for more viable production methods and producing standardized and safe drugs from these plants, thereby addressing MDGs 4 and 6.

18.1.4 Examples of Underutilized Plant Species of Great Economic Value

There are different classes of underutilized plant species, ranging from cereals, vegetables, legumes, fruits and nuts, root and tubers, etc. (Figures 18.1 and 18.2). Some underutilized vegetables are mentioned below.

18.1.4.1 *Talinum triangulare* (Water Leaf)

Water leaf is of Portulacaceae, a lenient mucilaginous leafy vegetable found in the tropics. It contains lutein and zeaxanthin as its carotenoids that act like a stimulant to improve the immune cells of the eyes. Eating this vegetable controls heart infections, blood stress, and cholesterol level. It also prevents diverticulosis and vision loss (Fasuyi, 2005). The plant is rich in minerals and amino acids and propagated mainly by cutting/division.

18.1.4.2 *Mucuna pruriens* (Velvet Bean)

Mucuna pruriens is a legume vegetable used by people of Nagaland. It is most times taken as substitute to coffee and it is dispersed by seed. Its quotations can be used as a poison antagonist for various snake bites. It has high protein content (23%–35%) together with digestibility when compared to pulses such as soybean and Lima bean (Gurumoorthi et al., 2003).

18.1.4.3 *Ipomoea aquatica* (Water Spinach)

Water spinach is an herbaceous aquatic or semi-aquatic trailing type with leaves used as vegetables in sauces and soups in some part of the world. It is propagated by seed and cutting. The plant is useful for the management of diabetes and liver problems. In Indonesia, it is used as a sleeping tablet by anemia patients due to high iron content. It can serve as scorpion venom antidote (Malalavidhane et al., 2001).

18.1.4.4 Amaranth

Amaranth has high food value than spinach and cabbage. It is used for consumption and commercial production in Africa with good market potential. It is produced with small inputs and capital risk for small-scale farmers. Based on the inter- and intraspecific diversity of cultivated amaranth, the crop is ideal for food production and nutrition security. Commercial seed companies know this market potential and include amaranth in their product portfolio.

18.1.4.5 Drumstick Tree

The *Moringa oleifera* is a grown species of the Moringaceae family widely grown in the tropics of Asia, Latin America, and sub-Saharan Africa.

18.1.4.5.1 Food Uses

The parts like leaves are used as salad and vegetables in soups and sauces with the fried seed. The roots are used as a condiment. The dried powder leaves are used as diet supplement for children and expectant women. It is added to soybean and groundnut paste to form an energy-dense supplemental food called ready-to-use food (RUF) which is used in the treatment of malnutrition.

18.1.4.5.2 Nutritional Content

The plant is rich in critical micronutrients, vitamins, and antioxidants. It can be grown easily with good processing properties and palatability. Maintaining moringa leaves at low temperature stabilizes nutrients and phytochemicals, while boiling the fresh leaves in water increases the bioavailable of iron.

18.1.4.5.3 Medicinal Uses

The crop is important to human health due to high content of reactive components like glucosinolates and isothiocyanates. These compounds inhibit mitosis and stimulate apoptosis. The plant extracts and porridges have antibiotic, anti-inflammatory, and hypoglycemic activities. In HIV/AIDS treatment, the powder can be used as an immune stimulant, while leaves and roots can be used for the treatment of abdominal tumors.

18.1.4.5.4 Agronomic and Horticultural Uses

Moringa is planted as a component of agroforestry systems for maintainable vegetable production. It can be intercropped with bean (*Cyamopsis tetragonoloba*), hot pepper (*Capsicum* spp.), cowpea (*Vigna unguiculata*), and onion (*Allium cepa*). The crop can be used as a livelihood for climbers such as yams (*Dioscorea* spp.) and beans (*Phaseolus* spp.). The leaves and twigs also serve as forage for animals.

18.1.4.5.5 Industrial Uses

The seeds contain 40% ben oil that fights rancidity, and good for cookery, lubrication, and as well as in the cosmetic industry. It is used in the production of biofuel due to its high octane number of 67, being highest among biodiesel fuels. The fiber of the trunk is appropriate for producing mats, cordage, and paper.

18.1.5 Potentials and Importance of Underutilized Plants

18.1.5.1 Food Security and Better Nutrition

Underutilized and neglected vegetables have low-input adaptation to agriculture and rich in nutrients. The erosion of these species can set consequences on food security and living standard of the poor, while their utilization gives better nutrition (Figure 18.4). Many unused vegetables have more pro-vitamin A and vitamin C than widely available commercial types. Utilization of neglected and underutilized vegetables is a good way to maintain a healthy diet and reduce micronutrient deficiencies like 'hidden hunger,' and other diet-related problems among vulnerable social groups in developing countries. They supplement diet-based staple crops through provision of some vitamins and minerals.

WW Examples of underutilized crops in Nigeria.

(Source: www.researchgate.net.)

18.1.5.2 Ecosystem Stability

There is a growing concern about crops that cannot survive in harsh environments like degraded vegetation, drought due to climate change, and land degradation. These species adapt better to stress and difficult environments, thereby playing a strategic role in the maintenance of diversity and stable environment. These species are maintained in niches by communities in crumbly ecosystems affected by salinization and desertification.

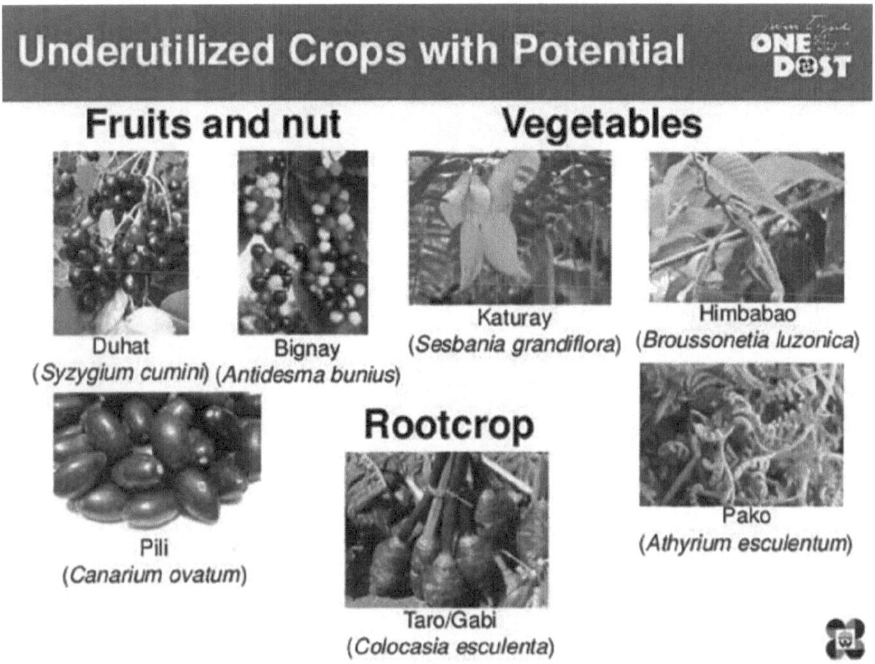

FIGURE 18.2 Examples of underutilized plant species.

(Source: www.slideshare.com.)

18.1.5.3 Economic Development

Neglected crops play vital part in the economy of the developing countries. They are deployed to solve the problems arising from malnutrition and poverty. Neglected crops contribute to the well-being of many tribal populations; being rich in vitamins, minerals, and high antioxidant activity which give rise to diversification of diet and micronutrients enrichment. Moreover, underutilized vegetables show resistance to stress and provide nutrient requirements of the vulnerable group.

18.1.5.4 New Tools for Using Biodiversity

New opportunities for the deployment of agro-biodiversity have emerged due to the availability of recent established tools to assess genetic diversity in the species and use of molecular markers and gene transfer techniques. The production of oil from threatened plants is also on the increase while biodiversity contribution to health, is also on the increase, despite the enormous progress made in synthetic chemistry and biotechnology (Ogunwusi et al., 2016).

18.1.5.5 Increased Income for Rural Poor

As a result of the growing plea from buyers in the advanced and emerging countries, there are new market niches for neglected and underutilized vegetables, due to their diversity and novelty in foods. Extra income is being generated by underprivileged farmers in a place where these crops have proportional advantages over staples or money-making crops because of the market opportunities. More so, the use of contemporary

FIGURE 18.3 Underutilized plants and income.

(Source: www.pinterest.com.)

technologies to change crops/other plants into assorted products by extending its shelf-life has created new prospects to market these species and their products (Figure 18.3).

Irrespective of their potentials in nutrient enrichment and composition, these plants attract little research attention. The potential of underutilized plants to contribute to industrial development is high and expanding. Their contributions to food security, commerce, and industry have been examined in various communities across the globe. Wild relatives of these crops have genes that are indispensable for improving cultivated crops or developing new strains to keep them from being overwhelmed by stresses (Figure 18.4).

18.1.6 Fats and Oils

Fats and oils could be sustainably produced from many underutilized industrial plants in Nigeria. Among the species of importance for oil production are *Pistacia vera* (Figure 18.5), *Dacryodes edulis*, *Elaeis guineensis*, *Irvingia gabonesis*, *Iwombolu*, *Ricinodendron heudelotii*, etc. (Okafor, 2012). The chemical composition of the

FIGURE 18.4 *Vigna subterranean* (Bambara groundnut).

(Source: www.pinterest.com.)

FIGURE 18.5 Pistachio oil and nuts (pistachio oil comes from the nuts of the *Pistacia vera* tree).

(Source: aocs.org.)

fruits, seeds, and leafy vegetables of these species have been reported by several authors. Okafor (1997) showed the oil content of *D. edulis* to be 66.87%, *Monodora myristica* – 42.0%, *Pentachletra macrophylla* – 58.15%, and *Plukenetia conophora* – 61.04%, respectively. The increasing role of fats and oils in the food and industrial sectors has recreated the need to develop appropriate strategies to harvest the potentials of these resources. With the current low price commanded by petroleum, oil from these sources is bound to assume greater importance industrially. A number of vegetable oil mills were established by government and private entrepreneurs in the 1960s. However, some of the mills do not have adequate raw materials.

The development of these natural resources will assist in the provision of raw materials to industries, thereby reducing expenditure on importation of fats and oils. Since fats and oil from the underutilized plants can be used as raw materials, strategies aimed at increasing their production level must focus on post-harvest technologies, especially in the area of storage, oil extraction, and application of new and emerging technologies (Talabi, 1990). Such a holistic approach will reduce dependence on foreign sources in the medium term and provide more sustainable systems of oil supply and utilization in the long run. Different sources of oil are suitable for different purposes, since each source has its unique characteristics. First are those who are merely for edible purposes; secondly, those suitable for both edible and industrial purposes; and thirdly, those suitable for only industrial purposes. Thus, categorization of the various sources of oil seeds is imperative for development purposes. For sustainable development of these oil sources, there is need for consistent government policy on fat and oil production from underutilized raw materials locally.

18.1.7 Fruit Juice and Beverages Production

Fruit juice connotes a product which is 100% liquid extract of the particular fruit obtained by mechanical expression (Aluko, 2013). Some of the underutilized plants have nutritional components which have been used for production of fruit juice and beverages locally. Some of these include *Artocarpous rigidus*, *Gynura procumbens*, *Gacinia atroviridis*, *Issungia gabonomis*, *Dalium guineense*, *Parkia biglobosa*, *Hibisicus sabdariffa*, and *Tamarindus indica* (Okafor, 2012) (Figure 18.6). Seeds of species such as *Treculia africana* have been used to produce nonalcoholic beverages in Nigeria, while *Spondia mombim* and *Diospyrous mespiliformis* have been used to produce alcoholic beverages. Their fruits are nutritious. The drinks produced from them contain vitamins and minerals, particularly vitamins C and B complex (RMRDC, 2003).

FIGURE 18.6 Underutilized fruit plants.

(Source: www.slideshare.com.)

The development of underutilized industrial plants for use in the food and beverage industry has become imperative as the fruit juice industry in Nigeria has traversed a variety of circumstances. While the first fruit juice factory was established in Nigeria in 1956, the factory became moribund after operating for 30 years. The industry did not enjoy rapid growth until 2002, when government banned importation of fruit juice retail packages into the country (Aluko, 2013). Among the major problems militating against the development of this sector are dearth of investment and absence of adequate varieties of raw materials. While the ban on fruit juice importation has led to the growth of the local fruit juice industry, the assumption that fruit juice concentrate importation will be phased out through expected increase in fruit production has remained a mere dream (Aluko, 2013). Thus, the incorporation of the underutilized plant species into the mainstream of raw materials for juice and beverages production in Nigeria has the potential of increasing varieties and activities in the sector.

18.1.8 Fiber Pulp Production

The major wood resources of the tropical forests of Nigeria are mainly mixed hardwoods. Apart from exhibiting heterogeneous properties, the fiber lengths of hardwoods are mostly tinier than those of softwoods (Oluwadare, 2010). These have placed considerable limitations on their utilization as pulpwood raw materials. Nevertheless, the increasing scarcity of softwoods supply to the pulp and paper industry has generated renewed interest in the use of hardwood and minor forest resources as raw ingredients for pulp and paper fabrication, and the innovation is yielding commendable results in most parts of the world. For instance, substitution of softwood with hardwood fibers in the paper production process is on the increase in the United States of America and Japan. Furthermore, this imperative has led development of Kenaf (*Hibiscus cannabinus*) as a viable substitute to softwoods in the production process of pulp and paper.

In Nigeria, considerable work has been done to incorporate indigenous hardwoods into the mainstream of pulpwood raw materials in order to reduce production cost and increase productivity and capacity utilization of the nation's integrated pulp and paper mills. However, this initiative has not been very successful in view of the heterogeneity of the properties of mixed hardwoods and their tendency to deposit. Various studies have, however, reported some underutilized plant species with long fiber characteristics. Examples of the plant species in this category are *Leucaena leucocephala*, *Sterculia setigera*, and *Sterculia oblonga* (Ogunwusi, 2003) (Figure 18.7). Ogunwusi (1991) reported the pulping properties of *Sterculia setigera* to be similar to those of long fiber pulp. The study observed that the fiber length of *S. setigera* increased from 2.47 mm in the heartwood zone to 2.50 mm in the transition zone and decreased to 2.26 mm in the sapwood area. The average length of fiber was reported to be 2.41 mm.

The properties of wood that makes it a suitable raw material for paper production is the fiber length. Fiber is a key element prompting strength improvement in paper (Dinwoodie, 1965). The differences between paper and pulp are due to the power of fiber features (Casey 1980). Thus, *S. setigera* may yield pulp with possessions reminiscent

FIGURE 18.7 *Leucaena leucocephala* (an underutilized plant used for pulp).

(Source: Semanticscholar.org.)

to long fiber. The mean alcohol/benzene soluble extractable content of *S. setigera* was reported to be 3.50% and within the range of 2%–4% which is suitable for pulp and paper making in hardwoods (Panshin and De zeew, 1980). Fibers play important role in the determination of the usefulness of wood species for several end uses.

18.1.9 Biodiversity and Underutilized Plant Species

Biodiversity in Nigeria is influenced by enormous geo-diversity. The variability in the climate and geographic features endows Nigeria with one of a very rich and diverse plant and animal species. Between the mangrove forests in the Southern part to the Sahel Savannah in the North, there are more than 400 food plants, less than 20 of which have been developed and handled in large tonnages. A study by the Federal Environmental Protection Agency (FEPA) in 1992 indicated that Nigeria possessed more than 5,000 plants and 22,000 species of animals. The study also identified about 200 species of lower plants and over 5,103 higher plants. With this number, Nigeria ranked 11th in Africa in terms of plant diversity. About 205 of the plant species are endemic to Nigeria, the 9th highest number among 42 countries. Of all these plant species, only a few numbers of these have been developed for use as industrial raw materials. This development has caused the country huge expenditure in terms of foreign exchange on the importation of products that can be produced locally from the plant species.

18.2 UNDERUTILIZED PLANT SPECIES AND FOOD SECURITY

Underutilized plant species contribute to food security and eradication of hunger as a prerequisite for peace and world security. It provides healthy and nutritious food thereby enhancing the livelihoods of farmers. Underutilized plants can buffer the economic and social shocks of the nation and provide solution during periods of stress and disasters.

Neglected crops play roles in the improvement of food security by:

i. Generating income for farmers;
ii. The risk of over-dependence on restricted crops is reduced;
iii. Raise sustainability of agriculture by reduction in fossil fuel-derived nitrogen fertilizers;
iv. Contributes to food quality and dietary diversity.

18.2.1 Cultivation of NUS and Role of Women in the Agricultural System

The roles of women in Nigeria have been improved through the cultivation of underutilized plants and it provides income opportunities to the rural and urban populace. About 60% of farmers from sub-Saharan African are females involved in farming activities to improve family life. Therefore, women are the main people in African farming system at both city and rural level. There are many economic benefits associated with NUS in different countries; amaranth is a very healthy green vegetable comparable to spinach and important for both household consumption and industrial purposes in Nigeria. The crop is produced with less capital risk by small-scale farmers and suitable for women's microcredit programs.

18.2.2 Challenges of Development of Underutilized Plants in Nigeria

Lack of attention and neglect by the government and research have remained the major constraints to the development of underutilized plant species in Nigeria. While information on some of these species are available in different research institutes and universities, little or no concerted efforts have been made to pool available information for development purposes. While the Nigeria Natural Medicine Development Agency has documented some of the plant materials used in medicine, no effort has been made to look at other areas of application based on available information. This lack of attention on the underutilized plant species limits the access to their optimal utilization potential. Well-trained African agricultural scientists can assist the farmer with new technologies by improving farming practices and developing new varieties as a way of promoting underutilized plant species in Africa (Dansi et al., 2012). Closely allied with this is the need for prioritization. Priority should be given on the basis of extent or level of current development through R&D, industrial potential, and socioeconomic contributions of the plants to local communities. For example, avocado pear is fast becoming known globally for its role as a plant with adequate nutritional potential and for its industrial utilization.

Focus on the development of such a plant species in Nigeria will not only promote socioeconomic development of communities where they are found, but also substantially improve foreign exchange earnings locally in view of the global demand for the plant and its products. For the promotion of such neglected and underutilized crops, it will be nice to have a national research and development program through the joint authority of the Ministry of Environment, Agriculture and Rural Development coupled with the Ministry of Science and Technology. Researchers, developers, producers, and finance institutions should be involved. Many of the underutilized species have multiple applications; and as such, belong to many categories of crops (GFAR, 1991).

The chain linking farmers up to final end users should be explored and exploited. This will provide revenues to rural communities and the mechanism that maintains the diversity of the species. Among the areas that will facilitate the development and underutilization of the plant species are new market opportunities. This will be promoted by the availability of biotechnological tools, which have the capability to alter useful plant species into products or extension of the shelf-life of fresh crops. This will aid commercialization and strengthen marketing systems of underutilized plants.

Some of the strategies that will eventually promote the development of the underutilized plant species in Nigeria, according to Global Forum on Agricultural Research (GFAR) (1999), include participatory research involving farmers, members of the local community, forest dwellers, scientists, and policy makers. This should be engendered and actively fostered among primary and secondary stakeholders. Allied with this is the need for promoting both in situ and ex situ management of the plant types. These will form the basis to develop or characterize the plants. Documentation of the plant species and dissemination of information to stakeholders would also play critical roles in the enhancement and utilization of the crops. Currently in Nigeria, there is no recent information on plant biodiversity of the country. The latest information obtained through a holistic and detailed study took place in the 1980s by the Federal Environmental Protection Agency (RMRDC, 1998). Thus, there is urgent need for a detailed study of plant bioresources of the country. Stock taking of available information is the most critical step to take in the field of documentation activities and provide necessary information for the development of national programs. Crop improvement activities should be initiated, maintained, and sustained. These should address storability and development of simple processing methods and tools. Research on market development should also be promoted. The development of new markets and gregarious exploitation of the old markets should be encouraged. This method was adopted for Roselle (*Hibiscus sabdariffa*) for eras in sub-Saharan Africa that finally became a known beverage in Europe due to modest marketing tactics (GFAR, 1999). Fortunately, the rural farmers subsist on some of these underutilized plant species. A national program directed toward their development will be applauded by farmers if they observe sincerity of purpose in such intentions.

18.2.3 Conclusion

Nigeria has a pool of plant biodiversity that can be adopted for fighting poverty locally. Some of these underutilized plant species could be useful industrial raw materials if developed in a comprehensive and sustainable manner. The development of the plant species will assist the country to navigate the problems it is currently facing in terms of foreign exchange earnings, unemployment, poverty alleviation, and underdevelopment.

REFERENCES

Aju, P. C. and Ezeibekwe I. O (2010). Understanding and appreciating the need for biodiversity conservation in Nigeria. *Journal of Medicinal Plants Research* 4(24), 2605–2608.

Andreas, W. E (2004). Potential of underutilized traditional vegetables and legume crops to contribute to food and nutritional security, income and more sustainable production systems. *Sustainability* 6, 319–335.

Aluko, B. A. (2013). Fruit Juice Industry in Nigeria. In P. A. Onwualu, A. A. Ogunwusi; A. E. Inyang and I. C. Olife (eds.) *Raw Materials Development for the Transformation of the Manufacturing Sector in Nigeria*, 31–44. Raw Materials Research and Development Council, Abuja, Nigeria.

Casey, J. P. (1980). *Pulp and Paper Chemistry and Chemical Technology*. Third Edition, Vol. I. Wiley Interscience Publication.

Dansi, A., Vodouhe, P., Azokpta, A., Yedomonhen, H., Assogba, P. and Loko, Y. L. (2012). Diversity of underutilised crop species of importance in Benin. *The Scientific World Journal* 10(1), 932–947.

Dinwoodie, J. M. (1965). The relationship between fibre morphology and paper properties: A review of literature. *Tappi* 48(8), 440–447.

FAO (2005). The State of Food Insecurity in the World 2005. Informative factsheet. Food and Agriculture Organization of the United Nations, Rome, Italy. 2–10.

Fasuyi, A.O. (2005). Nutritional potentials of some tropical leafy vegetables meals: Chemical characterization and functional properties of fresh and frozen vegetables. *Journal of Food Chemistry* 62(1), 59–64.

Frison, E. (2004). Biodiversity, nutrition and health: making a difference to hunger and conservation in the developing world. Keynote address to the 7th Meeting of the Conference of the Parties to the Convention on Biological Diversity, 1–30.

GFAR (1999). The Role of Underutilised Plant Species in the 21st Century. Global Forum on Agricultural Research. Washington, DC, USA. Unpublished.

Gurumoorthi, P., Pugalenthi, M., Janardhanan, K. (2003). Nutritional potential of five accessions of a South Indian tribal pulse Mucuna pruriens var. utilis; II Investigation on total free phenolics, tannins, trypsin and chymotrypsin inhibitors, phytohaemag glutinins, and in vitro protein digestibility. *Tropical and Subtropical Agroecosystems* 1, 153–158.

Malalavidhane, S., Wickramasinghe, S.M and Jansz, E.R (2001). An aqueous extract of the of the greenleafy vegetable Ipomoea aquatica is as effective as the oral hypoglycaemic drug tolbutamide in reducing the blood sugar levels of Wistar rats. *Phytotherapy Research* 15, 635–637.

NEST (1991). Nigeria's Threatened Environment. A National Profile. Nigeria Environment Study/Action Team (NEST), Ibadan.

Ogunwusi, A. A. (1991). Influence of Wood Quality and Pulping Variables on Pitch Deposits and on Paper Properties of 15 Nigerian Hardwoods. Ph.D. Thesis, University of Ibadan, Ibadan, Nigeria.

Ogunwusi, A. A. (2003). Wood Properties of *Sterculia setigera* Growing in the Savanna Belt of Nigeria. *Nigeria Journal of Forestry* 32(1), 50–55.

Ogunwusi, A. A. (2016). Role of New Agricultural Raw Materials in Industrial Development.

Okafor, J. C. (1997). Final Technical Report on the Project Development and Utilisation of Fast Disappearing and Underutilised Edible Woody Forest Species of South-eastern Nigeria.

Okafor, J. C. (2012). Potential of Plants in Health Promotion and Poverty Alleviation in Rural and Urban Communities in Nigeria. Being an Invited Keynote Paper Presented at the Moringa Workshop at Centre in Onitsha, 17–19 March, 2011.

Oluwadere, O. A. (2010). Wood Quality Studies in Plantain Grown Sterculia (*Sterculia setigera Del.*) in the Guinea Savanna, Nigeria. *Research Journal of Forestry* 5(1), 22–23.

Panshin, A. J. and DeZecuw, C. (1980). *Textbook of Wood Technology*. McGraw Hill Bk. Coy, 722–730.

RMRDC (1998). Report of the Multi-Disciplinary Task Force on Wood and Wood Products, RMRDC Publication, 203–207.

RMRDC (2003). Report of the Multi-Disciplinary Task Force on Wood and Wood Products, RMRDC Publication, 204–211.

Talabi, S. O. (1990). An Oilseed Industrial Strategy for Nigeria. Proceedings of the National Workshop on Policy.

19 The Place of Neglected Legumes in Human Nutrition and Food Security

Beckley Ikhajiagbe, Geoffrey Obinna Anoliefo, Doris I. Omoigui, and Zipporah Emilomo Omage
University of Benin

Edokpolor Osazee Ohanmu
University of Benin
Edo University

Saheed Ibrahim Musa
University of Benin
Admiralty University of Nigeria

CONTENTS

DOI: 10.1201/9781003178903-23

19.1 INTRODUCTION

A significant test towards agrarian sustainability includes the production of 70% more food crop for an extra 2.3 billion individuals by 2050 (FAO, 2009). In Nigeria, taking a gander at securing food, the degrees of produce development of brad harvests including cassava and beans are inadequate to satisfy future food needs without falling back on agrarian land extension. Moreover, is the inability of these staple crops to supply the necessary supplements suitable for advancement? The populace in Nigeria is unsurprising to ascend from around 200 million to 350 million by 2050. Will a nation confronted with a variety of pollution, marginal soil with little arable land available for expansion be able to feed a burgeoning populace?

19.2 GENERAL OUTLOOK ON GLOBAL HUNGER AND FOOD CRISIS WITH EMPHASIS ON AFRICA

Throughout the years, man has depended or domesticated a variety of crops that are currently named neglected and underutilized, either as food, fibre, grub, oil, or medicinal properties, of which their importance has been decreased in recent time. A wide scope of phrasing has been credited to characterizing it as: underutilized, conventional, forgotten, optional, new, or overlooked (Padulosi and Hoeschle-Zeledon, 2004). Neglected or underutilized legumes (NULs) are regularly portrayed as drought-resistant (Zeven, 1998), and this way demonstrate imperative in battling hunger. The global development of these traditional crops has kept on declining with its more prominent hereditary biodiversity and potential to improve food security. Over the world, a considerable lot of the plant species that are developed for food are underutilized while they assume a significant job in the food security, sustenance, and income of the provincial poor (Magbagbeola et al., 2010). Food uncertainty and hunger are brought about by numerous components which are regularly being interlaced with each other. In general, the main reasons for starvation are strife, weather, and absence of interest in agronomy. This way, food security persists and proceeds to be one of the major challenges affecting the present reality.

Global food security throughout the years has been obliged by a few variables remembering the considerable dependence for a couple of key staple crops, a circumstance that has nutritional, agronomic, financial, and environmental ramifications (Ebert, 2014). These days, just three major crops, such as rice, maize, and wheat, provide more than 60% of vitality intake worldwide (FAO, 1995). With the projected growth in the overall population, the demand for food has also increased (Baldermann et al., 2016). The absence of a decent variety of legumes is apparent because relying on less crops has bad implications for the routine of human consumption, which can lead to ailing health and diet-related illnesses. These ignored crops are critical particularly in human well-being and food security, and ought not to be neglected.

In Africa, the pace of malnutrition is intensifying at a distressing frequency. Lack of cash, dry season, and extreme weather reversing years of improvement with around 237million Africans been underfed compared to other regions. There are huge contrasts amongst islands in the predominance of extreme food uncertainty. Over 27.40% of populace in the region were marked as severely diet apprehensive in 2016, just about multiple times higher as some regions. Between 2014 and 2016, food uncertainty expanded with approximately 3% (FAO, IFAD, UNICEF, WFP and WHO, 2017). As indicated by World Food Program (WFP) (2020) Global Hotspots reported that millions of individuals in some Africa locale require life-sparing food help with the imminent months. In any case, this can be savaged if endeavours are taken to guarantee that NULs are developed to help the current mainstream crops.

19.3 POOR GDP IN AFRICA RESULTING FROM POOR INVESTMENT IN AGRICULTURE

Agricultural investment is an important necessity to create and arrange the rural segment in Africa. This will result to adequate rural production that guarantees food security for the developing populace. Africa is a large continent with 54 nations and its geographic size is colossal. At the degree of the continent, agribusiness brings up to 23% of GDP and is a main basis of livelihood by hosting about 55% of the dynamic populace in these nations. The farming area is basic to the accomplishment of poverty decrease and development in Africa. The fundamental challenges confronting food production is the long old traditional method of cultivating, undercapitalization, arrangement irregularities, minimized homestead size, and unfavourable environmental change (Garrity et al., 2012). The distributive impact of expanding portion of assets to farming is high. For example, an expansion of 1% in agrarian per capita GDP lessens poverty five times more than comparable increment in GDP per capital in different areas particularly among the centre poor (Alliance for a Green Revolution in Africa, 2014). Poor investment in agribusiness brings about poor GDP.

In Sub-Saharan Africa, agriculture is accountable for 60% of the anticipated advancement of poverty. Farming advancement not just directly affects pay and monetary development yet assumes a significant job in cultivating improvement in non-agrarian divisions of the economy (Lloyd, 1969). It is great to invest richly in agriculture so as to lessen poverty rate and increment the GDP in Africa. As per the report of FAO (2009a), total population is expanding at a quick rate and is anticipated to reach about 9.6 billion by 2050. Along these lines, the future global challenge is to cover the world's hunger through sustainable agriculture and food production. In order to meet the global food demands our attention ought to be on advancing the development and use of different crops which have been neglected and under-exploited however can possibly improve food and sustenance security particularly in Africa. In synopsis, how to improve the poor GDP because of neglected agrarian investment remains a massive task, which must be achieved by a crucial change of present food and agronomic frameworks.

19.4 FOOD PRODUCTIONS GAP

In meeting the food production gap of an escalating populace in Africa, the agricultural sector assumes an essential role in the achievement of poverty decline. Agribusiness in Africa is described by conventional method of cultivating, under-capitalization, approach irregularities, little and divided homestead size, and antagonistic environmental change impacts (Garrity et al., 2012). The minor homestead land, poor modernization, contaminated dry season, and urbanization has played a critical role in increased food production. Be that as it may, smallholder farmers developing under 2 hectares produce about 80% of Africa's agrarian yield (KPMG, 2013). The low agrarian production levels cannot adequately accommodate the rising populace, putting Africa as a major food importer. Hence a widespread food insecurity with more than 223 million underfed individuals and the figure is relied upon to ascend by around 132 million in 2050 (Alliance for a Green Revolution in Africa, 2014). Subsequently, positive correlations exist between undernourishment and output.

Malnourishment has antagonistic effect on labour output, health of families, and thus the society (Kariuki, 2011). Diao et al. (2007) found that inadequate nutrition prompts decrease in development pace of GDP *per capita* varying from 0.16% to 4.0%. The value included per labourer in the agribusiness is low in most African nations. The low profitability of agriculture represents most of rural Africans subsisting on under 1 USD every day (Rosen and Shapouri, 2012). The low profitability could be because of the poor transportation denting agricultural development attributing to the absence rural-urban linkages. The expense of transporting rural products in Africa is higher than 2 USD per ton for every kilometre (Nwuke, 2012). Thus, 85% of yield produced happens inside 6h foldaway time from the biggest metropolitan (African Development Bank, 2011).

19.5 CHALLENGES IN NUTRITIONAL GAP IN FOOD PRODUCTION

As a result of the swift industrialization and developing vulnerability, our food plans are experiencing some significant difficulties that is hindering to their longstanding sustenance and scopes in meeting malnutrition and nutrient objectives.

Firstly is the production gap in feeding of a burgeoning population (Ohanmu and Ikhajiagbe, 2018). Maybe, one of the inquiries raised is whether the developing universal populace have sufficient food to encounter its dietary needs? To take care of an added two billion individuals by 2050, global feed manufacturing should increase significantly (FAO, 2016a). Expanding the production of conventional staple crops is probably not going to satisfy this need; for instance, in key wetted farm plans, yield seems to be more than 70% of the profit potential. Depending on the yields solely will build up a significant space amongst food stream and market.

Environmental variation is another rising quandary in the pecuniary, well-being, and ecological scene of the region. African farming is principally susceptible to hostile climate change influences in view of excessive contact to variation in climate condition. In the meantime, near 4% of agrarian lands are moistened which shows overwhelming dependence on precipitation (United Nations Economic Commission for Africa, 2011). The adaptive capacities of farmers are inhibited by government neglect, vastly bureaucratic procedures of accessing credit with "throat-cutting" interest rates, scarce access to extension facilities, infrastructure deficits, among others (Africa Progress Panel, 2010; Bremner, 2012). The issue of terrestrial grab in Africa is rising and affecting property accessibility. This adds up to taking lands past the financial grasp of asset deprived farmers (Friis and Reenberg, 2010; Graham et al., 2008).

The low income resulting from cultivating activity is another factor. There is acceptable proof that numerous peasant farmers have turned increasingly more to off-farm income generating activities along with subsistence food cropping inferable from the low returns to labour in commercial crop production. Elmekki and Barker (1993) noticed that in Sudan, frequently more labourer families migrate looking for wage work. In 1975–1980, the total volume of agricultural export fell by 0.8% per annum, in 1980–85 by 2.9%, and in 1986–1988 by 2.5% (UNDP and World Bank, 1992). In Lesotho the agrarian export volume decreases in 1986–1988 were about 57% per annum. Other recognizable declines were Zambia 32%, Botswana 28%, Somalia 22%, Niger 14%, and Zaire 13%. Reduction in certain essential products in the international service markets has confirmed nearly as calamitous as drought.

Lastly is the nutritional value of the produce. The nutritional space between which crops sown and accessible, and which food is needed for healthy living. This entails growing the accessibility, entree to the healthy meals required for a balanced diet. Therefore, it is not enough to increase the output of essential produces which does not fast-track decreases in undernourishment as result of the deficient supply of nutrient-rich foods. In summary, the complications of the total challenges facing agrarian and food schemes involve an archetype alteration that highlights workable production and the ingestion of healthy meal.

19.6 ASSESSMENT OF THE INTERVENTIONS AT MEETING THE GAP

Over the most recent three decades farming improvement in Sub-Saharan Africa has relied upon government perspectives towards agriculture and the peasantry, government use, advancement, and venture strategies. For much of Africa, government is the topmost source of subsidy and sustenance for research and innovation, either in view of a constrained measure of private capital or as a result of the purposeful convergence of responsibility for undertakings in government hands. Numerous legislatures have focused their consideration on agricultural export, easier to tax than production for the home market and in a few nations the main foreign currency earner. Export crop decrease has been held to have financial and social downsides influencing salary distribution, food security, and the environ, which however has made noteworthy commitments to rural livelihoods and national monetary development with production esteems per hectare commonly those of fundamental food staples. It has also made some input to food sustenance by making the foreign exchange available either for food importation or the efforts required to upsurge domestic food growth (Maxwell, 1988).

In the past, exported crops brought in foreign asset in Africa, then interest weakened in late 1970s and 1980s when many of the foreign-owned estates and plantations were subdivided and transferred to African ownership or taken over by the state. In the early 1985, Africa boasted for around 14% of all foreign savings in less developed countries compared with 27% 10 years earlier (Riddell and Cockcroft, 1991). Currently the rate of return on investment in Africa is very low and both the investment totals and inflows to African agriculture are small. A few major plantations are left and a number of core schemes for smallholder cooperation, plus some agribusiness investment in food processing and exporting. Farmer involvement in large-scale agriculture has succeeded mostly where there persisted major commercial preferment by government selling agencies or by big trading organizations, often with extra sustenance in the form of seed or plant stock supply, agricultural advice, and subsidized production resources. Government promotion of smallholder reimbursement structures to generate higher and more lucrative smallholdings has had some success in producing a market reliant on smallholder class, particularly in Kenya (Clayton, 1964; King, 1977; Taylor, 1969).

In Nigeria and Ghana export firms often introduced retailing networks in the areas they wished to develop commercially in order to provide an incentive to produce cash crops, e.g. cashew nuts, but in many areas incentives were too poor for farmers lacking capital and modern tools and with access only to poorly developed produce markets. A major feature of the 1950s and 1960s was the creation of national marketing boards for export crops, with the objective of stabilizing prices and protecting national interests, and later, other forms of parastatal marketing organization, covering a wide range of produce. The latter were convenient for handling price controls, subsidies, and food aid and could act as counterpart agencies to which official aid could be tied (Mittendorf, 1988).

Government restraints on agricultural expansion may often be seen as cautious policy, by concentrating most of the funds into non-agricultural activity viewed as highly profitable or having better growth predictions, e.g. crude oil, or even by the taxation of commercial agriculture without regard to the effects of given levels of taxation on production. It may also be the by-product of other policies, e.g. manipulation of the exchange rate to keep imported goods' prices low, but to the disadvantage of agricultural exports. It is not without relevance that few governments have sought to promote agricultural services or agricultural education or to change the popular image of farming as an activity associated with rural backwardness and poverty. In some countries, agriculture was simply neglected (FAO, 1992). A major part of the problem has been that several countries in Africa are too small to afford a significant level of expenditure on agricultural research and development (Lipton, 1977). In others, however, attempts have been made to promote agriculture through research, development, advisory and extension services, the creation of markets, and the subsidizing of farm inputs. In 1987 government expenditure on agriculture in Sub-Saharan Africa averaged 7.4% of total expenditure, compared with an agricultural contribution of 33% of GDP, 79% of female employment, and 66% of male (World Bank, 1989; UNDP and World Bank, 1992). Only 6 countries of 21 reporting spent more than 10%, namely, Guinea-Bissau with 18.7% (the highest), Togo, Madagascar, Malawi, Kenya, and Ethiopia. All were low-income economies and four were extremely poor with Gross National Product (GNP) *per capita* of $210 or less by 1990, so that absolute levels of agricultural expenditure were low.

Government's agronomic restructuring and new contributions unaided are not sufficient to raise the level of agrarian impact needed, even if it is theoretical that agriculture's requirements are wholly understood. Agronomic output cannot be significantly improved without changes in African economies. Without assurance in the exporting sector, with no constancy and value in the process of farmer's market, and without food prices which inspire farmers to cultivate more food, there will be no interest in advancing agrarian development beyond the immediate needs of the farmers themselves. Attempts to transform whole economies and to reform economic management are becoming more widespread mainly as a result of the initiatives of the International Monetary Fund (IMF) and the World Bank in their structural adjustment and stabilization programmes. Due to some extent on economic recession, many African countries in the 1980s turned to the IMF and the World Bank for help. At the end of 1989, about 33 Sub-Saharan African states had adopted some sort of structural adjustment programme and 17 agreed to loosen their trading regimes (Mansoor and Inotai, 1991).

19.7 NEGLECTED/UNDERUTILIZED LEGUMES IN MAIN STREAM AGRICULTURE

Generally, crops are separated into two core groups: staple crops and underutilized crops/legumes (NULs). NULs typically belong though not restricted to the non-staple crops. NULs can be referred to as leguminous crops that gained slight consideration or have remained wholly disregarded by agriculturist, scientists, plant breeders, and lawmakers. They comprise of the open or semi-cultivated plants and non-timber species that are not naturally exported as supplies (Padulosi et al., 2013), mainly altered to homegrown settlements. These old-style diets have been in use for centuries, but have become progressively abandoned in current times due to more productive, lucrative, and enhanced crops substituted them in the agrobusiness. This cumulative oversimplification of agronomy and food classifications has led to NULs performing a marginal role in the present farming and food schemes. Again, their discernment as "food of the poor" often pillories them, making an impediment for producing and consuming them.

Amongst NULs, varieties that are nutritionally dense, environmentally resilient, frugally viable, and within range or pliable provide the possibility to tackle starvation and undernourishment. Improving nutritional and production assortment is key in addressing malnutrition from a food scheme perception. Agricultural diversity is an essential element in the NULs sustainable productivity that could contribute to enhancing food accessibility, affordability, and steadiness. According to Garn and Leonard (1989), of the 300,000–500,000 plant types that exist, about 30,000 are acknowledged as consumable plant varieties, and over 7,000 crops are either cultivated or farm for food, and/or obtained from the wild during the time of humankind.

Although NULs are not explored, they propose great chances for mitigating poverty, famine, and undernourishment. NULs provide a vital part in the contest against famine and are a significant resource for agronomy and rural development (FAO, 2012a). From time immemorial, NULs have been a major source for food and supplementary usages on a large scale, and in some nations are still popular particularly within marginal or small farmers in rural zones. Many of these crops are sold locally and handfuls were fortunate to make their way to export niche of the world (Akinnifesi et al., 2008).

19.8 UNDERUTILIZED LEGUMES OF AFRICA

Legumes are to a great extent proteinous plants from the family Fabaceae. They are a common source of proteins in African food diet. Africa has a huge variety of indigenous legumes, extending from huge rainforest trees to little yearly herbs. A number of them either have their significant centres of assorted variety in Africa or are restricted to specific regions of Africa as discussed below.

19.8.1 *LABLAB PURPUREUS*

Description: The common name is hyacinth bean, bonavist bean, field bean, and Egyptian bean. *L. purpureus* is a climbing, hot-seasonal, or short-lived perennial with a dynamic taproot. It has a dense, herbaceous stem that grows up to 3 ft, and vines extending to 25 ft from the plant (Valenzuela and Smith, 2002). The leaves are trifoliate and long-stemmed. The flowers may be blue, white, or purple which is variety dependent. Seedpod is 4–10 cm long (Cook et al., 2005; Venezuela and Smith, 2002), flatten, smooth, and pointy containing 2–4 seeds. Seeds colouration ranges from white, pale brown, cream, dark brown, red, black, or mottled which is variety dependent.

Distribution: *L. purpureus* is a traditional food crop that is believed to emerge in Africa (Cook et al., 2005) or India (Murphy and Colucci, 1999). It has been effectively cultivated in the Southern United States of Texas, Georgia, Florida, Puerto Rico, and as far north as the Great Lakes and Canada. It grows from sea level up to 6,500 ft. (FAO, 2012b).

19.8.1.1 Ethnobotany

L. purpureus has been used in the Philippines and China as a stimulant in reducing fever and pomposity, stimulate digestion, and as an anti-spasmodic (Stuart, 2011). In Namibia, the roots were used in treating heart conditions (Pennacchio et al., 2010).

19.8.1.2 Uses

Commercial crop*: L. purpureus* is grown as a pulse crop in Asia, Africa, and the Caribbean. It is consumed as a green vegetable (green bean, pod, leaf). Protein isolate obtained from the bean can be included as food additive for improving cake quality (Maass et al., 2010).

Forage: L. purpureus is used as forage, hay, and silage. As forage, it is often grown with sorghum or millet. The leaf is very edible making good hay for cattle and goat, but the stem is not while the seeds are fairly edible. Overall, it is one of the most eatable legumes for animals (Valenzuela and Smith, 2002).

19.8.1.3 Pests and Potential Problems

Largely, *L. purpureus* is less vulnerable to root diseases than cowpea (Cook et al., 2005), but usually affected by root-knot nematode infection (Valenzuela and Smith, 2002). It may develop diseases like bacterial wilt (*Xanthomonas* sp.) and powdery mildew (Cook et al., 2005).

19.8.2 *CANAVALIA ENSIFORMIS*

Description: *C. ensiformis* is a yearly plant with climbing or profuse growth forms. The plant is woody and possesses long tap root. The leaves, wedge-shaped at the base, tapering to the tip are 10 cm wide and 20 cm long with three egg-shaped leaflets. The 2.5 cm long flowers are rose-coloured, purplish, or white and have a red base. The sword-shaped seedpod is 30 cm long and 3.8 cm wide. Seeds are smooth and white with a brown seed mark which is approximately one-third the length of the seed. The roots possess nodules which fix nitrogen.

Distribution: The plant is seen in the tropics and subtropics, countries, and South Pacific – largely in farming. It is cultivated in the southwestern United States and Hawaii. *Canavalia* is a pantropical genus thought to have originated in the New World based on the large genetic diversity of species in the fossil record (Sauer and Kaplan, 1969).

Ethnobotany: The leaves spread on leafcutter anthills to eliminate ants (Bunch and ECHO staff, 1985). In Nigeria, *C. ensiformis* seed is used as an antibiotic and antiseptic (Olowokudejo et al., 2008). Archaeologically, *C. ensiformis* is used by native as nutrition and forage in droughty regions of Arizona and Mexico.

19.8.2.1 Uses: The Crop Has a Variety of Use as Listed Below

Commercial crop: Outside of the United States both young pods and green seeds are eaten as a vegetable. Seeds are also used as a coffee substitute. The mature bean contains potentially harmful saponins, cyanogenic glycosides, terpenoids, alkaloids, and tannic acid (Udedibie and Carlini, 1998) and must be cooked before eating. There is also pharmaceutical interest in the use of *C. ensiformis* as a source for the anticancer agent strigonelline and canavanine (Morris, 1999). Jack bean seed has been promoted in developing nations as a potential source of affordable and abundant protein. It has 29.0% protein content (Adebowale and Lawal, 2004).

Soil biofumigant: Incorporating jack bean dry matter into the soil has reduced root galling in tomatoes and increased tomato plant height and weight (Morris and Walker, 2002). Jack bean has been successfully used to increase yield and to suppress nematode (*Pratylenchus zeae*) populations and root necrosis when intercropped with maize (Arim et al., 2006). Farmer Participatory Research (FPR) in Uganda showed that jack bean intercropped with banana, coffee, sweet potatoes, or cassava is a preferred green manure among farmers (Fischler and Wortmann, 1999). However, research by McIntyre et al. (2001) has shown that intercropping banana with jack bean had no beneficial effect on nematode populations.

Aquaculture: Detoxified jack bean seed has been used successfully as a high protein fish meal substitute in tilapia aquaculture (Martinez-Palacios et al., 1988).

Phytoremediation: Jack bean can be grown in soils with high lead concentration and has potential to be used for restoration of lead-contaminated soils (Pereira et al., 2010).

Pests and Potential Problems: Jack bean has no significant pests or diseases (NAS/NRC, 1979). The toxicity of the seeds makes them unpalatable to insects and protects them from insect attack (Oliveira et al., 1999).

19.8.3 *Sphenostylis stenocarpa*

Description: The African yam bean (AYB) (*Sphenostylis stenocarpa* Hochst. Ex A. Rich) family Leguminosae is a neglected and underutilized leguminous plant and the most economically important among the seven species of *Sphenostylis* (Potter D., 1992). It is one of the most important tuberous legumes. It is grown for its eatable tubers, which resembles extended sweet potatoes, and its seeds that are confined in rigid and tough 20–30 long pods. It is mainly used as food and also to feed animals.

The AYB is a perpetual climber; about 1–3 m tall, normally grow as yearly. The leaves are trifoliate with oval leaflets (length: 2.7–13 cm; width: 0.2–5.5 cm). The inflorescence is a raceme that exhibits an acropetal mode of floral maturation (Adewale et al., 2012). The noticeable flowers are mauvish-pink, purple, or greenish-white in colour, about 2.5 cm in length, and borne on stout auxillary peduncles. The seed pods are linear, flat, with both margins raised 25–30 cm long and 1–1.5 cm broad containing 20–30 seeds which may be ellipsoid, rounded, or truncated and show considerable variation in size and colour; the largest are usually about 1 cm long and 0.7 cm wide. Seed colour may vary from creamy-white or brownish-yellow to dark brown, sometimes with black marbling, and there appear to be a number of types according to the seed colour. The plant produces small spindle-shaped tubers, about 5–7.5 cm long.

Local usage and in Africa and Nigeria: *S. stenocarpa* is a useful source of nutrients for many African communities with a nutritional value comparable to that of soybean, although the cooking time for the yam bean is much longer (4–6 h). It is rich in protein and starch and an important source of calcium and amino acids. It contains amino acids that are important for the development in early pre-school and school aged children and also those required for adults (Oshodi et al., 1995). The seeds and tubers represent the primary food for human consumption. The seeds may be eaten alone or in soups and are commonly served with yam, maize, or rice. The tubers are eaten either raw or cooked. It has potential utility in producing feed for livestock. The leafy vegetation remaining from the harvest provides useful fodder. It is very beneficial to livestock due to its protein content and the massive quantities produced. *S. stenocarpa* is a species that fixes nitrogen and is used for soil restoration as green manure.

Distribution and ecological footprint: *S. stenocarpa* are native to tropical west and central Africa and is cultivated in southern and eastern Africa (Ohanmu et al., 2017). It originated in Ethiopia. The crop is developed in Nigeria essentially for its seed, likewise developed for tuber part of Africa. Albeit AYB is dark from a general perspective, it exists in various nations. In wide terms, the development region reaches out from tropical West Africa to Sudan, Eritrea, and northern Ethiopia. From that point it stretches out southward to Congo in the west and Zanzibar in the east. Anyway, the best significance, without a twofold, is in West Africa, principally Nigeria and its quick neighbours. It flourishes on unfathomable, wabbly sandy and loamy soil with virtuous biological content and drainage system. It nurtures well in areas where yearly rainfall ranges between 800 and 1,400 mm, with temperatures comprised of 19°C and 27°C (Ecoport, 2009).

Pest and diseases: Contagious ailments detailed are fine moulds (due to *Oidium* sp.), which is parasitized by *iCercospora capsici;* leaf spot (brought about by *Phoma* sp.), and stem rust (brought about by *Aecidium* sp.). Infection mosaics have likewise been accounted for. Vermin have not been characterized in detail yet incorporate Orthopterous and Lepidopterous bugs. Leaf moving caterpillars and leaf excavators have been portrayed as making genuine harm to the foliage, and thrips harm the blossoms. Nematodes may assault the root framework prompting decrease in yield.

19.8.3.1 Economic Significance

1. It has huge potential for food security in Africa.
2. Insecticidal and medicinal uses – The plant is not vulnerable to pests and diseases related to most legumes (Duke, 1981). This quality is due to the intrinsic lectin in the crop seed.
3. Stable yields across wide environments.

19.8.4 *Mallotus subulatus*

Description: *Mallotus subulatus*, from the family Euphorbiaceae, is an NUL found mainly in the lowland grasslands of tropical Africa. It is usually collected from the open for local use as medicine. *M. subulatus* is a dioecious tree that grows up to 5 m tall with opposite, simple, and entire leaves with a 1–5.5 cm long petiole and 6–8 mm persistent tabulate stimulate. The leaf blade is oblong to ovate and its base rounded while its apex is acute to abominate; the leaf is also clangorous with an ancillary raceme inflorescence up to 5 cm long containing unisexual and regular flowers. The petals and disk are absent but the pedicure is short; the male flowers are in small fascicles, with obovoid calyx, and four free lobed cream stamens while the female flowers are solitary, with three re-curved sepals that are densely Stella, hairy, and cream in colour. The ovary is superior, globe, and 3-lobed. The fruit is a 3-lobed capsule, 16 mm in diameter covered with slightly hooked bristles; the seeds are ovoid and three in number (Schmelzer et al., 2008).

Uses: In some African countries such as Ghana, the infusion of the roots of *Mallotus subulatus* with the seeds of *Aframomum melegueta* is taken using an enema to treat lower back pains. In Gabon, the leaves are made into a pulp and used for head massage to treat headache and nervous breakdown. In Congo, the bark sap is rubbed into local scarification in case of side pain and lower back pains. In Nigeria, the roots, leaves, and fruits are ground and taken in water to treat dysentery and stomach ache. The wet leaves are pulped and applied to wounds to stop bleeding. Even in certain parts of Nigeria like the south, the powdered seeds of *M. subulatus* are applied to the faces of young men and girls as a mask in Igbo.

Distribution and ecological footprint: This legume has its native range in the lowland forests and wastelands of West and West-Central Africa, in countries such as Sierra Leone, Cameroon, Nigeria, Gabon, and Ghana.

Ecological significance: As a result of *Mallotus subulatus* being a neglected and underutilized legume, it lacks some necessary information regardless of its great potential stemming from its high medicinal properties (Schmelzer et al., 2008). Also, for consumption purposes, *M. subulatus* has a high protein content (Ade-Omowaye et al., 2015).

19.8.5 *VIGNA RACEMOSA*

Description: *Vigna racemosa* is a species of climbers in the family Fabaceae discovered in 1928. It is native to most of Tropical Africa and usually found in the savannah. It is locally known as *ewe* in Nigeria (Bolarinwa et al., 2015). *Vigna racemosa* is a perennial climbing herb of 0.5–7.0 m long. Its stem is usually shielded with long, spreading, bristly ferruginous hairs. The leaflets are three and are mostly oblong or ovate or lanceolate, with an acute or acuminate apex and a sub-acute, rounded or truncate base, sparsely to densely pilose on both surfaces; the petiole is 1.5–11 cm long while the rachis 0.5–1.5 cm long. The stipules are lanceolate, multinerved, prolonged, and bilobed at the base; the lobes are unequal, 2–4 mm long, and narrowly triangular. The peduncles are pubescent, especially toward the apex with 1.5–7 cm long rachis consisting of 4–13 nodes, and internodes of 2–11 mm long; the ovary consists of 6–9 ovules and the fruit is a linear-cylindrical, S-curved, pubescent pod with a short curved beak; the seed is small with both the hilium and aril eccentric (Mackinder et al., 2001).

Uses: *V. racemosa* is used as a source of food (a primary protein source) for man and animals (Cobbina, 1992). It is also used in the treatment of nacho-pharyngeal infections. In the Northern part of Nigeria, where the popular non-alcoholic beverage named "Kununzaki" is made, *V. racemosa* is used to produce it. *Vigna racemosa* exhibits many excellent nutritional attributes and contains high protein content and sulphur amino acids, which enables it to bridge up the protein deficiency gap at a low cost than other crops (Chopra, 1975). It can also be used as cover crops, fodder, for soil conservation and soil fertility maintenance.

Distribution and ecological footprint: This plant is native to several African countries such as Angola, Benin, Cameroon, Central African Republic, Chad, Congo, Gabon, Gambia, Guinea, Mali, Namibia, Niger, Nigeria, Senegal, Sudan, Togo, and Uganda. It has also been introduced to Colombia, New South Wales, and Honduras. *Vigna racemosa* is found mostly in the grasslands, rocky outcrop with wet flushes, and thin soil and forest-savanna mosaics.

19.8.6 *Vicia faba*

Description: Faba bean is one of the oldest crops grown by man (Fouad et al., 2013). It is the most important pulse globally after beans, peas, and chickpeas (Filippetti and Ricciardi, 1993). It is widely cultivated and known for its high protein content which is between 25% and 40% (Matthew and Marcellos, 2003) and it is a rich source of carbohydrates, vitamins, fibre, and minerals (Revilla, 2015). The Stem: Strong, hollow, tetragonal erect stem. Stipules: Very small, with a distinct black spot on it's down side. Root: Robust tap root, profusely branched secondary roots. Leaves: first young leaves bear two leaflets. Flower: white with a hint of pink at the top and a very distinct, satin black spot on both wing petals. Other flower colours include brown, violet, red, among others. Seed: About 2–3 seeds per pods are projected per inflorescence. This plant can produce far more than 12 pods, dispersed across more than six nodes.

Plant uses: Faba bean is consumed for its rich source of protein. In Nigeria, the seeds are boiled, seasoned, and eaten with an addition of oil. It is used as fodder, animal meal.

Ecologically, it contributes nitrogen (N) to the system by biologically fixing N_2. This can be associated with the *Rhizobium leguminosarum* found in its roots. It reduces fossil energy consumption as it can be used as a bio fuel. It is a source of food and feed rich in protein. As an autotroph, it is important in the food chain, as herbivores consume it. It produces oxygen during photosynthesis. It is important in biodiversity.

19.9 KIDNEY BEANS

The kidney bean is named after its kidney-like shape. Kidney beans are referred to as "common beans" and this is probably because they all are derived from a common bean ancestor that originated in Peru. It is also called French bean, String bean, and Frijoles (Singh, 1999; Pandey, 2003). Kidney beans are the best identified and commonly grown species of the genus *Phaseolus* (Shukla and Misra, 1979; Purseglove, 1988). Kidney Beans were brought to Europe in the 16th century by Spanish and Portuguese explorers returning from their voyages to the New World and it was introduced into Africa and Asia (Nassar et al., 2010). Like other beans they are highly proteinous. They are found in different colours. The colour of kidney beans varies from very light red to very dark and nearly purple.

19.10 NOMENCLATURE

Kingdom: Plantae
Superdivision: Spermatophyta
Division: Magnoliophyta
Class: Magnoliopsida
Subclass: Rosidae
Order: Fabales
Family: Fabaceae
Genus: *Phaseolus*
Species: *Phaseolus vulgaris*

Description: It is a herb (annual herbaceous). It possesses an erect, green, ribbed, cylindrical, strong stem. The leaves are of two types: the first two basal prophylls leaves are simple, opposite arrangement, green, cordate with entire margins, long petiolated, reticulate-pinnate venation, and the midrib that divides the blade into two equal parts (Nassar et al., 2010). The other leaves are compound. Here, leaves are divided into leaflet (trifoliate), pinnately compound with the terminal petiolule longer than the two lateral ones. The leaves are alternate, green, and ovate in shape with entire margin. The leaves are shielded with small hairs and reticulate-pinnate venation. The midrib divides the blade of the apical leaflet into two equal sides, but the lateral ones are unequal-sided. The leaf has two small stipules at its base. The apical leaflet has two small stipels and each of the lateral ones has only one stipel. The compound leaf is described by Nassar et al. (2010) in agreement with that documented by Purseglove (1988) and by Decoteau (2000).

The flowers are zygomorphic, perigenous, bisexual, and whitely coloured (Corolla having five petals) with green, tubular, hairy calyx. The fruits are green in colour, turns into yellow brown when matured. Pods are slender, curved or somewhat straight, long with prominent beaks. There are usually 4–5 seeds in a pod. The seeds are white in colour, relatively large in size, kidney in shape, usually white. Embryo is relatively large and dicotyledonous. Endosperm is absent (Nassar et al., 2010). The plant has a short taproot with lateral roots which usually bears nodulated outgrowth containing nitrogen-fixing bacteria.

19.11 USES

In tropical Africa, common bean is primarily produced and consumed as a pulse. It is eaten with carbohydrate-rich meals, such as those based on maize, to add diversity and flavour to the meal. It is the major protein source in various countries, e.g. in Rwanda, Burundi, and Kenya. It is mashed and made into soup. The leaves are sometimes eaten as vegetables and the crop residues are often used as animal feed. In Mali, a powder of carbonized seeds is applied to wounds.

Ecologically, the plant is important as it biologically fixes nitrogen. It serves as food for herbivores, and therefore, it is an important part of the food chain. It plays a role in biodiversity. It is a green plant and therefore, it photosynthesizes, producing oxygen for humans and animals.

19.12 WINGED BEANS (*PSOPHOCARPUS TETRAGONDOLOBUS* L.)

Description: The winged beans is a climbing enduring delivering new development every year from shallow diligent roots yet for ideal outcomes the winged bean is treated as a yearly harvest. The sinewy roots are various with the primary laterals running evenly close to the dirt surface; following a couple of months they normally become thickened and tuberous, close to the base of the plant; however, this doesn't happen in all strains. The roots are ordinarily vigorously nodulated. Plants in Malaysia may convey up to 440 huge knobs each and their new weight can arrive at 800 kg/ha. A solitary knob may weigh 0.6 g and have a width of up to 1.2 cm. The stem is tolerably thick, marginally furrowed and notched, and can arrive at 3–3.6 m in stature whenever given help. The leaves are trifoliate, on long firm petioles; the handouts are applaud, 7.5–15 cm long with the terminal one generally longer than the laterals and appended to the petiole by a checked pulvinus. The inflorescence is borne on an axillary raceme up to 15 cm long, with 2–10 blossoms which might be blue, white, or lilac. It has been accounted for that fertilization in certain species is by honey bees, and in their nonattendance case set is exceptionally low. The cases are four-sided, with qualities serrated wings running down the four corners. They contain 5–20 seeds which can fluctuate in shading from white, through differing shades of yellow and dark coloured to dark, and may likewise be mottled.

Ecology and distribution: The winged bean is thought to have started in Africa (Madagascar or Mauritius) and to have spread to Asia, and is currently developed for the most part as a market garden crop in southern India, Burma, Malaysia, New Guinea, Indonesia, the Philippines, China, and Thailand, and to a lesser degree in Africa, primarily in Ghana and Nigeria, and in the West Non-mainstream players.

Pests and diseases: At the point when developed in blended market garden culture or moving farming the winged bean is commonly liberated from genuine vermin and maladies however every so often assaulted by caterpillars, leaf excavators, grasshoppers, insect bugs, and nematodes. Presumably the most genuine infection, which happens in Papua.

19.12.1 *Cajanus cajan*

Description: The pigeon pea (*Cajanus cajan*) also known as pigeon pea, red gram, or tur is a perennial legume from the family Fabaceae (Ministry of Environment, forestry and climate charge, 2016). It originates from India and can be referred to as Congo pea. It is one of the most widely cultivated legumes because of its eatable seeds in the tropics and subtropics. Pigeon pea grows fast, strong, usually resilient, and drought resistance (Bekele-Tesemma, 2007). Pigeon pea is a vastly divided shrub with a woody base, slim stems, and trifoliate leaves. It is a quick-growing legume with a lengthy tap root (to 2 m) and a mass of fibrous roots. It grows up to 4 m (13 ft) in height but is not tolerant to low temperature or frost. Young stems are angled and pubescent.

Uses: The crop is mostly grown in the southern and central regions of the country and is primarily used for human consumption (Ayenan et al., 2017). Pigeon pea is a food crop (dried peas, flour, or green vegetable peas). When added to cereals, pigeon peas provide a balanced meal and hence are favoured by nutritionists as an essential ingredient for balanced nourishment. The dried peas might be grown temporarily, and then prepared for a flavour different from the green or dried peas. Sprouting also enhances the digestibility of dried peas via the reduction of indigestible sugars that would otherwise remain in the cooked peas (Akporhonor et al., 2006).

Pigeon pea has several uses in livestock feeding. The leaves and pods are valued and tasty protein-rich supplements. Leaves can be used to substitute alfalfa in ruminant foods in places where alfalfa are not cultivated. In most cases, the whole seeds and its by-products are used to feed livestock (Phatak et al., 1993). Pigeon pea is integrated in the cropping systems mainly in alley cropping for soil fertility restoration (Aihou et al., 2006; Versteeg and Koudokpon, 1993). It functions as a nitrogen-fixing cover/forage crop, as erosion control, as shade for other vegetables and herbs. The stems of pigeon pea are used as firewood and the dry stalks can be burned or used as thatch or in weaving baskets. It is useful in pest management (Atachi et al., 2009).

Distribution and ecological footprints: The source of *Canjanus cajan* is either from Eastern Africa or India (Ecocrop, 2016; Van der Maesen, 1989). The plant tilling is about 3,000 years ago (Mallikarjuna et al., 2011; Van der Maesen, 1989). Of recent, it has become a subtropic and pantropic species principally suited for rain-fed cultivation in semi-arid regions due to its deep tap root, heat resistance, and quick-growing nature (Mallikarjuna et al. 2011). It grows best in places where temperature ranges from 20°C to 40°C and deprived of hoarfrost (FAO, 2016a), and where annual precipitation is above 625 mm.

19.12.1.1 Economic Significance

It can be measured of highest relevance in food security in areas with limited rainfall and that are predisposed to droughts (Crop Trust, 2014). Useful in the field of medicine – Morton (1976) lists many folk medicinal uses for pigeon pea. In India and Java, the young leaves are applied to sores. Indochinese claim that the powdered leaves help expel bladder stones. Salted leaf juice is taken for jaundice. Leaves are also used for toothache, mouth wash, sore gums, child-delivery, and dysentery. Scorched seed added to coffee are said to alleviate headache and vertigo. Fresh seeds are said to help incontinence of urine in males, while immature fruits are believed of use in liver and kidney ailments (Duke, 1981).

19.12.2 *Phaseolus lunatus*

Description: *Phaseolus lunatus* L. (Lima bean) is a tropical and subtropical legume nurtured for its palatable seeds. There are uninhabited and habited types of *Phaseolus lunatus*, commonly denoted to as *Phaseolus lunatus* var. *silvester* Baudet and *Phaseolus lunatus* var. *Lunatus*, respectively. Lima beans (*Phaseolus lunatus*) belong to the family Fabaceae. Common names include: butter bean, lima bean, sugar bean, Madagascar bean, and java bean. As a crop, *Phaseolus lunatus* is the second most significant legume of the genus *Phaseolus* on the world (Maquet et al., 1999). It is an important source of protein for rural populations in South America and Africa (Lioi et al., 1998). In Nigeria, Lima bean is cultivated mainly for the dry seeds. Like other grain legumes, it is an important source of vegetable protein and it also improves soil fertility. It is well adapted to the humid rainforest environment of southern Nigeria. Lima bean (*Phaseolus lunatus* L.) is still regarded worldwide as a minor crop in spite of its ubiquity, diversity, and yield potential.

Uses: *P. lunatus* is cultivated primarily for its seeds, which are used in soups, salads, stews, and as a vegetable. Particularly in Asia and Africa, immature sprouts, leaves, and pods are also consumed. In the Philippines and Africa, dry seeds are used to produce high protein bean flour for enriching bread or noodles. Despite the great potential of the crop, it is highly underutilized in Nigeria and is known as "kapala" (Yoruba), "ukpa" (Igbo) in Southwestern and South-eastern Nigeria, respectively, where the seeds are commonly consumed (Ezeagu and Ibegbu, 2010; Seidu et al., 2014). Seeds and leaves are valued for their astringent qualities and consequently used as a diet for fever in traditional Asian medicine.

Despite the great potential of this crop, it is highly underutilized in the country and it has not received much attention in term of crop improvement and so local cultivars are still being grown by the farmers. Lima bean has not received the benefit of intensive research programme as devoted to cowpea and soybean (Lyman et al., 1985). It is cultivated in only about 4% land area devoted to grain legumes production in south-west Nigeria with no improved technology targeted towards the production of the crop and so yield is low. It is usually intercropped with cassava, maize, yam, cocoyam, and pepper. It is mainly produced for consumption as only about 35% of the grain produced is sold (Saka et al., 2004). There is, therefore, urgent need for the improvement of this crop.

After harvesting the pods, vines are occasionally fed to cattle. Seeds are maybe used as fodder. The leaves and stems may be turned into hay or silage. The silage contains 27.3% dry matter and up to 14.2% consumable nutrients. Lima bean can be used as cover cropping or biomanure. The high fibre content in lima beans prevents blood sugar levels from rising too rapidly after eating them due to the presence of those large amounts of absorption-slowing compounds in the beans, and the high soluble fibre content. Soluble fibre absorbs water in the stomach, forming a gel that slows down the absorption of the bean's carbohydrates. They can therefore help balance blood sugar levels while providing steady, slow-burning energy, which makes them a good choice for people with diabetes suffering with insulin resistance. The protein, carbohydrate, and mineral element contents of the seed are 24%, 61%, and 15%, respectively. One of the limitations to the eating of lima beans, like other beans, is that certain varieties require a longer time to cook or soak than others (Giami, 2001).

19.12.3 *Vigna luteola*

Vigna luteola (wild cowpea) is widely known in North America. The plant is a native, persistent, herbaceous vine that has trailing or twining branches that span 3 ft in length or longer. Formerly referred to scientifically as *Vigna repens*, it possesses numerous names such as dear pea, wild cowpea, and hairypod cowpea. Being a member of the subfamily papilionoideae of the Fabaceae family, *Vigna luteola* can be seen in tropical zones of different continents. Previously inherent in the New World, the plant was first cultivated in Ethiopia before spreading across the globe (Austin, 2004). *Vigna luteola* is non-specific in its *Rhizobium* requirement. It is day-neutral and flowers throughout the year.

Uses: The flowers of *Vigna luteola* are eaten as a boiled vegetable in Ethiopia and Malawi, and the roots are chewed for the sweet juice. Both leaves and flowers have phytomedicinal uses in some parts of Africa including Ethiopia, Malawi, and Nigeria.

Vigna luteola is most often considered a weed for crops due to its abundance (Pomilio and Zallocchi, 1989). However, the plant is used as a pasture plant and as a ground cover in many countries, such as Ghana, Zambia, and Australia. However, its short lifespan and vulnerability to insects and frost can make it ineffective (Grubben and Denton, 2004). *V. luteola* is a productive and palatable species that can be used as a short-term (±3 years) legume in pasture, particularly in damp areas. It is one of the best legumes for wet conditions and one of the best pioneer plants in such situations.

Distribution and ecological footprints: *Vigna luteola* can be found in tropical areas on many continents. Formerly native to the New World, the plant was brought into cultivation in Ethiopia and is now spread around the world (Austin, 2004). It has also been found in Zimbabwe, Senegal, Ethiopia, Egypt, and Malawi (Friiset al., 1998; Grubben and Denton, 2004).

19.12.4 *Vigna reticulata*

Vigna reticulata is a wild species whose distribution is limited to Africa. It is a leguminous species belonging to section Reticulatae, subgenus *Vigna*. Subtribe Phaseolinae, subfamily Papilionoideae of family Leguminosae (Aliyu et al., 2008).

Ecological distribution and footprint: Widespread in tropical Africa from Sierra Leone to Angola, Ethiopia, Mozambique, and Zimbabwe (Conservatoire et Jardin botaniques, 2012).

Ecology: Inundated places; hollows: edge of forest; grassland; woodland; grassy woods; miombo woodland; rocky and sandy places; Savannah; termite mounds; also as a weed: fallow land, cultivations; 0–2,400 m alt. (Conservatoire et Jardin botaniques, 2012).

19.12.5 *Vigna heterophylla*

Description: *Vigna heterophylla* is a legume with herbaceous stem, twining, densely clothed with deflexed bristly ferruginous hairs. Stipulus small, ovate (Baker, 1875). It belongs to the group Dicot; Family Fabaceae; Genes *Vigna;* Species *V. heterophylla* (A. Rich.) (Darbyshire et al, 2015).

Distribution: The plant is well distributed in Ethiopia, Kenya, Tanzania, Uganda, Cameroon Central African Republic, Rwanda, Zaire, and Zambia.

Economic value: It is interestedly written that NULs present great chances for combating poverty, starvation, and undernourishment (Padulosi et al., 2014). Also, wild relatives of legumes also present undeniable potential genetic resources that could be used for crop improvement (Dwivedi et al., 2008). However, the social and practical uses of non-domesticated wild and under-exploited *Vigna* species are almost not documented (Tomooka et al., 2014).

Pest and diseases: Many *Vigna* genus in the wild being under-exploited have been stated with promising agricultural characteristics such as disease resilience (Oyatomi et al., 2016), vital nutrients and mineral components (Difo et al., 2015; Marconi et al., 1997), and nutraceuticals (Bhat and Karim, 2009). Domesticated legumes do not show as much resistance to diseases and pest infestation (Difo, 2019). Studies show that domesticated legume crop production is usually challenged by biotic (pest and diseases) and abiotic (heat, frost, drought, and salinity), edaphic factors (associated with soil nutrient deficits) (Ojiewo et al., 2018). The study helped farmers' awareness on the usefulness of wild legumes that needs the attention of researchers (Difo, 2019).

19.12.6 *Vigna marina*

Description: *Vigna marina* is a prostrate, creeping vine and a perennial legume with stems 2–3 m long and often purplish-striped (Beach Protection Authority of Queensland, 2011). It is also known as the beach pea, nanea (Starr and Starr, 2011). It is a species of legume in the family Fabaceae (the peas).

Ecology and distribution: *Vigna marina* grows at the vegetation line on sandy beaches, frontal dunes, and beach ridges near the seashore (Description of Major Dune Plants, 2011). *Vigna marina* is rarely found inland on sea cliffs and dry shrubby slopes to about 395 ft (Native Plants Hawai'i, 2011). With *Vigna marina* being a salt tolerant species, it is found on tropical beaches around the world (United States Department of Agriculture, 2020). As presented on Table 19.1, it is found in the following distributional range:

Pests and diseases: This vining groundcover is prone to ants, spider mites, leaf miners, and seed weevils (Native Plants Hawai'i, 2011)

Economic impact: *V. marina* is regarded as a nuisance weed in the Pacific. It is also promising to act as cover and forage crop especially for saline conditions, but more research is needed to optimize its use (*Vigna marina* (PROSEA) – Plant Use English, 2016). Nonetheless, the economic impact of *V. marina* is unknown, believed to be insignificant.

19.12.7 *Vigna vexillata*

Vigna vexillata (zombi pea, wild cowpea) is a variable, perennial climbing plant found in regions such as Ethiopia, Nigeria, and Venezuela (Palaniswami and Peter, 2008). Perennial climbing or trailing herb 3–6 m long, from a tuberous, spindle-shaped woody rootstock, stems sparsely to very densely, covered with brown hairs (Conservatoire et Jardin botaniques, 2012). It is a twinning vine widely distributed in tropical Africa, Asia, and Australia, which produces large, thickened edible tubers.

TABLE 19.1
Distribution of *Vigna marina*

Region	Distribution
Africa	EAST TROPICAL AFRICA: Tanzania (coast)
	WEST-CENTRAL TROPICAL AFRICA: Cameroon (coast), Congo (coast), Democratic Republic of the Congo (coast)
	WEST TROPICAL AFRICA: Benin (coast), Ghana (coast), Liberia (coast), Nigeria (coast)
	SOUTH AFRICA: South Africa (Kwazulu-Natal (coast))
	WESTERN INDIAN OCEAN: Madagascar (coast), Seychelles (coast)
Pacific	NORTHWESTERN PACIFIC: Micronesia
	SOUTH-CENTRAL PACIFIC: French Polynesia
	SOUTHWESTERN PACIFIC: Vanuatu
South America	CARRIBEAN: Cuba (coast), Jamaica (coast), United States, Puerto Rico (coast)
	CENTRAL AMERICA: Panama (coast)
	BRAZIL: Brazil (Bahia (coast))
Asia-Tropical	INDIAN SUBCONTINENT: India, Maldives (coast), Sri Lanka (coast)
	PAPUASIA: Papua New Guinea (coast); Solomon Islands
	INDO-CHINA: Thailand (coast), Vietnam (coast)
	MALESIA: Indonesia (coast), Malaysia (coast). Philippines (coast)

Source: United States Department of Agriculture (2020).

V. vexillata is also regarded as a pasture cover crop, green manure, and erosion control plant (Garba and Pasqu et, 1998).

Ecology and distribution: *Vigna vexillata* can be found in grassland with scattered trees; thickets, like in the Somali Acacia-Commiphora bushlands; abandoned cultivations; Secondary formations; meadows, sunny places; grassy inundated plain (Conservatoire et Jardin botaniques, 2012). *V. vexillata*, sometimes known as "wild cowpea" is considered to be closely related to *Vigna unguiculata* (L) Walp. (Bisht et al., 2005; Delgado-Salinas et al., 1993). In Africa, *V. vexillata* is generally found between 1 and 2,300 in altitude, except for a variety of *vexillata*, which is found as high as 3,860 m altitude (Maxted et al., 2004). *V. vexillata* can also be found in the following distribution range (Table 19.2):

Pest and diseases: Varieties of *Vigna vexillata* are resistant to infestation and damage by the pod-sucking bug *Clavigralla tomentosicollis* than the cowpea varieties (Chiang and Singh, 1988). *V. vexillata* seeds are known to be resistant to cowpea weevil, one of the main pests of cowpea (Garba and Pasquet, 1998).

Economic importance: *Vigna vexillata* is a unique plant that can be used as forage for animal food; environmentally as a soil improver and vegetables for human consumption (United States Department of Agriculture, 2020). The plant is a very important food crop in several areas such as Namibia, where it is commonly harvested from the wild for local use (Conservatoire et Jardin botaniques, 2012). The root is consumed raw or cooked. The tubers are considered rich in protein and boiled or roasted as well as the young leaves, young pods, and seeds cooked and eaten as vegetable (National Research Council, 2002).

TABLE 19.2
Distribution of *Vigna vexillate*

Region	Distribution
Africa	NORTHEAST TROPICAL AFRICA: Chad, Ethiopia, Somalia, Sudan
	EAST TROPICAL AFRICA: Kenya, Tanzania, Uganda
	WEST-CENTRAL TROPICAL AFRICA: Burundi, Cameroon, Democratic Republic of the Congo, Equatorial Guinea, Rwanda
	WEST TROPICAL AFRICA: Cote D'Ivoire, Gambia, Ghana, Guinea-Bissau, Liberia, Mali, Niger, Nigeria, Senegal, Sierra Leone, Togo
	SOUTH TROPICAL AFRICA: Angola, Malawi, Mozambique, Zambia, Zimbabwe
	SOUTHERN AFRICA: Botswana, Eswatini, South Africa
Asia-Tropical	INDIAN SUBCONTINENT: Bangladesh, Bhutan, India, Nepal, Sri Lanka
	INDO-CHINA: Cambodia, Laos, Myanmar, Thailand, Vietnam
South America	CENTRAL AMERICA: Belize, Costa Rica, Guatemala, Honduras, Nicaragua, Panama
	NOTHERN SOUTH AMERICA: French Guiana, Guyana, Suriname, Venezuela
	BRAZIL: Brazil
	WESTERN SOUTH AMERICA: Bolivia, Colombia, Ecuador
Asia-Tropical	PAPUASIA: Indonesia (Papua), Papua New Guinea

Source: United States Department of Agriculture (2020).

19.13 UTILIZATION OF NUL IN MEETING FOOD AND NUTRITIONAL REQUIREMENTS

The obliteration of malnutrition, famine, and all kinds of hunger is fundamental to the Sustainable Development Goals (SDGs). The SDGs emphasize precisely on putting an end to hunger, attaining food safety, improving nutrition, and encouraging sustainable agrarian by 2030. Achieving "Zero Hunger" suggests leaving no stone unturned vis-à-vis deprivation and all sorts of undernourishment. To guarantee that people have access to appropriate nourishing food now and tomorrow necessitates extensive advancement of sustainable agronomy and preservation of biodiversity enlarged asset in farming.

Many of these NULs grown for food, fibre, fodder, oil, and as sources of traditional medicine play a major role in the subsistence of local communities and frequently are of special social, cultural, and medicinal value. With good adaptation to often marginal lands, they constitute an important part of the local diet of communities by providing valuable nutritional components, which are often lacking in staple crops (Jain and Gupta, 2013).

Traditional crops classically do not meet contemporary ethics for consistency and other features as they have been abandoned by breeders from the private and public sectors (Stamp et al., 2012). Thus, they tend to be less competitive in the marketplace compared with commercial cultivars. Currently NUL food sources ranging from minor grains and pulses, root and tuber crops, and fruits and vegetables to non-timber forest products have the potential to make a substantial contribution to food and nutrition security, protect against internal and external market disruptions and climate uncertainties, and lead to better ecosystem functions and services, thus enhancing sustainability (Keatinge et al., 2010).

A wider use of neglected and undervalued crops and species, either intercropped with main staples in cereal-based systems or as stand-alone crops, would provide multiple options to build temporal and spatial heterogeneity into uniform cropping systems, thus enhancing resilience to biotic and abiotic stress factors and ultimately leading to a more sustainable supply of diverse and nutritious food (Keatinge et al., 2010). As source of essential vitamins, micronutrients, protein, and other phytonutrients, underutilized legume crops such as mungbean have the potential to play a major role in strategies to attain nutritional security. Apart from their superior nutritional qualities, commercial, medicinal, and cultural value some of these produces do not need high agricultural efforts. Others can be cultivated in peripheral soils and simply intercropped or alternated with essential crops, and can basically suit integrated schemes like agroecology. NULs are also vital for sustainable food production as they reduce the impact of production systems on the environment. In all, due to their high adaptation to marginal circumstances, various NULs have the exceptional capability to bear or survive stresses (Ikhajiagbe et al., 2019; (Ohanmu and Ikhajiagbe, 2018), and also can make production schemes more sustainable.

19.14 EFFORTS AT ENHANCING ACCEPTABILITY OF NEGLECTED OR UNDERUTILIZED LEGUMES

19.14.1 Biotechnology

In an offer to proffer a strong and sustain food security, the world's present food scheme should be improved as it is defenceless and depends on a tight scope of food things. Current cultivating plans favour monocultures and institutionalized methodologies; this is quicker and progressively helpful for makers yet may destabilize food safety, sustenance, and biodiversity. Depending on a little division of staple harvests for our eating regimen have critical ramifications for food safety and sustenance. The absence of hereditary assorted variety inside the genetic supplies of the predominant developed yields leaves agrarian frameworks defenceless to vermin, infections, and other abiotic stresses. Rediscovering and development of ignored yields is one approach to lessen over-dependence on a couple of significant crops. Rural maintainability depends on a sound collaboration among farming and nature. This snare of natural associations works at three progressive degrees of hereditary assorted variety: agro-environments, decent variety among species (between explicit assorted variety), and inside species (intra-explicit assorted variety).

Intra-explicit assorted variety is the least complex type of decent variety that would forestall an overwhelming situation, for example, the Irish potato starvation, that resulted to over one million individuals kicks to the bucket due to famine right around two centuries back. Minimizing dismissed plants imperils agrarian biodiversity and compromises the maintainability of our diet frameworks. Regularly, NULs can increment farming supportability by decreasing the requirement for outer sources of info, for example, inorganic manures and pesticides. Coordinating promising NULs in cultivating frameworks can lessen the development of irritations and sicknesses when they are developed in pivot with the fundamental yields.

Contingent upon their attributes, NULs can likewise build soil ripeness, forestall soil disintegration, decrease vanishing, and smother weed development. NULs are regularly less requesting of their condition, thus offer more grounded versatility to environmental change. They can be progressively impervious to biotic burdens and give reliable gathers under horrible climatic settings or drained soils. For example, DeFries et al. (2016) found out in Central India that little millets are able to withstand climate fluctuations. This contrasted with rice, which was increasingly touchy to precipitation changeability. Canahua (*Chenopodium pallidicaule*), NUL Andean plant that is a piece of spinach and firmly identified with quinoa, in excess of 200 assortments. It is amazingly ice open minded and has high healthy benefit as far as protein substance and dietary fibre levels near quinoa (*Chenopodium quinoa*) (Giuliani et al., 2012). This versatile limit is a key characteristic of numerous NULs. They are developed in zones where significant crops needing high data sources come up short, and they to be progressively impervious to nearby nuisances and maladies. Furthermore, a few NULs in Nigeria have been reported (Ikhajiagbe et al., 2013; Ikhajiagbe, 2016). In this way, NULs give a place of refuge when the climate conditions are negative, or when outside information sources become inaccessible, unreasonably expensive, or unsafe to the earth.

19.15 SYNERGY WITH MAINSTREAM CROPS

During the 1940s and mid-1950s, Nigeria didn't need to battle with the issue of food instability. The noteworthy impact of the decay like a few nations of the world, the country's economy is feeling the brunt of the increasing expense of food things, particularly the ascent in the costs of staple foods. Altogether, the cost of rice has expanded by over 100% since 2006. It is informational to take note of that Nigeria requires 2.5 million metric huge amounts of rice every year while neighbourhood rice production is not exactly a large portion of a million metric tons for each year. With these figures as discharged by Minister of Agriculture and Water Resources, Nigeria is shy of 2,000,000 tons of rice, which it needs to source from different nations. It is assessed that Nigeria spent an incredible $2 billion dollars bringing in around 6,000,000 tons of wheat, $750 million on rice, $700 million on sugar, and $500 million on milk and other dairy items. As things stand now, Nigeria is probably going to spend considerably more. Subsidizing for preparing sustenance masters at all degrees of government should hence be a need. Mediations planned for improving weight control plans and raising degrees of food should join general well-being, sustenance instruction, and dietary systems. Agriculturalists need to work with nutritionists to distinguish shortages in neighbourhood diets and micronutrient admissions. Food-based mediations will be increasingly compelling when joined by network well-being programmes, promotion of dietary security programme through yields, aquaculture and animals expand the production of supplement rich food sources for direct utilization just as salary age. It is critical to make smallholder farmers and buyers in both urban and provincial zones of poor nations mindful of the advantages of preserving and utilizing Neglected and underutilized Species and to make researchers and policymakers mindful of the need to both secure and advance Neglected and underutilized Species which analysts, farmers, buyers, and getting the message across to approach creators implies giving proof that rationing and utilizing Neglected and Underutilized Species can profit networks, add to human sustenance and well-being, and bring benefits both broadly and universally disregard and underutilize Neglected and Underutilized Species since they are uninformed of their financial and healthy benefit. Agribusiness-based intercessions to improve sustenance security and tight the "food gap", to improve the likelihood that increments underway or buying power do prompt quickened decreases in lack of healthy sustenance, agricultural advancement programmes must incorporate sustenance security (notwithstanding food security) in their goals. In a rural setting, food security alludes to the "quality" part of food production, utilization, and use by all people in a family unit. While food security may build the all-out amount of vitality accessible for utilization, just sustenance security can ensure the quality and assorted variety of food vital for ensuring and advancing great nutritional status and well-being (Thompson and Amoroso, 2011). Similarly, as improving food security can be thought of as far as narrowing the gap among present and potential production yields, improving sustenance security can be thought of as far as narrowing the "food gap" between current food consumption examples and admission designs that are ideal regarding large scale and micronutrient content. Narrowing the food gap implies improving dietary decent variety through expanding accessibility and access to the foods fundamental for a sound eating regimen, and expanding the real admission of those food sources.

Key strategy suggestions for narrowing the food gap incorporate, consolidating unequivocal sustenance goals into agrarian research plans, advancement strategies and projects, building institutional and specialized limit at national and decentralized level, and advancing the idea of "food touchy" or "sustenance upgrading" agriculture locally and all around. Join express sustenance targets into farming advancement arrangements. It must be stressed in the process that the enhancement of sustenance outcomes, rather than, ultimately, the expansion of production to attain national independence, is necessary. Extra significant procedures incorporate adjustment measures for environmental change, advancement of expert poor food, and agrarian exchange arrangements that are helpful for sustenance security and empower positive overflows into nearby markets, and advancement of social well-being nets that diminish hurtful methods for dealing with stress among the most defenceless. The acknowledgement that sustenance stays caught in a "low need cycle" (Nattalicchio et al., 2009) in numerous nations is likewise picking up grasp.

19.16 COMPARISON BETWEEN NUL AND MAINSTREAM CROPS

NULs have significant features that include however not constrained to the accompanying. They are:

1. Associations with the traditional inheritance of their homes of inception.
2. Homegrown and conventional produces developed with indigenous information for which appropriation, science, development, and utilizations are inadequately recorded.
3. Adaptation to explicit agro-biological specialties, severe atmospheres, and marginal land.
4. Feeble or no proper seed source and duplication organizations.
5. Locally significant relationship with traditional utilizations and conventional food.
6. Produced in conventional invention frameworks with next to zero outside information sources.
7. Little consideration given by national research, extension administrations, dogma and decision makers, innovation suppliers, and purchasers.
8. Hardly signified in ex situ assortments.
9. In spite of having beneficial and therapeutic characteristics or additional numerous utilizations, they are not commonly appealing to the general population and private divisions.

19.17 CONCLUSION

NULs have numerous potentials in the ever-expanding population, decreasing all types of ailing health and upgrading economical agriculture. They have a key task to carry out in manageable farming because of their dexterity and flexibility to low-enter frameworks, resilience to climate fluctuations, and high healthy benefits. There is a requirement for expanded venture and purposeful endeavours to create and improve Food Sufficiency for Farmers (FSF) along the worth chain, utilizing economical farming.

REFERENCES

African Development Bank. (2011). Programmes for Infrastructure Development in Africa. 461–464.

Abitogun, A. S. and Olasehinde, E. F. (2012). Nutritional Evaluation of seed and characterization of crude jack bean (*Canavalia ensiformis*) oil. *IOSR Journal of Applied Chemistry*, *1*(6), 36–40. https://doi.org/10.9790/5736-0163640.

Adebowale, K. O., and Lawal, O. S. (2004). Comparative study of the functional properties of bambarra groundnut (*Voandzeia subterranean*), jack bean (*Canavalia ensiformis*) and mucuna bean (*Mucuna pruriens*) flours. *Food Research International*, *37*(4), 355–365. https://doi.org/10.1016/j.foodres.2004.01.009.

Ade-Omowaye, B. I. O., Tucker, G. A., and Smetanska, I. (2015). Nutritional potential of nine underexploited legumes in southwest Nigeria. *International Food Research Journal*, *22*(2), 798–806. https://www.researchgate.net/publication/281738665_Nutritional_potential_of_nine_underexploited_legumes_in_Southwest_Nigeria.

Adewale, B. D., Dumet, D. J., Vroh-Bi, I., Kehinde, O. B., Ojo, D. K., Adegbite, A. E., and Franco, J. (2012). Morphological diversity analysis of African yam bean (*Sphenostylis stenocarpa* Hochst. ex A. Rich.) Harms and prospects for utilization in germplasm conservation and breeding. *Genetic Resources and Crop Evolution*, *59*(5), 927–936. https://doi.org/10.1007/s10722-011-9734-1.

Africa Progress Panel. (2010). Raising agricultural productivity in Africa: options for action and the role of subsidies. http://www.cpahq.org/cpahq/cpadocs/Agricult%20Productivity.pdf.

Aihou, K., Sanginga, N., Vanlauwe, B., Diels, J., Merckx, R., and Van Cleemput, O. (2006). Soil factors limiting growth and establishment of pigeon pea (*Cajanus cajan* (L.) Millsp. In: Farmers' fields in the derived savannah of the Benin. *Bulletin de La Recherche Agronomique Du Bénin*, *52*, 12–21.https://www.google.com/url?sa=t&source=web&rct=j&url=http://www.slire.net/download/716/articles_n_52_de_aihou_.pdf&ved=2ahUKEwiV4__FutPuAhVYTBUIHQFcDeUQFjAAegQIARAB&usg=AOvVaw3HlyjEcdDQoNbv25trbPj2.

Akinlade, J. A., Farinu, G. O., Taiwo, A. A., Aderinola, O. A., Adebayo, T. A., Ojebiyi, O. O., and Olaniran, O. A. (2007). Agronomic and nutritive evaluation of jack beans (*Canavalia ensiformis*) for fodder in the derived savannah zone of Nigeria. *International Journal of Agricultural Research*, *2*(12), 1059–1063. https://doi.org/10.3923/ijar.2007.1059.1063.

Akinnifesi, F. K., Leakey, R. B., Ajayi, O. C., Sileshi, G., Tchoundjeu, Z., Matacala, P., and Kwesiga, F. (2008). *Indigenous Fruit Trees in the Tropics: Domestication, Utillization and Commercialization* (1st ed.).CABI Publishing, Wallingford. 438p.

Akporhonor, E. E., Egwaikhide, P. A., and Eguavoen, I. O. (2006). Effect of sprouting on invitro digestibility of some locally consumed leguminous seeds. *Journal of Applied Sciences and Environmental Management*, *10*(3), 55–58. http://www.bioline.org.br/request?ja06052.

Aliyu, B., Akoroda, M. O., and Padulosi, S. (2008). Variation with *Vigna reticulata* Hooker FII. *Nigerian Journal of Genetics*, *15*(1), 1–8. https://doi.org/10.4314/njg.v15i1.42266.

Alliance for a Green Revolution in Africa. (2014). *Africa agriculture status report 2014: Climate change and smallholder agriculture in sub-Saharan Africa*. https://ccafs.cgiar.org/resources/publications/africa-agriculture-status-report-2014-climate-change-and-smallholder.

Arim, O. J., Waceke, J. W., Waudo, S. W., and Kimenju, J. W. (2006). Effects of *Canavalia ensiformis* and *Mucuna pruriens* intercrops on *Pratylenchus zeae* damage and yield of maize in subsistence agriculture. *Plant and Soil*, *284*(1–2), 243–251. https://doi.org/10.1007/s11104-006-0053-9.

Atachi, P., Dannon, E. A., and Rurema, D. G. (2009). Trap cropping and intercropping of pigeonpea (*Cajanus cajan* L. Mill sp.) in pest management of cowpea (Vigna unguiculata L. Walp.) in southern Benin: Competing risk and pest status in pod attack. *Annales Des Sciences Agronomiques*, *9*(1). https://doi.org/10.4314/asab.v9i1.43226.

Austin, D. F. (2004). *Florida Ethnobotany*. Taylor & Francis, Abingdon. 952p.

Ayenan, M. A. T., Danquah, A., Ahoton, L. E., and Ofori, K. (2017). Utilization and farmers' knowledge on pigeonpea diversity in Benin, West Africa. *Journal of Ethnobiology and Ethnomedicine*, *13*(1), 37. https://doi.org/10.1186/s13002-017-0164-9.

Baker, C. (1875). *The international plant names index collaborators (2019)*. Available online:https://doi.org/10.15468/uhllmw (accessed 28 March 2020).

Baldermann, S., Blagojević, L., Frede, K., Klopsch, R., Neugart, S., Neumann, A., Ngwene, B., Norkeweit, J., Schröter, D., Schröter, A., Schweigert, F. J., Wiesner, M., and Schreiner, M. (2016). Are neglected plants the food for the future? *Critical Reviews in Plant Sciences*, *35*(2), 106–119. https://doi.org/10.1080/07352689.2016.1201399.

Baudoin, J. P. (2006). *Phaseolus lunatus* L. Record from Protabase. Brink, M., and Belay, G. (Eds). PROTA, Wageningen.

Beach Protection Authority of Queensland. (2011). Description of major dune plants. *Coastal Sand Dunes : Their Vegetation and Management*, *4*(14). https://web.archive.org/web/20110321005422/http://www.derm.qld.gov.au/register/p00290aa.pdf.

Bekele-Tesemma, A. (2007). *Profitable agroforestry innovations for eastern Africa: Experience from 10 agroclimatic zones of Ethiopia, India, Kenya, Tanzania, and Uganda*. World Agroforestry Centre (ICRAF), Nairobi. 385p.

Bhat, R., and Karim, A. A. (2009). Exploring the nutritional potential of wild and underutilized legumes. *Comprehensive Reviews in Food Science and Food Safety*, *8*(4), 305–331. https://doi.org/10.1111/j.1541-4337.2009.00084.x.

Bisht, I. S., Bhat, K. V., Lakhanpaul, S., Latha, M., Jayan, P. K., Biswas, B. K., and Singh, A. K. (2005). Diversity and genetic resources of wild Vigna species in India. *Genetic Resources and Crop Evolution*, *52*(1), 53–68. https://doi.org/10.1007/s10722-005-0286-0.

Bolarinwa, I. F., Olaniyan, S. A., Abdul-Hammed, M., and Oke, M. O. (2015). Production and quality evaluation of Kunun-zaki (A Nigerian fermented cereal beverage) from millet and *Vigna racemosa* blends. *Journal of Chemical and Pharmaceutical Research*, *7* (9), 347–352.

Bremner, J. (2012). *Population and Food Security: Africa's Challenges*. Population Reference Bureau. https://assets.aspeninstitute.org/content/uploads/files/content/docs/ee/Population_Reference_Bureau_Population_and_Food_Security_Africa_Bremner.pdf.

Bunch, R. and ECHO Staff. (1985). *TN #10 Green manure crops*. ECHOcommunity. https://www.echocommunity.org/en/resources/e7e1575d-f183-4882-af9c-bc0ec189dbbd.

Burkill, H. M., and Dalziel, J. M. E. (1985). *The Useful Plants of West Tropical Africa (Vol. 1)*. Royal Botanic Gardens, Richmond. 976p.

Chiang, H. S., and Singh, S. R. (1988). Pod hairs as a factor in *Vigna vexillata* resistance to the pod-sucking bug, *Clavigralla tomentosicollis*. *Entomologia Experimentalis et Applicata*, *47*(2), 195–199. https://doi.org/10.1111/j.1570-7458.1988.tb01136.x.

Chivenge, P., Mabhaudhi, T., Modi, A., and Mafongoya, P. (2015). The potential role of neglected and underutilised crop species as future crops under water scarce conditions in sub-saharan Africa. *International Journal of Environmental Research and Public Health*, *12*(6), 5685–5711. https://doi.org/10.3390/ijerph120605685.

Chopra, K. and Swamy, G. (1975). *An Analysis of Demand and Supply of Pulses in India 1951–1971*.Sterling Publishers, New Delhi.

Clayton, E. (1964). *Agrarian Development in Peasant Economies: Some Lessons from Kenya*. Pergamon, Oxford. 172p.

Cobbina, J. (1992). Herbage yield and soil fertility restoration potential of some tropical forage legumes. In: K. Mulongoy, M. Gueye, and D. S. C. Spencer (Eds.). *Biological Nitrogen Fixation and Sustainability of Tropical Agriculture*. Wiley, New York. pp. 455–462.

Conservatoire et Jardin botaniques. (2012). *CJB – African plant database – Detail*. Les Conservatoire et Jardin Botaniques Ville de Genève. https://www.ville-ge.ch/musinfo/bd/cjb/africa/details.php?langue=an&id=195837.

Cook, B.G., Pengelly, B.C., Brown, S.D., Donnelly, J.L., Eagles, D.A., Franco, M.A., Hanson, J., Mullen, B.F., Partridge, I.J., Peters, M., and Schultze-Kraft, R. (2005). Tropical Forages: An interactive selection tool. Web Tool. Brisbane, Australia: CSIRO, Queensland, Australia: Department of Primary Industries and Fisheries, Cali, Colombia: CIAT and Nairobi, Kenya: ILRI.

Crop Trust. (2014). Pigeonpea: Food for drought. Retrieved from https://www.croptrust.org/impact-story/pigeon-pea-food-for-drought/.

Darbyshire, I., Pickering, H., Kordofani, M., Farag, I., and Candiga, R. (2015). *The Plants of Sudan and South Sudan: An Annotated Checklist*. University of Chicago Press, Illinois. 420p.

Decoteau, D. R. (2000). *Vegetable Crops*. Prentice Hall, New Jersey. 465p.

DeFries, R., Mondal, P., Singh, D., Agrawal, I., Fanzo, J., Remans, R., and Wood, S. (2016). Synergies and trade-offs for sustainable agriculture: Nutritional yields and climate-resilience for cereal crops in Central India. *Global Food Security*, *11*, 44–53. https://doi.org/10.1016/j.gfs.2016.07.001.

Delgado-Salinas, A., Bruneau, A., and Doyle, J. J. (1993). Chloroplast DNA phylogenetic studies in new world phaseolinae (Leguminosae: Papilionoideae: Phaseoleae). *Systematic Botany*, *18*(1), 6–17. https://doi.org/10.2307/2419784.

Diao, X., Hazell, P., Resnick, D., and Thurlow, J. (2007). The role of agriculture in development: Implications for sub-Saharan Africa. Research Report Number 153, International Food Policy Research Institute (IFPRI).

Difo, V. H. (2019). *Wild vigna* legumes. Encyclopedia. Retrieved February 6, 2021, from https://encyclopedia.pub/181.

Difo, V. H., Onyike, E., Ameh, D. A., Njoku, G. C., and Ndidi, U. S. (2015). Changes in nutrient and antinutrient composition of *Vigna racemosa* flour in open and controlled fermentation. *Journal of Food Science and Technology*, *52*(9), 6043–6048. https://doi.org/10.1007/s13197-014-1637-7.

Duke, J. A. (1981). *Handbook of Legumes of World Economic Importance*. Springer, New York. 345p.

Dwivedi, S., Perotti, E., and Ortiz, R. (2008). Towards molecular breeding of reproductive traits in cereal crops. *Plant Biotechnology Journal*, *6*(6), 529–559. https://doi.org/10.1111/j.1467-7652.2008.00343.x.

Ebert, A. (2014). Potential of underutilized traditional vegetables and legume crops to contribute to food and nutritional security, income and more sustainable production systems. *Sustainability*, *6*(1), 319–335. https://doi.org/10.3390/su6010319.

Ecoport (2009). Ecoport | Feedipedia. (2009). Retrieved from https://www.feedipedia.org/node/1635.

Ecocrop (2016). FAO, Rome, Italy | Feedipedia. (2016). Retrieved from https://www.feedipedia.org/node/21869.

Elmekki, A., and Barker, J. (1993). Politics and agrarian crisis in Sudan. *Institute of Development Studies Bulletin*, *24*(3), 68–74.

Ezeagu, I. E., and Ibegbu, M. D. (2010). Biochemical composition and nutritional potential of ukpa: a variety of tropical lima beans (*Phaseolus lunatus*) from Nigeria - a short report. *Polish Journal of Food and Nutrition Sciences*, *60*(3), 231–235.

FAO. (1992). The state of food and agriculture 1991. Rome: FAO. Agriculture Series No. 24.

FAO. (1995). *Dimensions of need – Staple foods: What do people eat?* Dimensions of need. http://www.fao.org/3/u8480e/u8480e07.htm.

FAO. (2009). High level expert forum – how to feed the world in 2050. Food and Agricultural Organization of the United Nations, Rome.

FAO. (2012a). Neglected crops need a rethink – can help world face the food security challenges of the future (also available at www.fao.org/news/story/en/item/166368/icode/).

FAO. (2012b). Grassland species index. *Lablab purpureus*. http://www.fao.org/ag/AGP/AGPC/doc/Gbase/DATA/Pf000047.HTM (accessed 6 June 2012).

FAO. (2016a). Agroecology profile 'Integrating diverse grain legume for increased land productivity on small farms in Malawi'. Rome. [Cited 4 October 2016]. http://www.fao.org/agroecology/knowledge/practices/en/.

FAO. (2016b). Grassland index. A searchable catalogue of grass and forage legumes. FAO, Rome, Italy.

FAO, IFAD, UNICEF, WFP and WHO. (2017). The state of food security and nutrition in the world 2017: Building resilience for peace and food security. FAO, Rome, Italy.

Filippetti, A., and Ricciardi, L. (1993). Faba beans. In: G. Kalloo and B. O. Bergh (Eds.). *Genetic Improvement of Vegetable Crops*. Pergamon, Oxford. pp. 355–385.

Fischler, M., and Wortmann, C. S. (1999). Green manures for maize—bean systems in eastern Uganda: Agronomic performance and farmers' perceptions. *Agroforestry Systems*, *47*, 123–138. https://doi.org/10.1023/a:1006234523163.

Fouad, M., Mohammed, N., Aladdin, H., Ahmed, A., Xuxiao, Z., Shiying, B., and Tao, Y. (2013). *Genetic and Genomic Resources of Grain Legume Improvement: 5. Faba Bean*. Elsevier, Netherlands. pp. 113–136.

Friis, C., and Reenberg, A. (2010). *Land Grab in Africa: Emerging Land System Drivers in a Teleconnected World*. GLP-IPO. https://www.researchgate.net/publication/282847606_Land_grab_in_Africa_Emerging_land_system_drivers_in_a_teleconnected_world.

Friis, I., Vollesen, K., and Kongelige Danske videnskabernesselskab. (1998). *Flora of the Sudan-Uganda border area east of the Nile*. Kgl. Danske VidenskabernesSelskab, Denmark. 855p.

Gamborg, O.L., Miller, R. and Ojima, K. (1968) Nutrient requirements of suspension cultures of soybean root cells. *Experimental Cell Research*, *50*, 148–151.

Garba, M., and Pasquet, R. S. (1998). Isozyme diversity in Vigna vexillata (L.) A. Rich (Fabaceae) complex. *South African Journal of Botany*, *64*(3), 163–175. https://doi.org/10.1016/s0254-6299(15)30861-9.

Garn, S. M., and Leonard, W. R. (2009). What Did Our Ancestors Eat? *Nutrition Reviews*, *47*(11), 337–345. https://doi.org/10.1111/j.1753-4887.1989.tb02765.x

Garrity D., Dixon, J. and Boffa, J. (2012). Understanding African Farming Systems: Science and Policy Implications. Invited paper, Food Security in Africa: bridging research into practice, 30 November, 2012, Sydney, Australia, pp. 6–12.

Giami, S. Y. (2001). Quality attributes of three new improved lines of Nigerian lima beans (*Phaseolus lunatus* L. Walp.). *Plant Foods for Human Nutrition*, *56*(4), 325–333. https://doi.org/10.1023/a:1011820907356.

Giuliani, A., Hintermann, F., Rojas, W., and Padulosi, S. (Eds.). (2012). *Biodiversity of Andean grains: Balancing market potential and sustainable livelihoods*. Bioversity International, Rome. 192p. https://www.bioversityinternational.org/uploads/tx_news/Biodiversity_and_Andean_grains__balancing_market_potential_and_sustainable_livelihoods_1635.pdf.

Graham A., Aubry, S., Kunnemann R., and Suarez, S. M. (2008). CSD Monitoring 2009–2010 "Advancing African Agriculture" (AAA): *The Impact of Europe's Policies and Practices in African Agriculture and Food Security*, pp. 17–24.

Grubben G. J. H., and Denton, O. A. (2004) *Plant Resources of Tropical Africa: Vegetables*. PROTA Foundation, Wageningen. 667p.

Harouna, D. V., Venkataramana, P. B., Matemu, A. O., and Ndakidemi, P. A. (2019). Assessment of water absorption capacity and cooking time of wild under-exploited vigna species towards their domestication. *Agronomy*, *9*(9), 509. https://doi.org/10.3390/agronomy9090509.

Ikhajiagbe, B. (2016). Possible adaptive growth responses of *Chromolaena odorata* during heavy metal remediation. *Ife Journal of Science*, *18*(2), 403–411. Retrieved from https://www.ajol.info/index.php/ijs/article/view/144863.

Ikhajiagbe, B., Anoliefo, G. O., Oshomoh, E. O., and Airhienbuwa, N. (2013). Changes in heavy metal contents of a waste engine oil polluted soil exposed to soil pH adjustments. *British Biotechnology Journal3*(2), 158–168. https://doi.org/10.9734/BBJ/2013/2374.

Jain, S. M., and Gupta, S. D. (Eds.). (2013). *Biotechnology of Neglected and Underutilized Crops*. Springer, Berlin/Heidelberg. 459p.

Kariuki, J.G. (2011).*The Future of Agriculture in Africa*. Boston University Press, Massachusetts. 39p.

Keatinge, J. D. H., Waliyar, F., Jamnadas, R. H., Moustafa, A., Andrade, M., Drechsel, P., Hughes, J. '. A., Kadirvel, P., and Luther, K. (2010). Relearning old lessons for the future of food-by bread alone no longer: diversifying diets with fruit and vegetables. *Crop Science*, *50*(S1), S51–S62. https://doi.org/10.2135/cropsci2009.09.0528.

King, R. (1977). *Land Reform: A World Survey*. Routledge, New York. 462p.

KPMG.(2013). *Agriculture in Africa*. Canada Square, London.

Lioi, L., Lotti, C., and Galasso, I. (1998). Isozyme diversity, RFLP of the rDNA and phylogenetic affinities among cultivated Lima beans, *Phaseolus lunatus* (Fabaceae). *Plant Systematics and Evolution*, *213*(3–4), 153–164. https://doi.org/10.1007/bf00985196.

Lipton, M. (1977). Why poor people stay poor: a study of urban bias in world development. London: Temple Smith. --, 1985 The place of agricultural research in the development of Sub-Saharan Africa. *Institute of Development Studies bulletin*, *16*(3), 14–23.

Lloyd, P. J. (1969). The Role of "Agricultural Surpluses" in Economic Development. *The Developing Economies*, *8*(1), 39–51. https://doi.org/10.1111/j.1746-1049.1970.tb00547.x.

Lyman, J. M., Baudoin, J. P., and Hidalgo, R. (1985). Lima bean (*Phaseolus lunatus*). In: R. J. Summerfield and E. H. Roberts (Eds.). *Grain Legume Crops*. Collins, London. pp.477–519.

Maass, B. L., Knox, M. R., Venkatesha, S. C., Angessa, T. T., Ramme, S., and Pengelly, B. C. (2010). *Lablab purpureus*—A crop lost for Africa? *Tropical Plant Biology*, *3*(3), 123–135. https://doi.org/10.1007/s12042-010-9046-1.

Mackinder, B., Pasquet, R., Polhill, R., and Verdcourt, B. (2001). *Vigna racemosa G. Don Hutch. and Dalz. [family LEGUMINOSAE]*. JSTOR. https://plants.jstor.org/stable/10.5555/al.ap.flora.fz2338.

Magbagbeola, J. A. O., Adetoso, J. A., and Owolabi, O. A. (2010). Neglected and underutilized species (NUS): A panacea for community focused development to poverty alleviation/poverty reduction in Nigeria. *Journal of Economics and International Finance*, *2*(10), 208–211.

Marconi, E., Ruggeri, S., and Carnovale, E. (1997). Chemical evaluation of wild underexploited Vigna spp. seeds. *Food Chemistry*, *59*(2), 203–212. https://doi.org/10.1016/s0308-8146(96)00172-0.

Mallikarjuna, N., Saxena, K. B., and Jadhav, D. R. (2011). Cajanus. In: Chittaranjan Kole (Ed.). *Wild Crop Relatives: Genomic and Breeding Resources - Legume Crops and Forages*. Springer, Berlin/Heidelberg. pp. 21–33.

Mansoor, A., and Inotai, A. (1991). Integration efforts in sub-saharan Africa: failures, results and prospects - a suggested strategy for achieving efficient integration. In: Chibber, A. and Fischer, S. (Eds.). *Economic Reform in Sub-Saharan Africa*. World Bank, Washington, DC. pp. 217–232.

Maquet, A., Vekemans, X., and Baudoin, J.-P. (1999). Phylogenetic study on wild allies of lima bean, *Phaseolus lunatus* (Fabaceae), and implications on its origin. *Plant Systematics and Evolution*, *218*(1–2), 43–54. https://doi.org/10.1007/bf01087033.

Martinez-Palacios, C. A., Galván Cruz, R., Olvera Novoa, M. A., and Chávez-Martinez, C. (1988). The use of jack bean (*Canavalia ensiformis* Leguminosae) meal as a partial substitute for fish meal in diets for tilapia (*Oreochromis mossambicus* Cichlidae). *Aquaculture*, *68*(2), 165–175. https://doi.org/10.1016/0044-8486(88)90239-6.

Matthews, P., and Marcellos, T. (2003). Faba bean. Agfact P4.2.7, 2nd ed, 1–24.

Maxted, N., Mabuza-Diamini, P., Moss, H., Padulosi, S., Jarvis, A., and Guarino, L. (2004). An ecogeographic study. African vigna. *Plant Genetic Resources*, *5*(1), 55–56.

Maxwell, S. (1988). Cash crops in developing countries. *IDS Bulletin*, *19*(2), 1–4. https://doi.org/10.1111/j.1759-5436.1988.mp19002001.x.

McIntyre, B. D., Gold, C. S., Kashaija, I. N., Ssali, H., Night, G., and Bwamiki, D. P. (2001). Effects of legume intercrops on soil-borne pests, biomass, nutrients and soil water in banana. *Biology and Fertility of Soils*, *34*(5), 342–348. https://doi.org/10.1007/s003740100417.

Ministry of Environment, Forestry and Climate Change. (2016). Biology of *Canjanus cajan* (Pigeon pea), Series of Crop Specific Biology Documents published by the Genetic Engineering Appraisal Committee of the Government of India.

Mittendorf, H. J. (1988). Strategy of a marketing infrastructure in Africa. *Entwicklung Und Laendlicher Raum*, *22*(1), 6–9. https://agris.fao.org/agris-search/search.do?recordID=DE90Q0041.

Morris, J. B. (1999). Legume genetic resources with novel "value added" industrial and pharmaceutical use. In: J. Janick (Ed.). *Perspectives on New Crops and New Uses*. ASHS Press. pp. 196–201. http://www.hort.purdue.edu/newcrop/proceedings1999/v4-196.html (accessed 22 Aug. 2012).

Morris, J. B., and Walker, J. T. (2002). Non-traditional legumes as potential soil amendments for nematode control. *Journal of Nematology*, *34*(4), 358–361. http://naldc.nal.usda.gov/download/18902/PDF (accessed 7 Jan. 2013).

Morton, J. F. (1976). The pigeon pea (*Cajanus cajan* Millsp.), a high protein tropical bush legume. *HortScience*, *11*(1), 11–19.

Murphy, A. M., and Colucci, P. E. (1999). A tropical forage solution to poor quality ruminant diets: A review of *Lablab purpureus*. *Livestock Research for Rural Development*, *11*(2), 96–113. https://lrrd.cipav.org.co/lrrd11/2/colu112.htm.

NAS/NRC. (1979). *Tropical Legumes: Resources for the Future*. Report by an ad hoc advisory panel of the Advisory Committee on Technology Innovation, Board on Science and Technology for International Development, Commission on International Relations, National Academy of Sciences and the National Research Council, Washington, DC.

Nassar, R.M., Ahmed, Y.M., & Boghdady, M.S. (2010). Botanical Studies on Phaseolus vulgaris L. morphology of vegetative and reproductive growth. *International Journal of Botany*, *6*(3): 323-333.

Natalicchio, M., Garrett, J., Mulder-Sibanda, M., Ndegwa, S., and Voorbraak, D. (2009). *Carrots and Sticks: The Political Economy of Nutrition Reform*. World Bank, Washington, DC, 64p.

National Research Council. (2002). *Tropical Legumes: Resources for the Future*. The Minerva Group Incorporated, Colorado. 340p.

Native Plants Hawai'i. (2011). *Native Plants Hawaii – Viewing Plant: Vigna marina*. http://nativeplants.hawaii.edu/plant/view/Vigna_marina.

Nwuke, K. (2012). *Unlocking the African Moment: Rural infrastructure in Africa*. Semantic Scholar. https://www.semanticscholar.org/paper/Unlocking-the-African-Moment-%E2%80%93-Rural-infrastructure-Nwuke/902a31a85b9f9417f19fa610e56adc721bdaf7e6#citing-papers.

Ohanmu, E. O., Ikhajiagbe, B., and Anoliefo, G. O. (2017). Assessment of growth and yield responses of African yam bean (*Sphenostylis stenocarpa*) to cadmium pollution. *Nigeria Journal of Life Science*, *7*(2), 184–198.

Ohanmu, E. O., and Ikhajiagbe, B. (2018). Enzymatic and non-enzymatic response of *Sphenostylis stenocarpa* to cadmium stress. *Asian Journal of Applied Sciences*, *11*(3), 125–134. https://doi.org/10.3923/ajaps.2018.125.134.

Ojiewo, C., Monyo, E., Desmae, H., Boukar, O., Mukankusi-Mugisha, C., Thudi, M., Pandey, M., Saxena, K., Gaur, M. and Chaturvedi, K., … Varshney, R. K. (2018). Genomics, genetics and breeding of tropical legumes for better livelihoods of smallholder farmers. *Plant Breeding*, *138*(4), 487–499. https://doi.org/10.1111/pbr.12554.

Oliveira, A. E. A., Sales, M. P., Machado, O. L. T., Fernandes, K. V. S., and Xavier-Filho, J. (1999). The toxicity of jack bean (*Canavalia ensiformis*) cotyledon and seed coat proteins to the cowpea weevil (*Callosobruchus maculatus*). *Entomologia Experimentalis et Applicata*, *92*(3), 249–255. https://doi.org/10.1046/j.1570-7458.1999.00544.x.

Olowokudejo, J. D., Kadiri, A. B., and Travih, V. A. (2008). Ethnobotanical survey of herbal markets and medicinal plants in Lagos Nigeria. *Ethnobotanical Leaflets*, *12*, 851–865. Retrieved from https://ir.unilag.edu.ng/handle/123456789/5514.

Oshodi, A. A., Ipinmoroti, K. O., Adeyeye, E. I., and Hall, G. M. (1995). Amino and fatty acids composition of African yam bean (*Sphenostylis stenocarpa*) flour. *Food Chemistry*, *53*(1), 1–6. https://doi.org/10.1016/0308-8146(95)95778-5.

Oyatomi, O., Fatokun, C., Boukar, O., Abberton, M., and Ilori, C. (2016). Screening wild Vigna species and cowpea (*Vigna unguiculata*) landraces for sources of resistance to *Striga gesnerioides*. In: B.V. Ford-Lloyd, M. E. Dulloo, & N. Maxted (Eds.), Enhancing crop genepool use: Capturing wild relative and landrace diversity for crop improvement. CABI, Wallingford. pp. 27–31. https://doi.org/10.1079/9781780646138.0027.

Palaniswami, M. S., and Peter, K. V. (2008). Tuber and root crops (Vol. 9). New India Publishing Agency, New Delhi. 514p.

Pandey, B. P. (2003). *Economic Botany* (6th ed.). S Chand and Group, New Delhi. 680p.

Padulosi, S., and Hoeschle-Zeledon, I. (2004). Underutilized plant species: What are they? *LEISA-LEUSDEN*, 5–6. https://lib.icimod.org/record/11453#:~:text=The%20use%20of%20the%20term, has%20not%20been%20fully%20realised.

Padulosi, S., Thompson, J., and Rudebjer, P. (2013). *Fighting poverty, hunger and malnutrition with neglected and underutilized species*. Bioversity International, Rome. 60p. http://www.new-ag.info/en/book/review.php?a=3168.

Padulosi, S., Amaya, K., Jäger, M., Gotor, E., Rojas, W., and Valdivia, R. (2014). A holistic approach to enhance the use of neglected and underutilized species: the case of andean grains in Bolivia and Peru. *Sustainability*, *6*(3), 1283–1312. https://doi.org/10.3390/su6031283.

Pennacchio, M., Jefferson, L. V., and Havens, K. (2010). *Uses and Abuses of Plant-Derived Smoke: Its Ethnobotany as Hallucinogen, Perfume, Incense, and Medicine*. Oxford University Press, England. 264p.

Pereira, B. F. F., Abreu, C. A., Herpin, U., Abreu, M. F., and Berton, R. S. (2010). Phytoremediation of lead by jack beans on a Rhodic Hapludox amended with EDTA. *Scientia Agricola*, *67*(3), 308–318. https://doi.org/10.1590/s0103-90162010000300009.

Phatak, S. C., Nadimpalli, R. G., Tiwari, S. C., Bhardwaj, H. L. (1993). Pigeon peas: Potential new crop for the southeastern United States. In: J. Janick and J.E. Simon (Eds.). *New Crops*. Wiley, New Jersey. pp. 597–599.

Pomilio, A. B., and Zallocchi, E. M. (1989). Two new kaempferol isorhamninosides from *Vigna luteola*. *Journal of Natural Products*, *52*(3), 511–515. https://doi.org/10.1021/np50063a008.

Potter, D. (1992). Economic botany of *Sphenostylis* (Leguminosae). *Economic Botany*, *46*(3), 262–275. https://doi.org/10.1007/bf02866625.

Purseglove, J. W. (1988) *Tropical Crops: Dicotyledons*. Wiley, New Jersey. 719p.

Revilla, I. (2015). Impact of thermal processing on faba bean (*Vicia faba*) Composition. In: V. Preedy (Ed.). *Processing and Impact on Active Components in Food.* Academic Press, Cambridge, pp. 337–343. https://doi.org/10.1016/B978-0-12-404699-3.00040-8.

Riddell, R., and Cockcroft, L. (1991). Foreign direct investment in sub-saharan Africa. In: J. M. Underwood and I. Husain (Eds.). *African externalfinance in the 1990s.* World Bank, Washington, DC, pp. 139–149. https://www.semanticscholar.org/paper/Foreign-direct-investment-in-Sub-Saharan-Africa-Cockcroft-Riddell/dbe4c86bfb4fb2c21370d6097ffb73253b8351cc.

Rosen, S., and Shapouri, S. (2012). *Factors Affecting Food Production Growth in Sub-Saharan Africa.* Available from http://www.ers.usda.gov/amber-waves/2012-september/factors-affecting-food-production.aspx#.U3ukI_ldWSo.

Saka, J. O., Ajibade, S. R., Adeniyan, O. N., Olowoyo, R. B., and Ogunbodede, B. A. (2004). Survey of underutilized grain legume production systems in the southwest agricultural zone of Nigeria. *Journal of Agricultural & Food Information*, *6*(2–3), 93–108. https://doi.org/10.1300/j108v06n02_08.

Sauer, J., and Kaplan, L. (1969). Canavalia Beans in American Prehistory. *American Antiquity*, *34*(4), 417–424. https://doi.org/10.2307/277739.

Schmelzer, G. H., Gurib-Fakim, A., Arroo, R., Bosch, C. H., de Ruijter, A., Simmonds, M. S. J., Lemmens, R. H. M. J., and Oyen, L. P. A. (2008). *Plant resources of tropical Africa: Medicinal Plants* (Vol. 11). PROTA Foundation, Wageningen. 869p. https://www.researchgate.net/profile/Randolph_Arroo/publication/317510018_Plant_Resources_of_Tropical_Africa_111_Medicinal_plants_1/links/593ff602458515a62187b871/Plant-Resources-of-Tropical-Africa-111-Medicinal-plants-1.pdf.

Seidu, K. T., Osundahunsi, O. F., Olaleye, M. T., and Oluwalana, I. B. (2014). Chemial composition, phytochemical constituents and antioxidant potentials of lima bean seeds coat. *Annals Food Science and Technology*, *15*(2), 288–298.

Shukla, P., and Misra, S. P. (1979). *An Introduction to Taxonomy of Angiosperms.* Penguin Random House, New York. 533p.

Singh, S. P.(1999). *Common Bean Improvement in the Twenty-First Century.* Springer, Netherlands. 405p.

Snedecor, G. W., and Cochran, W. G. (1982). *Statistical Methods.* (7th ed.). Iowa State University Digital Press, Iowa. 507p.

Stamp, P., Messmer, R., and Walter, A. (2012). Competitive underutilized crops will depend on the state funding of breeding programmes: an opinion on the example of Europe. *Plant Breeding*, *131*(4), 461–464. https://doi.org/10.1111/j.1439-0523.2012.01990.x.

Starr, F., and Starr, K. (2011). *Vigna marina – Plants of Hawaii – Starr Environmental.* Starr Environmental. http://www.starrenvironmental.com/images/search/?q=Vigna+marina.

Stuart, G. (2011). Stuartx change. *Philippine alternative medicine.* http://www.stuartxchange.org/AltMed.html (accessed 26 July 2012).

Taylor, D. R. F. (1969) Agricultural change in Kikuyuland. In: Thomas, M. F. and Whittington, G. W. (Eds.). *Environment and Land Use in Africa.* Methuen Publishing Ltd, London. pp. 463–493.

Thompson, B., and Amoroso, L. (2011). *Combating Micronutrient Deficiencies: Food-Based Approaches.* CABI Publishing, Wallingford. 430p. https://doi.org/10.1079/9781845937140.0000.

Tomooka, N., Naito, K., Kaga, A., Sakai, H., Isemura, T., Ogiso-Tanaka, E., Iseki, K., and Takahashi, Y. (2014). Evolution, domestication and neo-domestication of the genus Vigna. *Plant Genetic Resources*, *12*(S1), S168–S171. https://doi.org/10.1017/s1479262114000483.

Udedibie, A. B. I. (1990). Nutritional evaluation of jackbean (*Canavalia ensiformis*) for the Nigerian poultry industry. *AMBIO*, *19*(8), 361–365. https://www.semanticscholar.org/paper/Nutritional-evaluation-of-jackbean-(Canavalia-for-Udedibie/68b1a7ae614a407719b3b5cddfd8eb74bbcc10c2.
Udedibie, A. B. I., and Carlini, C. R. (1998). Questions and answers to edibility problem of the *Canavalia ensiformis* seeds – A review. *Animal Feed Science and Technology*, *74*(2), 95–106. https://doi.org/10.1016/s0377-8401(98)00141-2.
Udry, C. (2009). *Networks, Local Institutions and Agriculture in Africa: Notes Toward a Research Program*. http://www.econ.yale.edu/~cru2/pdf/agdev.pdf.
UNDP and World Bank. (1992). *African Development Indicators*. World Bank, Washington, DC.
United Nations Economic Commission for Africa. (2011). *Climate change and agriculture in Africa: Analysis of knowledge gaps and needs*. UNECA Working Paper 7, pp. 6–7.
United States Department of Agriculture. (2020). *Vigna marina* Germplasm Resources Information Network (GRIN). Agricultural Research Service (ARS). Available online: https://www.usda.gov (accessed 28 March, 2020).
Valenzuela, H., and Smith, J. (2002). Lablab (Green Manure Crops). *Sustainable Agriculture GM-7*, 1–3. https://scholarspace.manoa.hawaii.edu/handle/10125/12737.
van der Maesen, L. J. G. (1989). *Cajanus cajan* (L.) Millsp. In: L. J. G. van der Maesen, and S. Somaatmadja, (Eds.). *Plant Resources of South-East Asia: I. Pulses*. Pudoc, Wageningen. pp. 39–42.
Vargas-Ayala, R., Rubiano, J. A., and Saavedra, E. D. (2000). *Sustainable and Integrated Management of Plantain-parasitic nematodes based on an intercropping system with tropical legumes*. Abstracts of XXXI annual meeting of ONTA, Auburn, Alabama.
Versteeg, M. N., and Koudokpon, V. (1993). Participative farmer testing of four low external input technologies, to address soil fertility decline in Mono province (Benin). *Agricultural Systems*, *42*(3), 265–276. https://doi.org/10.1016/0308-521x(93)90058-a.
Vigna marina (PROSEA) –PlantUse English. (2016, April 27). In Plant Net. https://uses.plantnet-project.org/en/Vigna_marina_(PROSEA).
World Bank. (1989). *World Development Report*. Oxford University Press, New York.
World Food Programme. (2020, January 6). *WFP Global Hotspots 2020 | World Food Programme*. https://www.wfp.org/publications/wfp-global-hotspots-2020.
Zeven, A. C. (1998). Landraces: A review of definitions and classifications. *Euphytica*, *104*(2), 127–139. https://doi.org/10.1023/a:1018683119237.

20 Antiprotozoal Activity of Some Medicinal Plants against *Entamoeba histolytica*, the Causative Agent of Amoebiasis

Charles Oluwaseun Adetunji
Edo State University Uzairue

Oyetunde T. Oyeyemi
University of Medical Sciences

CONTENTS

20.1 INTRODUCTION

Amoebiasis is a parasitic disease instigated by *Entamoeba histolytica*, a protozoan parasite with worldwide distribution (Gonzales et al., 2019). It causes high illness and death and has become an important foremost public health challenges being considered the third parasitic disease of medical importance after schistosomiasis and malaria (Pinilla et al., 2008). Amoebic colitis or extraintestinal abscess develops

DOI: 10.1201/9781003178903-24

in about 40–50 million individuals diseased with the cystic form of *E. histolytica* resulting in 100,000 deaths annually (Choudhuri and Rangan, 2012). Amoebic colitis is a major etiology of severe diarrhea in developed and underdeveloped countries most especially among children with the age of 5 (Shirley et al., 2018). The regions of the world with highest burden of amoebiasis include Asia, the tropical and sub-Saharan regions of Africa, and South and Central America (Shirley et al., 2018). The prevalence of amoebiasis which varies in different population strata is determined by the socioeconomic status, sanitary situations, and the investigative test usage (Gonzales et al., 2019).

Entamoeba histolytica infection can be acquired through diverse means. However, the maximum common method of spread is by ingestion of food or water polluted with cysts of the parasite (Stanley, 2003; Shirley 2018, Adetunji and Anani, 2021a,b,c; Adetunji et al., 2021; Dwivedi et al., 2021; Olaniyan and Adetunji, 2021; Mishra et al., 2020; Thangadurai et al., 2021; Anani et al., 2020; Islam et al., 2021). In high-income countries, returning travelers or recent migrants from pervasive regions, homosexuals, and long-standing persons are most affected groups (Salit et al., 2009; Herbinger, 2011; Shirley 2018). There is increasing evidence of amoebiasis concomitant existence with HIV infection (Hung et al., 2012) and an important diagnostic consideration for bloody diarrhea (Petri, 2010).

Diverse drugs are currently being used to manage amoebiasis, and treatment is always guided by diagnosis and severity of the illness (Martínez-Castillo et al., 2018). Metronidazole (MTZ), which is one of the mainstay drugs used for the treatment of amoebiasis, is known to exterminate the invasive trophozoites of *E. histolytica* but does not cause damage to cysts (Martínez-Castillo et al., 2018). The utilization of MTZ has been documented to demonstrate a higher adverse effect while there are several growing evidences that *E. Histolytica* could develop resistance to MTZ. These has necessitated the search for alternative anti-amoebic therapies (Orozco et al., 2009). Medicinal plants have been favored as new sources of therapies against infectious agents owing to their long history of use by humans. Many plant-derived bioactive compounds are known to be safe for human consumption thus offering a promising potential in new drug discoveries against amoebiasis. This chapter reviews diverse plant materials that have been studied for their anti-amoebic properties over the decades with a view to discovering future new therapies for amoebiasis.

20.2 GENERAL OVERVIEW OF *ENTAMOEBA HISTOLYTICA*, THE CAUSATIVE AGENT OF AMOEBIASIS

20.2.1 Morphology and Identification

Detection and identification of *Entamoeba histolytica* and its discrepancy from *Entamoeba dispar* is a vital goal in identification (Calderaro et al., 2006). The morphological description of *Entamoeba* was dated back to as early as 1925 during which two structurally identical species with quadri-nucleated cysts assessing 10 μm or more in diameter were described (Brumpt, 1925). By then the two species were indistinguishable owing to the absence of a means to differentiate them. The identification of *E. dispar*, the nonpathogenic species which requires no treatment in infected individuals, necessitated the need for alternative and improved detection methods (WHO, 1997) with capacity to differentiate between the two amoebas (Nuñez et al., 2001). It took the work of Sargeaunt and Williams (1978) on isoenzyme typing to be able to differentiate *E. histolytica* (the species which are pathogenic in nature) from *E. dispar* (the species which are nonpathogenic). This pioneering work of Sargeaunt and colleague had opened more opportunities to develop new tools and methods to differentiate between the two species.

Molecular based methods have helped to differentiate between these species thus solving some of the sensitivity and specificity deficiencies attributed to the classical microscopy utilized for recognition of protozoan pathogens (Pechangou et al., 2015). Multiplex polymerase chain reaction (mPCR), restriction fragment length polymorphism (RFLP), real-time PCR (RT-PCR), and gene sequencing are among the molecular tools developed for the identification of *Entamoeba* species (Samie et al., 2008; Bruijnesteijn van Coppenraet et al., 2009). Other possibility such as the use of Matrix-assisted laser desorption ionization-time of flight *mass spectrometry* (MALDI-TOF MS) to characterize *Entamoeba* species-specific amino acids and proteins has been explored. This proteomics approach has been acclaimed to offer a better method of differentiating between *Entamoeba* spp. compared to molecular methods because it is less protracted and less costly (Calderaro et al., 2015).

20.2.2 Transmission Cycle

The life cycle of *E. histolytica* is direct and homoxenous. Two forms exist in the parasite's transmission cycle; the infective cyst and the invasive trophozoite. Transmission occurs via feces-contaminated hands, consumption of cyst-infested foods and water (Shirley et al., 2018). Exposure to fecal matter during sexual escape among homosexual men is another route of transmission (Hung et al., 2012). Household insects like houseflies and cockroaches have also been implicated in contamination of foods with cysts of *E. histolytica* (Oyeyemi et al., 2016). Following ingestion of cyst, the cystic wall is thrown off for emergence of trophozoites which transfer to the large intestine and undergo binary fission to generate more cysts (Martínez-Castillo et al., 2018). Trophozoites may multiply to attain great densities and cumulative, a practice thought to stimulate encystation or virulence (Faust and Guillen, 2012). This behavior is similar to what is observed in some pathogenic bacteria like *Staphylococcus aureus,* which at low population express proteins. This could facilitate cell adherence and aggregation, whereas at high population, this ability becomes suppressed and the bacteria stimulate secretion of toxins and proteases essential for spreading (Waters and Bassler, 2005). Although *E. histolytica* secrete many proteins and cell activators (Zaki et al., 2006), there are no evidence that the process of encystation and pathogenicity in the protozoan is triggered by quorum sensing.

The trophozoites and cysts are voided in feces into the environment. Cysts are often released into the environment by asymptomatic individuals. The morphology of the cysts give them the ability to withstand harsh environmental conditions for weeks, whereas trophozoites lack such ability. This morphological advantage has further offered them the ability to resist gastric acidity and chlorine (Martínez-Castillo et al., 2018). The cysts can lose viability by boiling. The environmental stability and low infectious dose of cysts of *E. histolytica* can result in development of outbreaks (Escolà-Vergé et al., 2017).

20.2.3 Virulence Factors

What triggers the penetration of *E. histolytica* into the intestinal mucosa to induce amoebic colitis or propagation of the parasite to some other organs including the liver where it induces abscess is still largely unknown. The host and parasite genetic compositions, immune modulation, diet, and concomitant infections have all been suggested as factors that can influence the pathogenicity of *E. histolytica* (Meyer et al., 2016). A triad of protein families—amoebapores, galactose/N-acetyl D-galactosamine-inhibitable Gal/GalNAc-lectins, and cysteine peptidases (CPs)—have been implicated in amoebic liver abscess formation (Bracha et al., 2003; Matthiesen et al., 2013). The occurrence of common homologues of these probable virulence factors in *Entamoeba dispar* has however given room for more debate on whether additional factors are involved.

E. histolytica synthesizes prostaglandin E_2 (PGE_2) through a cyclooxygenase like enzyme (Belley and Chadee, 2000). *E. histolytica* PGE_2 unlike some other virulence factors did not suppress the host immune system but relatively bound EP4 receptors on colonic epithelial cells and triggered neutrophil attraction, protein production, and IL-8 mRNA expression (Meyer et al., 2016). It also increased the permeability of the intestinal mucosa wall thus resulting in diarrhea in intestinal amoebiasis (Lejeune et al., 2011). Although the parasite generates great levels of PGE_2 in the company of arachidonic acid, it can also excite colonic epithelial cell and macrophages to produce PGE2 as part of the pro-inflammatory rejoinder elicited by the parasite (Sanchez-Ramirez et al., 2004).

Several other factors that contribute to the parasite processes involved in virulence have been identified. These include actin- Bisphenol S (BPs), myosin IB, CaBP1—calcium-binding protein 1 (motility), KERP1-lysine- and glutamic acid-rich protein, β1FNR—beta-1 integrin-like fibronectin receptor (adherence), arginase, peroxiredoxin, MLIF—monocyte locomotion inhibitor factor (cytotoxicity and host defense escape), ruberythrin, thioredoxin reductase, FeSOD—iron-superoxide dismutase, KRiP-lysine-rich protein, LgL—light subunit of Gal/GalNAclectin (stress and glucose starvation response), and URE3-BP—upstream regulatory element 3 binding protein (calcium signaling) (Faust and Guillen, 2012).

20.3 TREATMENT OPTIONS AND CHALLENGES IN CONTROL OF AMOEBIASIS

Anti-amoebic drugs are so designed to attack *E. histolytica* occurrence in different sites within the host. These are grouped into luminal, tissue, and mixed amoebicides (Bansal et al., 2006). MTZ is the mainstay drug often applied for the management of *E. histolytica* illness. Other effective drugs include secnidazole, ornidazole, and tinidazole. Paromomycin, a broad-spectrum drug for management of some other luminal parasites like *Giardia intestinalis* and *Cryptosporidium* spp., is another very important anti-amoebic drug. Others include diloxanidefuroate, a second line of treatment after MTZ in symptomatic patients, emetine, diiodohydroxyquin, and chloroquine (Bansal et al., 2006; Farthing, 2006).

MTZ and tinidazole, which are derivatives of nitroimidazole, are used against the invasive trophozoites but with effect on amoebic cysts. Although, cysts within the host do not pose significant threat as the carriers are asymptomatic. However, excretion of cysts into the environment is dangerous being the parasite infective form. Some typical examples of drugs that could be applied for their effective treatment includes Diloxanide furoate, paromomycin, and iodoquinol (Li and Stanley, 1996) are the alternative treatment of asymptomatic cyst carriers (Anderson et al., 2004). The use of iodoquinol and its analog iodochlorhydroxyquin has been discouraged due to report of its involvement in causing myelo-optic neuropathy after continuous usage for a longer period of time. This side effect has however not been recounted with the prescribed amount utilized to manage protozoan infections (Farthing, 2006). Chloroquine, a known antimalarial, could be administered along MTZ/emetine for the management of hepatic amoebiasis (Bansal et al., 2006). The use of dehydroemetine and emetine as anti-amoebiasis has been grossly limited owing to their potential cardiovascular and gastrointestinal adverse effects (Farthing, 2006).

MTZ has also been implicated in causing loss of appetite, metallic flavor, diarrhea, and nausea due to high dosage and long therapy requirement (Stanley, 2003). Nitroimidazoles are known to induce genetic toxicity by the positioning of their CH_3 and NO_2 groups which causes DNA damage (Boechat et al., 2015). MTZ becomes biotransformed into nitroso intermediates which can form adducts in the DNA or impede the thioredoxin reductase-generating reactive oxygen species, causing oxidative cell damage (Elizondo et al., 1996; Leitsch et al., 2007). The drug can also cross the blood-brain barrier leading high level of noxiousness to brain cells. These drugs utilize different mechanisms of action in killing protoplasmic organelles of *E. histolytica* (Ravdin and Petri, 1995). Aside the drug-associated toxicity of the common anti-amoebic drugs, they also faced the problem of parasite resistance to them (Borst and Ouellette, 1995).

20.4 IN VITRO AND IN VIVO ANTIPROTOZOAL ACTIVITY FROM DIFFERENT PLANT EXTRACTS

Attentions are now drawn to the use of plant extracts as remedy for several diseases including those of bacterial, fungal, and parasitic origin. Several workers have conducted studies to evaluate the antiprotozoal efficacy of medicinal plants. The observations from these studies suggest discovery of new chemotherapy for amoebiasis in no distance time.

Herrera-Martínez et al. (2016) evaluated the influence of ethyl acetate extract and methanolic extract obtained from *Adenophyllum aurantium* against *E. histolytica* trophozoites. The result of the in vivo trial carried on hamsters administered with 2.5 or 5 mg from a root ethyl acetate extract after infesting them with *E. histolytica* prevented the growth of amoebic liver carbuncles up to 48.5%–89.0% on the animals tested, correspondingly. Adhesion to fibronectin and erythrophagocytosis were 28.7% and 37.5% prevented by the root ethyl acetate extract obtained from *Adenophyllum aurantium*. Also, the transmission electron microscopy analysis carried out revealed a decrease in the number of vacuoles without any adverse effect on cell damage after treating the *E. histolytica* with ethyl acetate extract of root of *Adenophyllum aurantium*. The modes of action of the plant extract were suggested to be linked to its capability to affect the arrangement of actin cytoskeleton structuration and inhibit the contractile rings, but the modes of action don't depend on the RhoA activation pathways and reactive oxygen species (Herrera-Martínez et al. 2016). The study recommended that the plant extract could be exploited for the management of amoebiasis diseases, but further experiment will be necessary on the toxicity study of thiophenes. This will go along the way to preventing the adverse effects that might be experienced after taking the extract from the *Adenophyllum aurantium* (L.), which has been confirmed as a usual infusion used for the management of amoebiasis (Herrera-Martínez et al. 2016). Sawangjaroen et al. (2006) carried out studies on the anti-amoebic potential of 12 Thai medicinal plants constituted into about 39 extracts which have been reported to be utilized by AIDS patients in southern Thailand. The effects of these extracts were tested against *E. histolytica* strains HTH-56:MUTM and HM1:IMSS in an in vitro experiment after incubation for 24 hours with concentrations that range from 31.25 to 1,000 μg/ml. Some of the plant used during the study included *Gynura pseudochina* (L.) DC., *Piper betle* L., *Spilanthes acmella* (L.) Murray, *Piper chaba* Hunter, *Murraya paniculata* (L.) Jack, *Eclipta prostrata* (L.) L., *Boesenbergia pandurata* (Roxb.) Schltr., *Alpinia galanga* (L.) Willd., *Acanthus ebracteatus* Vahl, Willd., *Barleria lupulina* Lindl., *Zingiber zerumbet* (L.) Roscoe ex Sm., and *Coccinia grandis* (L.) Voigt. The activities of these plants varied with *Z. zerumbet* (IC_{50} 196.9 μg/ml) and *M. paniculata* (IC_{50} 116.5 μg/ml) categorized as being "moderately active." These plants however were less potent than MTZ used as the standard drug with IC_{50} of 1.1 μg/ml. The study showed that there is a need to evaluate the efficacy of the plants as prophylaxis and in the systemic therapy of *E. histolytica*. The potential dual activities (anti-amoebiasis and anti-HIV) of these plants are highly desirable as eradication of *E. histolytica* in AIDS patients can help to minimize mortality and morbidity associated with immune compromise (Sawangjaroen et al., 2006; Prinsloo et al., 2018; Chinsembu, 2019).

Mayaa and Alasadiy (2015) evaluated the effect of cold aqueous of peels of *Citrus aurantium* and *Solanum melongena* against *E. histolytica* and *Giardia lamblia* parasite isolated from stool specimens of patients with diarrhea. The cold aqueous of the plants were tested at concentrations ranging from 50% to 100%. It was revealed that a concentration 100% exhibited the largest inhibitory activity on the two parasites. Moreover, the median lethal dose for bitter orange peel *C. aurantium* was 7.75 g/kg body weight while that of eggplant *S. longena* was 7.25 g/kg body weight, respectively.

Elizondo-Lu´evano et al. (2018) evaluated the effect of methanolic extract *of Argemone mexicana* against strain HM1-IMSS of *E. histolytica* in an in vitro assay. The result obtained showed that the plant extract concentration to inhibit the growth of the parasite by 50% (IC_{50}) was 78.39 μg/ml. The result obtained shows that the concentration of fractions C and D at 125 μg/ml obtained from methanolic extract demonstrated lethality percentages of 72.39% and 79.09%, respectively. The active constituents responsible for the biological activities of fractions C and D were detected using Ultra performance liquid chromatography - tandem mass spectrometer

(UHPLC-MS) which showed that the Tandem mass spectrometry (MS/MS) spectra obtained represent berberine and jatrorrhizine, respectively (Rubio-Pina et al., 2013; Xool-Tamayo et al., 2017).

Garbi et al. (2018) assessed the effects of petroleum ether and methanolic extracts of *Anogeisus leiocarpus* against *Giardia lamblia* and *E. histolytica*, respectively, in an in vitro assay. The concentration of the crude extract tested ranged from 125 to 1,000 ppm. The result obtained showed that 1,000 ppm methanolic extract of *A. leiocarpus* exhibited the highest inhibition against *E. histolytica* with 70.06% mortality within 72 h while the lowest activity 61.35% was observed in petroleum ether extract at the same concentration within 72 h. The positive control experimental group treated with 312.5 μg/ml MTZ recorded 83.42% parasite inhibition rate within 72 h of exposure.

In the recent times, medicinal plants research has adopted proteomics approach to elucidate bioactive compounds with antiprotozoal potential. Velázquez-Domínguez et al. (2013) isolated incomptine A from sesquiterpene lactone of the heliangolide type from the leaves of *Decachaeta incompta*. The incomptine A was tested in an in vitro assay against *E. histolytica* trophozoites. They explored the application of proteomics techniques centered on mass spectrometry (ESI-MS/MS) and two-dimensional gel electrophoresis study. Their study showed that 21 *E. histolytica* proteins were differentially articulated in reaction to incomptine A when it was applied for the management of *E. histolytica*. Moreover, the following glycolytic enzymes were downregulated including fructose-1,6-biphosphate aldolase, enolase, and ferredoxin oxidoreductase, respectively. The modes of action of incomptine A was further validated using electronic microscopy which revealed the upsurge in the number of glycogen granules. The study showed that incomptine A has an inhibitory consequence on *E. Histolytica* by changing its energy metabolism. It was suggested medicinal plants could serve as new source of a bioactive molecule containing incomptine A that could serve as new alternative therapy with antiprotozoal activity for the management of human amoebiasis which has been reported to affect many people in the developing countries (Velázquez-Domínguez et al. 2013).

Bolaños et al. (2014) evaluated the anti-amoebic activity of flavonoid (–)-epicatechin derived from *Geranium mexicanum* HBK in an in vitro assay against *E. histolytica* trophozoites. It was observed that the *E. histolytica* cytoskeleton proteins showed discrepancy abundance when exposed to (–)-epicatechin action. Also the result obtained indicate that the presence of (–)-epicatechin decreased cytolysis, adhesion and enhanced the rate of phagocytosis and migration as well as impede *E. histolytica* trophozoites development. Therefore, these facts showed that (–)-epicatechin portends the capability to distress virulence features of this human pathogen. The study showed that flavonoid (–)-epicatechin demonstrated an unusual influence on the trophozoites growth which might depend on the concentration of the active compound. The compound therefore can be a good candidate for drug development for MTZ-resistant *E. histolytica* strain.

Bolaños et al. (2015) evaluated the mode of action exhibited by flavonoid kaempferol isolated from *Helianthemum glomeratum*, a widespread Mexican therapeutic plant used for the treatment of *E. histolytica* trophozoites. A two-dimensional gel-based proteomic investigation was performed to substantiate the influence of the modified proteins that could describe the phenotypical variations detected in trophozoites after it was subjected to treatment. The trophozoites viability and growth was impaired by kaempferol at 27.7 μM. Moreover, kaempferol reduced the rate of adhesion, enhanced the rate of phagocytic activity and migration, but did not have any influence on the cytolytic and erythrocyte binding potential of *E. histolytica*. Consistently, the proteomic investigation has shown that myosin II heavy chain, cytoskeleton proteins actin, and cortexillin II were upregulated in rejoinder when kaempferol was applied as a treatment. The mode of action of kaempferol showed that it could be linked to the deregulation of proteins closed to the cytoskeleton which altered invasion mechanisms.

Martínez-Castillo et al. (2018) wrote a widespread assessment on the utilization of natural flavonoids that are polyphenolic compounds in nature as potential replacement to some synthetic drugs (nitroimidazoles, iodoquinol, diloxanidefuroate, paromomycin) used for the management/ treatment of *E. histolytica*. They identified a need to quest for a substitute and harmless natural drug because of the high level of resistance, cerebral toxicity, mutagenic effects experienced when these synthetic drugs were taken. Some of the highlighted natural and safe compounds were quercetin, -(−)epicatechin, and kaempferol. The modes of action through which most of these natural compounds exhibit their anti-amoebic activity include downregulation and upregulation of glyceraldehyde-phosphate dehydrogenase, fructose-1,6-bisphosphate aldolase, pyruvate:ferredoxin oxidoreductase, and their capability to cause structural alteration on the outer surface of amoebas which includes cytoskeletal protein reorganization and chromatin condensation. The authors also signify that there is a need to critically evaluate the bioavailability, necessary dosage, route of administration, and specific molecular targets of these natural drugs. It was established that the presence of flavonoids as an active constitute could be used as a permanent replacement for many drugs which are no longer effective to combat to *E. histolytica* due to enhancement in the level of resistance by the parasite.

Quintanilla-Licea et al. (2014), evaluated an in vitro assay on the antiprotozoal action of methanolic extracts of 32 plants used by Northeast Mexican traditional for the management of *E. histolytica* trophozoites. Their result showed that 18 out of all the 32 tested extracts exhibited a significant inhibitory effect on the tested strain of *E. histolytica*. Moreover, it was detected that six plant extracts exhibited approximately 80% antiprotozoal activity as well as 91.54% and 90.50% growth prevention at a dose of 150 μg/ml. Their study showed that 18 out of the 32 Mexican traditional medicinal plants screened could be utilized for the management of dysentery and diarrhea. Moreover, the extract from *Ruta chalepensis* Pers and *Lippia graveolens* Kunth showed significant antiprotozoal activity using a probit analysis with a 95% confidence level during the screening of the bio-guided fractionation. Therefore, the application of carvacrol and chalepensin could be notable pharmacological drugs that could be used for the handling of amoebiasis and significant neglected illnesses (Arfa et al., 2006; Xu et al., 2008; Baser et al., 2008).

Garza-Gonzalez et al. (2017) evaluated the antiprotozoal activity of *Thymus vulgaris* against *Entamoeba histolytica* trophozoites, *Trichomonas vaginalis*, and *G. lamblia*. The in vitro assay was performed using the methanolic extract of the *Thymus vulgaris* and its fraction in growth cultures containing the tested parasites.

The inhibitory effect of the crude extract was evaluated on the third day throughout the logarithmic growing phase.

Asarinin was isolated from *Zanthoxylum liebmannianum* leaves while the ethanolic extract exhibited an inhibitory effect against *E. histolytica* with IC50: 35.45 µg/ml (Arrieta et al., 2001). Lezama-Davila et al. (2007) isolated two active compounds containing melilotoside and narcissin from the aerial part of *Teloxys graveolens*. The IC_{50} of melilotoside and narcissin were 12.5 and 17.2 µg/ml correspondingly.

Pan et al. (2012) isolated (–)-epicatechin from aerial parts of *Rubus coriifolius*. The methanol:dichloromethane extract from *R. coriifolius* exhibited inhibitory effect against *E. histolytica* with the IC_{50} of the plant extract being 11.6 µg/ml. Moreover, Cruz-Paredes et al. (2013) isolated (–)-epicatechin from the root of *Geranium mexicanum*. The dichloromethane–MeOH extract obtained from the plant showed an inhibitory effect against *E. histolytica* and the IC50 of the plant extract was 1.9 µg/ml.

Ackers and Mirelman (2006) isolated incomptine A from the leaf of *Decachaeta incompta*. The inhibitory influence of the dichloromethane extract was IC50=2.6 µg/ml. Aguirre-Cruz et al. (1990) isolated two active compounds from *Salvia polystachya* aerial parts which included linearolactone and polystachynes A, B, and D. The acetone extract containing linearolactone showed an inhibitory effect against *E. histolytica* with IC50 22.9 µM. The IC50 of polystachynes A, B, and D ranged from 117.0 to 160.6 µM.

Arrieta et al. (2001) isolated benzyl glucosinolate from the root of *Lepidium virginicum*. The inhibitory concentration to inhibit the growth of *E. histolytica* by 50% was 100.1 µg/ml in methanolic extract of the plant. Calzada et al. (2003) isolated carvacrol from *Lippia graveolens* and chalepensin from *Ruta chalepensis* as the main active compounds from these plants. The IC50 of carvacrol and chalepensin were 44.3 and 45.95 µg/ml, respectively. Alanís et al. (2003) isolated thiophenes from the root of *A. aurantium*. The ethyl acetate extract obtained from the root of the plant exhibited inhibitory action against *E. histolytica* with IC_{50} 230 µg/ml.

20.5 SOME PHYTOCHEMICALS WITH ANTIPROTOZOAL ACTIVITY AND THEIR STRUCTURAL ELUCIDATION STUDIES

Herrera-Martínez et al. (2016) validated the active components present in 14.36 g of ethyl acetate extract of *A. aurantium* (L.) using nuclear magnetic resonance spectroscopy. The following compounds 5-(40 0-Hydroxy-10 0-butynyl)-2–20-bithiophene and α-Terthienyl which were thiophenes in nature were present in the ethyl acetate extract of the plant. This might be accountable for the anti-amoebiasis action recorded when applied against *E. histolytica*.

Elizondo-Lu´evano et al. (2018) utilized HPLC analysis, mass spectrometry, ^{1}H-NMR, and ^{13}C-NMR experiments to evaluate the active compounds accountable for the anti-amoebicidal activity present in the methanolic extract of *Argemone mexicana*. The most active compound was obtained from fractions C and D which were later analyzed using UHPLC-MS which showed a crucial signal at *m/z* 336 and 338. The fraction C and D represent berberine and jatrorrhizine, respectively. The fragment of an ion at *m/z* 336 resulted into *m/z* 321, 320, and 292 while the ion observed at *m/z* 338 were 323, 322, and 294. The fragmentation of the ion at these two points corresponds to those reported for berberine and jatrorrhizine when tested by MS/MS spectra. Furthermore,[1] H-NMR and ^{13}C-NMR were later used to characterize the nature of the alkaloids mentioned above. The signal obtained from fraction D corresponded to the presence of berberine while the presence of jatrorrhizine could not be detected due to the low sensitivity of the tested compound. Therefore, there is a need to carry out a further experiment on the detection of a chromatographic peak that resulted into these active signals and to further understand the mechanism of action of these compounds.

Bautista et al. (2014) tested the inhibitory effect of 8-acyl and 8-alkyl incomptine A analogs against *G. lamblia and E. histolytica*. They reported the effect of the amount and the various lengths of alkyl groups, incomptine A (1), and acyl antiprotozoal sesquiterpene lactone against the tested protozoans.

Quintanilla-Licea et al. (2014) isolated and performed structural elucidation of the dynamic compound accountable for the antiprotozoal action of methanolic extracts of 32 plants used in Northeast Mexican traditional using a bio-guided assay and 1D- and 2D-NMR analyses. The bioassay-guided fractionation showed that the active compound responsible for the antiprotozoal activity against the tested strain of *E. histolytica* during their study were carvacrol and chalepensin, respectively.

Garza-Gonzalez et al. (2017) utilized thin-layer chromatography for the separation of active components responsible for the antiprotozoal activity against *T. vaginalis* trophozoites, *E. histolytica*, and *G. lamblia*. The fractionation of the crude extract was performed by polarity-based partitioning followed by the checking of purity which was carried out using thin-layer chromatography. The result showed that the three main bioactive components present were oleanolic, ursolic, and betulinic acids.

Several scientists have established the antiprotozoal potential of numerous plants most especially against *Entamoeba histolytica*. This biological activity could be linked to the presence of several phytochemical components. Typical example of such bioactive compounds that could be applied in the management of *Entamoeba histolytica* that has been established to possess biological activities includes kaempferol, Quercetin, apigenin, galangin, (−)-epigallocatechin, pinocembrin, narigenin, Chaparrin (a simaroubolidane), steroidal saponins, ellagic acid, α-linoleic, flavonoid juglamin, linoleic as shown in Table 20.1.

20.6 THE MECHANISM OF ACTION OF SOME PHYTOCHEMICALS RESPONSIBLE FOR THE ANTIPROTOZOAL ACTIVITIES

Numerous scientists have documented the modes of action of carvacrol most especially those involved in the antiprotozoal and the antimicrobial action (Ultee et al., 1999; Ultee et al., 2002; Arfa et al., 2006; Xu et al., 2008; Baser, 2008). Also carvacrol proteins have been validated to exhibit several antimicrobial activity and antiprotozoal action against the biological membranes of pathogenic microorganism including protozoa. It possesses the capability to diminishing the intracellular adenosine triphosphate (ATP) pool through adequate decreasing ATP fabrication and enhancing ATP hydrolysis. It has been established that decrease in transmembrane electric capability could be the driving force of ATP fabrication that could improve

TABLE 20.1
Influence of Biologically Active Phytochemical Constituents Present in Medicinal Plants against *Entamoeba histolytica*

S/N	Medicinal Plants	Phytochemical Constituents	Mode of Action	References
1.	*L. graveolens*	Kaempferol, Quercetin, apigenin, galangin, (−)-epigallocatechin, pinocembrin, narigenin	Inhibited the growth of *E. histolytica*	González-Güereca et al. (2007), Lin et al. (2007), Calzada et al. (1999)
2.	*Castela texana*	Chaparrin (a simaroubolidane)	Concentration of 100 μg/mL derived from the active compound produced 100% growth inhibition against *E. histolytica*	Calzado-Flores et al. (1986)
3.	*Agave lechugilla*	Steroidal saponins	Occurrence of steroidal saponins with anti-inflammatory and antiprotozoal activity	Hernández et al. (2005), Romero-González et al. (2006), Monterrosas-Brisson et al. (2013)
4.	*Mentha spicata*	Terpenes	Antiprotozoal activity	Lawrence (2006)
5.	*Opuntia ficus-indica*	Neobetanin, flavonoids, alkaloids indicaxanthin	Antiprotozoal activity	Saleem et al. (2006)
6.	*Juglans mollis*	Ellagic acid, α-linoleic, flavonoid juglamin, linoleic	Antiprotozoal activity	Cruz-Vega et al. (2008), Shah et al. (2014)
7.	*Ruta chalepensis* Pers	Crude extract	Antiprotozoal activity with 90.50% growth inhibition at a concentration of 150 μg/mL with IC_{50} values of 60.07 μg/ml)	Quintanilla-Licea et al. (2014)
8.	*Lippia graveolens* Kunth	Rude extract	Antiprotozoal activity (91.50% growth inhibition at a concentration of 150 μg/ml with IC_{50} values of 59.14 μg/ml)	Quintanilla-Licea et al. (2014)

proton permeability of the membrane, maintenance of cytoplasmic pH and osmotic pressure. Furthermore, it has been documented the function of potassium (K^+) in the stimulation of the cytoplasmic enzymes. The removal of K^+ out of the cell is a proof that destabilization of the membrane has occurred (Ultee et al., 1999). The biologically active constituents available in essential oil have been validated to exhibit their action on pathogenic protozoans most especially with the presence of 70% of carvacrol (Baser, 2008; Machado et al., 2010). It was discovered that the addition of essential oil led to ultrastructural changes in the appearance of the trophozoite which prompted availability of membrane blebs, deformations of the trophozoite, irregular dorsal, roundly shaped, and ventral surface, internalization of ventral disc and flagella, while electrodense precipitates in the nuclei and cytoplasm. Their result indicated that essential oils could cause autolysis of the protozoa by several process related to the loss of osmoregulation induced by changes in the plasmatic membrane (Machado et al., 2010).

MfotieNjoya et al. (2014) utilizes the leaves of *Codiaeum variegatum* in the treatment of bloody diarrhea that affected the majority of populations in Cameroon. The authours also validated that the presence of biologically active constituents available in the essential oils of *C. variegatum* that could be utilized for the management of *Entamoeba histolytica*. They also established the modes of activity against *Entamoeba histolytica*. The mechanism of action was established using bioassay-guided screening of the numerous fractions and several aqueous extracts from *C. variegatum* leaves against trophozoites of *Entamoeba histolytica*. Their experiment validates the potential of anti-amoebic activity of the extracts. The anti-amoebic action of the fraction was validated through morphological appearances of induced autolysis in trophozoites entailing lysis and cell rounding. The authors also utilized differential gene expression evaluation with the aids of high-throughput RNA sequencing which established the anti-amoebic action through targeting ceramide. The experiment validated the bioactive lipid that are involved in the interruption of the biochemical processes that occurs the inside cell membrane which entails apoptosis, differentiation, cell growth arrest, and proliferation of the authors established that the process involved in the ceramide biosynthesis pathway which influenced its anti-amoebic action against *E. histolytica*.

20.7 FUTURE RESEARCH PERSPECTIVES IN THE CONTROL OF AMOEBIASIS

This chapter has highlighted various in vivo and in vitro antiprotozoal activity from different plant extracts. Therefore, there is a need to carry out structural elucidation in order to isolate the most active compound available in these plants. Also, some further study will be necessary in order to establish the modes of action through which most of these active compounds execute their modes of action against *Entamoeba histolytica*.

REFERENCES

Adetunji, C.O., Anani, O.A., and Panpatte, D. (2021). Mechanism of actions involved in sustainable ecorestoration of petroleum hydrocarbons polluted soil by the beneficial microorganism. In: Panpatte, D.G., and Jhala, Y.K. (eds.), *Microbial Rejuvenation of Polluted Environment. Microorganisms for Sustainability*, vol 26. Springer, Singapore. https://doi.org/10.1007/978-981-15-7455-9_8

Adetunji, C.O., and Anani, O.A. (2021a) Recent advances in the application of genetically engineered microorganisms for microbial rejuvenation of contaminated environment. In: Adetunji C.O., Panpatte D.G., Jhala Y.K. (eds.), *Microbial Rejuvenation of Polluted Environment. Microorganisms for Sustainability*, vol 27. Springer, Singapore. https://doi.org/10.1007/978-981-15-7459-7_14

Adetunji, C.O., and Anani, O.A. (2021b) Utilization of microbial biofilm for the biotransformation and bioremediation of heavily polluted environment. In: Panpatte, D.G., and Jhala, Y.K. (eds.), *Microbial Rejuvenation of Polluted Environment. Microorganisms for Sustainability*, vol 25. Springer, Singapore. https://doi.org/10.1007/978-981-15-7447-4_9

Adetunji, C.O., and Anani, O.A. (2021c) Bioremediation of polythene and plastics using beneficial microorganisms. In: Adetunji C.O., Panpatte D.G., and Jhala Y.K. (eds.), *Microbial Rejuvenation of Polluted Environment. Microorganisms for Sustainability*, vol 27. Springer, Singapore. https://doi.org/10.1007/978-981-15-7459-7_13

Anani, O.A., Mishra R.R., Mishra P., Enuneku A.A., Anani G.A., and Adetunji C.O. (2020) Effects of toxicant from pesticides on food security: Current developments. In: Mishra, P., Mishra, R.R., and Adetunji, C.O. (eds.), *Innovations in Food Technology*. Springer, Singapore. https://doi.org/10.1007/978-981-15-6121-4_22

Ackers, J. P. and Mirelman, D. (2006). Progress in research on Entamoeba histolytica pathogenesis. *Current Opinion in Microbiology* 9:367–373.

Aguirre-Cruz, M. L., Valadez-Salazar, A. and Muñoz, O. (1990). In vitro sensitivity of Entamoeba histolytica tometronidazole. *Archivos de InvestigaciónMédica* 1:23–26.

Alanís, A. D., Calzada, F., Cedillo-Rivera, R. and Meckes, M. (2003). Antiprotozoal activity of the constituents of Rubuscoriifolius. *Phytotherapy Research* 17(6): 681–682.

Anderson, K. A., Brooks, A. S., Morrison, A. L., Reid-Smith, R. J., Martin, S. W., Benn, D. M, and Peregrine, A. S. (2004). Impact of Giardia vaccination on asymptomatic Giardia infections in dogs at a research facility. *Canadian Veterinary Journal* 45: 924–930.

Arfa, A. B., Combes, S., Preziosi-Belloy, L., Gontard, N. and Chalier, P. (2006). Antimicrobial activity of carvacrol related to its chemical structure. *Letters in Applied Microbiology* 43(1): 41–54.

Arrieta, J., Reyes, B., Calzada, F., Cedillo-Rivera, R. and Navarrete, A. (2001). Amoebicidal and giardicida compounds from the leaves of Zanthoxylum liebmannianun. *Fitoterapia* 72: 295–297.

Bansal, D., Malla, N. and Mahaja, R. C. (2006). Drug resistance in amoebiasis. *Indian Journal of Medical Research* 123:115–118.

Baser, K. H. C. (2008). Biological and pharmacological activities of carvacrol and carvacrol bearing essential oils. *Current Pharmaceutical Design* 14: 3106–3120. doi: 10.2174/138161208786404227.

Bautista, E., Calzada, F., López-Huerta, F. A., Yépez-Mulia, L. and Ortega, A. (2014). Antiprotozoal activity of 8-acyl and 8-alkyl incomptineA analogs. *Bioorganic and Medicinal Chemistry Letters* 24 (15): 3260–3262.

Belley, A. and Chadee, K. (2000). Production of prostaglandin E2 by Entamoeba histolytica via a novel cyclooxygenase. *Archives of Medical Research* 31: 74–75.

Boechat, N., Carvalho, A. S., Salomão, K., Castro, S. L., Araujo-Lima, C. F., Mello, F.V., Felzenszwalb, I., Aiub, C. A., Conde, T. R., Zamith, H. P., Skupin, R. and Haufe, G. (2015). Studies of genotoxicity and mutagenicity of nitroimidazoles: demystifying this critical relationship with the nitro group. *Memórias do Instituto Oswaldo Cruz* 110(4): 492–499.

Bolaños, V., Díaz-Martínez, A., Soto, J., Rodríguez, M. A., López-Camarillo, C., Marchat, L. A. and Ramírez-Moreno, E. (2014). The flavonoid (–)-epicatechin affects cytoskeleton proteins and functions in Entamoeba histolytica. *Journal of Proteomics* 111: 74–85.

Bolaños, V., Díaz-Martínez, A., Soto, J., Marchat, L. A., Sanchez-Monroy, V. and Ramírez-Moreno, E. (2015). Kaempferol inhibits Entamoeba histolytica growth by altering cytoskeletal functions. *Molecular and Biochemical Parasitology* 204(1): 16–25.

Borst, P. and Ouellette, M. (1995). New mechanisms of drug resistance in parasitic protozoa. *Annual Review of Microbiology* 49: 427–460.

Bracha, R., Nuchamowitz, Y. and Mirelman, D. (2003). Transcriptional silencing of an amoebapore gene in Entamoeba histolytica: molecular analysis and effect on pathogenicity. *Eukaryot Cell* 2(2): 295–305.

Bruijnesteijn van Coppenraet, L. E., Wallinga, J. A., Ruijs, G. J., Bruins, M. J. and Verweij, J. J. (2009). Parasitological diagnosis combining an internally controlled real-time PCR assay for the detection of four protozoa in stool samples with a testing algorithm for microscopy. *Clinical Microbiology and Infection* 15(9): 869–874.

Brumpt, E. (1925). Etude sommaire de l'"Entamoebadispar" n. sp. Amibe a kystesquadrinuclees, parasite de l'homme. *Bulletin of the New York Academy of Medicine (Paris)* 94:943–952.

Calderaro, A., Gorrini, C., Bommezzadri, S., Piccolo, G., Dettori, G. and Chezzi, C. (2006). Entamoeba histolytica and Entamoebadispar: comparison of two PCR assays for diagnosis in a nonendemic setting. *Transactions of the Royal Society of Tropical Medicine and Hygiene* 100: 450–457.

Calderaro, A., Piergianni, M., Buttrini, M., Montecchini, S., Piccolo, G. and Gorrini, C. (2015). MALDI-TOF Mass Spectrometry for the Detection and Differentiation of Entamoeba histolytica and Entamoeba dispar. *PLoS ONE* 10(4): e0122448. doi: 10.1371/journal.pone.0122448.

Calzada, F., Meckes, M. and Cedillo-Rivera, R. (1999). Antiamoebic and antigardial activity of plant flavonoids. *Planta Medica*. 65:78–80. doi: 10.1055/s-2006-960445.

Calzada, F., Velazquez, C., Cedillo-Rivera, R. and Esquivel, B. (2003). Antiprotozoal activity of the constituents of Teloxys graveolens. *Phytotherapy Research* 17: 731–732.

Calzado-Flores, C. C., Segura-Luna, J. J., Domínguez, X. A. and García-González, S. (1986). Castelatexana: Cernimiento de suactividad antiamibiana. *Archivos de investigacion medica (Mex.)* 17:127–134.

Chinsembu, K. C. (2019). Chemical diversity and activity profiles of HIV-1 reverse transcriptase inhibitors from plants. *Revista Brasileira de Farmacognosia* 29(4): 504–528.

Choudhuri, G. and Rangan, M. (2012). Amebic infection in humans. *Indian Journal of Gastroenterology* 31 (4):153–62.

Cruz-Paredes, C., Bolívar, B. P., Gómez-Velasco, A., Juárez, Z. N., Sánchez, A. E., Hernández, L. R. and Bach, H. (2013). Antimicrobial, antiparasitic, anti-Inflammatory, and cytotoxic activities of Lopezia racemosa. *The Scientific World Journal*. 237–438. doi: 10.1155/2013/237438.

Cruz-Vega, D. E., Verde-Star, M. J., Salinas-González, N., Rosales-Hernández, B., Estrada-García, I., Mendez-Aragón, P., Carranza-Rosales, P., González-Garza, M. T. and Castro-Garza, J. (2008). Antimycobacterial activity of Juglans regia, Juglans mollis, Carya illinoensis and Bocconia frutescens. *Phytotherapy Research* 22: 557–559. doi: 10.1002/ptr.2343.

Dwivedi, N., Dwivedi, S., and Adetunji, C. O. (2021) Efficacy of microorganisms in the removal of toxic materials from industrial effluents. In: Adetunji C.O., Panpatte D.G., and Jhala, Y.K. (eds.), *Microbial Rejuvenation of Polluted Environment. Microorganisms for Sustainability*, vol 27. Springer, Singapore. https://doi.org/10.1007/978-981-15-7459-7_15

Elizondo-Luévano, J. H., Castro-Ríos, R., Sánchez-García, E., Hernández-García, M. E., Vargas-Villarreal, J., Rodríguez-Luis, O. E. and Chávez-Montes, A. (2018). In vitro study of antiamoebic activity of methanol extracts of argemone Mexicana on trophozoites of entamoeba histolytica hm1-IMSS. *Canadian Journal of Infectious Diseases and Medical Microbiology*, Article ID 7453787, 8 pageshttps://doi.org/10.1155/2018/7453787.

Elizondo, G., Gonsebatt, M. E., Salazar, A. M., Lares, I., Santiago, P., Herrera, J., Hong, E., and Ostrosky-Wegman, P. (1996). Genotoxic effects of metronidazole. *Mutation Research* 370(2): 75–80.

Escolà-Vergé, L., Arando, M., Vall, M., Rovira, R., Espasa, M., Sulleiro, E., Armengol, P., Zarzuela, F., and Barberá, M. J. (2017). Outbreak of intestinal amoebiasis among men who have sex with men, Barcelona (Spain), October 2016 and January 2017. *Euro surveillance: Bulletin Europeen sur les maladies transmissibles = European Communicable Disease Bulletin*, 22(30) 30581. https://doi.org/10.2807/1560-7917.ES.2017.22.30.30581

Farthing, M. J. (2006). Treatment options for the eradication of intestinal protozoa. *Nature Clinical Practice, Gastroenterology and Hepatology* 3 (8): 436–445.

Faust, D. M. and Guillen, N. (2012). Virulence and virulence factors in Entamoeba histolytica, the agent of human amoebiasis. *Microbes and Infection* 14: 1428–1441.

Garbi, M. I., Kabbashi, A. S., El-BadriElamin, O., Koko, W. S., Dahab, M. M. and Elshikh, A. A. (2018). In Vitro gardicidal and amoebicidal activity of Anogeissus leiocarpus leaves extracts. *Horticulture International Journal* 2(5):232–235. doi: 10.15406/hij.2018.02.00057.

Garza-Gonzalez, J. N., Vargas-Villarreal, J., Verde-Star, M. J., Rivas-Morales, C., Oranday-Cardenas, A., Hernandez-Garcia, M. E., De La Garza- Salinas, L. and Gonzalez-Salazar, F. (2017). Antiprotozoal activity of a Thymus vulgaris methanol extract and its fractions. *Health*. 9(7): 1081–1094. https://doi.org/10.4236/health.2017.97079.

Gonzales, M. L. M., Dans, L. F. and Sio-Aguilar, J. (2019). Antiamoebic drugs for treating amoebic colitis. *Cochrane Database of Systematic Reviews* 1, CD006085. doi: 10.1002/14651858.CD006085.pub3.

González-Güereca, M. C., Soto-Hernández, M., Kite, G. and Martínez-Vázquez, M. (2007). Antioxidant activity of flavonoids from the stem of the Mexican oregano (Lippia graveolens H.B.K. var. berlandierischauer). *Revista fitotecnia Mexicana* 30:43–49.

Herbinger, K. H., Fleischmann, E., Weber, C., Perona, P., Löscher, T. and Bretzel, G. (2011). Epidemiological, clinical, and diagnostic data on intestinal infections with Entamoeba histolytica and Entamoeba dispar among returning travelers. *Infection* 39(6): 527–535. doi: 10.1007/s15010-011-0155-z.

Hernández, R., Lugo, E. C., Díaz, L. and Villanueva, S. (2005). Extracción y cuantificaciónindirecta de las saponinas de Agave lechuguilla Torrey. *E-Gnosis* 3: 1–9.

Herrera-Martínez, M., Hernández-Ramírez, V. I., Hernández-Carlos, B., Chávez-Munguía, B., Calderón-Oropeza, M. A. and Talamás-Rohana, P. (2016). Antiamoebic activity of Adenophyllum aurantium (L.) strother and its effect on the actin cytoskeleton of Entamoeba histolytica. *Frontiers in Pharmacology* 7:169. doi: 10.3389/fphar.2016.00169.

Hung, C. C., Ji, D. D., Sun, H. Y., Lee Y. T., Hsu, S. Y., Chang, S. Y., Wu, C. H., Chan, Y. H., Hsiao, C. F., Liu, W. C., and Colebunders, R. (2008). Increased risk for Entamoeba histolytica infection and invasive amebiasis in HIV seropositive men who have sex with men in Taiwan. *PLOS Neglected Tropical Diseases* 2(2): e175. doi: 10.1371/journal.pntd.0000175.

Hung, C. C., Chang, S. Y. and Ji, D. D. (2012). Entamoeba histolytica infection in men who have sex with men. *Lancet Infectious Diseases* 12:729–736

Islam, S., Thangadurai, D., Adetunji, C. O., Nwankwo, W., Kadiri, O., Makinde, S., Michael, O. S., Anani, O. A., and Adetunji, J. B. (2021) Nanomaterials and nanocoatings for alternative antimicrobial therapy. In: Kharissova O.V., Martínez L.M.T., Kharisov B.I. (eds.), *Handbook of Nanomaterials and Nanocomposites for Energy and Environmental Applications*. Springer, Cham. https://doi.org/10.1007/978-3-030-11155-7_3-1

Lawrence, B. M. (Ed.). (2006). *Mint: The Genus Mentha (1st ed.)*. CRC Press. https://doi.org/10.1201/9780849307980.

Leitsch, D., Kolarich, D., Wilson, I. B., Altmann, F., and Duchêne, M. (2007). Nitroimidazole action in Entamoeba histolytica: a central role for thioredoxin reductase. *PLOS Biology* 5: e211.

Lejeune, M., Moreau, F. and Chadee, K. (2011). Prostaglandin E2 produced by Entamoeba histolytica signals via EP4 receptor and alters claudin-4 to increase ion permeability of tight junctions. *The American Journal of Pathology* 179: 807–818.

Lezama-Davila, C. M., Isaac-Márquez, A. P, Zamora-Crescencio, P., Uc-EncaladaMdel, R., Justiniano-Apolinar, S. Y., del Angel-Robles, L., and Satoskar, A. (2007). Leishmanicidal activity of Pentalinon andrieuxii. *Fitoterapia* 78(3): 255–257.

Li, E. and Stanley, S. L. (1996). Amebiasis. *Gastroenterology Clinics of North America* 25: 471–492.

Lin, L. Z., Mukhopadhyay, S., Robbins, R. J. and Harnly, J. M. (2007). Identification and quantification of flavonoids of Mexican oregano (Lippia graveolens) by LC-DAD-ESI/MS analysis. *Journal of Food Composition and Analysis* 20:361–369 doi: 10.1016/j.jfca.2006.09.005.

Machado, M., Dinis, A. M., Salgueiro, L., Cavaleiro, C., Custódio, J. B. A., and Sousa, M. C. (2010). Anti-Giardia activity of phenolic-rich essential oils: effects of Thymbra capitata, Origanum virens, Thymus zygis subsp. sylvestris, and Lippia graveolens on trophozoites growth, viability, adherence, and ultrastructure. *Parasitology Research* 106:1205–1215.

Martínez-Castillo, M., Pacheco-Yepez, J., Flores-Huerta, N., Guzmán-Téllez, P., Jarillo-Luna, R. A., Cárdenas-Jaramillo, L. M., Campos-Rodríguez, R. and Shibayama, M. (2018). Flavonoids as a natural treatment against Entamoeba histolytica. *Frontiers in Cellular and Infection Microbiology* 8: 209 doi: 10.3389/fcimb.2018.00209.

Matthiesen, J., Bär, A. K., Bartels, A. K., Marien, D., Ofori, S., Biller, L., Tannich, E., Lotter, H., and Bruchhaus, I. (2013). Overexpression of specific cysteine peptidases confers pathogenicity to a nonpathogenic Entamoeba histolytica clone. *mBio* 4(2), e00072-13. https://doi.org/10.1128/mBio.00072-13.

Mayaa, F. A. K. and Alasadiy, Y. D. K. (2015). Effect of cold aqueous plant extract (Citrus aurantium and Solanum melongena) against the Giardia lamblia parasite and Entamoeba histolytica in vitro and measuring the LD50 for extracts in Al-Muthanna province. *Research Journal of Pharmaceutical, Biological and Chemical Sciences* 6(6): 1241–1252.

Mishra, P., Mishra, R. R., & Adetunji, C. O. (Eds.). (2020). *Innovations in Food Technology: Current Perspectives and Future Goals.* Springer Nature Singapore Pte Ltd. 2020. Springer, Singapore. Pp. 143-162. 1st Edition. DOI: 10.1007/978-981-15-6121-4

Meyer, M., Fehling, H., Matthiesen, J., Lorenzen, S., Schuldt, K., Bernin, H., Zaruba, M., Lender, C., Ernst, T., Ittrich, H., Roeder, T., Tannich, E., Lotter, H., and Bruchhaus, I. (2016). Overexpression of differentially expressed genes identified in non-pathogenic and pathogenic Entamoeba histolytica clones allow identification of new pathogenicity factors involved in amoebic liver abscess formation. *PLOS Pathogens* 12(8): e1005853. doi: 10.1371/journal.ppat.1005853.

MfotieNjoya, E., Weber, C., Hernandez-Cuevas, N. A., Hon, C. C., Janin, Y., Kamini, M. F., Moundipa, P. F. and Guillén, N. (2014). Bioassay-guided fractionation of extracts from Codiaeum variegatum against Entamoeba histolytica discovers compounds that modify expression of ceramide biosynthesis related genes. *PLoS Neglected Tropical Diseases* 8(1): e2607. https://doi.org/10.1371/journal.pntd.0002607.

Monterrosas-Brisson, N., Arenas-Ocampo, M. L., Jiménez-Ferrer, E., Jiménez-Aparicio, A. R., Zamilpa, A., Gonzalez-Cortazar, M., Tortoriello, J., Herrera-Ruiz, M. (2013). Anti-inflammatory activity of different Agave plants and the compound Cantalasaponin-1. *Molecules* 18: 8136–8146. doi: 10.3390/molecules18078136.

Nuñez, Y. O., Fernández, M. A., Torres-Nuñez, D., Silva, J. A., Montano, I., Maestre, J. L. and Fonte, L. (2001). Multiplex polymerase chain reaction amplification and differentiation of Entamoeba histolytica and Entamoeba dispar DNA from stool samples. *American Journal of Tropical Medicine and Hygiene* 64:293–297.

Orozco, E., Marchat, L. A., Gómez, C., López-Camarillo, C., Pérez, D. G. (2009). *Drug Resistance Mechanisms in Entamoeba histolytica, Giardia lamblia, Trichomonas vaginalis, and Opportunistic Anaerobic Protozoa, in Antimicrobial Drug Resistance*, Springer. pp. 549–559.

Oyeyemi, O. T., Agbaje, M. O. and Okelue, U. B. (2016). Food-borne human parasitic pathogens associated with household cockroaches and houseflies in Nigeria. *Parasite Epidemiology and Control* 1(1):10–13.

Olaniyan, O.T., and Adetunji, C.O. (2021) Biochemical role of beneficial microorganisms: An overview on recent development in environmental and agro science. In: Adetunji, C.O., Panpatte, D.G., and Jhala, Y.K. (eds.), *Microbial Rejuvenation of Polluted Environment. Microorganisms for Sustainability*, vol 27. Springer, Singapore. https://doi.org/10.1007/978-981-15-7459-7_2

Pan, L., Lezama-Davila, C. M., Isaac-Marquez, A. P., Calomeni, E. P., Fuchs, J. R., Satoskar, A. R. and Kinghorn, A. D. (2012). Sterols with antileishmanial activity isolated from the roots of *Pentalinon andrieuxii*. *Phytochemistry* 82:128–135. doi: 10.1016/j.phytochem.2012.06.012.

Pechangou, N. S., Upninder, K., Kapil, G., Rakesh, S. and Moundipa, F. P. (2015). Molecular differentiation of *Entamoeba* spp. isolated from Cameroonian human immunodeficiency virus (HIV) infected and uninfected patients. *The Journal of Parasitology and Vector Biology* 7(7): 139–150.

Petri, W. A. and Haque, R. (2010). Entamoeba species, including amoebiasis. In: Mandell G.L., Bennett J.E., Dolin R. editor (s). *Mandell, Douglas, and Bennett's Principles and Practice of Infectious Diseases*. 7th Edition. Philadelphia, PA: Churchill Livingstone Elsevier, 3411–3425.

Pinilla, A.E.; López, M.C. and Viasus, D.F. (2008). History of the Entamoeba histolytica protozoan. *Revista medica de Chile* 136(1): 118–124.

Prinsloo, G., Marokane, C. K. and Street, R. A. (2018). Anti-HIV activity of southern African plants: Current developments. phytochemistry and future research. *Journal of Ethnopharmacology* 210: 133–155. doi: 10.1016/j.jep.2017.08.005.

Quintanilla-Licea, R., Mata-Cárdenas, B. D., Vargas-Villarreal, J., Bazaldúa-Rodríguez, A. F., Kavimngeles-Hernández, I., Garza-González, J. N., and Hernández-García, M. E. (2014). Antiprotozoal activity against Entamoeba histolytica of plants used in northeast Mexican traditional medicine. Bioactive compounds from Lippia graveolens and Rutachalepensis. *Molecules (Basel, Switzerland)* 19(12): 21044–21065. https://doi.org/10.3390.

Ravdin, J. I. and Petri, W. A. (1995). Entamoeba histolytica (amoebiasis). In: Mandell G.L., Bennett J.E., Dolin R., editors. *Mandell, Douglas and Bennett's Principles and Practice of Infectious Disease*. New York: Churchill Livingstone; pp. 2395–2408.

Romero-González, J., Peralta-Videa, J. R., Rodríguez, E., Delgado, M. and Gardea-Torresdey, J. L. (2006). Potential of Agave lechuguilla biomass for Cr(III) removal from aqueous solutions: Thermodynamic studies. *Bioresource Technology* 97: 178–182. doi: 10.1016/j.biortech.2005.01.037.

Rubio-Pina, J. and Vazquez-Flota, F. (2013). Pharmaceutical applications of the benzylisoquinoline alkaloids from Argemone mexicana L. *Current Topics in Medicinal Chemistry* 13(17): 2200–2207. doi: 10.2174/15680266113139990152.

Saleem, M., Kim, H. J., Han, C. K., Jin, C. and Lee, Y. S. (2006). Secondary metabolites from Opuntia ficus-indica var. Saboten. *Phytochemistry* 67: 1390–1394. doi: 10.1016/j.phytochem.2006.04.009.

Salit, I. E., Khairnar, K., Gough, K. and Pillai, D. R. (2009). A possible cluster of sexually transmitted Entamoeba histolytica: genetic analysis of a highly virulent strain. *Clinical Infectious Diseases* 49(3): 346–53. doi: 10.1086/600298.

Samie, A., Obi, C. L., Bessong, P. O., Houpt, E., Stroup, S., Njayou, M., Sabeta, C., Mduluza, T. and Guerrant, R. L. (2008). Entamoeba histolytica: genetic diversity of African strains based on the polymorphism of the serine rich protein gene. *Experimental Parasitology* 118(3): 354–361.

Sanchez-Ramirez, B., Ramirez-Gil, M., Vazquez-Moctezuma, I., Ramos-Martinez, E. and Talamas-Rohana, P. (2004). Entamoeba histolytica: induction of cyclooxygenase-2 expression during amoebic liver abscess formation in hamsters (Mesocricetus auratus). *Experimental Parasitology* 106(3–4): 119–125.

Sargeaunt, P. G. and Williams, J. E. (1978). Electrophoretic isoenzyme patterns of Entamoeba histolytica and Entamoeba coli. *Transactions of the Royal Society of Tropical Medicine and Hygiene* 72(2): 164–166. https://doi.org/10.1016/0035-9203(78)90053-6.

Sawangjaroen, N., Phongpaichit, S., Subhadhirasakul, S., Visutthi, M., Srisuwan, N. and Thammapalerd, N. (2006). The anti-amoebic activity of some medicinal plants used by AIDS patients in southern Thailand. *Parasitology Research* 98(6): 588–592. doi: 10.1007/s00436-005-0119-2.

Shah, T. I., Sharma, E. and Ahmad, G. (2014). Juglans regia Linn: a phytopharmacological review. *World Journal of Pharmaceutical Science* 2: 364–373.

Shirley, D. T., Farr, L., Watanabe, K. and Moonah, S. (2018). A review of the global burden, new diagnostics, and current therapeutics for amebiasis. *Open Forum Infectious Diseases* 5(7): ofy161. https://doi.org/10.1093/ofid/ofy161.

Stanley, S. L. (2003). Amoebiasis. *Lancet* 361 (9362): 1025–1034.

Thangadurai, D., Dabire, S. S., Sangeetha, J., Said Al-Tawaha, A. R. M., Adetunji, C. O., Islam, S., Shettar, A. K., David, M., Hospet, R., and Adetunji, J. B. (2021) Greener composites from plant fibers: Preparation, structure, and properties. In: Kharissova, O.V., Martínez, L.M.T., and Kharisov, B.I. (eds.), *Handbook of Nanomaterials and Nanocomposites for Energy and Environmental Applications*. Springer, Cham. https://doi.org/10.1007/978-3-030-11155-7_21-1

Ultee, A., Kets, E. P. W. and Smid, E. J. (1999). Mechanisms of action of carvacrol on the food-borne pathogen Bacillus cereus. *Applied and Environmental Microbiology* 65:4606–4610.

Ultee, A., Bennik, M. H. J. and Moezelaar, R. (2002). The phenolic hydroxyl group of carvacrol is essential for action against the food-borne pathogen Bacillus cereus. *Applied and Environmental Microbiology* 68:1561–1568. doi: 10.1128/AEM.68.4.1561-1568.2002.

Velázquez-Domínguez, J., Marchat, L. A, López-Camarillo, C., Mendoza-Hernández, G., Sánchez-Espíndola, E., Calzada, F., Ortega-Hernández, A., Sánchez-Monroy, V. and Ramírez-Moreno, E. (2013). Effect of the sesquiterpene lactone incomptine A in the energy metabolism of Entamoeba histolytica. *Experimental Parasitology* 135(3):503–510. https://doi.org/10.1016/j.exppara.2013.08.015.

Waters, C. M. and Bassler, B. L. (2005) Quorum sensing: cell-to-cell communication in bacteria. *Annual Review of Cell and Developmental Biology* 21:319–346.

WHO (1997). UNESCO Report of a consultation of experts on amoebiasis. *Weekly Epidemiological Record* 72:97–99.

Xool-Tamayo, J. F., Monforte-González, M., Rubio-Piña, J., Mirón-López, G. and Vázquez-Flota, F. (2017). Early developmental onset of alkaloid biosynthesis in Mexican poppy (Argemone Mexicana L) Papaveraceae. *Phytochemistry Letters* 20: 300–305. doi: 10.1016/j.phytol.2016.12.020.

Xu, J., Zhou, F., Ji, B. P., Pei, R. S. and Xu, N. (2008). The antibacterial mechanism of carvacrol and thymol against *Escherichia coli*. *Letters in Applied Microbiology* 47: 174–179. doi: 10.1111/j.1472-765X.2008.02407.x.

Zaki, M., Andrew, N. and Insall, R. H. (2006). Entamoeba histolytica cell movement: a central role for self-generated chemokines and chemorepellents. *Proceedings of the National Academy of Sciences of the United States of America U.S.A.* 103: 18751e18756.

21 Harnessing the Underexploited Plant Species for Economic Growth in Nigeria

Ifeoma B. Enweani-Nwokelo and
Ugoy Sonia A. Ogbonna
Nnamdi Azikiwe University

Grace I. Okwu
Ambrose Alli University

Onyekachukwu O. Enweani
University of Nigeria

CONTENTS

DOI: 10.1201/9781003178903-25

21.1 INTRODUCTION

Nigerian climate is very favourable to numerous plant species which are endowed with numerous medicinal and antimicrobial properties. Estimates of over 65,000 species of flowering plants that have medicinal properties which are attributed to abundant quantity of phytochemicals have been reported (Akinpelu and Kolawole 2004; Farombi et al., 2005). Some plants also contain some natural dyes which can be used for dying fabrics and staining of microorganisms and diseased tissues. Currently most of the dyes used in Nigeria and possibly overseas for staining of bacterial, fungal, diseased tissues are synthetic (Ochei and Kolhatkar, 2008). There are disadvantages in the use of synthetic dyes as they may cause skin allergies on humans upon exposure and produce toxic waste (Goodarzian and Ekrami, 2010). Some plant materials have also been explored in the production of organic growth media that would be used for the culture of microorganisms (Oledibe, 2017).

Micronutrients are very essential for the well-being of humanity especially Nigerians. A lot of desirable micronutrients are embedded in indigenous Nigerian plants and plant-based products. Some have been found to control diarrhoea and malnutrition (Enweani, 1998). In this chapter, we bring to bear some underutilized Nigerian plant-based constituents and plant-based products which have been long neglected but have enormous potentials in contributing to the economic growth of Nigeria and improving well-being of Nigerians especially women and children. We took a look at some plant materials that can be used for medicinal and antimicrobial purposes which we had undertaken studies on. *Garcinia kola*, *Vitex doniana*, *Lantana aculeata*, *Lawsonia inermis*, *Cnestis ferruginea*, *Pterocarpus soyauxii*, *Curcuma longa*, *Hibiscus sabdariffa*, *Bixa orellana*, *Lupinus arboreus* are some of the Nigerian plants of economic value that are underexploited which are reported in this write up.

21.2 ANTIMICROBIAL POTENTIALS OF SOME UNDEREXPLOITED PLANTS STUDIED BY OUR RESEARCH TEAM

Quantitative analysis, screening and nutritional composition of these underutilized plants are shown in Table 21.1. Various phytochemicals screened for included alkaloids, saponins, tannins, flavonoids, amongst others.

21.2.1 *Garcinia kola*

Bitter kola called *Garcinia kola Heckel* {akinu (Igbo), orogbo (Yoruba) grows in Nigeria. It produces a reddish yellowish or orange-coloured fruit containing 2–4

TABLE 21.1
Proximate Analysis for *D. dumetorum, N. latifolia, V. doniana* (%)

S/N	Plant Product (g)	Moisture Content	ASH	Crude Protein	Crude Fibre	Ether Extract	Dry Matter	Carbohydrate
1	*D. dumetorum* (raw)	7.85	3.59	9.34	2.51	0.42	90.28	76.29
		7.01	3.97	9.28	2.44	0.38	90.30	76.92
	Mean/SD	7.43 ± 0.42	3.78 ± 0.19	9.31 ± 0.03	2.48 ± 0.03	0.40 ± 0.02	90.29 ± 0.01	76.61 ± 0.31
2	*D. dumetorum* (soaked)	9.41	2.81	9.01	3.02	0.51	89.82	75.24
		9.31	2.88	9.02	3.11	0.48	89.71	75.20
	Mean/SD	9.32 ± 0.09	2.85 ± 0.03	9.02 ± 0.01	3.07 ± 0.04	0.50 ± 0.01	89.77 ± 0.05	75.22 ± 0.02
3	*D. dumetorum* (boiled)	9.26	2.14	16.32	9.42	1.45	90.74	61.41
		9.74	2.18	16.40	9.31	1.52	90.26	60.85
	Mean/SD	9.50 ± 0.24	2.16 ± 0.02	16.36 ± 0.04	9.37 ± 0.05	1.49 ± 0.03	90.50 ± 0.24	61.13 ± 0.28
4	*N. latifolia* (raw)	6.05	4.67	21.23	6.07	10.72	90.14	51.26
		6.33	4.77	21.42	6.14	10.59	90.18	50.75
	Mean/SD	6.19 ± 0.14	4.57 ± 0.2	21.33 ± 0.09	6.11 ± 0.03	10.66 ± 0.07	90.16 ± 0.02	51.01 ± 0.26
5	*V. doniana* (raw)	16.66	11.55	8.20	1.50	16.12	90.34	45.97
		16.46	11.50	8.24	1.58	16.19	90.34	46.03
	Mean/SD	16.56 ± 0.1	11.53 ± 0.05	8.22 ± 0.02	1.54 ± 0.04	16.16 ± 0.03	90.34 ± 0	46.00 ± 0.03

Source: Ogbonna et al. (2019).

seeds (Iwu et al., 1987). Different parts of the plant contain phytochemical, minerals and antioxidants which are responsible for its medicinal and pharmacological functions (Ogunmoyole et al., 2012). Bitter kola is used as snack and stimulant due to high content of caffeine in the seeds. In folk medicine, the plants' parts are utilized to treat infection caused by Gram positive bacteria, Ebola virus infections, flu, dysentery and diarrhoea (Iwu, 1993). Ezeanya and David (2013) reported that the plant's bark, seeds and stem are traditionally used in the treatment of throat infections, inflammation of the respiratory tract and acute fever. *Garcinia kola* nut has been shown in previous researches to possess some antibacterial properties against *S. aureus* and *E. coli* (Amalu et al., 2014). Alkaloids possess analgesic, anti-plasmodic and bactericidal properties. Saponins are found to neutralize some enzymes in the intestine which can become harmful. This builds the immune system and promises wound healing.

Tannins have been reported to hasten the healing of wounds and inflamed mucous membrane (Okwu and Okwu, 2004). Calcium fluxes (minerals) are important mediators of hormonal effects on target organs through several intracellular signalling pathways as stated by FAO/WHO (1998). Magnesium functions as a co-factor of many enzymes. These enzymes are involved in energy metabolism, protein synthesis, RNA (ribonucleic acid) and DNA (deoxyribonucleic acid) synthesis. They also maintain the electrical potential of the nervous tissue and cell membranes. Important roles are played by them in the metabolism of calcium (AL-Ghamdi et al., 1994).

Terpenoids show great potential in treatment of pathogenic microorganisms. They have exhibited antibacterial activity against *E. coli. Staphylococcus, Pseudomonas aeruginosa, Proteus mirabilis, Klebsiella pneumonia* (Pierra et al., 2011). Methicillin resistant *S. aureus, Staph. epidermidis* (Santo et al., 2008), *Listeria monocytogenes* (Nero and Moneira, 2010), *Aspergillus flavus* (Leandro and Varga, 2012), *Bacillus cereus and E. coli* were inhibited by methanolic seed extract of *G. kola* (Jackie et al., 2014). The antimicrobial screening of the crude ethanol extract of *G. kola* seeds was shown to have significant activity against some clinical bacterial isolates including *S. aureus, Strep. mutans* and *Strep. viridans* with mean zones of inhibition in comparable range with that of gentamicin (2.5 μg/ml) used as a reference standard. This was reported by Ajayi et al. (2014).

G. kola mesocarp revealed alkaloids, saponins, flavonoids, tannins and cardiac glycoside as constituent parts. Morabandza et al. (2013) reported that its mesocarp also contain alkaloids, anthocyanins, quinines and anthraquinones except cardiac glycoside which was found in Nigerian species. Its phytochemical composition indicated that *G. kola* mesocarp has good chemotherapeutic value. Antibacterial activity of methanolic extract of *G. kola.* mesocarp showed some promising results. Mean zone diameter of 200 mg/ml up to 6.25 mg/ml of *G. kola* exhibited good antibacterial activity against all Gram positive test isolates (*M. morgani, E. coli and P. mirabilis*) as reported by Olise et al. (2019). Mean zone diameter for those test organisms ranged from 17.6 ± 1.28, 10.5 ± 0.55 and 8.0 ± 0.44 mm, respectively, as compared to mean zone diameter of 5 ug/l on the three bacterial isolates recorded as 19.8 ± 0.72, 19.8 ± 0.72 and 17.3 ± 1.85 mm, respectively. *Garcinia kola* stem bark has been shown to contain a complex mixture of phenolic compounds such as tannins, guttiferin (Etkin, 1981), bioflavonoids, xantherenes, benzopherone, kolaflavonone and Garcinia flavonone (Iwu and Igboko, 1982) all of which have antimicrobial activity. Akintowa and

Essien (1990) reported that *Garcinia kola* exhibits purgative, antibacterial and antiviral activity.

In line with all the potentials inherent in *Garcinia kola* tree, it is hereby advocated for increased production at commercial scale and pharmaceutical companies to scale up in the area of utilizing it as a drug component in the treatment of target disease conditions. Prof. Maurice Iwu's products (Interceed Health Products, IHP) are examples of such initiatives in the right direction. *G. kola* could be of benefit as a contraceptive and is widely consumed as a stimulant and improves respiratory activities (Olise et al., 2019).

21.2.2 *Vitex doniana*

Vitex doniana (black plum) is a perennial tropical fruit bearing tree widely distributed in West Africa and the high rainfall areas and found in every part of Nigeria. It is commonly referred to as 'mbembe' or 'uchakoro'(Igbo), 'Orinola'(Yoruba) and 'Dinyar'(Hausa). Phytochemical constituents of the plant include glysosides, flavonoids, alkaloids and essential fatty acids (Arokiyaraj et al., 2009). This plant grows wild and produces a lot of fruits in Nigeria. Fruit drinks have been produced by Okigbo (2003) from *Vitex doniana* fruit pulp. Flavonoids confers antioxidant activity on tissues of humans. They are water-soluble antioxidant and free radical scavengers which prevent oxidative cell damage and have strong anticancer activity. Flavonoids also lower risk of heart diseases (Okwu and Okwu, 2004).

Previous studies have demonstrated that 70% of ethanolic, methanolic and ethyl acetate extracts of stem bark of *V. doniana* have proven activity against *S. aureus, E. coli, B. subtilis, S. typhi* at varying concentrations. Stem bark of *V. doniana* also inhibited sporulation and mycelia development of *A. flavus, A. parasiticus, A. terreus, A. ochraceus, A. nidulans and A. fumigates* (Ouattara et al., 2013; Lagnika et al., 2012).

In our study, the mean zone diameter of test bacterial isolates using concentrations of 200 µg/ml of methanolic extract of *V. doniana* was 2.3 ± 0.635 for *B. subtilis* and *S. aureus.* Extracts of different parts of the plant maybe valuable in the treatment of dysentery and other gastro enteric infections. MIC of different parts of *V. doniana* against *B. subtilis, E. coli, P. mirabilis, S. aureus* ranged between 50 and 200 mg/ml. There was no activity in respect of fungal isolates tested (Ogbonna, 2017). Progesterone-like compounds in the fruit has been suggested as the probable cause of a transient reduction in reproductive functioning in female olive baboons if they consume large amounts of *V. doniana* fruits (Ogbonna, 2017).

Aqueous extract of the stem bark has been shown to produce a dose-dependent hypotensive effect. The leaves and root are anodyne (World Agroforestry Centre, 2015). Juice extracted from young mashed tender leaves of the plant is reported to be squeezed into the eyes in order to treat conjunctivitis and other eye disease conditions. A paste from the mashed tender leaves and bark are also applied to wounds and burns. Pulverized bark solution is taken to treat colic and control bleeding after child birth. Twigs are used as chewing sticks. This has potential of extracts being used in toothpaste production. Dried seeds yield about 30% oil which has low saponification values. This can be used for skin cream, resin and paint production (Agro Forestry

Centre 2015). Traditional healers employ *Vitex doniana* alone or in a combination with stem bark of *Adansonia digitata* in the treatment of diarrhoea, leprosy and dysentery (Adjanohoun et al., 1989). The plant's leaves are used as antiseptic and antidiabetic. In Mali, aerial parts are utilized as diuretic, tonifiant, aphrodiasiac and bactericide (Goetz, 2006).

21.2.3 *Lantana aculeata*

Lantana aculeata is a flowering ornamental plant. Different parts of *L. aculeata* are reported to possess essential oils, flavonoids, alkaloids, oligosaccharides, phenolic compounds, saponins, steroids, glycosidestritepenes, sesquiterpenoides and tannins as major phytochemical groups (Bhakta and Ganjewala, 2009; Kensa, 2011).

In previous studies (Srivastar and Singhm 2011) *L. aculeata* had been found to possess antifungal, anti-proliferative and antimicrobial activities. Leaves are used to treat cuts, ulcers, swelling, eczema, tumour, high blood pressure, measles, fever, sores, fevers, cold. Oil *extracted from Lantana* is used in the treatment of skin itches and as an antiseptic for wounds. Decoctions prepared from the plant are applied externally in case of leprosy and scabies. A 20 mg/ml dose of *L. aculeate* exhibited significant antifungal activity against *Alternaria* spp. and white and brown rot fungi (Srivastar and Singh 2011; Tripathi et al., 2009).

Extracts of *L. aculeata* exhibited antimicrobial activity against *S. aureus*, *P. vulgaris*, *P. aeruginosa*, *V. chlolerae*, *E. coli* and two multi-resistant strains of *E. coli* and *S. aureus* (Barreto et al., 2010). In our recent study, the MIC of the leaf extract indicated the range in respect of test bacterial isolates (*B. subtilis*, *S. aureus*, *S. pneumoniae*, *E. coli*, *P. mirabilis*) to be from 3.125 to 12.51 mg/ml. Mean zone diameter of the ethanol extracts (200 mg/ml) in respect of the test bacterial isolates were between 7.0 ± 1.73 and 9.6 ± 2.517 mm (Ogbonna, 2017).

Major leaf and flower essential oil constituents of *L. aculeata* as reported by Andersson and Dobson (2003) is Sesquiterpenes with mainly B-caryphyllene, zingiberene, humulene, arcurcumene, gemacrene-D and bisabolene.

21.2.4 *Lawsonia inermis* (Itenra)

It is a perennial shrub and medicinal plant that is widely distributed across Nigeria (Emin and Mehmet, 2012). It was reported by Babu and Subhasree (2009) that *L. inermis* extracts could be a potential alternative to standard antimicrobials that have become less effective or ineffective against certain pathogenic microorganisms. The plant contains substances such as *lawsonia* (oxynapthoquinone being the principal colouring matter), Gallic acid, mucilage, traces of alkaloid, anthopyanins. Leaf extracts also contain phytochemicals such as *phytosterol*, tannins, steroidal compounds and flavonoids (Raja et al., 2013). Extracts of *L. inermis* leaf, flower, seed, stem, bark and roots have been found to exhibit antioxidant, antidiabetic, hepatoprotective, hypoglycaemic, antimicrobial, anticancer and wound healing properties (Hema et al., 2010). Gull et al. (2013) reported that methanol aqueous, chloroform and acetone leaf extracts inhibited the activity of *S. aureus*, *B. subtilis*, *S. typhi*, *S.*

epidermis, *E. coli*, *Kleb.* species and *Shigella* species. We observed in our study (Ogbonna, 2017) that MIC exhibited by *L. inermis* leaf against the test bacterial isolates was 2.5–12.5 mg/ml and the zone diameter (200 mg/ml) ranged from 3.25–0.35 to 15.6 ± 1.15 mm. *Lawsonia inermis* roots are taken as means of enhancing fertility in women. It also serves as a potent medicine for gonorrhoea. A decoction of them is considered to be diuretic and pectoral for bronchitis (Olise et al., 2019).

Fresh bruised leaves are used in Malaysia as poultices to relieve burning sensation of the feet, treat beriberi, skin disease, circumcision wounds, boils and distension of the stomach. Decoction can also be gargled to treat gum boils or prescribed to relieve abdominal pains after childbirth (Hema et al., 2010).

Lawsonia inermis leaf extract exhibited absolute toxicity against *Microsporum gypseum* and *Trichophyton mentagrophytes*. Fungitoxicity of the extract remained unaltered at high temperature on autoclaving and after long storage (Singh and Pandeyi, 1989). *Lawsonia inermis* ethanol leaf extract exhibited significant antifungal effect against phytopathogenic fungi. Ethanol extract could be used as an alternative source of antifungal agents for protection of plants or crops against fungal infections as stated by Anwar et al. (2007).

21.2.5 *Cnestis ferruginea*

Cnestis ferruginea is a shrub or climber of deciduous forest widely dispersed in West Africa including Nigeria. It bears orange-red fruits with velveting hairs on the follicle (Burkill, 1995). Common names of the plant in Nigeria include 'Furaamanya/ Otito' (Hausa); 'Okpunkita'; 'Amunkita' (Igbo);'Akaraoje/ Bonyinbonyin'(Yoruba); 'Ukpo-ibieka'(Edo) and 'Usiereebua' (Efik) (Burkii, 1995).

The plant has an ornamental value and is used across tropical Africa to treat diverse ailments in traditional African medicine (Ishola and Ashorobi, 2007). In Ivory Coast, a paste of the root bark is rubbed on the forehead for headache and with the addition of ash of burnt bark of Aubrevillei is given as an appetite stimulant in cases of illness (Adjanohoun and Ake, 1972). The leaves are laxative and are taken by Yorubas as decoction. A leaf decoction is given in Zaire to treat bronchitis and as an abortifacient (Troupin, 1952). In Ivory Coast, leaf sap is placed on the eyelids, instilled into the eyes for the treatment of eye infections. The leaf or the root is used for dysmenorrhoeal (Oliver, 1990). A paste of young leaves compounded with a little wood ash is let down with water and administered as an enema for heartburn. The roots are recognized in Zaire as a purgative. *Cnestis ferruginea* has anti-hepatotoxic effect which is assumed to be related to glutathione-mediated detoxification as well as free radical suppressing activity. Considerable antioxidant activities were found in the plant extracts (Akharaiyi et al., 2012). Treatment with the extract reversed the hyperandrogenaemia and attenuated the irregular oestrous cycle in polycystic ovarian syndrome (PCOS)-induced rats. Saponins and flavonoids present in the plant are considered responsible for the clinical benefits of *C. ferruginea* roots in the management of PCOS (Yakubu and Ibiyo, 2013).The leaves contain highly potent hypoglycaemic principle. This could be a potential source for isolation of new orally active anti-hyperglycaemic compounds for attenuating secondary complications of diabetes which include atherosclerosis, liver and renal dysfunction (Adisa et al., 2010). They could enter into remedies for skin infections in Nigeria. It has been demonstrated that it has antibacterial effects on *Sarcinalutea* and *Staphylococcus*. The roots are used for skin ointment by women in Benin Province.

The fruit contains a soft juicy, somewhat bitter and acid pulp. Teeth cleaning and whitening are achieved by rubbing fruit extract on the teeth, which leaves a refreshing taste in the mouth (Aubreville, 1959). Fruit juice is applied to help in wound healing. The fruit with ground seeds using spirit or boiled in wine produces a remedy for snake bite in Nigeria. From our study, 200 mg/ml of *C. ferruginea* ethanolic fruit extract demonstrated antifungal activity on all the test fungal isolates (*A. niger, C. albicans, C. albicans* ATTC 10231) revealing mean zone diameter of between 5.3 ± 1.155 and 12.3 ± 1.128 mm as compared to 9.0 ± 1.000 and 20.3 ± 1.528 mm of 50 µg/ml Miconazole. MIC of *C. ferruginea* ethanolic extract against tested isolates ranged from 25 to 200 µg/ml for bacterial isolates and 25 to 50 µg/ml for fungal isolates tested.

21.2.6 *Pterocarpus soyauxii*

It is a high rain forest tree. The plant stem bark is greyish brown to brown in colour, scales off in fine irregular scales and contains a red sap. It is commonly known as African padauk/narra/mukkwa and has a high hygroscopic stability and good natural durability due to its extractives (Dean et al., 1980). Various parts of *P. soyauxii* extracts, containing different phytochemicals and minerals, are used in Africa for the treatment of different diseases (Okafor, 1999). Plant-based products in medicine are useful. It is due to the presence of bioactive substances including alkaloids, tannins, flavonoids, phenolic compounds, steroids, resins and other secondary metabolites which they contain. These are capable of producing definite physiological action in the body. Banerjee and Mukherjee (1981), Bezuidenhoudt et al. (1987) and Kiec-swierczynska et al. (2004) stated that *Pterocarpus soyauxii* contains various compounds such as bioflavonoids (santalin A, santarubins A and B), isoflavonoids (pterocarpin, formononetin and prunetin), an isoflavone, quinine (claussequinone), isoflavanes (vestitol and mucronulatol), tannins, ascorbic acid, glucosides, triterpenes, xanthanes. Previous studies (Kiec-swierczynska et al., 2004, Olise et al., 2019) have shown that *P. soyauxii* possess antimicrobial activity. In African traditional medicine, leaves, stem, wood, bark, flowers and seeds are used in the treatment of diabetes, hypertension, intestinal parasites, renal and cutaneous diseases. In Gabon, bark is used in diuretic while in Nigeria, fresh leaves are used as food (Okafor, 1999).

The ethanolic fresh and older leaves extract of this plant inhibited *E. coli, S. typhi, S. flexneri, S. aureus, K. pneumoniae* and *C. albicans* with zone of inhibition ranging from 3.00 to 11.75 mm (Osuagwu and Akomas, 2013). In our study, the mean zone diameter of tested bacterial isolates at 200 mg/ml of methanolic extract of *P. soyauxii* stem extract was in range of 3.6 ± 0.577–12.03 ± 3.5/mm in respect of *S. pneumoniae, S. aureus, B. subtilis, K. pnemoniae, P. aeruginosa*. MIC of this plant stem extract was in the range of 1.502–50 mg/ml in respect of the test bacterial isolates (Olise et al., 2018).

21.2.7 *Pterocarpus santallinoides, Smilax kraussiana, Rauvolfia vormitoria, Alchornea cordifolia, Cassytha filiformis*

Pterocarpus santallinoides, *Smilax Kraussiana* and *Rauvolfia vormitoria* had been used extensively in Africa as phytomedicine (Emencheta et al., 2019). Various parts of these three Nigerian plants revealed antimicrobial activities against the test microorganisms with the mean inhibition diameter ranging from 2 ± 0 to 9 ± 0 mm. The range of the MIC recorded for the extracts and fractions that were effective was 6.25–100 mg/ml. The MBC values ranged from 12.5–100 mg/ml. *Escherichia coli* and *Salmonella typhi* were susceptible to the fractions.

Antibacterial activities of three plant extracts (*Alchornea cordifolia*, *Cassytha filiformis and Pterocarpus santilinoides*) are utilized in traditional/folklore medicine for the treatment of diarrhoea in Nigeria. This was evaluated by employing three Gram negative bacterial isolates (*Salmonella typhi*, *Shigella dysenteriae* and *Escherichia coli*) and one Gram positive bacterial isolate (*Staphylococcus aureus*) from stool sample of diarrhoeic patients visiting a tertiary hospital in Nigeria (Obaji et al., 2019, 2020).

The MIC value of methanol extracts of *A. cordifolia*, *C. filiformis* and *P. santalinoides* against drug-resistant strains of *S. typhi*, *Sh. dysenteriae*, *E. coli* and *Staph. aureus* ranged from 3.125 to 12.5 µg/ml, while the combination of *A. cordifolia* and *P. santalinoides* has largely synergistic effect on all the test isolates. The polyherbal suspension involving A. *cordifolia* and *P. santalinoides* in combination showed excellent synergistic activity against all the test isolates (Obaji et al., 2019, 2020).

21.2.8 *Curcuma longa*

A study conducted by Kim et al. (2005) reported that ethyl acetate extract of *C. longa* exhibited antibacterial activities, anti-inflammatory effect, anti-helminthic property, anti-cholesterol action and anticancerous activity. Anti-inflammatory effect was said to have been linked with the inhibition of TNF-α IL-8 monocyte inflammatory proteins–I, IL-IB. It has also been reported to block cytokine-induced transcription of leukocyte adhesion molecule ICAM-I, VCAM-I and E-selection (Gupta and Gosh 1999). Curcumin inhibits platelet aggregation induced by ADP epinephrine or collagen. Curcumin is also described to lessen the uptake of cholesterol from the gut, increase the high-density lipoproteins (HDL) and reduce the low-density lipoproteins. It can also inhibit the peroxidation of serum low-density lipoproteins which leads to artherosclerotic lesions.

21.2.9 *Hibiscus sabdariffa*

Leaves of roselle are very rich in vitamin C and anthocyanin and retain a high composition even after cooking. *Hibiscus sabdariffa* has anti-hypertensive, anticancer and hepatoprotective activities. Stem has been used in folk medicine as a diuretic, mild laxative and treatment for cardiac and nerve diseases as well as cancer. It contains flavonoids (hibiscitin and hibiscetin) (Olise et al., 2018).

21.2.10 *Bixa orellana*

In our traditional/folklore medicine, pulp surrounding the seed of Bixa plant is widely used to treat burns, bleeding, dysentery, gonorrhoea, constipation and fever. The seeds have expectorant properties. The roots aid in digestion and cough suppressant. They are employed in the treatment of prostrate disorders and internal inflammation, obesity, high cholesterol, arterial hypertension, cystitis, renal insufficiency and to eliminate the uric acid (Olise et al., 2018).

Other studies included the activities of different parts of *Moringa oleifera* against some clinical pathogenic fungal isolates from sputum samples of some HIV positive individuals (Onuoha et al., 2015). Activities of different parts of *Napoleona imperalis* against some Nigerian dermatophytic isolates and their relationship with some haematological parameters were considered by Odeyemi et al. (2015). Phytochemical and antifungal activities of different parts of three Nigerian herbs were studied by Okwuosa et al. (2012). Enweani et al. (2005) examined the ascorbic acid and microbiological analyses of extra-cotyledonous deposits of Pride of Barbados (*Caesalpina pulcherima*) at various temperatures. Enweani et al. (1998, 2004) also studied the phytochemical constituents of *Anona muricata* and *A. squamosa* and the potential to be used as complementary source of electrolyte replacement in children suffering from acute watery diarrhoea. These studies revealed that these plants contain various phytochemicals which exhibited antimicrobial activities and abundant antioxidants amongst other constituents.

21.3 NIGERIAN UNDEREXPLOITED PLANTS AS SOURCES OF DYES/STAINS

All tissues and their constituent cells are mainly transparent and colourless. Different structures cannot be easily distinguished from each other when examined with the conventional light microscope. Histological techniques used for the distinction of tissue components commonly cause two changes in the tissue, either an alternation of contrast or an alternation in colour. These changes can be achieved by staining methods. Successful staining methods are both specific and sensitive. Specifically or selectivity is the ability to discriminate between individual tissue components and to colour one or a few of these, leaving others unstained. There was toxicity-induced haematological alterations after acute and chronic administration of tartrazine (E102) in Albino rats (Awvioro et al., 2007).

In microscopy, staining is an auxiliary technique used to enhance contrast in the microscopic image. Stains and dyes are usually employed in life sciences to highlight structures in biological tissues to be viewed under different microscopes. Biological stains are prepared from dyes (Baker et al., 1998). Medicinal plants contain dyes. Various phytochemical compounds detected in plants are known to have beneficial importance in industrial and medical sciences (Wagini et al., 2014).

Dyeing of tissues by stains (natural or synthetic) is dependent on binding to tissues. Ionic bonding which involves electrostatic attraction between oppositely charging ions is the most important form of bonding in histological staining.

Several structures are selectively stained with various natural and synthetic dyes. Acidity, alkalinity and mordant have been reported to affect same stains. *Curcuma longa*, *Hibiscus sabbdariffa* were used as a counter stain in place of eosin (Bassey et al. 2011). These dyes are acidic in nature. *Curcuma longa* served as one of best counter stains with yellow colouration. It has already been established that acidic tissue elements (nucleus) would have affinity for basic stains while cytoplasm, which is basic in nature, would have affinity for acid dyes. *Bixa orellana* seed extract could serve as a better alternative or complementary stain to eosin especially when used with acetic acid (Olise et al., 2019).

Lawsonia inermis extract could be used as a primary stain because it is naturally acidic. But when needed to be used as a counter stain in order to stain the nucleus, then ammonia (pH11.6) has to be added to the plant extract to increase pH to alkalinity. *Lawsonia inermis* contains a principal colouring matter called lawsone (2-hydroxy-1.4 hapthaquinone) which is used as a natural dye (Raja et al. 2013). It can be inferred that these natural dyes from plant extracts can serve as a source of stains for microorganisms in the presence of appropriate mordant. It is used as a raw material for natural hair dyes, dyes for textile and tattoo (Kok et al., 2005). Aqueous and ethanolic extracts of *Lawsonia* were used to stain epidermal cells of Allium cepa L. (Aguru et al., 2015). In our findings (Olise et al., 2018), nucleic staining with *L. inermis* was achieved as primary stain in place of haematoxylin. Cytoplasm portrayed moderate intensity of either pale pink or pink colouration. Kaempferolmonoglucoside and luteolin (flavones) are present in *L. inermis* leaf extract. They appear as a yellow crystalline solid which is derivative of the yellow colour exhibited by *L. inermis* leaf extract.

Garcinia kola mesocarp stains red and is an acidic stain that should be allowed to stay shorter than with eosin procedure. Ink are also derived from the dried fruits, young leaves and bark of *Vitex doniana* subjected to boiling and it serves as dye for clothing (Olise et al., 2018). *Pterocarpus soyauxii* contains about 13% extractives, which were abundantly used as a dye-source in the 19th century (Deon et al., 1900). Attempts have been made to investigate the feasibility of employing the natural dye extracted from *P. soyauxii* as a sensitizer for Titanium dioxide thin films. This is because without complex synthetic and purification process Titanium dioxide is most commonly used as a photo anode in dye-sensitized solar cells which is non-toxic and possess good optical and electrical properties, thus can be efficiently sensitized by a dye (Rajalakshmi et al., 2013).

Plant extracts studied by our research team (Olise et al., 2018) as potential stains on clinical bacterial and fungal isolates and diseased human tissues indicated acidic pH range of 4.2–6.4. Acidic pH is needed to create affinity for cytoplasm being basic naturally (Baker et al. 2001). This explains why the extracts acted as counter stains. Stains derived from ethanolic extracts of *C. ferruginea* fruit and methanolic stem extract of *P. souyauxii* exhibited stronger affinity for fungal than bacterial isolates tested. Extracts of *C. ferruginea* and *P. soyauxii* plant parts which were treated with inorganic acid increased the colour intensity from red to deep red. Malvidin is an anthocyanin (flavonol) found in *P. soyauxii*. Malvidin is responsible for the reddish colour as exhibited by the plant (Jing, 2006).

In our study, anthocyanins (polyphenols) were extracted from *V. doniana*. They are the most naturally occurring important group of water-soluble pigments as reported by Harborne (1998). They are responsible for various colours exhibited by fruits and vegetables including the red, blue, orange, purple or orange colour of many fruits and vegetables (Seeram, 2008). They can form flavylium cations which distinguishes it from other flavonoids (Mazza, 2007). Ethanolic extract of *G. kola* mesocarp, methanolic stem bark extract of *P. soyauxii*, ethanolic fruit extract of *C. ferruginea* and ethanolic leaf extract of *L. inermis* were substituted as counter stain in Gram staining technique in respect of *S. aureus* and *E. coli*, *C. albicans* and *A. niger*. The fungal isolates were properly stained which compared favourably with standard lactophenol staining technique. Extract of *P. soyauxii* was also used to stain human appendix tissue which compared favourably with standard staining procedure of Harris Haematoxylin; Gills Haematoxylin and Eosin; Mayers Haematoxylin; Gills Haematoxylin (Olise et al., 2018).

Rhizome of *Curcuma longa* is used majorly as dye to colour products such as tanned leather, cotton cloth thread and palm fibres as a golden yellow colour in West Africa (Prasad et al., 2011). Rhizome of *Curcuma longa* is used mainly as a direct dye that has affinity for cellulosic fibres. They do not require mordant to colour products such as tanned leather, cotton, cloth, thread and palm fibres as a yellow colour in West Africa (Prasad et al., 2011). It is useful as a histological stain for counter stain for cytoplasmic staining. Avwioro et al. (2007) and Bassey et al. (2011) successfully used the dye to stain collagen fibres and blood cell. Our recent studies had shown that *C. longa* was effective in staining of the cytoplasm as a counter stain giving yellow colouration. Henna leaves aqueous extract had been applied as a counter stain in Gram staining procedure (Chukwu et al., 2011).

Summarily, extracts of *Curcuma longa*, *P. soyauxii*, *C. ferruginea* and other plant materials studied compared favourably with primary or secondary stains with or without mordant of standard Gram staining techniques and H and E.

21.3.1 *Bixa orellana*

Ethanolic extracts of *Bixa orellana* seeds is used to colour lips as a dye. Its dye is used to colour textile, fabrics, weapons, silks, wool in India. Our study portrayed aqueous and ethanolic extracts of *B. orellana* mixed with acetic acid and used as a counter stain in place of eosin, giving an excellent yellowish green cytoplasmic colouration (Olise et al., 2018).

21.3.2 *Lipinus arboreus*

A bright yellow dye is derived from the flowers and roots of *L. arboreus*.

21.3.3 *Hibiscus sabdariffa*

Hibiscus sabdariffa red calyces are used as food colouring. Aqueous extract of roselle with various modifications stained testicular tissue of a rabbit (Egbujo et al., 2008). Diagnostic results were obtained (Ihuma et al., 2012) when methanolic /

extracts of *H. sabdariffa* was applied to stain tissues with fungal infections as well as histological sections from the small intestine. Formalin-fixed paraffin embedded in tissue sections were stained using *H. sabbdariffa* extracts mixed with ferric chloride and glacial acetic acid (Bernard, 2008). *Hibiscus sabbdariffa* could serve as a primary stain without mordant giving the cytoplasm of the tissues a pink colouration without staining the nucleus. Purified acidic part of *H. sabdariffa* was used in place of eosin (Emman, 2000). Extracts of *H. sabdariffa* and *Azadirachata indica* had been employed in the staining of the ova of intestinal parasites. Advantages of dyes from natural sources over synthetic ones cannot be overemphasized. Apart from the toxic nature of synthetic dyes, high cost of production and importation makes it imperative for us to search for an alternative good source of dyes. These shows there are enormous natural resources in plant products in Nigeria which could be scaled up to save lean foreign reserves for the nation.

21.4 SOURCING OF MEDIA FOR CULTIVATION OF MICROORGANISMS FROM NIGERIAN PLANT PRODUCTS

A culture medium may either be liquid or gelled substance that support the growth of microorganisms under laboratory conditions. A culture medium suitable for growth of microorganisms is formulated with the consideration of a carbon source, nitrogen source such as protein, availability of enzymes, vitamins, mineral elements such as phosphorus and sulphur, suitable pH, inorganic salt and water. The cost of media importation into the country has sky rocketed with the dwindling economy. This has encouraged continuous search for alternative, readily available and cheaper local plant materials for fungal culture.

Previous studies in the search for alternative sources of media have not fully succeeded. In an attempt to compound the best alternative medium, the following was formulated from local plant materials. A proper combination of *Zea mays*, *Dioscorea dumentorum* and *Glycine max* yielded a good fungal medium with a good gelling property, solidity and consistency which compared favourably with standard Sabouraud Dextrose Agar (SDA). The pH of the formulated fungal medium was 5.15 while that of standard SDA was 5.05 after autoclaving. Mean colony diameter, estimation of sporulation and aerial mycelial growth of test *Aspergillus niger* and *Penicillium* species compared favourably to same parameters assessed using SDA (Oledibe, 2018).

21.5 NUTRITIONAL BENEFITS OF SOME UNDEREXPLOITED PLANT - DERIVED FOOD PRODUCTS IN NIGERIA

Plant – based food product was formulated from some Nigerian underexploited plant sources by our research team (Ogbonna et al., 2019). The composite food products were formulated from *Dioscorea dumentorum* (Plate 21.1), *Musa acuminate* and *Vitex doniana* (Plate 22.3). The composite plant products contained crude protein, carbohydrate, ether extract, moisture, ash, crude fibre and dry matter (Tables 21.1–21.3).

PLATE 21.1 *Dioscorea dumetorum* tuber.

TABLE 21.2
Phytochemical Composition of *D. dumetorum, N. latifolia, V. doniana* (mg/100 g)

S/N	Plant Product (g)	Saponin	Tannin	Alkaloid	Hydrocyanide	Phenol
1	*D. dumetorum* (raw)	0.41	1.11	2.11	1.85	0.81
		0.41	1.02	2.15	1.88	0.80
	Mean/SD	0.41 ± 0.0	1.07 ± 0.05	2.13 ± 0.04	1.87 ± 0.03	0.81 ± 0.01
2	*D. dumetorum* (soaked)	2.55	0.71	1.24	1.45	-
		2.58	0.72	1.28	1.48	-
	Mean/SD	2.57 ± 0.03	0.72 ± 0.005	1.26 ± 0.04	1.47 ± 0.03	-
3	*D. dumetorum* (boiled)	0.32	0.34	0.84	1.12	0.20
		0.40	0.38	0.92	1.14	0.14
	Mean/SD	0.36 ± 0.08	0.36 ± 0.04	0.88 ± 0.02	1.13 ± 0.02	0.17 ± 0.06
4	*N. latifolia* (raw)	0.61	2.10	0.11	0.31	-
		0.64	2.15	0.17	0.32	-
	Mean/SD	0.63 ± 0.01	2.13 ± 0.02	0.14 ± 0.04	0.32 ± 0.005	-
5	*V. doniana* (raw)	0.51	1.81	1.51	1.81	0.07
		0.58	1.88	1.55	1.82	0.08
	Mean/SD	0.55 ± 0.03	1.85 ± 0.03	1.53 ± 0.02	1.82 ± 0.005	0.08 ± 0.005

Source: Ogbonna et al. (2019).

TABLE 21.3
Micronutrient Composition of *D. dumetorum, N. latifolia and V. doniana* (mg/100 g)

S/N	Plant Product (g)	Calcium	Magnesium	Potassium	Sodium	Phosphorus	Copper
1	*D. dumetorum* (raw)	58.7	15.89	62.14	-	ND	0.09
		58.5	15.88	63.02	-	ND	0.08
	Mean/SD	58.6/58.6 ± 0.1	15.89/15.89 ± 0.1	62.58/62.58 ± 0.44	-	ND	0.09 ± 0.01
2	*D. dumetorum* (soaked)	56.70	14.14	60.51	-	ND	0.04
		56.41	14.45	60.52	-	ND	0.08
	Mean/SD	56.56/±0.14	14.30/±0.15	60.52/±0.1	-	ND	0.06/±0.02
3	*D. dumetorum* (boiled)	42.63	32.11	48.12	30.41	35.04	2.52
		42.52	32.18	48.20	30.52	34.39	2.82
	Mean/SD	42.58/±0.06	32.15/±0.03	48.16±0.04	30.47±0.05	34.72±0.33	2.67±0.15
4	*N. latifolia*	114.01	28.41	45.21	17.81	32.30	12.04
		110.21	28.45	45.22	17.31	32.14	12.08
	Mean/SD	112.11 ± 0.10	28.43 ± 0.02	45.22 ± 0.01	17.56/±0.25	32.22/±0.08	12.06/±0.02
5	*V. doniana*	34.92	20.28	16.77	ND	20.44	3.07
		34.70	20.20	16.75	ND	20.40	3.11
	Mean/SD	34.81/±0.11	20.24/±0.04	16.76/±0.01	ND	20.42/±0.02	3.09/±0.02

Source: Ogbonna et al. (2019).

They also contained various proportions of phytochemicals including saponin, tannin, alkaloid, hydrocyanin and phenol. *Vitex doniana* fruits are very rich in vitamins tested including $B_{1,}$ $B_{2,}$ B_6, C and provitamin. Vitamin composition of the formulated composite products yielded vitamins $B_{1,}$ B_2, B_6, C, E and provitamin. Enhanced micronutrient values were recorded for the formulated composites in respect of calcium, magnesium, potassium, sodium, phosphorus and copper.

Nauclea latifolia fruit, leaf and bark were prepared separately and herbal fruit drink made from the different parts of the plant. It was observed by our research team that hot water extract (infusion) of different plant parts of *N. latifolia* served a nutritional healthy drink with lots of antioxidants and vitamins (Ogbonna et al., 2019). In Nigerian folklore medical practices, decoctions from different plant parts are used in the treatment of diabetes, fever, indigestion, cough, analgesic, antimicrobial and antiparasitic. It is also used as a chewing stick. The proximate analysis values of raw *N. latifolia* fruit (Plate 22.2) consisted of 6.19% ± 0.14% moisture, 4.57% ± 0.2% ash, 21.33% ± 0.09% crude protein, 6.11% ± 0.03% crude fibre, 10.66% ± 0.07% ether extract, 90.16% ± 0.02% dry matter and 51.01% ± 0.2% carbohydrate content (Ogbonna et al., 2019).

The phytochemical composition of *N. latifolia* from our study revealed that the fruit contains 2.13 ± 0.02 (mg/100) saponin, 2.13 ± 0.02 tannin, 0.14 ± 0.04 alkaloid,

PLATE 21.2 *Nauclea latifolia* ripe fruit.

PLATE 21.3 *Vitex doniana* ripe and unripe fruit.

0.32 ± 0.00 hydrogen cyanide and phenol was not detected. Micronutrient composition of *N. latifolia* was 112.11 ± 0.10 calcium, 28.43 ± 0.02 magnesium, 45.22 ± 0.01 potassium, 17.56 ± 0.5 sodium, 32.22 ± 0.08 phosphorus, 12.06 ± 0.02 copper.

The two fruits (*V. doniana* and *N. latifolia*) are very rich in vitamins and make excellent healthy fruit drink which is a natural source of micronutrient to Nigerians especially women and children (Plate 21.3).

Fruit juice of *Anona muricata* and *Anona squamosais* an excellent natural juice. It contains micronutrients and electrolytes available to women and children. It contains potassium, sodium, iron, magnesium, calcium, chloride and bicarbonate. It also contains carbonate, phosphorus, zinc and copper (Enweani et al., 1998; 2004).

Edible mushroom has been processed into powdery forms to make it accessible to Nigerians considering its numerous health benefits. It contains a lot of vitamins, lovastatin (which lowers cholesterol in the blood) (unpublished data).

21.6 ECONOMIC DEVELOPMENT OF NIGERIA FROM PERSPECTIVE OF HARNESSING UNDEREXPLOITED PLANTS

Nigerians spent N334.3 billion to import foodstuff and tobacco in 6 months as reported in http://nairametrics.com 2019/10/03. Nigeria imports of pharmaceutical products was US$606.31 million during 2018, according to United Nations COMTRADE database on international trade as reported by Nigeria imports of pharmaceutical products-data (https://tradingeconomics.com). Historical chart and statistics was updated in June 2020 (Figure 21.1).

Nigeria depends so much on drugs imported into the country and this is very worrisome as COVID-19 pandemic rages. Some of the drugs imported into the country include analgesics, vitamins, antiretrovirals, antimalarials, antibiotics including tablets, capsules and syrups. Others include antitussive syrups, infusions, antacids, antiseptics/ disinfectants and injectables. About 70% of Nigeria's drug usage are

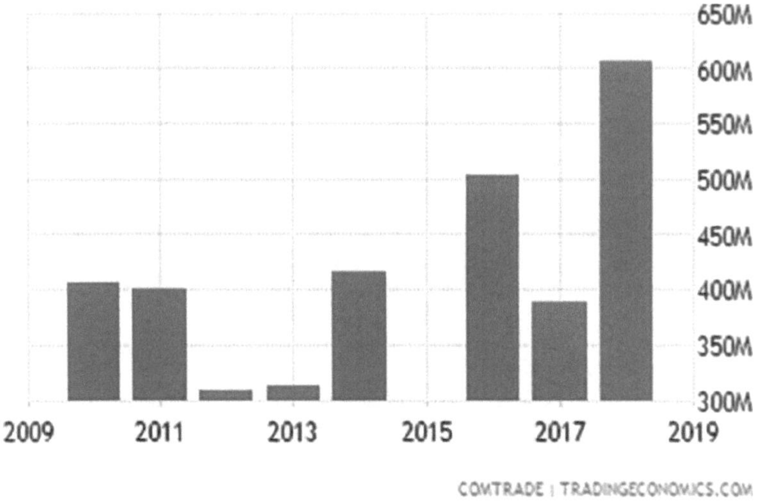

FIGURE 21.1 Nigeria imports of pharmaceutical products.

imported from other nations mainly China and India. In addition, Nigeria relies on imported equipment in addition to active pharmaceutical ingredients used in drug manufacturing.

In this period of COVID-19 pandemic that is ravaging people of the world, Nigeria will have to battle with additional challenges because air travels are not possible due to lockdown coupled with economic recession. It is pertinent that national drug sufficiency and drug security is planned for and addressed to combat diseases in order to reduce death rate and cater for other healthcare needs. Counterfeit and substandard pharmaceutical products, at times, are imported. This poses threat to the economy and security of Nigeria. About 115 pharmaceutical companies are established in Nigeria. Most of them buy active pharmaceutical ingredients from other manufacturing sources, then formulate into finished drugs products.

There are lots of researches in the higher institutions and research institutes in Nigeria where there are lots of published works on phytochemical composition, antimicrobial properties, proximate analysis of numerous indigenous plants which have been proved to be source of bioactive ingredients in the treatment of ailments that Nigerians do suffer from. The major challenge is lack of pharmaceutical industry – University/Research Institutes Partnership that is channelled to solving our local health problems.

Other important challenges tackled by the pharmaceutical companies in Nigeria which have adversely affected them include power outage, poor incentives on the part of the government, weak technology and engineering base, lack of industrial network and supply chain with high taxation (https://theconversation.com).

Currently, there is a National Agency for Food and Drug Administration and Control in Nigeria (NAFDAC) policy of Five plus Five-Year Validity (Migration to Local Production). This addresses local production of essential medicines. This is geared towards reduction of the quantity of pharmaceutical products that are imported into Nigeria. At long term, it will encourage essential medicines being produced within Nigeria (NAFDAC 2019).

Drugs that are consumed in Nigeria include pharmaceutical dosage forms. These are syrups, ointments, tablets, capsules. About 70% of them, in 2011, with capacity utilization of local manufacturing facilities, were operating at about 40% of their capacity (Pharmaceutical Sector Profile, 2011). It is likely that appropriate fiscal and non-fiscal incentives, with the new policy in Nigeria, can increase local pharmaceutical production. This will develop the domestic pharmaceutical industry. Growth of pharmaceutical industries in Nigeria should be put in its right perspective. The bulk of healthcare expenditure, in Nigeria, is mainly out-of-pocket expenses. Growth in consumption should translate into higher healthcare spending.

The rise in non-communicable diseases (such as heart diseases and diabetes) present opportunities for pharma companies to position themselves as long-term partners to the government providing access to much needed medicines. Nigeria could contribute between $1.9 billion and 2.2 billion to pharma sales growth, 55% of it from prescription drugs over a period of 10 years. However, Nigeria is a complex market. Companies need to address short-term economic setback and drop rooted structural challenges they can take advantage of growth in the long term.

Medical tourism, especially to India and South Africa, cost more than $1 billion in 2013. A survey of pattern of antibiotic use in a tertiary hospital in Nigeria revealed that 21%–26% of the hospital budget was spent on antibiotics. It was observed that high-cost broad-spectrum antibiotics were commonly used. This scenario makes it more pertinent for local sourcing of antimicrobials from our Nigerian herbs rich in bioactive properties. For example, COVID-19 is a severe acute respiratory syndrome caused by Corona virus 2 (the new virus). Nigerian herbs like *Nauclea latifolia* is rich in antioxidants and it has anti-inflammatory, immunomodulatory, antimicrobial, antidiabetic amongst other properties which has been used in traditional medicine. Further exploration of this important herb is solicited for in the mitigation of COVID-19 pandemic. Another important herb is *Artemisia annua* which is being considered as possible treatments for COVID-19 and should be tested for efficacy and adverse effects.

Another useful plant-based fruit juices is *Anona muricata* and *A. squamosa*. The juices are useful in the control of diarrhoea in children under five years (Enweani et al., 1998, 2004). The moisture content of fruits is 70%. The proximate nutritional compositions include carbohydrate, glucose, proteins, lipids, ash, fibre and citrate. Juices also contain potassium, sodium, iron, magnesium, calcium, chloride and bicarbonate. They also contain carbonate, phosphorus, zinc and copper. These fruit juices meet and surpass the condition of ORT in replacement of fluid and electrolyte loss in children suffering from diarrhoea.

Specific deficiencies in respect of vitamins and micronutrients (e.g. zinc) needs to be replenished in addition to nutrients that provide energy (e.g. glucose). Potassium is contained in the *Anona* fruit juices which is advantageous as large potassium losses are associated with acute diarrhoea especially in infants. The base-deficit acidosis is corrected by citrate or bicarbonate function. The absorption of glucose has a linear relationship with the absorption capacity of the intestine. This is applicable irrespective of the causal agent of diarrhoea, and this forms the physiological basis of oral rehydration therapy where oral rehydration salt solution was applied. It is important to break the vicious cycle of malnutrition and diarrhoea by nutritional repletion during the management of persistent diarrhoea. Consumption of these fruit juices can alleviate the challenges of micronutrient depletion. Zinc has also been used as part of palliative medications in the management of COVID-19 infection. *Karin Karin* is a very common vegetable in Nigeria. It has been suggested that flavonoids obtainable in this vegetable enhance zinc absorption in the gut of humans. Nigeria needs to rely heavily on her natural endowed herbs for the management and treatment of her citizenry.

Prof. Mauruce Iwu, CEO, Niemeth and a Professor of Pharmacology and herbal medicine has made a lot of progress and positive impact through his numerous researches and patency in the use of Nigerian herbs to promote health and healthy living. He has been marketing a range of products. IHP of researches undertaken by Bioresources Development Group based in Abuja, Nigeria. IHP include Detox tea, Garcinia IHP, Immunovit, Erovit, Moringa tea, Bichap, Vernoniaoccimum, Ginger tea, Physogen tea, Hepavital tea, Phytolax tea, Virgin coconut oil, Ganoderma green tea and coffee, Syferol, amongst others. All these range of products are derived from Nigerian herbs. All these products emanated from Nigerian medicinal herbs and they boost immune system of humans for better health and well-being.

Another range of organic health products made from Nigerian medicinal plants is BESOMIK which include Ona meal, herbal fruit drink, herbal fruit tea, fruit juice, mushroom, amongst others. All these promote healthy well-being of Nigerians as well as providing micronutrients, antioxidants and antimicrobials. BESOMIK is a range of research products of Nnamdi Azikiwe University, Awka, Nigeria.

Huge sums of foreign exchange are spent in the importation of all these processed products into Nigeria. If we can channel our research interests and get these underutilized plant materials into finished processed products, foreign exchange will be saved. Export of those products will improve Nigerian foreign reserve which will enhance the nation's economy.

Nigerians should seize the opportunity of the Covid-19 pandemic which should be an opportunity for drug manufacturers to pressure government into doubling efforts to ensure viability of local drug manufacturing. Nigeria being blessed with so much medicinal plants should take opportunity to improve on her research on phytomedicine for disease management and improve drug security. In addition, there is abundance of micronutrients in the Nigerian herbs needed especially by women and children for health and well-being. Nigeria needs to take stock and look inwards to harvest and process enormous indigenous local herbs and make them available, accessible and affordable to treat and manage numerous ailments ranging from communicable to non-communicable diseases affecting Nigerians.

REFERENCES

Ajayi T.O., Moody J.O., Fukushi Y., Adeyemi T.A (2014). Antimicrobial Activity of Garcinia kola (Heckel) Seed Extracts and Isolated Constituents Against Caries-causing Microorganisms. *African Journal of Biomedical Research* Vol.17 No 3

Akinpelu, D.A. and Kolawole, D.O. (2004). Phytochemistry and antimicrobial activity of leaf extract of *Piliostigmatization* (Schum). *Science Focus* 7:64–70.

Akintonwa, A. and Essien, B. (1990). Productive effect of *Garcinia kola* seed extract against paracetamol-induced hepatoxicity in rats. *Journal of Ethropharmacology* 29:207–211.

Andersson S., and Dobson H.E.M (2003). Behavioral Foraging Responses by the Butterfly Hekiconius Melpomene to Lantana camara Floral Scent. *Journal of Chemical Ecology* 29: 2303–2318

Anwar, M.N., Begum J., Yusuf, M., Chowdhum, J.U., Khan, S and Nural, M. (2007) Antifungal activity of forty higher plants against phyto-pathogenic fungi. *Bangladesh Journal of Microbiology* 24(1)76–78.

Anyawu, N. C. and Arigbe-Osula, M. E. (2012). Pattern of antibiotic use in a tertiary hospital: pharmacotherapy. *Pharmacokinetics and Pharmacodynamics* 19(2): 195; https://ejhp.bmj.com>content.

Arokinyaraji, K., Perinbam, P.A., Agastian, R., and Motian, K. (2009). Phytochemical analysis and activity of vitex agnus-Catus. *International Journal of Green Pharmacy* 34: 162–164.

Avwioro, O.G., Onwuka, S.K., Moody, J.O., Agbedahunsi, J.M., Oduola T., Ekpo, O. E and Oladele, A.A (2007). *Curcuma longa* extract as a histological dye for collegen fibres and red blood cells. *Journal of Anatomy* 210(5): 600–603.

Babu P.D and Subhasree R.S (2009). Antimicrobial Activities of *Lawsonia inermis. Academic Journal of Plant Sciences* 2(4):231–232.

Banerjee, A and Mukherjee, A. K (1981). Chemical Aspects of Santalin as a Histological stain. *Biotechnic and Histochemistry* Vol. 56.

Barreto, F. S. (2010) Antibacterial activity of *Lantana camara* Linn and *Lantana monteviclensis* Bng Extracts from *Cariri-ceara* Brazil. *Journal of Young Pharmacists* 2(1):42–44.

Bassey, R.B., Oremosu, A.A and Osinubi, A.A.A (2012). *Curcuma longa*: staining effect on histomorphology of the testis. *International Journal of Morphology* 26(4): 927–930.

Bezuidenhoudt,C.B., Brandt, V.E. and Ferreira, D (1987). Flavonoids Analogues from Pterocarpus Species. *Phytochemistry* 26(2): 531–535.

Bharkta, D. and Ganjewala, D. (2009). Effect of leaf positions on total phenolics, flavonoids and proantho-cyanidin content and antioxidant activities in *Lantana camara* (L). *Journal of Scientific Research* 1(2):363–369.

Burkill, H.M. (1985). The use of plants of West Tropical Africa. Vol. 1 p. 1055.

Chukwu, O.O.C., Odu, C.E., Chukwu, O.I., Hafiz, N., Chidozie, V.N. and Onyimbo, I.A. (2011). Application of extrats of Henna (Lawsonia inermis) Leaves as a Counter Stain. *African Journal of Microbiology Research* 5(21): 3351–3356.

Egbujo, E. C., Adisa, O. J., and Yahaya, A.B. (2008). A Study of the Staining Effect of Roselle (*Hibiscus sabddriffa*) on the Histological Section of the Testis. *International Journal of Morphology* 26(4): 927–930.

Elekima, I., Nwachukwu, O. E., Nduka, N., Nwanjo, H.U., Ukwukwu, D. (2019). Biochemical and histological changes associated with azo food dye (Tartrazine) in male albino rats. *Asian Journal of Research in Biochemistry* 5(1): 1–14.

Emencheta, S.C., Enweani, I.B., Oli, A.N., Okezie, U.M. and Attama, A. A. (2019). Evaluation of antimicrobial activities of fractions of plant parts of *Pterocarpus santalliodes. Biotechnology Journal International* 23(3):1–11.

Enweani, I.B, Esebelahie, N.O, Obroku, J and Obi, C.L (1998). Use of soursop and swetsop juices in the management of diarrhea in children. *Journal of Diarrhea Diseases Research* 12(2):133–135.

Enweani, I. B., Obroku, J., Enahoro, T and Omoifo, C. (2004). The biochemical analysis of soursop (*Anona muricata L.*) and sweetsop (*A. squamosa* L.) and their potential use as oral rehydration therapy. *Journal of Food, Agriculture and Environment* 2(1):39–43.

Enweani, I.B., Prohp, T.R., Ainemiokhale, L.G., Asia, A. and Uzoaru, S.C. (2005). Ascorbic acid and microbiological analyses of extracotyledenous deposits of pride of barbados (*Caesalpina pulcherima*) at various temperatures. *Nigerian Journal of Biotechnology* 16(1):100–105.

Ezeanya, C.C., and Daniel E. O. (2012). Antibacterial Activity of Garcinia kola Seed and Leaf Extracts of some Selected Clinical Isolates. *Science Journal of Microbiology*; doi:10.7237/sjmb/298

Farombi, E.O., Adepoju, B.F, Ola-Davis, O.E and Emerole, G.O. (2005). Chemoprevention of alaflatoxin B_1-induced genotoxicity and hepatic oxidative damages in rats by Kolaviron, a natural bioflavonoid of garcinia kola seeds. *European Journal of Cancer Preview* 14:207–214.

Fostering Local Production of Essential Medicines in Nigeria- World Health Organization. *https://www.who.int>online_first* Publication: Bulletin of the WHO; Type: Perspectives Article ID:BLT.19:249508.

Fatokun O (2020): Fostering Local Production of Essential Medicines in Nigeria - Bulletin of the World Health Organization. https://www.who.int>online_first Type: Perspectives Article ID:BLT.19:249508 P. 1 - 5

Goet, Z. (2006). Traitment des Troubles de la Libido masculine. *Phytotherapy Clinique* 1:9–14.

Goodarzian, H. and Ekrami, E. (2010). Extraction of dye from madder plant (*Rubiatinctorium* L.) and Dyeing of Wool. *World Application of Science Journal* 9(4): 434–436.

Gull, M., Maria, S., Muhammed S.A. and Muhammed A.A. (2013). Phytochemical toxicological and antimicrobial evaluation of Lawsoniainermis extracts against clinical isolate of pathogenic bacteria. *Annals of Clinical Microbiology and Antimicrobiology* 12(1):1–36.

Gupta, B. and Ghosh, B.(1999). *Curcuma longa* inhibits DNA-α induced expression of adhension molecules on human umblical vein endothelial cells. *International Journal of Immunopharmocology* 21(11): 745–757.

Harborne J.B. (1998): Phytochemical Methods. A Guide to Modern Technique of Plant Analysis. Chapman and Hall Publishers, London p. 51–59

Hema, R., Kumaravel, S. and Gomath, N. (2010). Gas chromatography mass spectroscopic analysis of *Lawsonia inermis* Leaves. *New York Science Journal* 3(12):99–101.

Ishola, I.O. and Ashorobi, R. B. (2007). Anti-stress potential of aqueous root extract of *Cnestis feruginea*. *International Journal of Pharmacology* 3(3): 295–298.

Iwu, M.M. and Igboko, O. (1982). Flavonoid of *Garcinia kola* Seeds. *Journal of Natural Products* 45(5). doi: 10.1021/np50023a026.

Kensa, V.M. (2011) Studies of phytochemical screening and antibacterial activities of *Lantana camara* Linn. *Plant Sciences Feed* 1(5):74–79.

Kiec-Swierczynska M., Krecisz M. B., Swierczynka-Machura D., and Palczynski C.(2004). Occupational Allergic Contact Dermatitis Caused by Padouk wood (Pterocarpus soyauxii Taub). Contact Dermatitis 50: 384–385.

Kim, K.J., Yu, H.H., Cha, J.A., Seo, S.J and You, Y.O (2005). Antibacterial activity of *Curcuma longa* against methicillin resistant *S. aureus*. *Phylothers Research* 19(7): 599–604.

Lagnika, L., Amoussa M., Adjovi, Y and Sann, A.(2012). Antifungal, antibacterial, and antioxidant properties of *Adensonia digitate* and *Vitex doniana* from Benin Pharmocopeia. *Journal of Pharmacognosy and Phytotherapy* 4(4):44–52.

Morabandza, J., Ongoka, R.P., Matini, L., Epa, C., Nkounkou, L.C and Abena, A.A. (2013). Chemical Composition of the mesocarp of Garcinia kola Heckel (Clusiaceae) fruit. *Journal of Recent Science* 2(1)53–58.

NAFDAC (2019). Curbing Substandard, Falsified(SFs) and Counterfeit Medicines. Abuja: National Agency for Food and Drug Administration and Control, 2019. Available from: https://www.nafdac.gov.ng/curbing-sunstandard-falsified-sfs-andcounter-medicines Accessed June 29, 2020.

Nigeria Imports of Pharmaceutical Products (1996–2018). Data/2020 Forecast – Trading Economics. https://tradingeconomics.com Accessed June 29, 2020.

Nigeria Relies Heavily on Drug Imports. Why this sis worrying in the time of COVID-19-The Conversation. https://theconversation.com. Accessed June 29, 2020.

Obaji, M., Enweani, I.B, Oli, A.N (2020). A unique combination of *Alchorneacordifolia* and *Pterocarpussantalliodes* in the management of multidrug resistant bacterial isolates. *Asian Journal of Biomedical and Pharmaceutical Sciences* 10(69): 9–18.

Ochei, J. and Kolhatkar, A. (2008). *Microscopic Examination of Bacteria, Medical Laboratory Science Theory and Practice*, 7th Ed. Tata Mac Graw-Hill Publishing Company LTD, New Delhi. p. 1338.

Odeyemi, O., Enweani, I. B. and Alajekwu, O. U. (2019). Activity of different parts of *Nauclea Latifolia* and *Napoleona imperialis* against some Nigerian dermatophytic isolates and their relationship with some haematological parameters. A paper Presented at the Biotechnological Society of Nigeria 28th AGM and Scientific Conference Held ar Godfrey Okoye University, Enugu, August 26–30.

Ogbonna, U. S, Enweani, I. B, Umeaku, N. C and Okwu, G. I. (2019). Nutritional potential and microbial quality of food products from *Dioscorea dumentorium* fortified with *Musa acuminate* and *Vitex doniana*. A Paper Presented at the 6th International Interdisciplinary Conference IIC6. Bok of Abstract, p.32 (FSNLSTI_42) Which Held at the University of Eldoret, Kenya, September 2019.

Oli, A. N., Obaji, M. and Enweani, I. B. (2019). Combinations of *Alchorneacordifolia, Cassytha filiformis* and *Pterocarpus santalliodes* in diarrhoegenic bacterial Infections. *BMC Research Notes* 12:649–654.

Olise, N. A, Enweani, I. B, Urama, E. U, Oshim, I.O and Ngwoke, C. C. (2018). The use of extracts from Nigerian indigenous plants as staining techniques in bacteriology, mycology and histopathology. *International Journal of Plant Research* 8(1): 15–24. doi: 10.5923/j. plant 20180801.03.

Okeimute, P. A., Wokem, G. N., Enweani, I. B., Okosa, J. N, Okpe, A.N, Opara, C. E. and Moudebe C. N. (2020). Antimicrobial Effect of Garlic on *Staph. aureus* from Clinical Specimens in Madonna University Teaching Hospital, Elele, Nigeria. *International Journal of Recent Scientific Research* 11(3c): 37840–37845. doi: 10.24327/IJRSR.

Okigbo, R.N. (2003): Fermentation of black plum (*Vitex doniana sweet*) for production of wine. *Fruits (France)* 58: 6(6): 363–369.

Okwuosa, O. M. T. B., Chukwura, E. I., Chukwuma, G. O., Okwuosa, C. N., Enweani, I. B., Agbakoba, N. R., Chukwuma, C. M., Manafa, P. O. and Umedum, C. N. (2012). Phytochemical and antifungal activities of *Uvaria chamae* leaves and roots, *Spondias mombin* leaves and bark and *Combretum racemosum* leaves. *African Journal of Medical Sciences* 41 Suppl. 99–103.

Okwu, D. E and Okwu, M. E. (2004). Chemical composition of *Spondias mombin* plant parts. *Journal of Sustainable Agricultural Environment* 6: 140–147.

Omotayo, F. (2020). Fostering local production of essential medicines. Nigeria Publication: Bulletin of the WHO; Type: Perspective Article ID:BLT.19.249508, pp. 1–5.

Onuoha, V., Enweani, I. B and Ogbonna, U. S. A. (2015). Activities of different parts of *Moringa oleifera* against some clinical pathogenic fungi isolated from sputum samples of hiv positive individuals. A poster Presentation at the 19th ISHAM Congress Held at Melbourne, Australia. May 4–8. Poster No. 724.

Pattern of Antibiotic Use in a Tertiary Hospital in Nigeria- ResearchGate Winning in Nigeria: Pharma's Next. https://www.rFrontier/McKinsey. https://www.nckinsey.com>our-insights.

Pharmaceutical Sector Profile: Nigeria Vienna: United Nations industrial Development Organisation (2011). https://www.unido.org/sites/default/files/2011-04/NigeriaPharma %20Sector%20Profile0320ilEbookOpdf (accessed June 29, 2020).

Pharmacotherapy: Pharmacokinetics and Pharmaco Dynamics 19(2). https://ejhp.bmj.com>content.

Prasad, S., Aggarwal, B.B., Benzie, I.F.F and Walchtel-Galor, S. (2011). Tumeric, the golden spice for traditional medicine to modern medicine. Herbal Medicine: Biomolecule and Clinical Aspects. 2nd Edition.

Quattara, A., Coulibaly, A., Adima, A.A and Quattara, K.(2013). Exploration of the Antistaphylococci activity of *Vitex doniana* (Verbenaceae) stem bark extracts sch. *Academic of Journal of Pharmacy* 2: 21–24.

Raja, W., Ovais, M and Dubai, A. (2103). Phytochemical screening and antibacterial activity of Lawsoniainermis leaf extract. *International Journal of Microbiology Research* 4(1):33–36.
Singh, C and Pandeyi D.K (1989). Fungitoxic studies on bark extract of *Lawsonian inermis* against Ringworm Fungi Hindusthan. *Antibiotics Bulletin* 31: 1–2.
Srivastan, P. and Singh, C. (2011). Antifungal potential of two common weeds against plant pathogenic fungi-*Altemaria* spp. *Asian Journal of Experimental Biological Sciences* 2(525–528).

Part 5

Working Group 13

Medical Insect Biotechnology, Genomics and Molecular Genetics of Pests

Handling Editor: Dr. Abdulrazak Ibrahim

22 Exploiting Arthropod Midgut Components for Development of Interventions against Infectious Diseases

Oluwafemi Abiodun Adepoju, Bashiru Ibrahim, and Emmanuel Oluwadare Balogun
Ahmadu Bello University

CONTENTS

DOI: 10.1201/9781003178903-27

22.1 INTRODUCTION

More than 17% of all infectious diseases are vector-borne diseases killing more than 700,000 people annually (WHO, 2020a). One of the most common vectors of infectious diseases are arthropods. Arthropod vectors are mainly insects harboring parasites, bacteria, or viruses causing infectious diseases (Aksoy, 2000; WHO, 2020a). These insects acquire pathogens from infected hosts during a bloodmeal; the ingested pathogens pass through the insect's midgut into the hemocoel and finally to the salivary glands where they can infect a new host when an insect bites an uninfected person (Whitten et al., 2006). The digestive tract of insects is divided into the foregut, the midgut, and the hindgut. Mechanical breakdown of food and chemical digestion by salivary enzymes and enzymes regurgitated from the midgut take place in the foregut; the midgut is the site of production and secretion of digestive enzymes (de Sousa & Conte, 2013; Gullan & Cranston, 2014). Beyond serving digestive functions, the midgut of arthropods serves osmoregulatory and immune functions (Napoleão et al., 2019). Insects do not have adaptive immunity like vertebrates but rely on their innate immunity to fight pathogens (Yassine & Osta, 2010). The immune function of arthropod's midgut has attracted the attention of researchers and scientists working on transmission-blocking interventions against infectious diseases. Insect's midgut is home to various symbiotic microorganisms which also play vital roles in the development of their innate immunity (Douglas, 2011).

The midgut immunity of insects is due to the production of midgut-antimicrobial effector molecules by differential gene expression or endosymbionts (Caicedo et al., 2019). *Wolbachia* is an endosymbiont of arthropods which has been reported to modify the susceptibility of its host to arbovirus by different mechanisms (Yin et al., 2020).

Wolbachia confers on mosquito resistance to *Plasmodium* and viruses (Bian et al., 2013; Moreira et al., 2009; Pan et al., 2012). Presently the role of the midgut microbiota in insect immunity is poorly understood but highly important in the development of intervention strategies targeting the transmission of insect-borne pathogens (Nouzova et al., 2019). Understanding the role of the midgut microbiota in innate immunity of arthropods is essential to the creation of transgenic vectors for the control of infectious diseases. With increasing climate change concerns and its relevance to the expansion of foci for vector-borne diseases (Balogun et al., 2016; Ramirez, 2017), it is imperative for scientists to intensify efforts towards eco-friendly techniques for vector controls.

22.2 MEDICALLY IMPORTANT ARTHROPOD VECTORS

Mosquitoes and tsetse fly are hematophagous insects found in the same geographical areas in many parts of sub-Saharan Africa where they cause infections and devastating losses, both of lives and economically (Nikolaeva et al., 2020; Pal & Chakravarty, 2020; WHO, 2020a). Mosquitoes use the same feeding mechanisms with tsetse fly but do not transmit trypanosomes (Nok et al., 2002). The different susceptibility of vectors to infectious pathogens might be due to differences in the composition of their gut microbiota and/or presence of different anti-parasite effector molecules in their midguts (Yin et al., 2020). These anti-parasite effector molecules interfere with parasite development and transmission.

22.2.1 Mosquitoes

Mosquitoes are the most important vectors of infectious diseases in sub-Saharan Africa. They are hosts to diverse groups of microorganisms, including bacteria, fungi, and viruses (Gao et al., 2020). *Aedes* transmit viruses of Dengue fever, Chikungunya, Rift Valley fever, Yellow fever, Zika fever, and lymphatic filariasis parasite (WHO, 2020a). The *Anopheles* mosquito transmits the parasites causing malaria and lymphatic filariasis (WHO, 2020a). *Plasmodium falciparum* is responsible for the highest mortality due to malaria especially in children under 5 years and women getting pregnant for the first time (Nikolaeva et al., 2020).

22.2.2 Tsetse Flies

Tsetse flies are the vectors for trypanosomes which cause Human and Animal African trypanosomiasis that is endemic in 36 countries of sub-Saharan Africa and is responsible for the loss of about USD4–5 billion annually (Pal & Chakravarty, 2020; Ruiz-Postigo et al., 2012). Tsetse flies are monophagous, feeding only on vertebrate blood; therefore, they rely on other symbiotic microorganisms for additional diet to complement their restricted diet (Soumana et al., 2013). Tsetse flies establish symbiotic relationships with several microorganisms including bacteria and viruses which influence their vectorial competence. Some endosymbionts of tsetse flies can be stably transmitted vertically from mother to offspring in laboratory and wild-type strains of tsetse flies (Wang et al., 2013).

Tsetse flies and mosquitoes coexist in the same geographical areas and feed through the same feeding mechanism, but mosquitoes do not transmit trypanosome and do not serve as its reservoir. Nok et al. (2002) carried out an investigation on the factors responsible for the resistance of *Anopheles* to trypanosome infection and identified a 45-kDa trypanocidal sialoglycoprotein (gp45) in the mosquito midgut. This protein is thought to inhibit the procyclic developmental stage of trypanosomes in mosquitoes and therefore responsible for their refractoriness to trypanosome infection. They concluded on the possibility of using gp45 in the development of transgenic tsetse for trypanosomiasis control.

22.2.3 Ticks

The emergence and increasing prevalence of tick-borne diseases is a subject of public health concern (Madison-Antenucci et al., 2020). Ticks are non-insect arthropod vectors, which have four pairs of legs and lack clearly segmented bodies which distinguish them from insects. There are two major families of ticks: the hard ticks (Ixodidae), covering about 650 species, and the soft ticks (Argasidae), covering about 150 species. The soft tick, *Ornithodoros moubata*, is the vector of the relapsing fever in Africa. Ticks are parasitic blood-sucking arthropods; they feed on the bloods of animals and humans thereby transmitting infectious pathogens. Other human diseases transmitted by ticks include Rocky Mountain spotted fever, Q fever, and Lyme disease. Ticks are responsible for spreading several diseases affecting domestic animals and livestock, causing huge economic losses to farmers around the world (WHO, 1997).

22.3 THE MALARIA BURDEN AND INTERVENTIONS

Malaria is one of the most devastating diseases worldwide. Globally, there were an estimated 229 million cases of malaria and 409,000 deaths in the year 2019 according to the World Malaria Report (WHO, 2020b). Nigeria had the highest burden of malaria cases (27% of all cases) and deaths (23% of all deaths) in 2019 (WHO, 2020b). In malaria endemic areas, control strategies involve the use of chemical compounds like insecticides targeting the vector or antimalarial drugs targeting the parasite (Yassine & Osta, 2010). Standard vector control interventions use insecticide-treated bed nets, mosquito sprays, and medication therapies to reduce human infection (Billingsley, 1994). A major weakness of these strategies is the emergence of insecticide-resistant mosquitoes and drug-resistant parasites. Currently there are no effective vaccines against the malaria parasite. Current intervention strategies are now focusing on blocking the transmission of the malaria parasite from the mosquito vector to the human host (Mathias et al., 2012; Nikolaeva et al., 2015, 2020).

22.4 TARGETING MIDGUT STAGE OF THE MALARIA PARASITE'S LIFE CYCLE

The malaria parasite is most vulnerable during its developmental stage in the mosquito midgut (Wang & Jacobs-Lorena, 2013). The mosquito midgut serves a special role in mosquito immunity and is the most trying environment for *Plasmodium*

(Whitten et al., 2006). During the passage of *Plasmodium* from the midgut to the salivary glands, it is recognized and labeled by mosquito recognition factors; the labeled parasites are destroyed by mosquito killing factors in the midgut and other surrounding tissues (Whitten et al., 2006).

Following a bloodmeal from an infected host, gametocytes in the ingested blood undergo several developmental stages in the mosquito: the gametocytes are fertilized to form zygotes, which in turn develop into ookinetes; the ookinetes finally form oocysts which exit the midgut through its epithelia tissues into the hemocoel. The oocysts must transverse two epithelia tissues in the mosquito: the midgut and the salivary glands for successful infection and transmission to occur (Wang & Jacobs-Lorena, 2013).

The ability of the malaria parasite to develop successfully in the mosquito midgut and reach the salivary gland is a necessary requirement for its transmission from mosquitoes to humans. During a bloodmeal from a malaria-infected host, the *Anopheles* ingests about 10,000 gametocytes out of which only about 1,000 successfully develop into ookinetes. Less than five of these ookinetes develop into sessile oocysts even in mosquito strains that are susceptible to infection (Shahabuddin & Costero, 2001; Sinden, 1999; Sinden & Billingsley, 2001). Oocysts that successfully traverse the midgut epithelia tissue release thousands of sporozoites into the hemocoel which find their ways to the salivary glands. Infection is initiated in a new host through the bite of mosquitoes harboring these sporozoites.

The efficiency of the mosquito-midgut immunity in neutralizing and reducing the number of the malaria parasites is an attractive subject to researchers working on developing interventions targeting transmission blocking of the malaria parasite. This strategy will not only be useful in the control of malaria infection but several other diseases transmitted by mosquitoes (Abraham & Jacobs-Lorena, 2004).

22.5 GENETIC MANIPULATION OF VECTOR COMPETENCE

About 30–40 of anopheline mosquitoes that are natural vectors for human malaria have been found to be amenable to genetic manipulation (Coutinho-Abreu et al., 2010). Insects possess anti-pathogens effector genes whose products contribute to the elimination of infectious pathogens. Identifying suitable effector genes that can be genetically manipulated without fitness cost to the vector is a prerequisite for the development of transgenic vectors (Wang & Jacobs-Lorena, 2013). Effector genes are differentially expressed in certain organs, including the midguts of insects. These can be hijacked for the development of transgenic vectors that are refractory to infection.

22.6 GENETIC MANIPULATION STRATEGIES

Transgenesis and paratransgenesis are two novel genetic manipulation strategies used to alter the competence of disease vectors. These approaches have been used on laboratory strains of mosquitoes and their gut microbiota in the development of transgenic mosquitoes that are refractory to malaria infection (Wang & Jacobs-Lorena, 2013).

22.6.1 Transgenesis

Transgenesis involves the direct manipulation of the vector itself for the expression of effector genes whose products deliver anti-pathogen effector molecules (Wang & Jacobs-Lorena, 2013). Transgenic *Anopheles stephensi* strain genetically manipulated for expression of the salivary gland and midgut peptide 1 (SM1) which binds to a putative mosquito-midgut ookinete receptor to prevent ookinete invasion of the midgut has been reported to lose its vectorial competence to transmit the malaria parasite or cause infection (Ghosh et al., 2001; Ito et al., 2002).

The genome of insect vectors can be genetically manipulated for differential expression of anti-parasite effector molecules in specific tissues, like the midgut. An ideal effector molecule should interfere with parasite development and/or transmission without affecting the performance of the vector (Wang & Jacobs-Lorena, 2013). Certain anti-*Plasmodium* genes in mosquitoes are refractory to genetic manipulation and may impose fitness load on the mosquito. The fitness load imposed on mosquitoes by refractory genes is one area that must be painstakingly considered in the development of transgenic mosquitoes (Corby-Harris et al., 2010). An ideal candidate gene for genetic manipulation must not impose fitness cost on the vector.

22.6.1.1 Targeting Midgut Effector Molecules for Interventions

Anti-*Plasmodium* effector molecules are classified based on their mechanisms of action. A first class of mosquito-midgut effector molecules include attacins, defensins, gambicin, and cecropins, which play important roles in the innate immunity of mosquitoes (Herrera-Ortiz et al., 2011; Kim et al., 2004; Kokoza et al., 2010; Vizioli et al., 2001). These molecules kill invading parasites without affecting the mosquito. A second class of effector molecules act by interacting with the parasite. In mosquitoes, this class of effector molecules binds to the surface of the ookinete or sporozoite, or to secreted proteins (de Lara Capurro et al., 2000; Isaacs et al., 2011). Their interaction with the parasite prevents it from invading the midgut of the mosquito. *Plasmodium* enolase-plasminogen peptide, EPIP, prevents invasion of the midgut by binding to ookinete's surface (Ghosh et al., 2011). Other antimicrobial peptides have been documented in the literature (de Lara Capurro et al., 2000; Isaacs et al., 2011; Yoshida et al., 2001). Thirdly, some effector molecules interact with mosquito midgut or salivary gland epithelia to mediate their effect. An example is SM1, a 12-amino acid peptide that binds putative receptors that are found on the luminal surface of mosquito midgut and the basal surface of the salivary gland epithelia. SM1 exerts blocking effects on ookinete and sporozoite invasion (Ghosh et al., 2001). A list of several anti-*Plasmodium* effector molecules was presented by Wang and Jacobs-Lorena (2013).

22.6.1.2 Mosquito-Midgut Proteases in Transmission-Blocking Vaccines (TBV)

Midgut proteases act on components of the bloodmeal and may offer some defense against pathogenic organisms in the ingested blood (Oliveira et al., 2009). Carboxypeptidase A has been studied for the development of Transmission-Blocking Vaccine (TBV) to target the sexual stage of malaria parasite in mosquitoes. Mosquito-midgut carboxypeptidase A is a promising candidate for TBV. Exopeptidases such as aminopeptidase N and carboxypeptidase B are other midgut enzymes that have been listed to have potentials in the development of TBV (Lavazec et al., 2007; Mathias et al., 2012; Raz et al., 2013; Venkat Rao et al., 2017).

22.6.1.3 Targeting Mosquito-Midgut Protein for Intervention

Certain mosquito-midgut proteins will either stimulate or inhibit the invasion of the midgut by *Plasmodium* ookinetes. Some of these molecules are attractive targets for transmission-blocking interventions. Strategies exploiting the differential expressions of anti-*Plasmodium* molecules in transgenic vectors can achieve transmission blocking in mosquitoes. Epithelial serine protease (ESP) and cactus proteins are mosquito-midgut proteins that enhance the entry of *Plasmodium* ookinetes into the midgut lamina. It is not known if the knockdown of *ESP* or *cactus protein* genes in mosquitoes will generate transgenic mosquitoes that are refractory to *Plasmodium* infection. Thioester-containing protein 1 (TEP1) and serine protease inhibitor 6 (SRPN6) on the other hand have been shown to inhibit the entry of ookinetes into the midgut lamina (Abraham et al., 2005; Garver et al., 2012, 2009; Molina-Cruz et al., 2012; Rodrigues et al., 2012; Zou et al., 2011). Transgenic mosquitoes expressing TEP1 or SRPN6 or any other immune-related genes have potential to inhibit the development and transmission of the malaria parasite and other pathogens transmitted by mosquitoes. A collection of molecules involved in vector-pathogen interaction in malaria infection is available in Sreenivasamurthy et al. (2013).

22.6.1.4 Mosquito-Midgut Immune-Related Gene Families

Immune-related genes products inhibit the development and transmission of parasites. Some of these genes are expressed in the midgut of arthropods. Modification of mosquito immune-related genes can cause a decline in its vectorial competence. *Akt* signaling in mosquito midgut has been shown to disrupt parasite development with a concurrent reduction in human infection. Bloodmeal-induced *Akt* expression has been associated with the refractoriness of mosquitoes to *Plasmodium* infection (Corby-Harris et al., 2010). In the same manner, overexpressing the immune deficiency pathway-mediated transcription factor *Rel2* makes mosquitoes resistant to *Plasmodium* infection (Dong et al., 2011). Species or strain differences may influence the natural expression of immune-related genes and the susceptibility of vectors to pathogens (Behura et al., 2014; Behura et al., 2011; Chauhan et al., 2012).

22.6.1.5 Midgut Barriers: Midgut Infection Barrier (MIB) and Midgut Escape Barrier (MEB)

Aedes aegypti has been shown to differentially express the immune-related genes that determine the midgut infection barrier (MIB) and midgut escape barrier (MEB) (Black IV et al., 2002; Caicedo et al., 2019). Pathogens ingested in the bloodmeal would have to penetrate the midgut of mosquitoes for successful infection and transmission to take place. The midgut of the mosquito represents an important barrier to infection. MIB and MEB play important roles in determining the vector competence of mosquitoes (Black IV et al., 2002). Genes involved in determining these barriers are potential targets for the development of transmission-blocking interventions (Caicedo et al., 2019).

22.6.2 Expression of Immune-Related Genes in Mosquito Midgut

In a knockdown experiment using RNA interference, silencing *Gram-negative binding protein, keratinocyte-lectin*, and *Cathepsin-b* in *Ae. aegypti* dengue-susceptible (Cali-S) strain resulted in reduced number of mosquitoes with transmissible dengue virus to a significant amount, viral titers in the salivary glands were also greatly reduced (Caicedo et al., 2019). These genes involved in refractoriness of mosquitoes to infection can be explored in the development of transgenic mosquitoes for control of infectious diseases.

22.6.2.1 Expressing Anopheline Midgut Trypanocidal Protein gp45 in Tsetse Fly

Despite that mosquitoes and tsetse flies are found in the same areas and use the same feeding mechanisms, they have different susceptibility to infectious pathogens. Mosquitoes do not transmit trypanosomes and tsetse flies do not transmit *Plasmodium*. The presence of a mosquito-midgut protein gp45 that kills trypanosomes with high efficiency has been reportedly linked to the resistance of *Anopheles* to trypanosomes (Nok et al., 2002). This anopheline protein has potential in the development of transgenic tsetse fly for biological control of trypanosomiasis. Transgenic tsetse flies expressing gp45 will be able to kill trypanosomes in their midgut without transmitting the infection. Transgenic tsetse will lower the burden of trypanosomiasis in endemic areas.

22.6.2.2 Exploiting Gut Immunity in Ticks

Blood meal especially from infected hosts exposes ticks to both pathogenic and non-pathogenic organisms. Ticks respond to assaults from these organisms by the innate immunity of their midgut. Antimicrobial activities of the midgut components from extracts of American dog ticks *Dermacentor variabilis* (Say) against *Micrococcus luteus* have been reported (Sonenshine et al., 2005). Ticks use hemoglobin as the main nutrient for their development and reproduction (Fernando & Fischer, 2020). It is believed that the by-products of hemoglobin digestion play an important role in the midgut immunity of *Dermacentor variabilis* (Sonenshine et al., 2005).

Defensin is an antimicrobial peptide that has been detected in the midguts of *Stomoxys calcitrans, Anopheles gambiae, Glossina morsitans*, and several other hematophagous insects including some hard and soft ticks (Boulanger et al., 2002; Munks et al., 2001; Nakajima et al., 2002; Nakajima et al., 2001; Richman et al., 1996). A study revealed the differential expression of defensins and lysozyme in the midgut of the soft tick, *Ornithodoros moubata*, following blood feeding (Nakajima et al., 2001). Like other hematophagous insects, blood feeding stimulates differential expression of antimicrobial genes in ticks. Defensin has been reported in certain tissues including the midguts, hemolymph, fat bodies and hemocytes of both (soft) and ixodid (hard) ticks as well as the synganglia of ixodid ticks (Ceraul et al., 2007; Ceraul et al., 2003; Nakajima et al., 2002). *D. variabilis* challenged with *R. montanensis*, an obligate natural intracellular bacterium of ticks, were shown to differentially express the antimicrobial genes *defensin-1, defensin-2*, and *lysozyme* in their midgut tissues and fat body (Ceraul et al., 2007).

The bacterium rickettsiae transmitted by ticks encounters different antimicrobial peptides in the tissues of infected ticks. These infected tissues express antimicrobial genes. Ceraul et al. (2007) carried out an investigation on differential expression of antimicrobial genes in the midgut and fat bodies of fed and unfed ticks. They reported that defensin-1 was expressed almost exclusively in the hemocytes, while defensin-2 was expressed mostly in the ovary, midgut, and fat body; and lysozyme had the highest expression in hemocytes and midgut tissues.

A new *defensin-like* encoding gene, *HE defensin*, which codes for hemolymph defensin has been identified from the hemolymph expressed sequence tags (EST) database of *H. longicornis*; this gene has been shown to be upregulated in blood-feeding ticks. HE defensins showed antiviral activities *in vitro* against Langat virus making it a potential therapeutic molecule against tick-borne diseases, especially the flaviviruses (Talactac et al., 2017).

22.6.3 Paratransgenesis

Paratransgenesis is the genetic modification of vector microbiota and associated microorganisms for the delivery of anti-pathogen effector molecules. This has been successfully experimented in mosquitoes. Anti-*Plasmodium* effector molecules are able to kill *Plasmodium* and limit malaria transmission (Wang & Jacobs-Lorena, 2013). This strategy has been proposed for the development of interventions for several diseases transmitted by insect arthropods.

22.6.3.1 Paratransgenesis in Mosquitoes

Most disease-causing pathogens transmitted by insects depend on their hosts to complete their lifecycle. *Plasmodium* absolutely depends on the mosquito midgut for its development and survival. Since the same midgut milieu is shared by *Plasmodium* and several other microbial symbionts capable of delivering anti-*Plasmodium* molecules which can inhibit parasite growth and cause death, paratransgenesis takes advantage of this phenomenon (Hurwitz et al., 2011). In mosquitoes, bloodmeal significantly increases the gut microbiota population by a factor of 100- to 1000-fold (Whitten et al., 2006). The mosquito midgut is the most challenging environment for the malaria parasite (Wang & Jacobs-Lorena, 2013). The gut microbiota produces several anti-*Plasmodium* effector molecules which effectively kills *Plasmodium* (Cirimotich et al., 2010; Wang & Jacobs-Lorena, 2013). The microbiota of mosquitoes can be genetically engineered for the development of transgenic mosquitoes.

22.6.3.2 Paratransgenesis in Tsetse Flies

The midgut microbiota of tsetse fly include *Wigglesworthia glossinidia*, *Sodalis glossinidius*, and *Wolbachia pipientis* (Wang et al., 2013). *Wigglesworthia* and *Sodalis* have opposing effects on trypanosomes in tsetse flies. *Wigglesworthia*-infected tsetse flies are refractory to trypanosomes infection while *Sodalis* infection in tsetse flies increase the susceptibility of tsetse flies to trypanosomes infection (Soumana et al., 2013; Wang et al., 2009; Welburn et al., 1993). *Wigglesworthia* has potential in the development of paratransgenic symbiont which can be used in tsetse fly for biological control of trypanosomiasis.

22.6.3.3 Paratransgenesis in Ticks

Attempts are being made to create paratransgenic symbionts and recombinant vectors able to express transgenes in ticks. The challenge in creating transgenic ticks is in part due to their months- to year-long generation times. There are already breakthroughs in the generation of aposymbiotic ticks and the reconstitution of ticks with transformed *R. buchneri* (Kurtti et al., 2016). Transgene expression using *Agrobacterium tumefaciens* has been demonstrated in tick cells (Machado-Ferreira et al., 2015). This strategy will open up new opportunities in using recombinant technology to deliver antimicrobial effector molecules in ticks for the control of tick-borne diseases. Little is known about the immune system of ticks and their endosymbionts. For successful transgenesis, the transgenic vector should not impose a fitness cost on the tick and must be able to establish and colonize the tick's midgut.

22.6.4 Basic Requirements for Paratransgenesis

According to Wang and Jacobs-Lorena (2013), there are five basic requirements for paratransgenesis: there must be a stable symbiotic relationship between the transgenic symbiont and the vector; the symbiont must be amenable to in vitro culture and can be genetically manipulated; the products of effector genes should not impair symbiont and vector fitness; these products should be soluble for secretion to target site for interaction with the pathogen; there must also be an efficient method of introducing the engineered symbiont into the field.

The problems of resistant parasites and fitness cost of paratransgenesis to either the transgenic microorganism or the vector are important areas for consideration in developing paratransgenic-based interventions (Lee et al., 2019; Wang & Jacobs-Lorena, 2013). To circumvent the problems associated with parasite resistance to effector molecules, paratransgenesis strategies using multiple effector molecules that act through different mechanisms have shown some successes (Wang & Jacobs-Lorena, 2013). Another potential problem to the use of genetically modified organisms is that different species or strains of asymbiont or vector may need to be genetically engineered separately. Although, inhibition of *Plasmodium* development in different mosquito strains showed that *A. gambiae* which is native to Africa showed similar results with *A. stephensi* which is of Asian origin. In the same way, *P. falciparum*, a human parasite, showed similar results with the rodent parasite *P. berghei*. This is an indication that paratransgenesis may show good results in other medically important mosquito species (Wang & Jacobs-Lorena, 2013).

22.6.5 Establishing Symbiont Infection in the Gut

A successful paratransgenic symbiont must be able to evade its host gut immunity to establish infection and multiply (Wang et al., 2013). Understanding the host's gut immunity and how it affects genetically modified symbionts is important for successful paratransgenesis. A finding reported a strategy for protecting genetically engineered *E. coli* modified to express *Sodalis* outer membrane protein (OmpA) from tsetse midgut immunity. OmpA confers on *Sodalis* the ability to form biofilm. Biofilm formation is very important in determining if *Sodalis* will establish infection in tsetse fly. The biofilm protects *Sodalis* from the effect of tsetse midgut immune system and is important for its ability to colonize tsetse's midgut (Maltz et al., 2012). *Sodalis*-infected tsetse flies are highly susceptible to trypanosome infection (Soumana et al., 2013; Welburn et al., 1993). Genetic modification of symbionts to express OmpA can ensure their stability and establishment in their vector host.

22.7 THE MIDGUT MICROBIOTA OF ARTHROPOD VECTORS

Endosymbionts play important roles in the physiology, ecological adaptation, and evolution of insects. The gut microbiota gives insects nutritional advantages, tolerance to environmental disturbances, and complement their immune system (Weiss & Aksoy, 2011). The gut microbiota can be genetically engineered by paratransgenesis to increase the refractoriness of insect vectors to infectious pathogens with the aim of limiting disease transmission to humans (Chang et al., 2003; Durvasula et al., 1997; Favia et al., 2007; Rao et al., 2005; Riehle & Jacobs-Lorena, 2005).

22.7.1 Mosquitoes Midgut Microbiota

Paratransgenesis is an attractive strategy in the development of transmission-blocking interventions because microbial symbionts can be vertically transmitted from parents to offspring in insect vectors (Bartholomay & Michel, 2018; Wang et al., 2013; Wang & Jacobs-Lorena, 2013). Stable vertical transmission of genetically modified symbionts is required for efficient and continuous control of infectious diseases using transgenic organisms. Gut microbiota has been shown to significantly reduce oocysts load in the midgut of mosquitoes. Antibiotics which kill the gut microbiota of mosquitoes have been demonstrated to increase the susceptibility of mosquitoes to *Plasmodium* infection (Dong et al., 2009).

Anti-*Plasmodium* microorganism *Enterobacter* bacterium strain Esp. Z has been shown to kill the malaria parasite by producing Reactive Oxygen Species which prevents midgut invasion in mosquitoes (Cirimotich et al., 2011). *Enterobacteriaceae* plays a protective role in the natural infection of mosquitoes with *Plasmodium.* A study has shown positive correlation between *Enterobacteriaceae* and reduction of *Plasmodium* infection in the midgut of *A. gambiae* (Baton & Ranford-Cartwright, 2005).

22.7.1.1 Wolbachia-Infected Aedes aegypti Has Reduced Susceptibility to Dengue Virus

Wolbachia pipientis are bacteria symbionts found in a large number of insects and are vertically transmitted from mother to offspring (Hilgenboecker et al., 2008). *Wolbachia* infection reduces parasite load in different species of mosquitoes and other species of arthropods limiting their vectorial competence for transmission of infectious pathogens (Bian et al., 2010; Iturbe-Ormaetxe et al., 2011; Kambris et al., 2009; Moreira et al., 2009; Werren et al., 2008).

Wolbachia-infected *Anopheles* are not found in natural infection. Although a study has shown the possibility of *Wolbachia* infection in *Anopheles* (Hughes et al., 2014), a stable infection has not been reported (Walker & Moreira, 2011). Introduction of paratransgenic *Wolbachia* Mel strain into *Aedes aegypti*, which is not a natural host of *Wolbachia*, reduced the susceptibility of *Aedes* to dengue virus (Carrington et al., 2018). The mechanisms through which *Wolbachia* enhances the immunity of mosquito is unknown. Infecting other mosquito species with *Wolbachia* has not been reported to be effective in malaria control. A report showed increased infection and transmission of *Plasmodium falciparum* in *Wolbachia*-infected *Anopheles* (Hughes et al., 2014). Similarly, increased infection of *Wolbachia*-infected *Ae. aegypti* with some non-medically important insect-specific flaviviruses has been reported (Amuzu et al., 2018).

22.7.1.2 Mosquito-Midgut Yeast Modulates the Immune System

The guts of mosquitoes play host to many species of yeast: *Candia, Pichia*, and *Wickerhamomyces anomalus* were identified as promising endosymbionts that could deliver antimalarial effector proteins to anopheline vectors (Ignatova et al., 1996; Ricci, Damiani, et al., 2011; Ricci, Mosca, et al., 2011; Wang & Jacobs-Lorena, 2013).

22.7.2 Tsetse Fly Midgut Microbiota

The midgut microbiota of tsetse fly is very important to their survival and/or the trypanosome they harbor (Lazaro et al., 2002; Lysyk et al., 1999). The endosymbionts of tsetse flies include *Wigglesworthia glossinidia*, *Sodalis glossinidius*, *Wolbachia pipientis*, and Salivary Gland Hypertrophy Virus (SGHV) (Wang et al., 2013). *Wigglesworthia* and *Sodalis* are both found intracellularly and extracellularly in different tissues of tsetse fly including the midgut (Balmand et al., 2013; Cheng & Aksoy, 1999).

22.7.2.1 Wigglesworthia-Infected Tsetse Flies Are Refractory to Infection with Trypanosomes

Wigglesworthia is a mutualistic endosymbiont that serves immunological function in tsetse fly and impacts its fertility (Pais et al., 2008). The absence of *Wigglesworthia* in the developmental larvae stage of tsetse fly has been linked to a severely compromised immune system in adult tsetse flies (Wang et al., 2013). *Wigglesworthia*-free tsetse are vulnerable to infection with *E. coli* K12 that are usually non-pathogenic and to trypanosomes (Wang et al., 2009; Weiss et al., 2012, 2011, 2013). *Wigglesworthia* interacts with the maternally transmitted tsetse fly immune milk protein, peptidoglycan recognition protein (PGRP-LB), to modulate the immune system of tsetse's offspring (Wang & Aksoy, 2012; Wang et al., 2009).

22.7.2.2 Tsetse Flies Harboring Sodalis Are Susceptible to Infection with Trypanosomes

Sodalis is a commensal endobacteria whose functional role in tsetse fly remains unclear. Tsetse fly harboring *Sodalis* was shown to be three times susceptible to trypanosomes infection compared to *Sodalis*-free tsetse through a mechanism that involves inhibition of midgut trypanocidal lectins (Soumana et al., 2013; Welburn et al., 1993). *Sodalis*-free tsetse flies and streptomycin-treated tsetse flies have been shown to be refractory to infection with trypanosomes (Dale & Welburn, 2001;

Welburn et al., 1993). A possible effect of gut microbiota on vectorial competence has been reported for *Glossina* endosymbionts (Welburn & Maudlin, 1999).

22.7.3 Ticks Midgut Microbiota

Gram-negative bacteria of the phylum Proteobacteria are the predominant species in tick midgut. These organisms modulate the reproductive fitness and vectorial competence of ticks. Tick's bacterial endosymbionts are vertically transmitted from parents to offspring, and function in their immunity by stabilizing the peritrophic matrix and maintaining the epithelial barrier integrity of their midgut (Narasimhan & Fikrig, 2015). The non-vectored symbiotic *Rickettsia peacockii* inhibit the infection of the tick *Dermacentor andersoni* with *Rickettsia rickettsii* and prevents its transmission by ticks (Burgdorfer & Anacker, 1981; Niebylski et al., 1997, 1999). *R. rickettsii*, the causative agent of Rocky Mountain spotted fever, is lethal to experimentally and transovarially infected Rocky Mountain wood ticks (Niebylski et al., 1999).

22.7.4 Other Immunity-Related Endosymbionts

Wolbachia spp. are endosymbionts found in the midgut of insects. They impact the reproduction of their host and protect them from viral infection and other disease pathogens (Soumana et al., 2013). *Wolbachia* sp. has been shown to inhibit the transmission of West Nile virus in *Culex sp.*(Soumana et al., 2013). *Serratia marcescens* has been reportedly linked to increased mortality in *Anopheles albimanus* and *G. pallidipes* flies (Gonzalez-Ceron et al., 2003; Poinar Jr et al., 1979).

22.8 EFFECT OF GUT BACTERIA ON PATHOGEN TRANSMISSION

Several gut bacteria species have been studied for their effect on pathogen transmission in infectious disease management. Some of these bacteria species stimulate or inhibit pathogen transmission through a mechanism involving the production and secretion of effector molecules. *Enterobacter* sp. found in the gut of the anopheline mosquito and *Rhodnius* sp. inhibit *Plasmodium falciparum* and *Trypanosoma cruzi* and reduce their ability to cause infection. *Serratia marcescens* and *Pseudomonas fluorescens* are other bacteria species found in the midgut of *Rhodnius* sp. that contribute to its refractoriness to *Trypanosoma cruzi* infection (Poinar Jr et al., 1979; Soumana et al., 2013). Infection of *A. gambiae* with *Serratia marcescens* reduced infection with rodent *Plasmodium* (Bando et al., 2013).

22.8.1 Recombinant Technology in the Development of Interventions

Escherichia coli bacteria transformed with recombinant plasmid carrying an anti-parasite gene can be used to express and display anti-*Plasmodium* molecules in the midgut of mosquitoes. A problem with the use of this strategy is the poor performance of attenuated *Escherichia coli* in the midgut of mosquitoes. In a study that used *E. coli* to express recombinant single-chain immunotoxin to inhibit malaria parasite development, the expressed proteins formed inclusion bodies in the mosquito midgut

(Riehle et al., 2007; Yoshida et al., 2001). Recombinant proteins may also attach to *E. coli* surface denying its diffusion to the mosquito-midgut target or parasite for action (Riehle et al., 2007). Ticks have been shown to digest or destroy *E. coli* within their midgut cells (Matsuo et al., 2004).

Pantoea agglomerans is the dominant bacteria in the natural microbiota of some mosquito species in Mali and Kenya (Straif et al., 1998). The problems associated with expressing recombinant protein using *E. coli* in mosquitoes has been addressed by using the *P. agglomerans* expression system. This bacterium has been genetically modified to produce anti-*Plasmodium* proteins using the HlyA system (Wang et al., 2012). The anti-*Plasmodium* molecules scorpine and $(EPIP)_4$ have been successfully delivered to mosquito midgut without any fitness cost to the mosquito or the transgenic expression bacteria (Bisi & Lampe, 2011; Wang et al., 2012).

22.9 FIELD APPLICATION OF TRANSGENIC TECHNOLOGY

For transgenic technology to be useful in the control of infectious diseases, the breakthroughs recorded in the laboratory should be effectively transferred and reproduced in the field. To successfully establish a genetically modified organism in the field, the transgenic organism must have competitive advantages over its wild-type counterparts (Wang & Jacobs-Lorena, 2013). An effective way of introducing and establishing transgenic organisms must be solved to be able to use this technology in the control of infectious diseases.

22.10 CONCLUSION

Although there are still controversies on the use of genetically modified organisms, intervention strategies using genetically modified organisms hold some promise in the control of infectious diseases. Strategies must be developed for the selective maintenance of transgenic organisms and simultaneous elimination of wild-type populations. The use of transgenic organisms after all safety concerns have been addressed is still subject to the acceptance of the public and approval by regulatory authorities.

REFERENCES

Abraham, E. G., & Jacobs-Lorena, M. (2004). Mosquito midgut barriers to malaria parasite development. *Insect Biochemistry and Molecular Biology 34*(7), 667–671. https://doi.org/10.1016/j.ibmb.2004.03.019.

Abraham, E. G., Pinto, S. B., Ghosh, A., Vanlandingham, D. L., Budd, A., Higgs, S., Kafatos, F. C., Jacobs-Lorena, M., & Michel, K. (2005). An immune-responsive serpin, SRPN6, mediates mosquito defense against malaria parasites. *Proceedings of the National Academy of Sciences, 102*(45), 16327–16332.

Aksoy, S. (2000). Tsetse–a haven for microorganisms. *Parasitology Today, 16*(3), 114–118.

Amuzu, H. E., Tsyganov, K., Koh, C., Herbert, R. I., Powell, D. R., & McGraw, E. A. (2018). Wolbachia enhances insect-specific flavivirus infection in Aedes aegypti mosquitoes. *Ecology and Evolution, 8*(11), 5441–5454.

Balmand, S., Lohs, C., Aksoy, S., & Heddi, A. (2013). Tissue distribution and transmission routes for the tsetse fly endosymbionts. *Journal of Invertebrate Pathology, 112*, S116–S122.

Balogun, E. O., Nok, A. J., & Kita, K. (2016). Global warming and the possible globalization of vector-borne diseases: a call for increased awareness and action. *Tropical Medicine and Health*, 44(1), 1-3.

Bando, H., Okado, K., Guelbeogo, W. M., Badolo, A., Aonuma, H., Nelson, B., Fukumoto, S., Xuan, X., Sagnon, N. F., & Kanuka, H. (2013). Intra-specific diversity of Serratia marcescens in Anopheles mosquito midgut defines Plasmodium transmission capacity. *Scientific Reports, 3*, 1641.

Bartholomay, L. C., & Michel, K. (2018). Mosquito immunobiology: The intersection of vector health and vector competence. *Annual Review of Entomology, 63*, 145–167. https://doi.org/10.1146/annurev-ento-010715-023530.

Baton, L. A., & Ranford-Cartwright, L. C. (2005). How do malaria ookinetes cross the mosquito midgut wall? *Trends in Parasitology, 21*(1), 22–28. https://doi.org/10.1016/j.pt.2004.11.001.

Behura, S. K., Gomez-Machorro, C., Debruyn, B., Lovin, D. D., Harker, B. W., Romero-Severson, J., Mori, A., & Severson, D. W. (2014). Influence of mosquito genotype on transcriptional response to dengue virus infection. *Functional and Integrative Genomics, 14*(3), 581–589.

Behura, S. K., Gomez-Machorro, C., Harker, B. W., Debruyn, B., Lovin, D. D., Hemme, R. R., Mori, A., Romero-Severson, J., & Severson, D. W. (2011). Global cross-talk of genes of the mosquito Aedes aegypti in response to dengue virus infection. *PLoS Neglected Tropical Diseases, 5*(11).

Bian, G., Joshi, D., Dong, Y., Lu, P., Zhou, G., Pan, X., Xu, Y., Dimopoulos, G., & Xi, Z. (2013). Wolbachia invades Anopheles stephensi populations and induces refractoriness to Plasmodium infection. *Science, 340*(6133), 748–751.

Bian, G., Xu, Y., Lu, P., Xie, Y., & Xi, Z. (2010). The endosymbiotic bacterium Wolbachia induces resistance to dengue virus in Aedes aegypti. *PLoS Pathogens, 6*(4).

Billingsley, P. F. (1994). Approaches to vector control: new and trusted. 2. Molecular targets in the insect midgut. *Transactions of the Royal Society of Tropical Medicine and Hygiene 88*(2), 136–140. https://doi.org/10.1016/0035-9203(94)90265-8.

Bisi, D. C., & Lampe, D. J. (2011). Secretion of anti-Plasmodium effector proteins from a natural Pantoea agglomerans isolate by using PelB and HlyA secretion signals. *Applied and Environmental Microbiology 77*(13), 4669–4675.

Black IV, W. C., Bennett, K. E., Gorrochótegui-Escalante, N., Barillas-Mury, C. V., Fernández-Salas, I., de Lourdes Muñoz, M. a., Farfán-Alé, J. A., Olson, K. E., & Beaty, B. J. (2002). Flavivirus susceptibility in Aedes aegypti. *Archives of Medical Research, 33*(4), 379–388.

Boulanger, N., Brun, R., Ehret-Sabatier, L., Kunz, C., & Bulet, P. (2002). Immunopeptides in the defense reactions of Glossina morsitans to bacterial and Trypanosoma brucei brucei infections. *Insect Biochemistry and Molecular Biology, 32*(4), 369–375.

Burgdorfer, W., & Anacker, R. L. (1981). Rickettsiae and rickettsial diseases. Conference on Rickettsiae and Rickettsial Diseases (1980: Rocky Mountain Laboratories).

Caicedo, P. A., Serrato, I. M., Sim, S., Dimopoulos, G., Coatsworth, H., Lowenberger, C., & Ocampo, C. B. (2019). Immune response-related genes associated to blocking midgut dengue virus infection in Aedes aegypti strains that differ in susceptibility. *Insect Science 26*(4), 635–648. https://doi.org/10.1111/1744-7917.12573.

Carrington, L. B., Tran, B. C. N., Le, N. T. H., Luong, T. T. H., Nguyen, T. T., Nguyen, P. T., Nguyen, C. V. V., Nguyen, H. T. C., Vu, T. T., & Vo, L. T. (2018). Field-and clinically derived estimates of Wolbachia-mediated blocking of dengue virus transmission potential in Aedes aegypti mosquitoes. *Proceedings of the National Academy of Sciences, 115*(2), 361–366.

Ceraul, S. M., Dreher-Lesnick, S. M., Gillespie, J. J., Rahman, M. S., & Azad, A. F. (2007). New tick defensin isoform and antimicrobial gene expression in response to Rickettsia montanensis challenge. *Infection and Immunity, 75*(4), 1973–1983.

Ceraul, S. M., Sonenshine, D. E., Ratzlaff, R. E., & Hynes, W. L. (2003). An arthropod defensin expressed by the hemocytes of the American dog tick, Dermacentor variabilis (Acari: Ixodidae). *Insect Biochemistry and Molecular Biology, 33*(11), 1099–1103.

Chang, T. L.-Y., Chang, C.-H., Simpson, D. A., Xu, Q., Martin, P. K., Lagenaur, L. A., Schoolnik, G. K., Ho, D. D., Hillier, S. L., & Holodniy, M. (2003). Inhibition of HIV infectivity by a natural human isolate of Lactobacillus jensenii engineered to express functional two-domain CD4. *Proceedings of the National Academy of Sciences, 100*(20), 11672–11677.

Chauhan, C., Behura, S. K., Debruyn, B., Lovin, D. D., Harker, B. W., Gomez-Machorro, C., Mori, A., Romero-Severson, J., & Severson, D. W. (2012). Comparative expression profiles of midgut genes in dengue virus refractory and susceptible Aedes aegypti across critical period for virus infection. *PloS one, 7*(10).

Cheng, Q., & Aksoy, S. (1999). Tissue tropism, transmission and expression of foreign genes in vivo in midgut symbionts of tsetse flies. *Insect Molecular Biology, 8*(1), 125–132.

Cirimotich, C. M., Dong, Y., Clayton, A. M., Sandiford, S. L., Souza-Neto, J. A., Mulenga, M., & Dimopoulos, G. (2011). Natural microbe-mediated refractoriness to Plasmodium infection in Anopheles gambiae. *Science, 332*(6031), 855–858.

Cirimotich, C. M., Dong, Y., Garver, L. S., Sim, S., & Dimopoulos, G. (2010). Mosquito immune defenses against Plasmodium infection. *Developmental and Comparative Immunology, 34*(4), 387–395. https://doi.org/10.1016/j.dci.2009.12.005.

Corby-Harris, V., Drexler, A., De Jong, L. W., Antonova, Y., Pakpour, N., Ziegler, R., Ramberg, F., Lewis, E. E., Brown, J. M., & Luckhart, S. (2010). Activation of Akt signaling reduces the prevalence and intensity of malaria parasite infection and lifespan in Anopheles stephensi mosquitoes. *PLoS Pathogens, 6*(7).

Coutinho-Abreu, I. V., Zhu, K. Y., & Ramalho-Ortigao, M. (2010). Transgenesis and paratransgenesis to control insect-borne diseases: current status and future challenges. *Parasitology International, 59*(1), 1–8.

Dale, C., & Welburn, S. (2001). The endosymbionts of tsetse flies: manipulating host–parasite interactions. *International Journal for Parasitology, 31*(5–6), 628–631.

de Lara Capurro, M., Coleman, J., Beerntsen, B. T., Myles, K. M., Olson, K. E., Rocha, E., Krettli, A. U., & James, A. A. (2000). Virus-expressed, recombinant single-chain antibody blocks sporozoite infection of salivary glands in Plasmodium gallinaceum-infected *Aedes aegypti. The American Journal of Tropical Medicine and Hygiene, 62*(4), 427–433.

de Sousa, G., & Conte, H. (2013). Midgut morphophysiology in Sitophilus zeamais Motschulsky, 1855 (Coleoptera: Curculionidae). *Micron, 51*, 1–8.

Dong, Y., Das, S., Cirimotich, C., Souza-Neto, J. A., McLean, K. J., & Dimopoulos, G. (2011). Engineered anopheles immunity to Plasmodium infection. *PLoS Pathogens, 7*(12).

Dong, Y., Manfredini, F., & Dimopoulos, G. (2009). Implication of the mosquito midgut microbiota in the defense against malaria parasites. *PLoS Pathogens, 5*(5).

Douglas, A. E. (2011). Lessons from studying insect symbioses. *Cell Host & Microbe, 10*(4), 359–367.

Durvasula, R. V., Gumbs, A., Panackal, A., Kruglov, O., Aksoy, S., Merrifield, R. B., Richards, F. F., & Beard, C. B. (1997). Prevention of insect-borne disease: an approach using transgenic symbiotic bacteria. *Proceedings of the National Academy of Sciences, 94*(7), 3274–3278.

Favia, G., Ricci, I., Damiani, C., Raddadi, N., Crotti, E., Marzorati, M., Rizzi, A., Urso, R., Brusetti, L., & Borin, S. (2007). Bacteria of the genus Asaia stably associate with Anopheles stephensi, an Asian malarial mosquito vector. *Proceedings of the National Academy of Sciences, 104*(21), 9047–9051.

Garver, L. S., Bahia, A. C., Das, S., Souza-Neto, J. A., Shiao, J., Dong, Y., & Dimopoulos, G. (2012). Anopheles Imd pathway factors and effectors in infection intensity-dependent anti-Plasmodium action. *PLoS pathogens, 8*(6).

Garver, L. S., Dong, Y., & Dimopoulos, G. (2009). Caspar controls resistance to Plasmodium falciparum in diverse anopheline species. *PLoS Pathogens, 5*(3).

Ghosh, A., Edwards, M., & Jacobs-Lorena, M. (2000). The journey of the malaria parasite in the mosquito: hopes for the new century. *Parasitology Today, 16*(5), 196–201.

Ghosh, A. K., Coppens, I., Gårdsvoll, H., Ploug, M., & Jacobs-Lorena, M. (2011). Plasmodium ookinetes coopt mammalian plasminogen to invade the mosquito midgut. *Proceedings of the National Academy of Sciences, 108*(41), 17153–17158.

Ghosh, A. K., Ribolla, P. E., & Jacobs-Lorena, M. (2001). Targeting Plasmodium ligands on mosquito salivary glands and midgut with a phage display peptide library. *Proceedings of the National Academy of Sciences, 98*(23), 13278–13281.

Gonzalez-Ceron, L., Santillan, F., Rodriguez, M. H., Mendez, D., & Hernandez-Avila, J. E. (2003). Bacteria in midguts of field-collected Anopheles albimanus block Plasmodium vivax sporogonic development. *Journal of Medical Entomology, 40*(3), 371–374.

Gullan, P. J., & Cranston, P. S. (2014). *The Insects: An Outline of Entomology*. John Wiley & Sons.

Herrera-Ortiz, A., Martínez-Barnetche, J., Smit, N., Rodriguez, M. H., & Lanz-Mendoza, H. (2011). The effect of nitric oxide and hydrogen peroxide in the activation of the systemic immune response of Anopheles albimanus infected with Plasmodium berghei. *Developmental & Comparative Immunology, 35*(1), 44–50.

Hilgenboecker, K., Hammerstein, P., Schlattmann, P., Telschow, A., & Werren, J. H. (2008). How many species are infected with Wolbachia?–a statistical analysis of current data. *FEMS Microbiology Letters, 281*(2), 215–220.

Hughes, G. L., Rivero, A., & Rasgon, J. L. (2014). Wolbachia can enhance Plasmodium infection in mosquitoes: implications for malaria control? *PLoS Pathogens, 10*(9).

Hurwitz, I., Fieck, A., Read, A., Hillesland, H., Klein, N., Kang, A., & Durvasula, R. (2011). Paratransgenic control of vector borne diseases. *International Journal of Biological Sciences, 7*(9), 1334.

Ignatova, E., Nagornaia, S., Povazhnaia, T., & Ianishevskaia, G. (1996). The yeast flora of blood-sucking mosquitoes. *Mikrobiolohichnyi Zhurnal (Kiev, Ukraine: 1993), 58*(2), 12–15.

Isaacs, A. T., Li, F., Jasinskiene, N., Chen, X., Nirmala, X., Marinotti, O., Vinetz, J. M., & James, A. A. (2011). Engineered resistance to Plasmodium falciparum development in transgenic Anopheles stephensi. *PLoS Pathogens, 7*(4).

Ito, J., Ghosh, A., Moreira, L. A., Wimmer, E. A., & Jacobs-Lorena, M. (2002). Transgenic anopheline mosquitoes impaired in transmission of a malaria parasite. *Nature, 417*(6887), 452–455.

Iturbe-Ormaetxe, I., Walker, T., & O'Neill, S. L. (2011). Wolbachia and the biological control of mosquito-borne disease. *EMBO Reports, 12*(6), 508–518.

Kambris, Z., Cook, P. E., Phuc, H. K., & Sinkins, S. P. (2009). Immune activation by life-shortening Wolbachia and reduced filarial competence in mosquitoes. *Science, 326*(5949), 134–136.

Kim, W., Koo, H., Richman, A. M., Seeley, D., Vizioli, J., Klocko, A. D., & O'brochta, D. A. (2004). Ectopic expression of a cecropin transgene in the human malaria vector mosquito Anopheles gambiae (Diptera: Culicidae): effects on susceptibility to Plasmodium. *Journal of Medical Entomology, 41*(3), 447–455.

Kokoza, V., Ahmed, A., Shin, S. W., Okafor, N., Zou, Z., & Raikhel, A. S. (2010). Blocking of Plasmodium transmission by cooperative action of Cecropin A and Defensin A in transgenic Aedes aegypti mosquitoes. *Proceedings of the National Academy of Sciences, 107*(18), 8111–8116.

Kurtti, T. J., Burkhardt, N. Y., Heu, C. C., & G, M. U. (2016). Fluorescent protein expressing Rickettsia buchneri and Rickketsia peacockii for tracking symbiont-tick cell interactions. *Veterinary Science, 3*(4), 34.

Lavazec, C., Boudin, C., Lacroix, R., Bonnet, S., Diop, A., Thiberge, S., Boisson, B., Tahar, R., & Bourgouin, C. (2007). Carboxypeptidases B of Anopheles gambiae as targets for a Plasmodium falciparum transmission-blocking vaccine. *Infection and Immunity, 75*(4), 1635–1642.

Lazaro, J., Nitcheu, J., Predicala, R. Z., Mangalindan, G. C., Nesslany, F., Marzin, D., Concepcion, G. P., & Diquet, B. (2002). Heptyl prodigiosin, a bacterial metabolite, is antimalarial in vivo and non-mutagenic *in vitro*. *Journal of Natural Toxins, 11*(4), 367–377.

Lee, W. S., Webster, J. A., Madzokere, E. T., Stephenson, E. B., & Herrero, L. J. (2019). Mosquito antiviral defense mechanisms: a delicate balance between innate immunity and persistent viral infection. *Parasit Vectors, 12*(1), 165. https://doi.org/10.1186/s13071-019-3433-8.

Lysyk, T., Kalischuk-Tymensen, L., Selinger, L., Lancaster, R., Wever, L., & Cheng, K.-J. (1999). Rearing stable fly larvae (Diptera: Muscidae) on an egg yolk medium. *Journal of Medical Entomology, 36*(3), 382–388.

Machado-Ferreira, E., Balsemão-Pires, E., Dietrich, G., Hojgaard, A., Vizzoni, V. F., Scoles, G., Bell-Sakyl, L., Piesman, J., Zeidner, N. S., Moerbeck, L., & Soares, C. A. (2015). Transgene expression in tick cells using Agrobacterium tumefaciens. *Experimental & Applied Acarology, 67*(2), 269–287.

Maltz, M. A., Weiss, B. L., O'Neill, M., Wu, Y., & Aksoy, S. (2012). OmpA-mediated biofilm formation is essential for the commensal bacterium Sodalis glossinidius to colonize the tsetse fly gut. *Applied. Environmental Microbiology, 78*(21), 7760–7768.

Mathias, D. K., Plieskatt, J. L., Armistead, J. S., Bethony, J. M., Abdul-Majid, K. B., McMillan, A., Angov, E., Aryee, M. J., Zhan, B., Gillespie, P., Keegan, B., Jariwala, A. R., Rezende, W., Bottazzi, M. E., Scorpio, D. G., Hotez, P. J., & Dinglasan, R. R. (2012). Expression, immunogenicity, histopathology, and potency of a mosquito-based malaria transmission-blocking recombinant vaccine. *Infection and Immunity 80*(4), 1606–1614. https://doi.org/10.1128/iai.06212-11.

Matsuo, T., Okoda, Y., Badgar, B., Inoue, N., Xuan, X., Taylor, D., & Fujisaki, K. (2004). Fate of GFP-expressing Escherichia coli in the midgut and response to ingestion in a tick, Ornithodoros moubata (Acari: Argasidae). *Experimental Parasitology, 108*(1–2), 67–73.

Molina-Cruz, A., DeJong, R. J., Ortega, C., Haile, A., Abban, E., Rodrigues, J., Jaramillo-Gutierrez, G., & Barillas-Mury, C. (2012). Some strains of Plasmodium falciparum, a human malaria parasite, evade the complement-like system of Anopheles gambiae mosquitoes. *Proceedings of the National Academy of Sciences, 109*(28), E1957–E1962.

Moreira, L. A., Iturbe-Ormaetxe, I., Jeffery, J. A., Lu, G., Pyke, A. T., Hedges, L. M., Rocha, B. C., Hall-Mendelin, S., Day, A., & Riegler, M. (2009). A Wolbachia symbiont in Aedes aegypti limits infection with dengue, Chikungunya, and Plasmodium. *Cell, 139*(7), 1268–1278.

Munks, R., Hamilton, J., Lehane, S., & Lehane, M. (2001). Regulation of midgut defensin production in the blood-sucking insect Stomoxys calcitrans. *Insect Molecular Biology, 10*(6), 561–571.

Nakajima, Y., Van Der Goes van Naters-Yasui, A., Taylor, D., & Yamakawa, M. (2002). Antibacterial peptide defensin is involved in midgut immunity of the soft tick, Ornithodoros moubata. *Insect Molecular Biology, 11*(6), 611–618.

Nakajima, Y., van Naters, A. v. d. G., Taylor, D., & Yamakawa, M. (2001). Two isoforms of a member of the arthropod defensin family from the soft tick, Ornithodoros moubata (Acari: Argasidae). *Insect Biochemistry and Molecular Biology, 31*(8), 747–751.

Napoleão, T. H., Albuquerque, L. P., Santos, N. D., Nova, I. C., Lima, T. A., Paiva, P. M., & Pontual, E. V. (2019). Insect midgut structures and molecules as targets of plant-derived protease inhibitors and lectins. *Pest Management Science 75*(5), 1212–1222. https://doi.org/10.1002/ps.5233.

Narasimhan, S., & Fikrig, E. (2015). Tick microbiome: the force within. *Trends in Parasitology, 31*(7), 315–323.

Niebylski, M. L., Peacock, M. G., Fischer, E. R., Porcella, S. F., & Schwan, T. G. (1997). Characterization of an endosymbiont infecting wood ticks, Dermacentor andersoni, as a member of the genus Francisella. *Applied and Environmental Microbiology, 63*(10), 3933–3940.

Niebylski, M. L., Peacock, M. G., & Schwan, T. G. (1999). Lethal effect of Rickettsia rickettsiion its tick vector (Dermacentor andersoni). *Applied and Environmental Microbiology, 65*(2), 773–778.

Nikolaeva, D., Draper, S. J., & Biswas, S. (2015). Toward the development of effective transmission-blocking vaccines for malaria. *Expert Review of Vaccines, 14*(5), 653–680. https://doi.org/10.1586/14760584.2015.993383.

Nikolaeva, D., Illingworth, J. J., Miura, K., Alanine, D. G. W., Brian, I. J., Li, Y., Fyfe, A. J., Da, D. F., Cohuet, A., Long, C. A., Draper, S. J., & Biswas, S. (2020). Functional Characterization and Comparison of Plasmodium falciparum Proteins as Targets of Transmission-blocking Antibodies. *Molecular and Cellular Proteomics, 19*(1), 155–166. https://doi.org/10.1074/mcp.RA117.000036.

Nok, A. J., Njoku, G., & Balogun, E. (2002). A 45-kDa midgut glycoprotein from Anopheles albimanus mosquito mediates the killing of trypanosomes. *Cell Biochemistry and Function: Cellular Biochemistry and Its Modulation by Active Agents or Disease, 20*(3), 257–262.

Nouzova, M., Clifton, M. E., & Noriega, F. G. (2019). Mosquito adaptations to hematophagia impact pathogen transmission. *Current Opinion in Insect Science*, *34*, 21–26. https://doi.org/10.1016/j.cois.2019.02.002.

Oliveira, F., Jochim, R. C., Valenzuela, J. G., & Kamhawi, S. (2009). Sand flies, Leishmania, and transcriptome-borne solutions. *Parasitology International*, *58*(1), 1–5. https://doi.org/10.1016/j.parint.2008.07.004.

Pais, R., Lohs, C., Wu, Y., Wang, J., & Aksoy, S. (2008). The obligate mutualist Wigglesworthia glossinidia influences reproduction, digestion, and immunity processes of its host, the tsetse fly. *Applied and Environmental Microbiology 74*(19), 5965–5974.

Pal, A., & Chakravarty, A. (2020). Disease resistance for different livestock species. *Genetics and Breeding for Disease Resistance of Livestock*, 271.

Pan, X., Zhou, G., Wu, J., Bian, G., Lu, P., Raikhel, A. S., & Xi, Z. (2012). Wolbachia induces reactive oxygen species (ROS)-dependent activation of the Toll pathway to control dengue virus in the mosquito Aedes aegypti. *Proceedings of the National Academy of Sciences, 109*(1), E23–E31.

Poinar Jr, G., Wassink, H., van der Linden, M., & van der Geest, L. (1979). Serratia marcescens as a pathogen of Tsetse flies (Glossina morsitans and Glossina pallidipes, biological control. *Acta Tropica* 36 (3):223–227.

Rao, S., Hu, S., McHugh, L., Lueders, K., Henry, K., Zhao, Q., Fekete, R. A., Kar, S., Adhya, S., & Hamer, D. H. (2005). Toward a live microbial microbicide for HIV: Commensal bacteria secreting an HIV fusion inhibitor peptide. *Proceedings of the National Academy of Sciences, 102*(34), 11993–11998.

Ramirez, B. (2017). Support for research towards understanding the population health vulnerabilities to vector-borne diseases: increasing resilience under climate change conditions in Africa. Infectious diseases of poverty, 6(1), 1-7.

Raz, A., Dinparast Djadid, N., & Zakeri, S. (2013). Molecular characterization of the carboxypeptidase B1 of Anopheles stephensi and its evaluation as a target for transmission-blocking vaccines. *Infection and Immunity, 81*(6), 2206–2216. https://doi.org/10.1128/iai.01331-12.

Ricci, I., Damiani, C., Scuppa, P., Mosca, M., Crotti, E., Rossi, P., Rizzi, A., Capone, A., Gonella, E., & Ballarini, P. (2011). The yeast Wickerhamomyces anomalus (Pichia anomala) inhabits the midgut and reproductive system of the Asian malaria vector Anopheles stephensi. *Environmental Microbiology, 13*(4), 911–921.

Ricci, I., Mosca, M., Valzano, M., Damiani, C., Scuppa, P., Rossi, P., Crotti, E., Cappelli, A., Ulissi, U., & Capone, A. (2011). Different mosquito species host Wickerhamomyces anomalus (Pichia anomala): perspectives on vector-borne diseases symbiotic control. *Antonie Van Leeuwenhoek, 99*(1), 43–50.

Richman, A., Bulet, P., Hetru, C., Barillas-Mury, C., Hoffmann, J., & Kafatos, F. (1996). Inducible immune factors of the vector mosquito Anopheles gambiae: Biochemical purification of a defensin antibacterial peptide and molecular cloning of preprodefensin cDNA. *Insect Molecular Biology, 5*(3), 203–210.

Riehle, M. A., & Jacobs-Lorena, M. (2005). Using bacteria to express and display anti-parasite molecules in mosquitoes: Current and future strategies. *Insect Biochemistry and Molecular Biology, 35*(7), 699–707. https://doi.org/10.1016/j.ibmb.2005.02.008.

Riehle, M. A., Moreira, C. K., Lampe, D., Lauzon, C., & Jacobs-Lorena, M. (2007). Using bacteria to express and display anti-Plasmodium molecules in the mosquito midgut. *International Journal for Parasitology, 37*(6), 595–603.

Rodrigues, J., Oliveira, G. A., Kotsyfakis, M., Dixit, R., Molina-Cruz, A., Jochim, R., & Barillas-Mury, C. (2012). An epithelial serine protease, AgESP, is required for Plasmodium invasion in the mosquito Anopheles gambiae. *PloS one, 7*(4).

Ruiz-Postigo, J. A., Franco, J. R., Lado, M., & Simarro, P. P. (2012). Human African trypanosomiasis in South Sudan: how can we prevent a new epidemic? *PLoS Neglected Tropical Diseases, 6*(5).

Shahabuddin, M., & Costero, A. (2001). Spatial distribution of factors that determine sporogonic development of malaria parasites in mosquitoes. *Insect Biochemistry and Molecular Biology, 31*(3), 231–240.

Sinden, R. (1999). Plasmodium differentiation in the mosquito. *Parassitologia, 41*(1–3), 139–148.

Sinden, R. E., & Billingsley, P. F. (2001). Plasmodium invasion of mosquito cells: hawk or dove? *Trends in Parasitology, 17*(5), 209–211.

Sonenshine, D. E., Hynes, W. L., Ceraul, S. M., Mitchell, R., & Benzine, T. (2005). Host blood proteins and peptides in the Midgut of the tick Dermacentor variabilis Contributeto bacterial control. *Experimental and Applied Acarology, 36*(3), 207–223.

Soumana, I. H., Simo, G., Njiokou, F., Tchicaya, B., Abd-Alla, A. M., Cuny, G., & Geiger, A. (2013). The bacterial flora of tsetse fly midgut and its effect on trypanosome transmission. *Journal of Invertebrate Pathology, 112 Suppl*, S89–S93. https://doi.org/10.1016/j.jip.2012.03.029.

Sreenivasamurthy, S. K., Dey, G., Ramu, M., Kumar, M., Gupta, M. K., Mohanty, A. K., Harsha, H. C., Sharma, P., Kumar, N.;, Pandey, A., Kumar, A., & Prasad, T. S. (2013). A compendium of molecules involved in vector-pathogen interactions pertaining to malaria. *Malaria Journal, 12*, 216. https://doi.org/10.1186/1475-2875-12-216.

Straif, S. C., Mbogo, C. N., Toure, A. M., Walker, E. D., Kaufman, M., Toure, Y. T., & Beier, J. C. (1998). Midgut bacteria in Anopheles gambiae and An. funestus (Diptera: Culicidae) from Kenya and Mali. *Journal of Medical Entomology, 35*(3), 222–226.

Talactac, M. R., Yada, Y., Yoshii, K., Hernandez, E. P., Kusakisako, K., Maeda, H., Galay, R. L., Fujisaki, K., Mochizuki, M., & Tanaka, T. (2017). Characterization and antiviral activity of a newly identified defensin-like peptide, HEdefensin, in the hard tick Haemaphysalis longicornis. *Developmental & Comparative Immunology, 68*, 98–107.

VenkatRao, V., Kumar, S. K., Sridevi, P., Muley, V. Y., & Chaitanya, R. K. (2017). Cloning, characterization and transmission blocking potential of midgut carboxypeptidase A in Anopheles stephensi. *Acta Tropica, 168*, 21–28. https://doi.org/10.1016/j.actatropica.2016.12.035.

Vizioli, J., Bulet, P., Hoffmann, J. A., Kafatos, F. C., Müller, H.-M., & Dimopoulos, G. (2001). Gambicin: a novel immune responsive antimicrobial peptide from the malaria vector Anopheles gambiae. *Proceedings of the National Academy of Sciences, 98*(22), 12630–12635.

Walker, T., & Moreira, L. A. (2011). Can Wolbachia be used to control malaria? *Memórias do Instituto Oswaldo Cruz, 106*, 212–217.

Wang, J., & Aksoy, S. (2012). PGRP-LB is a maternally transmitted immune milk protein that influences symbiosis and parasitism in tsetse's offspring. *Proceedings of the National Academy of Sciences, 109*(26), 10552–10557.

Wang, J., Weiss, B. L., & Aksoy, S. (2013). Tsetse fly microbiota: form and function. *Frontiers in Cellular and Infection Microbiology, 3*, 69.

Wang, J., Wu, Y., Yang, G., & Aksoy, S. (2009). Interactions between mutualist Wigglesworthia and tsetse peptidoglycan recognition protein (PGRP-LB) influence trypanosome transmission. *Proceedings of the National Academy of Sciences, 106*(29), 12133–12138.

Wang, S., Ghosh, A. K., Bongio, N., Stebbings, K. A., Lampe, D. J., & Jacobs-Lorena, M. (2012). Fighting malaria with engineered symbiotic bacteria from vector mosquitoes. *Proceedings of the National Academy of Sciences, 109*(31), 12734–12739.

Wang, S., & Jacobs-Lorena, M. (2013). Genetic approaches to interfere with malaria transmission by vector mosquitoes. *Trends in Biotechnology*, *31*(3), 185–193. https://doi.org/10.1016/j.tibtech.2013.01.001.

Weiss, B., & Aksoy, S. (2011). Microbiome influences on insect host vector competence. *Trends in Parasitology, 27*(11), 514–522.

Weiss, B. L., Maltz, M., & Aksoy, S. (2012). Obligate symbionts activate immune system development in the tsetse fly. *The Journal of Immunology, 188*(7), 3395–3403.

Weiss, B. L., Wang, J., & Aksoy, S. (2011). Tsetse immune system maturation requires the presence of obligate symbionts in larvae. *PLoS Biology, 9*(5).

Weiss, B. L., Wang, J., Maltz, M. A., Wu, Y., & Aksoy, S. (2013). Trypanosome infection establishment in the tsetse fly gut is influenced by microbiome-regulated host immune barriers. *PLoS Pathogens, 9*(4).

Welburn, S., Arnold, K., Maudlin, I., & Gooday, G. (1993). Rickettsia-like organisms and chitinase production in relation to transmission of trypanosomes by tsetse flies. *Parasitology, 107*(2), 141–145.

Welburn, S., & Maudlin, I. (1999). Tsetse–trypanosome interactions: rites of passage. *Parasitology Today, 15*(10), 399–403.

Werren, J. H., Baldo, L., & Clark, M. E. (2008). Wolbachia: master manipulators of invertebrate biology. *Nature Reviews Microbiology, 6*(10), 741–751.

Whitten, M. M., Shiao, S. H., & Levashina, E. A. (2006). Mosquito midguts and malaria: cell biology, compartmentalization and immunology. *Parasite Immunology, 28*(4), 121–130. https://doi.org/10.1111/j.1365-3024.2006.00804.x.

WHO. (1997). *Vector Control - Methods for Use by Individuals and Communities*. Retrieved 21st June 2020 from https://www.who.int/docstore/water_sanitation_health/vectcontrol/ch26.htm

WHO. (2020). *Vector-borne diseases*. Retrieved 17th June 2020 from https://www.who.int/news-room/fact-sheets/detail/vector-borne-diseases.

Yassine, H., & Osta, M. A. (2010). Anopheles gambiae innate immunity. *Cell Microbiol, 12*(1), 1–9. https://doi.org/10.1111/j.1462-5822.2009.01388.x.

Yoshida, S., Ioka, D., Matsuoka, H., Endo, H., & Ishii, A. (2001). Bacteria expressing single-chain immunotoxin inhibit malaria parasite development in mosquitoes. *Molecular and Biochemical Parasitology, 113*(1), 89–96.

Zou, Z., Souza-Neto, J., Xi, Z., Kokoza, V., Shin, S. W., Dimopoulos, G., & Raikhel, A. (2011). Transcriptome analysis of Aedes aegypti transgenic mosquitoes with altered immunity. *PLoS Pathogens, 7*(11).

23 Applications of Metagenomics in Vector Surveillance for Effective Prediction and Control of Mosquito-Borne Viral Disease Outbreaks

Nwadiuto (Diuto) Esiobu
Florida Atlantic University

Bamidele Oderinde and Njidda Gadzama
University of Maiduguri

Ihekerenma Okoli
Applied Biotech International Nigeria Ltd.

CONTENTS

DOI: 10.1201/9781003178903-28

23.1 INTRODUCTION

Viruses are obligate intracellular pathogens, hijacking host's transcriptional and metabolic processes for cloning themselves within compatible hosts and spreading in the population. Some of the most devastating diseases of humans, animals, and plants are caused by viruses. Yet they are primarily responsible for recycling dissolved organic matter in the oceans, which in turn drives the food chain and earth's homeostasis (Wilhelm and Suttle, 1999). Their ability to mutate and jump host boundaries has caused more destructive global pandemics than any other organism on earth. It is estimated that only a tiny fraction of viruses on earth has been discovered. Using metagenomics, new viruses would be proactively picked up and added to the list of viruses. The activities of arthropods whose habitation and feeding centers around humans, domestic/wild animals, plants, and sewage encourage diverse interactions that promote a wide variety of ecologies for a huge diversity of viruses (many of which are yet to be discovered) to hide-out in a dormant state. A genomic surveillance program will enable the understanding of the nature and survival of some of the known viruses and their mutation rates as the insect habits change in response to environmental factors. The epidemiology of arbovirus infection in man is influenced by the probability of contacts between the vector, the human population and, for many viruses, the amplifying vertebrate host such as birds (most arboviruses that cause encephalitis), monkeys (yellow fever virus, Kyasanu Forest Virus), horses (Eastern Equine Encephalitis Virus, Western Equine Encephalitis Virus, Venezuelan Equine Encephalitis Virus, West Nile Virus), sheep (Rift Valley Fever), pigs (Japan Encephalitis Virus), and rodents (in transmission of Lassa viral fever). Climate change and demographic behaviors are important determinants of virus reservoirs and, in recent decades, have aided several arboviral infection range to expand beyond traditional boundaries.

The majority of mosquito virus studies have focused on the detection of specific mosquito vectors (Millanby, 1956; Gadzama, 1977) and already known arboviruses or those capable of replicating in specific insect and vertebrate cell lines (Blanc and Gutierrez, 2015). Viruses present in mosquitoes can include viruses that are biologically or mechanically transmitted by these vectors, as well as other viruses that are not transmitted by mosquitoes but are drawn indiscriminately from host reservoirs. Many viruses cannot be cultured on cell lines in the laboratory, necessitating use of methods such as degenerate Polymerase Chain Reaction (PCR) and pan-viral microarrays which rely on the detection of highly conserved regions in known viral genomes for viral discovery (Wang et al., 2002). To circumvent these issues, recent studies have demonstrated the effectiveness of viral particle purification and shotgun sequencing (viral metagenomics) for describing novel viruses (Edwards and Rohwer, 2005; Delwart, 2007). Using viral metagenomics, novel viruses have been characterized from nasopharyngeal aspirates (Allander et al., 2005), fecal samples (Breitbart et al., 2003; Victoria et al., 2009), blood (Jones et al., 2005; Breitbart and Rohwer, 2005), and tissue samples such as lungs (Ng et al., 2009; Willner et al., 2009) and tumors (Ng et al., 2009).

23.2 ECOLOGY OF VECTORS IN VIRAL INFECTION

Alarming trends in the resurgence of vector-borne diseases are anticipated to continue unless more effective action is taken to address the variety of underlying causes. Social factors, anthropogenic environment modification, and/or ecological changes appear to be the primary drivers. The ecological dimension of vector-borne disease research and management is an important element because this issue is an essential ecological problem with biophysical, social, and economic dimensions. However, there is often a lack of clarity about the ecological dimension, the field of ecology (role, limitations), and related concepts pertinent to ecosystem approaches to health. An ecological perspective can provide foresight into the appropriateness of interventions, provide answers to unexpected vector control responses, and contribute to effective management solution in an ever-changing environment (Ellis and Wilcox, 2009).

Vector-borne diseases pose a significant public health problem today, with a number of diseases resurging in recent decades alongside newly emerging infectious diseases (Solinski et al., 2003). Some of these were effectively controlled just 50 years ago, but these previous hard-won gains are now threatened or have already been lost (Gubler, 2002, 2005). Dengue is perhaps the most striking example. From 1950 to 1959, less than 1,000 cases were reported worldwide (Guha-Sapir and Schimmer, 2005). Now an estimated 50–100 million cases occur annually. At least 20 other vector-borne diseases have also emerged during this time, having increased in incidence and/or expanding their geographical range. The correlation between the increase in prevalence rates and environmental degradation over the last century (due largely to anthropogenic effects) have fueled growing concerns and acceptance of the interdependence of One Health of man and the environment (Hassan et al., 2005; WHO, 2005).

The association of changes in anthropogenic environment and infectious diseases in particular has recently begun to attract attention and prominence in relation to policy (e.g., Millennium Ecosystem Assessment, 2005). The growing interest in ecosystem approaches to health is a response to an increasing recognition of "the inextricable links between humans and their biophysical, social, and economic environments and that these links are reflected in a population's state of health" (Lebel, 2003).

The disturbing trends in vector-borne disease emergence and reemergence underscore the critical need for governments to set up molecular or genomic surveillance systems which will not only produce early alarm and infection prediction patterns but will inform policy on vector management and environmental mitigation. The systemic nature of these changes and complexity of interactions among factors make simple, targeted, or "silver bullet" solutions ineffective, except for in the short term. This is a major national security risk no country should entertain after witnessing the capacity of viruses like the COVID19 to cause epidemics and devastating pandemics and economic downturn. An ecosystem-wide approach is highly advocated to provide sustainable simultaneous monitoring of both insect vector dynamics and the virobiont community structure over time. Such solutions span public health interests and areas of nature resource management and agricultural pest management

(Campbell-Lendrum and Molyneux, 2005). The cost of such a surveillance network, especially with declining costs of sequencing, pales in contrast to the enormous public health benefits to be gained.

23.3 NEED TO ESTABLISH DATABASE FOR THE IDENTITY AND GENOMIC PROFILE BASELINE OF ARBOVIRUSES AND THEIR ASSOCIATED VECTORS AND NATURAL DIVERSITY IN NIGERIA

In most developing nations, misdiagnosis of arboviral diseases is quite common because of overlapping symptoms with the widely prevalent malaria and typhoid fever. The expertise and infrastructural capacity to accurately diagnose the viral diseases is so poor (Kolawole et al., 2018) that none of the six geopolitical regions can boast of a standard routine detection laboratory. Unfortunately, the major challenge experienced during recent outbreaks of arbovirus infections in the country was the delay in outbreak confirmation which led to wider spread and panic. Atypical symptoms of arbovirus infections with other hyperendemic pathogens such as malaria and typhoid make it necessary that a surveillance molecular laboratory is established for prompt alert. In Nigeria, suspected outbreaks of arbovirus infections (which were not investigated using appropriate diagnostic facilities) have been reported. The recent rise in incidences of malaria/typhoid in most healthcare centers in Nigeria may actually signal emerging and remerging arbovirus infections. Generally, human practices such as increased global travel, accelerated urbanization resulting in increasing numbers of overcrowded, informal settlements, or 'shanty towns' characterized by low-grade housing, inadequate water supply and storage, and poor roads, poor sanitation and waste management services favored mosquito vector breeding sites in Nigeria. Misdiagnosis of arbovirus co-infections as malaria and/or typhoid, combined with a lack of virus surveillance and low awareness of arbovirus infections by healthcare providers, increases the potential for uncontrolled spread, under-reporting, under-recognition, and underestimation of arboviruses in most African countries including Nigeria. Before now, there was a dearth of skilled workforce but that situation is rapidly changing. In addition, there are now very well-equipped private laboratories like Applied Biotech International Nigeria Ltd in Wuye, Abuja and several public stations that could handle the monitoring.

Apart from public health reasons, the need for an ongoing vector-virus surveillance programs is critical for national and food security. Nigeria is by far the largest producer of cassava in the world and is endowed with expansive bioresources. The vulnerability of bioresources to decimation by a new virus makes it a very fragile asset for any country. Consequently, the country should take a holistic approach to crafting its bioeconomy strategies and how to protect and preserve it. Extensive globalization and travels allow rapid spread and introduction of plant and animal viruses. The spread of Zika to Miami in the United States in 2015 was mediated by travelers and accompanying mosquitoes. In all geopolitical regions of Nigeria, mosquitoes can thrive all year-round and could harbor a wide range of viruses. Detecting individual viruses will be extremely cumbersome and will not detect unknown and newly emerging viruses. That is why the proposed vector-virus metagenomics detection is the best approach. In addition, the recent release of genetically modified mosquitoes in neighboring countries for the control of malaria has the potential of recruiting new viruses in these modified vector hosts. No one could predict, a priori, the dynamics of the remnant viruses and their virus host range. Periodic sampling of key locations in the country will capture mosquitoes which will then be subjected to metagenomic sequencing protocols to determine all the viruses and the various mosquitoes in circulation at the time. The information will be fed into a national database called the Arbovirus and Vectors in Nigeria Database (AVIVEND).

23.4 EPIDEMIOLOGICAL BENEFITS AND PREDICTIVE POWERS OF SUCH AN APPROACH

Arboviruses surveillance program can monitor viruses in vertebrate hosts and/or arthropod vectors. The surveillance of virus in vertebrate hosts has several drawbacks relative to surveillance of arthropods, including the limited accessibility of hosts, small sample sizes of hosts, and variation of susceptibility of different vertebrate species to the same virus (van den Hurk et al., 2012). Surveillance of mosquitoes for circulating arboviruses can be used as an early warning system because it allows the detection of seasonal initiation or increase of virus circulation.

Over the years, studies have shown that public health interventions are cost-saving, both to health services as well as the wider economy. Furthermore, some are very rapid: fall's prevention interventions reported substantial returns within 6–12 months (Beard et al., 2006). Predictive power of genomic vector surveillance for arbovirus will not only prepare the healthcare frontiers but will also reduce mortality and morbidity from a deadly disease that is prevented.

One might reasonably expect equally rapid returns for preventive interventions from immunization, direct health care, smoking cessation, and nutrition and sentinel arbovirus surveillance.

Attempting to quantify the value and returns of genome-based surveillance in the short-run could be both challenging and misleading with significant implications for policy makers, whose assignments are often limited in time (typically 3–5 years). One of the most successful approaches is to develop a comprehensive strategic healthcare roadmap that outlines infrastructure and technical frameworks to guide the creation of databases and preventive surveillance. With the dwindling costs of genome sequencing and bioinformatics, such technical systems are now affordable. Partnerships between the Centers for Disease Control and Prevention, the Federal Ministry of Health, the Federal Ministry of Agriculture and Natural Resources, and the Federal Ministry of Science and Technology should create a solid collaboration with the private sector in public-private partnership (PPP) for this invaluable venture.. It is financially preferable for healthcare systems to aim at preventing ill-health rather than to treat diseases and manage the subsequent economic and social consequences. A number of studies have calculated the return on investment for individual prevention interventions; however, no systematic review has spanned the breadth of public health.

There is no known routine surveillance in place in Nigeria as most of the results are from planned research studies that are based on available resources targeted for academic publications, research grants, collaborative studies, or part of requirements for the award of degrees which are not routine. An example of such a study is the sporadic serology research by Oderinde et al. (2020), where a total of 200 serum samples were collected from patients exhibiting febrile illness who visited the State Hospital in Maiduguri for medical attention between March and April 2018. Sera were tested for Flavivirus RNA by a pan-flaviviral primer set using hemi-nested real-time PCR; 26 of the samples were positive and after sequencing with the cFD2 and MAMD primers to characterize the infections. Out of 15 samples sequenced, 13 aligned with WNV Lineage 1 perfectly matching West Nile Virus (WNV) Italian isolate 358 (FJ472944.1) and Indian isolate Gwl-01 2015 (MG516600.1). Interestingly, one additional sequence matched Zika Virus (ZIKV) instead, aligned with strain MR766 (MK105975.1). These results, though scanty, emphasize the need for facilities for differential diagnosis and early detection of arboviruses in not only our health institutions but also in our health and veterinary research institutes.

The subtropical climate of Florida which is somewhat similar to Nigeria's allows many mosquito populations to thrive; and for more than a 100 years, the state has been highly successful in systematically understanding the threat posed by mosquito-borne pathogens especially the arboviruses and in devising effective control measures. Through disease surveillance, based on mosquito biology, control strategies have been developed. Tabachnick (2016) provides an excellent review on these advances. Arbovirus surveillance programs, primarily based on *Culex*-borne diseases such as the Saint Louis Encephalitis Virus (SLEV), West Nile Virus (WNV), and Eastern Equin Encephalitis Virus (EEEV), integrate oviposition data of the adult female Culex mosquitoes, the age structure, and gonotrophic cycles of *Cx. nigripalpus* or *Cx. p. quinquefasciatus* populations, GIS data, rainfall, and flooding to develop precise risk assessment. In 2005, the Florida arbovirus sentinel chicken surveillance system used routine serotyping of the sentinel chicken flock to precisely detect and confine the transmission of the WNV in Pinellas County. In summary, many factors are used to create models for predicting mosquito disease outbreaks used for focal vector control efforts (Day et al., 2008). The State recently is aggressively funding incorporated molecular tracking and surveillance such as the one proposed here.

Recent advances in next-generation sequencing and bioinformatics will allow us to determine the diversity and richness of all and specific mosquitoes in any locality in one test. Integrating such data with output from our novel rapid impudence-based virus detection device will provide insight into the dynamics of vector communities, environment, and disease. A robust disease surveillance system should be applicable to all types of mosquito vectors (capture and detection), permit a simultaneous direct detection of specific etiologic agents in the sampled mosquito population, and be rapid.

This chapter makes a case for a versatile surveillance system that will identify various mosquito vectors and the viral pathogens they harbor in one metagenomic test. In addition, it will simultaneously determine the viral density and profiles (e.g., ZIKV) in mosquitoes. Over time, a database of the mosquito and viral/bacterial symbionts in a given ecology is established. Similar to metagenomics assays, vector community shifts can be precisely detected before disease transmission reaches epidemic levels. Genomic surveillance not only detects new mutations in the wild (such as the one that caused the COVID-19) before they jump host range,but allows health officials to predict outbreaks and mitigate impacts.

23.5 WHAT IS NEEDED TO SET UP A SURVEILLANCE

Building capacity to invest in both long-term and focal interventions increases the resilience of communities and reduces disease burden. Optimizing the spatial and temporal application of these interventions is also part of precision epidemiology. This process should be dynamic and adaptable according to epidemiological scenarios.

However, the three Vs are with us—The Viruses, their Vectors, and their Victims (human/animals)! Implementation of mosquito surveillance can give decision makers enough time to enact efficient measures for outbreak control (Gu et al., 2008). Mosquito-based virus surveillance programs are also important tools in the study of the eco-epidemiology and transmission of viruses by enabling identification of potential vectors of a given virus as well as the spatiotemporal dynamics of the virus. Mosquito surveillance can also be useful for identifying new arbovirus species of medical and veterinary importance (van den Hurk et al., 2012; Gu et al., 2008; Crabtree et al., 2009; Ochieng et al., 2013)

One of the key elements of the control of arbovirus transmission is early detection of virus activity or increased virus activity in vector populations. Surveillance programs designed to monitor these parameters provide an early warning system of increased risk of transmission and disease outbreak. Detection or accurate prediction of virus activity in vector populations and specific diagnosis of infection in the human or animal host are crucial components of effective control and treatment strategies and facilitate early warning of potential or existing outbreaks and initiation of vector management and/or vaccination programs.

One of the key arbovirus surveillance is the detection of virus in arthropod vector populations. Traditional methods include trapping vectors such as mosquitoes, ticks, and midges, identifying them to species level ,and analyzing vector pools for known viruses of interest. Virus isolation by inoculating mice, embryonated eggs, or colonized vector species has largely been replaced by in vitro methods (e.g., inoculation of arthropod and vertebrate cell lines) but is still the most effective means for monitoring virus activity for some viruses. However, the enhanced technologies for the specific detection of viral nucleic acid, such as multiplexed real-time PCR protocols, provide more rapid, sensitive, and specific approaches for detecting virus activity in vector populations. Furthermore, the recent application of next-generation sequencing technologies to rapidly analyze nucleic acid of unidentified viral isolates provides a revolutionary approach for the discovery and genetic characterization of new vector-borne viruses.

Another approach routinely used to monitor virus activity is sentinel animal surveillance. This is particularly useful for veterinary or zoonotic arboviruses where domestic animals (e.g., cattle, pigs, or chickens) can be effectively employed.

The serological monitoring of strategically positioned flocks or herds can provide early warning of virus transmission in specific region/communities. However, such herd-based systems are limited by the costs of maintaining the animals (particularly in remote areas), lack of specificity of serological assays, and the ability to only target viruses that infect the selected sentinel animal or those that are transmitted by vectors that feed on the sentinel host. To address these problems, a new approach (reviewed by A. F. van den Hurk and colleagues) has recently been applied to the specific detection of virus activity in mosquito populations, which is particularly useful for remote locations. The specific detection of viral RNA expectorated by infected mosquitoes feeding on a sugar-coated matrix has enabled these investigators to collect samples over prolonged periods in remote locations and specifically identify viruses carried by any mosquito feeding on the sugar bait. The potential application of this system to a wide variety of arbovirus surveillance scenarios is very promising.

Monitoring of arthropod vector populations is a very important tool in arbovirus surveillance, especially in detecting increase in known vectors or the introduction of new species into the community, e.g., recent outcomes of *Aedes albopictus* into Europe and Australia. Interestingly, new strategies such as satellite-based remote sensing of vector breeding sites and assessment of the risk of virus transmission based on proximity to human or animal habitation can rapidly provide highly useful data on a very large scale.

Accurate and timely diagnosis of arbovirus infections is also crucial to ensure appropriate patient management, for the reporting of virus activity in a region and to allow instigation of control strategies such as vector management, vaccination, and public awareness campaigns. Serological assays required for this purpose are not readily available in low resource country like Nigeria where facilities for differential diagnosis of arbovirus are lacking. Early research and development of point-of-care assays, particularly for yellow fever, dengue, and West Nile viruses, also provides a useful tool for the clinician or veterinarian treating the patient on presentation at the clinic (see the reviews by Hall et al., 2012).

23.6 CONTRAST COSTS WITH ENORMOUS PUBLIC HEALTH BENEFITS OF DISEASE PREVENTION, TARGETED GOVERNMENT EFFORTS

World Health Organization defined Health Promotion as "The process of enabling people to increase control over, and to improve, their health. It moves beyond a focus on individual behavior towards a wide range of social and environmental interventions."

Disease prevention differs from health promotion because it focuses on specific efforts aimed at reducing the development and severity of chronic diseases and other morbidities, while wellness is related to health promotion and disease prevention. Wellness is described as the attitudes and active decisions made by an individual that contributes to positive health behaviors and outcomes.

Prevention also embraces early detection efforts, such as screening at-risk populations with updated and differential routine facilities, as well as strategies for appropriate management of existing diseases and related complications. Active surveillance for diseases vectors and detection from environments before spread to humans are one of the best public health protection tools.

Targeted investments in prevention can achieve Benjamin Franklin's belief that "an ounce of prevention is worth a pound of cure" is thus borne out by the cost savings demonstrated, particularly when compared with recent returns for investment in health care.

Meaningful genomic-based arboviral research can be translated into effective policies and actions at local, national, and international levels, using public health evidence to illustrate the arguments. To reduce arboviral disease across the nation, we must rethink our healthcare system. It is essential to have a coordinated, strategic prevention approach that promotes healthy behaviors, expands early detection and diagnosis of disease, supports people of every age, and eliminates health disparities. With community-based public health efforts that embrace prevention with active mosquito—vector surveillance as a priority, we can become a healthier nation.

Research translation and promising research findings are relevant only when they reach the people they are designed to serve. Key scientific advances must be applied and evaluated, reflected in state and local health policies, and widely adopted as community practices across the country. We must support community-based prevention research to identify the causes of health inequities and the best ways to provide resources needed for health and access to high-quality preventive care and clinical services.

23.7 NATIONAL SECURITY RATIONALE FOR GENOMIC SURVEILLANCE

An examination of the nexus of security, insecurity, and health shows that security is a prerequisite for health. The many and varied ways that armed violence—including threats of armed violence—can affect people's health can be documented by formal studies; however, valuable data also exist in other reports, such as media reports. The health community needs to recognize that people's insecurity is a massive global health issue. The foreign policies of donor governments should incorporate recognition that documentation, analysis, and publication of data describing the impact of insecurity on people's health can lead to the creation of policies to enhance people's security (WHO, 2007). Ballistic trauma and its sequelae can only be negative in terms of the impact on health of the individual victims. But the often ignored is the potential magnitude of the multisectoral devastation of a novel viral pandemic like the COVID-19 can cause. The indirect effects include economic breakdown, loss of safety nets, fear, coercion, displacement, and deprivation of essentials such as clean drinking water, food, and health care. Fragile health infrastructure coupled with increases in the number of Internally Displaced Peoples and the associated social problems in Nigeria underscores the importance of proactive prevention strategies such as arbovirus mosquito surveillance.

Presently, various regions in Nigeria are faced with insecurity in diverse forms. The insurgency varies from Boko Haram, Armed bandits, Cattle rustlers, Fulani herdsmen, Kidnapping, and so on. These are great impediments to meaningful surveillance and research, especially in arbovirus-related research where insects and animals (both domestic and wild) need to be studied and sampled.

Insecurity generally disrupts development of nations and societies. Denney (2013) observed that the relationship between insecurity and underdevelopment is much stronger than the relationship between peace and development. That is, where there is conflict, there is often underdevelopment. The present insecurity situation in the country not only depresses genomic surveillance, but dwarfs any research and development efforts which are essential for nation's progressive development and economic sustainability.

23.8 CONCLUSION

Recent advances in the life sciences have drastically reduced the cost of gene sequencing, making metagenomic studies within reach of any government. A genomic-based surveillance will periodically capture mosquitoes and employ cutting-edge molecular techniques to identify mosquito taxonomic profiles and all the virus and bacterial symbionts they carry throughout the year. The data is stored in the cloud where artificial intelligence setups are used to mine it. Forward leaning leaders, especially in tropical countries where climatic conditions favor virus mutations and prolonged thriving of their mosquito hosts, should prioritize funding a routine genomic-based surveillance of vectors and their virobionts. Creating a baseline database allows for rapid tracking and mitigation of emerging outbreaks (captured during the surveillance) before widespread impact. Such a standing project is often executed through public-private partnerships in advanced economies for sustainability and costs a tiny fraction of the cost of mitigating public health crisis. In summary, genomic surveillance of vectors and their viruses will proactively protect national security from potential biowarfare, help prevent plant diseases (by extension), and preserve bioresources and national biological assets from insect-borne virus devastation. Moreover, the platform once created can be used to surveil against any microbial pathogen of humans and animals.

REFERENCES

Allander, T., Tammi, M. T., Eriksson, M., Bjerkner, A., Tiveljung-Lindell, A. and Anderson, B. (2005). Cloning of a human parvovirus by molecular screening of respiratory tract samples. *Proceedings of National Academy of Science, USA* 102(36): 12891–12896.

Beard, J., Rowell, D., Scott, D., van Beurden, E., Barnett, L, Hughes, K. and Newman, B. (2006). Economic analysis of a community-based falls prevention program. *Journal of Research Institute Public Health* 120:742–751.

Blanc, S. and Gutierrez, S. (2015). The specifics of vector transmission of arboviruses of vertebrates and plants. *Current Opinion in Virology* 15: 27–33.

Breitbart, M., Hewson, I., Felts, B., Mahaffy, J.M., Nulton, J., Salamon, P., Rohwer, F. (2003). Metagenomic analyses of an uncultured viral community from human faces. *Journal of Bacteriology* 185: 6220–6223.

Breitbart, M. and Rohwer, F. (2005). Method for discovering novel DNA viruses in blood using viral particle selection and shotgun sequencing. *Biotechniques* 39: 729–736.

Campbell-Lendrum, D. and Molyneux, D. (2005). Ecosystems and vector-borne disease control. In: Epstein, P., Githeko, A., Rabinovich, J., Weinstein, P., editors. *Ecosystems and Human Well-Being: Policy Responses. v. 2.* Washington, DC: Island Press, pp. 353–372.

Crabtree, M., Sang, R., Lutomiah, J., Richardson, J. and Miller, B. (2009). Arbovirus surveillance of mosquitoes collected at sites of active Rift Valley fever virus transmission: Kenya, 2006–2007. *Journal of Medical Entomology* 46(4): 961–964.

Day, J. F., Tabachnick, W. J. and Smartt C.T. (2015) Factors that influence the transmission of West Nile virus in Florida. *Journal of Medical Entomology* 52: 743–754. doi: 10.1093/jme/tjv076.

Delwart, E. L. (2007). Viral metagenomics. *Reviews in Medical Virology* 17: 115–131.

Edwards, R. A. and Rohwer, F. (2005). Viral metagenomics. *Nature Reviews, Microbiology* 3: 504–510.

Ellis, B. R. and Wilcox, B. A (2009). The ecological dimensions of vector-borne disease research and control. *Cad Saude Publica* 25: S155–S167. https://www.scielosp.org/article/csp/2009.v25suppl1/S155-S167/en/.

Gadzama, N. M. (1977). The distribution of adult mosquitoes at Ahmadu Bello University. *Savanna* 6(2): 179–189.

Gubler, D. J. (2002). The global emergence/resurgence of arboviral diseases as public health problems. *Archives of Medical Research* 33: 330–342.

Gubler, D. J. (2005). The emergence of epidemic dengue fever and dengue hemorrhagic fever in the Americas: A case of failed public health policy. *Review of Pan American Journal of Public Health* 17(4): 221.

Guha-Sapir, D. and Schimmer, B. (2005). Dengue fever: New paradigms for a changing epidemiology. *Emerging Themes Epidemiology* 2:1.

Gu, W, Unnasch, T. R., Katholi, C. R., Lampman, R. and Novak, R. J. (2008). Fundamental issues in mosquito surveillance for arboviral transmission. *Transactions of the Royal Society of Tropical Medicine and Hy*giene 102(8): 817–822.

Hall, R. A., Blitvich, B. J., Johansen, C.A. and Blacksell, S. D (2012). Advances in arbovirus surveillance, detection and diagnosis. *BioMed Research International*, 2012, Article ID 512969, 2 pages, https://doi.org/10.1155/2012/512969.

Hassan, R. M., Scholes, R. and Ash, N. (2005). *Ecosystems and Human Well-being: Current State and Trends: Findings of The Condition and Trends Working Group of the Millennium Ecosystem Assessment.* Washington, DC: Island Press.

Jones, M.S., Kapoor, A., Lukashov, W., Simmonds, P., Hecht, F., Delwart, E. (2005). New DNA viruses identified in patients with acute viral infection syndrome. *Journal of Virology* 79: 8230–8236.

Kolawole, O. M., Adelaiye, G., Ogah, J. I. (2018). Emergence and associated risk factors of vector borne west Nile virus infection in Ilorin, Nigeria. *Journal of Arthropod-Borne Diseases* 12(4): 341–350.

Lebel, J. (2003). Health: An ecosystem approach; the issue. Case studies, lessons and recommendations. In focus: an ecosystem approach. IDRC, Ottawa, ON Google scholar. https://idl-bnc-idrc.dspacedirect.org/handle/10625/30918.

Millamby, K. (1956). Mosquito populations at Ibadan in Nigeria. *Bulletin of Entomological Research* 47:125–130.

Ng, T.F.F., Suedmeyer, W.K., Wheeler, E., Gulland, F. and Breitbart, M. (2009). Novel anellovirus discovered from a mortality event of captive California Sea Lions. *Journal of General Virology* 90: 1256–1261.

Ochieng C., Lutomiah J., Makio A., Koka H., Chepkorir E, Yalwala S, Mutisya J, Musila L, Khamadi S, Richardson, et al. (2013). Mosquito-borne arbovirus surveillance at selected sites in diverse ecological zones of Kenya; 2007–2012. *Virology Journal* 10: 140.

Oderinde, B.S., Mora-Cardenas, E., Carletti, T., Baba, M.M., Marcello, A. (2020). Prevalence of locally undetected acute infections of Flaviviruses in North-Eastern Nigeria. *Virus Research 286*: 198060.

Smolinski, M.S., Hamburg, M.A. and Lederberg, J. (2003). *Microbial Threats to Health: Emergence, Detection, and Response.* Washington, DC: National Academies Press.

Tabachnick, W. J. (2016) Climate change and the arboviruses: Lessons from the evolution of the dengue and yellow fever viruses. *Annual Review of Virology* 3(1):125–145. doi:10.1146/annurev-virology-110615-035630.

van den Hurk, A. F, Hall-Mendelin, S., Johansen, C. A., Warrilow, D. and Ritchie, S. A. (2012). Evolution of mosquito-based arbovirus surveillance systems in Australia. *Journal of Biomedicine and Biotechnology* 2: 325659.

Victoria, J. G., Kapoor, A., Li, L., Blinkova, O., Slikas, B., Wang, C., Naeem, A., Sohail, Z. and Delwart, E. (2009) Metagenomic analyses of viruses in stool samples from children with acute flaccid paralysis. *Journal of Virology* 93 9: 4642–4651.

Wang, D., Coscoy, L., Zylberberg, M., Avila, P.C., Boushey, H.A., Ganem, D. and DeRisi, J. L. (2002). Microarray-based detection and genotyping of viral pathogens. *Proceedings of National Academy of Science USA* 99: 15687–15692.

Wilhelm, S. W. and Suttle, C.A. (1999) Viruses and nutrient cycles in the sea: Viruses play critical roles in the structure and function of aquatic food webs. *BioScience* 49 (10): 781–788, https://doi.org/10.2307/1313569.

Willner, D., Furlan, M., Haynes, M., Schmieder, R., Angly, F.E., Silva, J., Tammadoni, S., Nosrat, B., Conrad, D., Rohwer, F. (2009). Metagenomic analysis of respiratory tract DNA viral communities in cystic fibrosis and non-cystic fibrosis individuals. *Public Library of Science One* 4: e7370.

World Health Organization (2005). Ecosystems and human well-being: health synthesis. Geneva: World Health Organization; www.who.int.

Index

Note: **Bold** page numbers refer to tables and *italic* page numbers refer to figures.